MODERN TRENDS IN PHYSICS RESEARCH

To learn more about the AIP Conference Proceedings, including the Conference Proceedings Series, please visit the webpage **http://proceedings.aip.org/proceedings**

MODERN TRENDS IN PHYSICS RESEARCH

Second International Conference on Modern Trends
in Physics Research

MTPR-06

Cairo, Egypt   *6 – 11 April 2006*

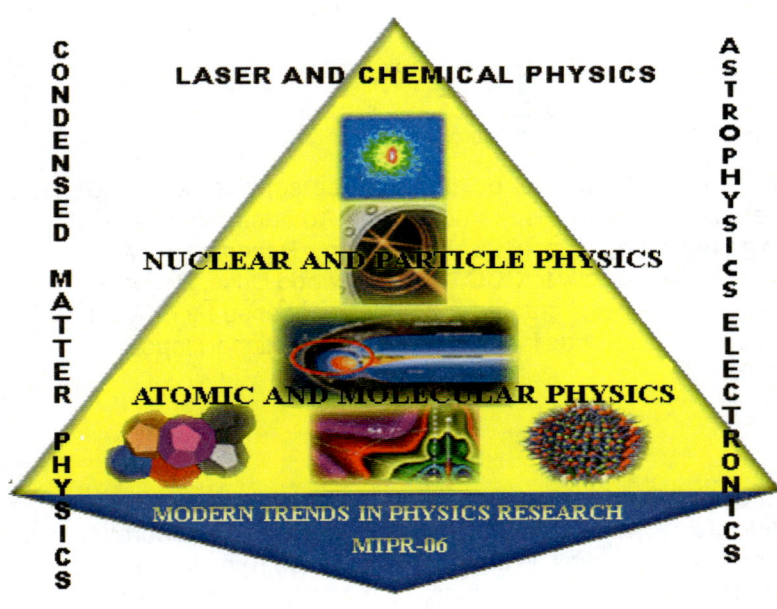

Editor
Lotfia El Nadi

AMERICAN
INSTITUTE
ᴏꜰPHYSICS

Melville, New York, 2007
AIP CONFERENCE PROCEEDINGS ■ VOLUME 888

Editor:

Lotfia El Nadi

Department of Physics
Faculty of Science
Cairo University
Cairo-Giza
EGYPT

E-mail: lotfianadi@gmail.com

L.C. Catalog Card No. 2006940927
ISBN 978-0-7354-0394-9
ISSN 0094-243X
Printed in the United States of America

CONTENTS

I. ATOMIC, MOLECULAR & CONDENSED MATTER PHYSICS
I-1- KEYNOTE AND PLENARY PAPERS

I-2—INVITED LECTURE PAPERS

I-3—CONTRIBUTING PAPERS

II. CHEMICAL PHYSICS, LASER & APPLICATIONS
II-1—KEYNOTE AND PLENARY PAPERS

II-2—INVITED LECTURE PAPERS

II-3—CONTRIBUTING PAPERS

II. NUCLEAR, PARTICLE PHYSICS & ASTROPHYSICS
III-1—KEYNOTE AND PLENARY PAPERS

III-2—INVITED LECTURE PAPERS

III-3—CONTRIBUTING PAPERS

ERRATA

Conference Participants on the stage attending the opening of MTPR-06 at Cairo University Main Hall on Thursday 6 April 2006

PREFACE

Modern Trends in Physics Research MTPR-06, the Cairo – Luxor conference, which took place from April 6-11, 2006, was the second of the International Conference series, to be carried out biannually by the Physics Department, Faculty of Science, Cairo University. The purpose of the conference was to get an advanced picture of the activities in some branches of the physics domain in the last two years after the MTPR-04 conference. The results presented at MTPR-06 conference were an astonishing display of new and extremely breakthrough experimental and theoretical discoveries in physics, its applied branches and technology aspects. This proceeding highlights the contributions presented at the conference. It provides some detailed account of the latest results in the fields of atomic, molecular, condensed matter, lasers, nuclear, particle and astrophysics. The papers and some review articles have been written by international and national experts in these fields and should serve as reference material to scientists already wishing to get into the aspects of modern physics and technology of the twenty first century.

An appreciable number of young researchers from over twelve Egyptian Faculties of Science and institutions, gathering around 180 participants attended and contributed to this conference. Experts from Argentina, China, Finland, France, India, Italy, Jordon, Kuwait, Pakistan and USA enriched the conference with innovative research of their Labs. They also encouraged possible cooperation between their groups and the Egyptian ones in the Advanced Technological Fields of Physics.

We hope that the intimacy and discussions encountered during the conference will truly mark impressive future collaboration between, not only Egyptian – Egyptian groups but also between Egyptian – International scientists. The approaches already going on, between Egyptian scientists and particularly French, American and Chinese experts declare the dawn of internationally motivated cooperation.

The conference proceedings contain ten sectors, the first consists of the honorary keynote overview of " Coaxing Cell to Cajoling Condensates: The Many Uses of Optical Forces" by Kristian Helmerson from NIST, Gaithersburg, USA. The other chapters cover each of the topics which formed the theme of the conference. Each chapter opens with the keynotes, plenary and invited presentation then followed by the contributed ones.

I am grateful to all authors who cared and provided excellent papers. I would like in particular to thank all international participants for their continuous advice, moral support and patience which greatly helped to rank the MTPR-06 conference as a successful realm of advanced physics research.

I wish to express my gratitude to Professor Hamdi Hassanien Dean of the Faculty of Science, to the authorities of both Cairo University and Helwan University and to the organizing committee for their generous support and encouragement which helped greatly to shape and perform the MTPR-06 conference.

In particular I would like to thank Dr. Galila Abdelatif Mehena for her outstanding help during the preparation of this proceedings.

I would like to thank my colleges Dr. Hussein A. Moniem, Dr. Magdy M. Omar and Mrs. Souzan Fekry for their efforts and advice before, during and after the conference.

Thanks are also extended to staff members of the Physics Department who helped directly or indirectly to the success of this conference.

I would like to extend my appreciation for the sincere devotion of Mr. Mohammed Hassan and the technical staff for the time and efforts they gave to finalize this proceedings.

The social and cultural features of the conference; through visits to important and cultural sites in Cairo, Giza and Luxor gave the conference a spatial flavor of mixing future, present and old civilization all together.

In conclusion it is worthy to assure the readers of this volume that through the days to come in my life, I am always ready to serve in spreading knowledge, promoting international cooperation and caring to upgrade the wonderful world of Physics Research.

For the readers of this volume I hope you will feel and use the excitement and enthusiasm to join our future MTPR-08 (2008) conference. The website for MTPR-08 is www.eun.eg/MTPR-08.html will appear by April 15 (2007) In Shaa ALLAH.

The Editor
Lotfia M. El Nadi
Laser Physics Lab.,
Faculty of Science,
Cairo University,
Egypt

Coaxing Cells to Cajoling Condensates: The Many Uses of Optical Forces

Kristian Helmerson

National Institute of Standards and Technology, Gaithersburg, Maryland, USA 20899-8424

Abstract. Optical forces, due to the interaction of light with matter, is described. Several applications of optical forces in atomic physics and biotechnology are given, based on work in the Laser Cooling and Trapping Group at NIST.

Keywords: Optical forces, laser cooling, atom trapping, atom optics, optical tweezers
PACS: 42.50.Vk, 32.80.Pj, 03.75.-b, 03.75.Be, 03.75.Dg, 03.75.Lm, 87.68.+z, 87.80.-y

INTRODUCTION

The application of optical forces crosses many disciplines. There are currently over 300 groups utilizing optical forces for a variety of applications from atomic, molecule and optical physics to biophysics to condensed matter physics. This proceeding is based on a lecture given in Cairo, Egypt on April 6, 2006 at the Modern Trends in Physics Research conference (MPTR-06). I will describe the basic principles of optical forces and then give specific examples of applications, drawing almost exclusively on the research going on in the Laser Cooling and Trapping Group at the National Institute of Standards and Technology in Gaithersburg, Maryland. This proceeding is not meant to be either a pedagogical, exhaustive or thorough review of the field. To do so would require, at the very least, the entire space of the conference proceedings. Instead, this contribution is intended to give the reader a flavor of what optical forces are and some of the possible applications.

OPTICAL FORCES

The interaction of light and matter inevitably involves the exchange of momentum. Most people have heard that light comes in quantized packets of energy called photons. The energy of the photon, E, is proportional to the frequency of the light or inversely proportional to the wavelength, λ, and is simply given by $E = hc/\lambda$, where h is the Planck constant. What is less well known is that the photon also carries momentum. The momentum of the photon, p, is simply E/c. Thus the momentum of the photon is also proportional to the frequency of the light or inversely proportional to the wavelength, and is simply given by $p = h/\lambda$. This is often written as $\vec{p} = \hbar \vec{k}$, where $\hbar = h/2\pi$ and $k = 2\pi/\lambda$ is the magnitude of the wavevector of the light that points in the direction of the propagation of the light.

CP888, *Modern Trends in Physics Research,*
Second International Conference on Modern Trends in Physics Research—MTPR-06,
edited by L. El Nadi

The momentum of the photon can be transferred to an object. Typically, when this happens, the wavelength of the light and hence the magnitude of its momentum does not change. Instead, the direction of the light changes and so the direction of its momentum also changes. By conservation of momentum, the object that changed the direction of the momentum of the light has to account for this change, hence the momentum of the object changes in the opposite direction. As an example, consider the situation in figure 1. A photon of momentum p, traveling from left to right impinges an atom of mass m, initially at rest. The photon is reflected by the atom and is now moving from right to left with momentum $-p$. The momentum of the photon has changed by $-2p$, hence by conservation of momentum, the atom has to be moving with velocity $v = 2p/m$. Such redistribution of photons is at the heart of the microscopic picture of the elementary process involved in optical forces.

$$p \qquad m \qquad\qquad -p \qquad mv = 2p$$

FIGURE 1. Diagram illustrating momentum transfer from photons to an atom. By conservation of momentum, any change in the photon momentum from the scattering process has to be reflected by a corresponding change in the atom's momentum.

Optical forces are often divided in to two categories - the scattering force and the dipole force. The scattering force, also called the spontaneous force or radiation pressure, is the force exerted on an atom by the incoherent scattering of photons. More precisely it is the force on an atom corresponding to absorption of a photon followed by spontaneous emission. The photon absorbed transfers $\hbar k$ of momentum to the atom. The spontaneous emission of the photon is symmetrically distributed and so the momentum transfer due to spontaneous emission, averaged over many absorption-emission events, is zero. Figure 2a illustrates this process with cold atoms from a Bose-Einstein condensate (this will be explained later). In this situation, the atoms absorb light from a laser beam traveling from right to left. As a result an atom gets a momentum kick of $\hbar k$ to the left. Those atoms subsequently emit photons, of momentum $\hbar k$ in a random direction, due to spontaneous emission. By conservation of momentum, the atoms recoil with opposite momentum. The spatial pattern of the atoms in figure 2a is a measurement of their momentum distribution. There is a ring pattern of atoms due to the random direction of the spontaneously emitted photon. (The distribution is nominally spherical in space, but looks like a ring when projected on a plane.) The center of the ring is displaced from the initial position of the atoms (indicated by the blue arrow) due to the momentum kick from the incident laser beam. For larger objects, we can associate the scattering force with the reflection of light, as illustrated in the right side of figure 2a

The dipole force, also called the gradient force or stimulated force, is the force exerted on an atom due to coherent redistribution of photons. The dipole force can be considered as arising from stimulated Raman events - the absorption and stimulated emission of photons. (Note that the absorption and stimulated emission cannot be

thought of as successive and independent events; their correlation is central to the proper understanding of the force.) Figure 2b illustrates this process where a sample of cold atoms interacts with two laser fields, one coming from the left and the other coming from the right. In this case the an atom can absorb a photon (and get a momentum kick of $\hbar k$) from the laser beam coming from the right and simultaneously be stimulated to emit a photon into the laser beam coming from the left (and recoil from the emitted photon's momentum of $\hbar k$). The atoms, which are initially at rest, that have undergone this process receive a net momentum kick of $2\hbar k$. We can also understand the dipole force in analogy to a driven, classical oscillator. A harmonically bound charge driven by an oscillating electric field, $\dot{E}$, has an oscillating dipole moment, $\dot{P}$, proportional to the driving field, hence $\dot{P} = \alpha\dot{E}$. The induced dipole is in phase with the driving field when driven below resonance and out of phase when driven above resonance. The energy of interaction between the dipole and field is $W = -\dot{P} \cdot \dot{E} = -\alpha E^2$ and the force is $\dot{F} = -\dot{\nabla}W = \alpha\dot{\nabla}E^2$. Since E^2 is proportional to the intensity, we see that the force exerted is proportional to the gradient of the intensity. Below resonance the energy is negative and the force is such that the oscillator will be drawn toward the more intense region of the laser field, while above resonance it will be drawn to the weaker part of the laser field. For bigger objects, we can associate the dipole force with the refraction of light, as illustrated in the right side of figure 2b.

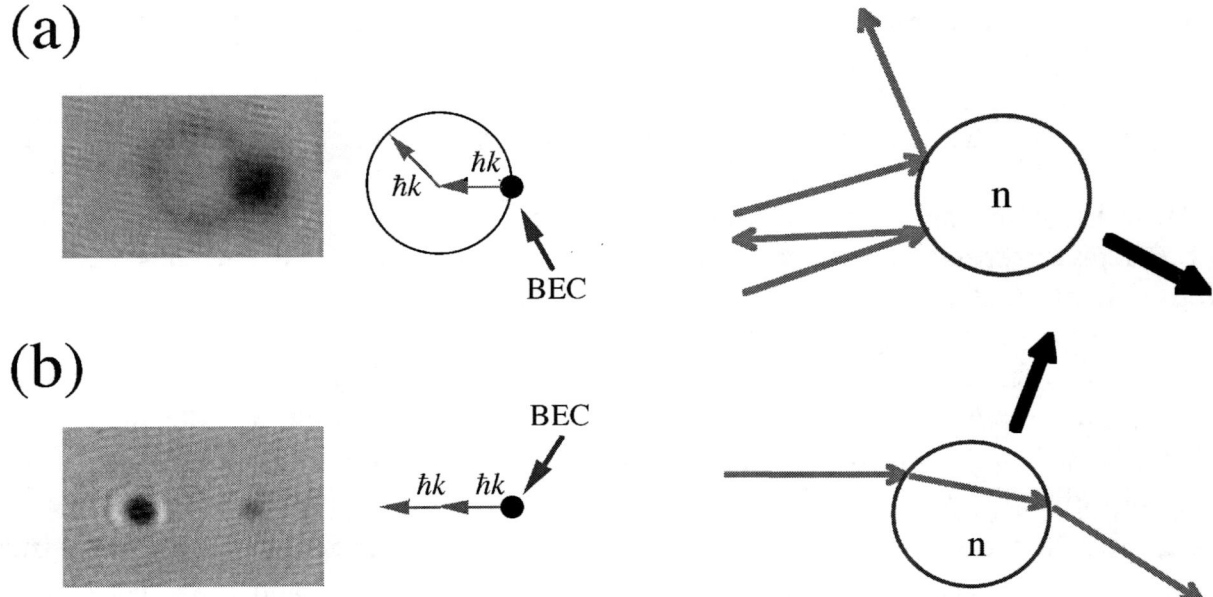

FIGURE 2. Diagrams and images representing the two types of optical forces. (a) The scattering force results from the absorption of a photon followed by spontaneous emission. The spontaneously emitted photon is typically in a random direction so, averaged over many scattering events, the net force on an atom is in the direction of the incident laser beam. For a larger particle, the scattering force is associated with the reflection of light. (b) The dipole force results from the simultaneous absorption and stimulated emission of photons. The next momentum transfer will be in the direction determined by the difference between the laser beams participating in the stimulated process. For a larger particle, the dipole force is associated with the refraction of light.

APPLICATIONS OF OPTICAL FORCES

Neutral atoms, unlike charged particles, are weakly coupled to static electric or magnetic fields. There is no good "handle" such as the charge to grab to easily manipulate the motion of an atom. Atoms do interact with resonant electromagnetic fields and in particular with resonant and near resonant laser fields. The history of optical forces can be traced to studies on the manipulation of atoms with light. Over the past 25 years, researchers have learned how to cool and trap atoms with lasers [1-3]. Optical forces are being used for many applications in many fields. In the Laser Cooling and Trapping Group of NIST, for example, among our various applications of optical forces are the manipulation of matterwaves, investigation of ultracold collisions, simulation of solid state systems, quantum information processing and studying biological molecules and membranes. In the following sections, I will use my own research to give examples of applications in two fields: atom optics and biotechnology.

Atom Optics

Atom optics, the manipulation of atoms with mirrors, beamsplitters and lenses in analogy to the manipulation of light, is a rapidly advancing field of research. Until recently, however, experiments have used thermal sources of atoms [4, 5], much as early experiments in optics used lamps. What was lacking was a coherent source of matter-waves similar to the laser for light. The creation of a Bose-Einstein condensate (BEC) of a dilute atomic gas has opened up the possibility of realizing a matter-wave source analogous to the optical laser.

Bose-Einstein Condensation

Bose-Einstein condensation is a phase transition in which, for a sufficiently high enough phase space density, a macroscopic fraction of bosonic particles occupy a single quantum state. For a cloud of atoms, the transition occurs when the density of the cloud is high enough and the temperature is low enough such that the average spacing between atoms is less than the thermal deBroglie wavelength. In figure 3, we show a simple picture of the process of Bose-Einstein condensation. Consider a cloud of atoms with a density n. The average spacing between the atoms is $d \approx n^{-1/3}$. The thermal deBroglie wavelength, λ_{dB}, is inversely proportional to the average speed of the atoms, hence $\lambda_{dB} \propto T^{-1/2}$. At high temperatures, the atoms have a short thermal deBroglie wavelength, so they can be considered as particles and essentially act as billiard balls. As the temperature decreases, the thermal deBroglie wavelength gets longer. The atoms behave like wavepackets and interference effects become evident. When the temperature is close to the BEC transition temperature, T_c, the thermal deBroglie wavelength is comparable to the average spacing between the atoms and hence the wavepackets begin to spatially overlap in phase. That is, the wavepackets, on average, begin to constructively interfere. As the temperature decreases below T_c more and more atomic wavepackets constructively interfere producing a macroscopic

matterwave. In other words, there is a macroscopic occupation of a single quantum state, typically the lowest energy state in a trap.

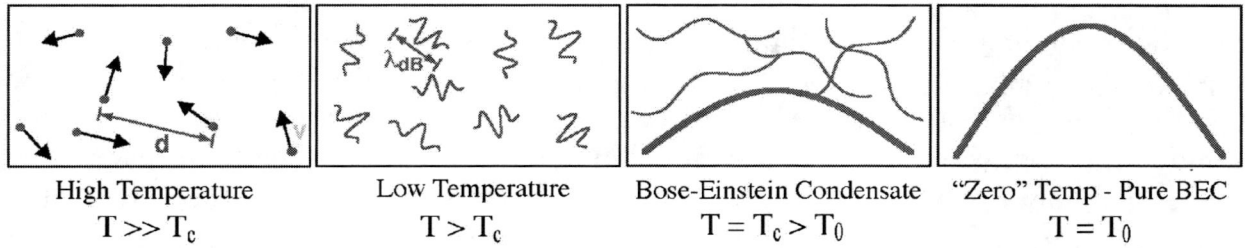

High Temperature	Low Temperature	Bose-Einstein Condensate	"Zero" Temp - Pure BEC
$T \gg T_c$	$T > T_c$	$T = T_c > T_0$	$T = T_0$

FIGURE 3. Simply picture of Bose-Einstein condensation (adapted from [8]).

The first Bose-Einstein condensates were realized in 1995. Since then, there are now over 50 groups, worldwide, producing condensates of a weakly interacting atomic gas. Although the details to realize experimentally an atomic BEC varies from laboratory to laboratory, they all basically follow three "simple" steps – laser cooling to collect a large number of cold atoms, holding the atoms in a conservative trapping potential, and evaporative cooling of the sample of atoms to reach the low temperatures and high densities necessary for BEC. In our laboratory, we produce a condensate of $\sim 10^6$ sodium atoms in a magnetic trap [6]. I will not describe the production of the BEC. Details on the various techniques can be found in refs. [7-9].

Atoms from a BEC are nearly the ideal, monochromatic source for atom optics. Many atom optical elements involve the interaction of the atoms with an optical field and the associated transfer of the photon momentum to the atoms. Because of the repulsive atom-atom interaction, which can be described by a mean field, the BEC swells to a size significantly larger than the ground state wavefunction of the harmonic trap confining the atoms [9-11]. The spatial extent of the resulting wavefunction can be several orders of magnitude larger than the optical wavelength. Hence the momentum width, given by the Heisenberg uncertainty principle, can be much less than the photon's momentum. Not all experiments will realize this reduced, intrinsic momentum width. The interaction energy may be converted to kinetic energy when the atoms are released from the trap. Nonetheless, the resulting additional momentum spread, due to the atom-atom interaction, can still be significantly less than the momentum of a single photon.

Diffraction By Optical Standing Waves

When an atomic beam passes through a periodic optical potential formed by a standing light wave, it diffracts similar to the diffraction of light by a periodic grating. The diffraction can be divided into two regimes, normal and Bragg diffraction. Both diffraction processes can be thought of as arising from the simultaneous absorption of a photon from one laser beam of the optical standing wave and stimulated emission of a photon due to the counterpropagating laser beam. This necessarily means that the momentum transfer to the atomic beam by the optical standing wave is quantized in units of $2\hbar k$, twice the momentum of a single photon. The atoms in a BEC are essentially, initially at rest. A situation similar to the passage of an atomic beam

through the standing wave can be achieved by exposing the condensate to a pulsed, optical standing wave.

In normal diffraction, the condensate atoms are exposed to a non-adiabatic pulse of an optical standing wave. For short interaction times such that the atoms do not move appreciably along the direction of the standing wave (the Raman-Nath regime), the standing wave potential can be considered as a thin phase grating that modifies the atomic deBroglie wave with a phase modulation that varies sinusoidally in space. The phase modulation comes about due to the AC Stark shift, which is the shift in the energy level from the interaction with a light field. (The AC Stark shift is proportional to the intensity of the light, which varies sinusoidally for an optical standing wave.) An atom, initially with zero momentum, is projected, by this phase modulation, onto states with momenta $p_n = n2hk$, ($n = 0, \pm1, \pm2,...$). Energy conservation is satisfied by the spread in energies associated with the non-adiabatic "turn-on" and "turn-off" of the standing wave [4]. We've used normal diffraction of a BEC for measuring the coherence length of the condensate [12] and for demonstrating the matterwave analog of the optical Talbot effect [13]. We've also investigated normal diffraction beyond the Raman-Nath regime [14].

Bragg Diffraction

A particularly useful technique that we have developed to coherently transfer momentum to the BEC is Bragg diffraction. In Bragg diffraction, the condensate atoms are exposed to an adiabatic pulse of an optical standing wave, and energy conservation must be explicitly satisfied in the interaction between the atoms and the light field. This is illustrated in the diagram of figure 4. The energy difference of the atom after the change of momentum of $2\hbar k$ must come from the photon field. For an atomic beam, this is typically accomplished by choosing the angle of incidence such that the atoms see a differential Doppler shift between the two counterpropagating laser beams comprising the standing wave. In the case where we start with BEC essentially at rest, this differential Doppler shift can be created by moving the standing wave with respect to the atoms. We create our moving standing wave by having a frequency difference δ between the two counterpropagating waves that make up the standing wave. In the presence of this moving standing wave, an atom initially at rest will simultaneously absorb photons from the higher frequency laser beam and be stimulated to emit photons into lower frequency beam, acquiring a unidirectional momentum of $2n\hbar k$ in the process. In order to satisfy energy conservation, the detuning δ must be chosen such that $\hbar\delta = n4E_{rec}$, where $E_{rec} = \hbar^2 k^2/2M$ is the recoil energy, the final energy that an atom initially at rest would have after absorption of a single photon of wavelength λ. Figure 4 shows images of the spatial distribution of atoms that have been Bragg diffracted. The images of the distribution of atoms is taken after some time of flight (TOF) period, such that moving atoms will be displaced from their initial position. Hence these images are essentially a map of the momentum distribution of the atoms. The top image shows a BEC initially at rest. In the bottom

image, a large fraction of the atoms have been displaced because they have acquired a momentum of $2\hbar k$ from the Bragg diffraction process.

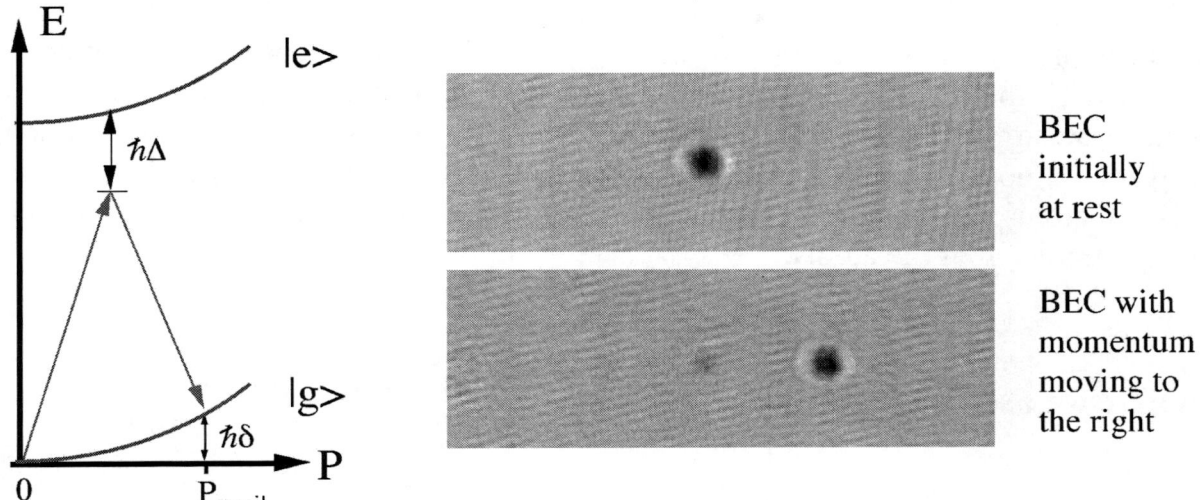

FIGURE 4. Diagram illustrating energy and momentum conservation for Bragg diffraction. Note that the frequency difference, δ, of the laser beams has to be equal to the kinetic energy change of the atom after P_{recoil} of momentum is transferred. Images of atoms before and after Bragg diffraction, demonstrating unidirectional momentum transfer.

We have observed 1st-order ($n = 1$) Bragg diffraction with 100% diffraction efficiency and up to 6th-order, with a momentum transfer of $\sim12\hbar k$ (corresponding to a velocity of 35 cm/s) with 15% efficiency [6]. Because the Bragg diffraction process does not involve a change in the internal state of the atom, it can be applied to both trapped and untrapped atoms. We have also shown that Bragg diffraction is momentum selective. If the bandwidth of a Bragg pulse is narrower than the Doppler width associated with the velocity spread of a BEC, only the fraction of the BEC within the bandwidth of the Bragg pulse will diffract. The velocity selectivity of the Bragg diffraction can be used to measure the intrinsic momentum distribution of a trapped BEC [15].

Atom Lasers

The macroscopic occupation of the ground state of a trap by a BEC is similar to the occupation of a single mode of an optical cavity by photons. One can, in fact, describe a coherent source of matterwaves from a BEC in direct analogy to an optical laser. Such a source of matterwaves is, for lack of a better term, referred to as an "atom laser". Figure 5 illustrates the analogy. In an optical laser, the macroscopic light field is confined in an optical cavity, typically formed by two mirrors. For a macroscopic matterwave, such as a BEC, the "cavity" is the trapping potential. In an optical laser, there is an externally pumped gain medium, which is the source of photons for building up the macroscopic field. For an atom laser, the pumping mechanism and gain medium are the thermal atoms and evaporative cooling. Finally, in order to extract the light from the optical cavity, some sort of output coupler is used. Typically,

one of the mirrors of the cavity is partially transmitting, which results in a directional laser beam. To realize an atom laser an atom output coupler is needed. One way to do this is to suddenly turn off the trapping potential, releasing all of the atoms. This output is equivalent to a cavity-dumped optical laser. The first demonstration of an output coupler for slowly extracting atoms from a BEC was reported by Ketterle's group at MIT in 1997 [16]. They used coherent, rf-induced transitions to change the internal state of the atoms from a trapped state to an untrapped one. This method, however, did not allow the direction of the output coupled atoms to be chosen. The extracted atoms fell under the influence of gravity and expanded because of the intrinsic repulsion of the atoms.

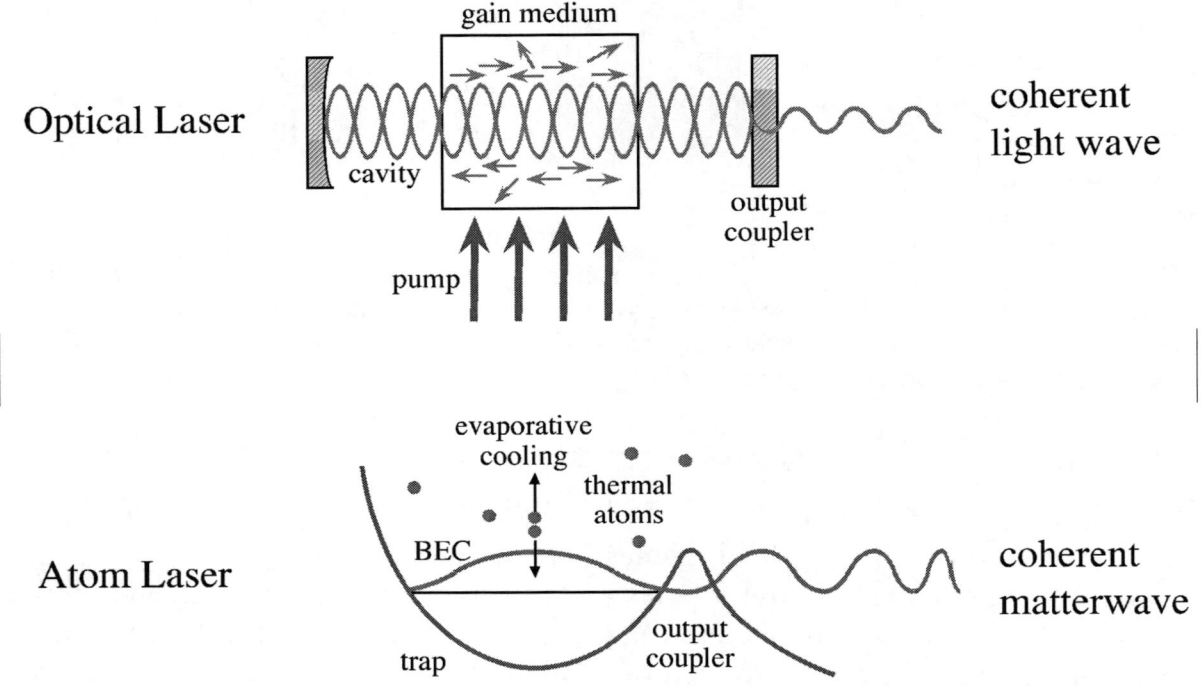

FIGURE 5. Atom laser optical laser analogy.

At NIST, we have developed a highly directional method to couple out a variable fraction of a condensate [17]. We use stimulated Raman transitions to coherently transfer trapped condensate atoms in the F=1, m=−1 magnetic sublevel to the untrapped F=1, m=0 sublevel, while giving them a momentum kick. This is similar to the process of Bragg diffraction of atoms discussed earlier; however, output coupling involves a stimulated Raman transition between different internal, as well as external, states. The frequency difference between the lasers includes the Zeeman energy between the two magnetic sublevels, in addition to the change in kinetic energy of the atoms.

Figure 6 is an image of the distribution of atoms 1.6 ms after applying 140 Raman output coupling pulses. (For reasons having to do with the geometry of our experiment [17], we used pulses rather than continuous application of the Raman beams.) In the time between two subsequent Raman pulses each output coupled wavepacket, with a velocity kick of 6 cm/s, moves only 2.9 µm. These pulses strongly overlap because

this 2.9 μm spatial separation is much smaller than the ~50 μm size of the condensate, therefore the output coupled atoms form a continuous matter wave.

The momentum kick from the Raman output coupling process produces a highly directional beam of atoms. Unlike the other output couplers demonstrated 16, 18, 19], which rely on gravity to determine the direction of the beam of atoms, the direction of our beam can be chosen by a suitable orientation of the Raman laser beams. In fact, the beam of extracted atom shown in Fig. 6 is perpendicular to gravity.

Figure 6. Demonstration of a highly directional, well collimated atom laser. Bose-Einstein condensed atoms in the m = −1 magnetic sublevel are confined in a magnetic trap. A stimulated optical Raman transition is used to couple atoms to the m = 0 magnetic sublevel which is not affected by the magnetic fields. The two photon Raman process imparts a $2\hbar k$ momentum kick to the transferred atoms causing them to leave the region of the condensate. The orientation of the laser beams determines the direction of the momentum transfer, which in our case is perpendicular to gravity (into the page).

Nonlinear Atom Optics

The advent of the laser as an intense, coherent light source enabled the field of nonlinear optics to flourish. The interaction of light in materials, whose index of refraction depends on the intensity, has led to effects such as multi-wave mixing of optical fields to produce coherent light of a new frequency, and optical solitons, pulses of light that propagate without dispersion. Nonlinear optics now plays an important role in many areas of science and technology. With the experimental realization of Bose-Einstein condensation and the atom laser, we now have an intense source of matterwaves analogous to the source of light from an optical laser. This has led us to the threshold of a new field of physics: nonlinear atom optics [20]

The analogy between nonlinear optics with lasers and nonlinear atom optics with Bose-Einstein condensates can be seen in the similarities between the equations that govern each system. For a condensate of interacting bosons, in a trapping potential V, the macroscopic wave function Ψ satisfies a nonlinear Schrödinger equation [11],

$$i\mathsf{h}\frac{\partial \Psi}{\partial t}=\left(-\frac{\mathsf{h}^2}{2M}\nabla^2+V+g|\Psi|^2\right)\Psi, \tag{1}$$

where M is the atomic mass, g describes the strength of the atom-atom interaction ($g > 0$ for sodium atoms), and $|\Psi|^2$ is proportional to atomic number density.

Four-Wave Mixing of Matterwaves

The nonlinear term in Eq. (1) is similar to the third-order susceptibility term, $\chi^{(3)}$, in the wave equation for the electric field describing optical four-wave mixing. We therefore expect that if three coherent matterwaves are spatially overlapped with the appropriate momenta, a fourth matterwave will be produced due to the nonlinear interaction, analogous to optical four-wave mixing. In contrast to optical four-wave mixing, the nonlinearity in matterwave four-wave mixing comes from atom-atom interactions, described by a mean-field interaction term $g|\Psi|^2$; there is no need for an external nonlinear medium.

Using the atoms from a BEC, we have observed such four-wave mixing of matterwaves [21]. In our four-wave mixing experiment, we used optically induced Bragg diffraction to create three overlapping wavepackets with appropriately chosen momenta. When the three wavepackets spatially separated, a fourth wavepacket, due to the wave-mixing process, was observed (see Fig. 7, right).

The process of four-wave mixing of matterwaves (and also optical waves) can be thought of as Bragg diffraction off of a matter grating. In this picture, two of the matterwaves interfere to form a matterwave grating. The third wave can Bragg diffract off of this grating, giving rise to the fourth wave. An alternative picture of four-wave mixing is in terms of stimulated emission. In this picture it is helpful to view the four-wave mixing process in a reference frame where the process looks like degenerate four-wave mixing; that is, all of the waves have the same energy.

In four-wave mixing, both energy and momentum (corresponding to phase matching) must be conserved. Since atoms, unlike photons, cannot be created out of the vacuum we have the additional requirement, for matter waves, of particle number conservation. (If we included the rest mass of the atom, particle number conservation is contained in energy conservation.) Given these three conditions, one can show that the only four-wave mixing configurations possible with matter waves are those that can be viewed in some frame of reference as degenerate four-wave mixing. This is also the geometry of phase conjugation. Fig. 7 (left) shows the four-wave mixing geometry for matterwaves viewed in the degenerate or phase conjugation frame.

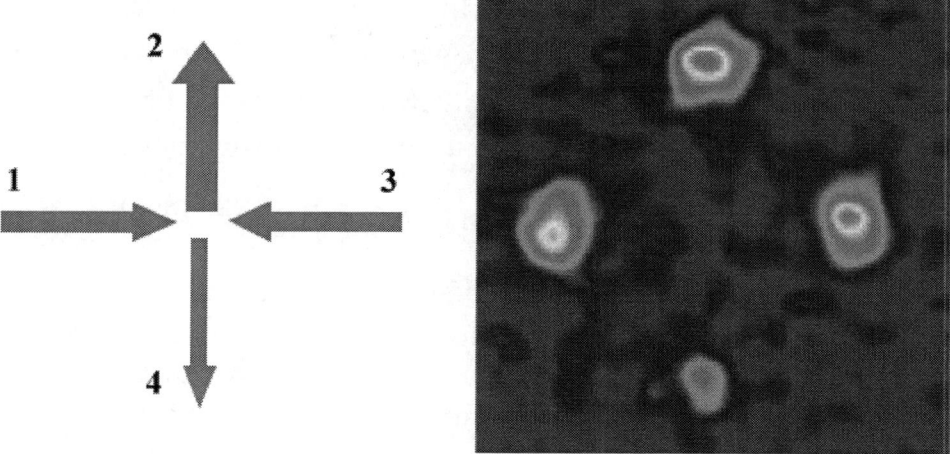

Figure 7. The process of four-wave mixing of matterwaves can always be transformed to a reference frame where the mixing process is degenerate (all of the waves have the same energy; left). The nonlinear term describing the mean-field, s-wave interaction of the atoms is responsible for the four-wave mixing. Atoms from waves 1 and 3 scatter off each other and go off back-to-back. The scattering process can be stimulated by wave 2, so that it is more likely that one of the scattering pairs goes into this wave. By momentum conservation, wave 4 is created. The small cloud of atoms in the image on the right is the fourth wave generated by four-wave mixing of matterwaves.

In the picture of four-wave mixing as arising from stimulated emission, atoms in waves 1 and 3 can be considered as undergoing an elastic collision. The scattering process results in atoms going off back-to-back in order to conserve momentum, but at some arbitrary angle with respect to the incident direction. (The scattering process is typically s-wave and the outgoing waves can be considered spherical.) In the presence of wave 2, however, this scattering process can be stimulated. There is an enhanced probability that one of the atoms from the collision of waves 1 and 3 will scatter into wave 2. (This probability is enhanced by the atoms in wave 2.) Because of momentum conservation, the enhanced scattering of atoms into wave 2 results in an enhanced number of atoms in wave 4. In this picture, it is obvious that the four-wave mixing process removes atoms from waves 1 and 3 and puts them into waves 2 and 4.

Solitons

Solitons are stable, localized waves that propagate in a nonlinear medium without spreading. They may be either bright or dark, depending on the details of the governing nonlinear wave equation. A bright soliton is a peak in the amplitude, while a dark soliton is a notch with a characteristic phase step across it. Equation (1), which describes the weakly interacting, zero-temperature BEC, also supports solitons. The solitons propagate without spreading (dispersing) because the nonlinearity balances the dispersion; for Eq. (1) the corresponding terms are the nonlinear interaction $g|\Psi|^2$, and the kinetic energy $-\left(\hbar^2/2M\right)\nabla^2$, respectively. Our sodium condensate only supports dark solitons because the atom-atom interactions are repulsive [22] ($g > 0$).

A distinguishing characteristic of a dark soliton is that its speed, v_s, is less than the Bogoliubov speed of sound $v_0 = \sqrt{gn/M}$, where n is the unperturbed condensate density. The soliton speed can be expressed either in terms of the phase step δ ($0 < \delta \leq \pi$), or the soliton "depth" n_d, which is the difference between n and the density at the bottom of the notch [22]:

$$v_s/v_0 = \cos(\delta/2) = \sqrt{1 - n_d/n}. \tag{2}$$

For $\delta = \pi$ the soliton has zero velocity, zero density at its center, a width on the order of the healing length and a discontinuous phase step [22]. As δ decreases the velocity increases, approaching the speed of sound. The solitons are shallower and wider, with a more gradual phase step. They travel opposite to the direction of the phase gradient. Because a soliton has a characteristic phase step, optically imprinting a phase step on the BEC wavefunction should be a way to create a soliton [23, 24].

A three-dimensional image of an arbitrarily complex object can be constructed by sending light, with sufficient spatial coherence, through the appropriate phase and/or amplitude mask. This is the basic principle behind physical optics, which includes wave phenomena like diffraction and holography. Diffraction can be achieved with a periodic phase and/or amplitude mask; while a more complicated mask is needed to construct a complex holographic image. In each case, the mask modifies the incoming wave and subsequent propagation produces the desired pattern of light. This idea can be readily adapted to atom optics, especially when the "incoming" matter wave is from a highly coherent source such as a Bose-Einstein condensate.

We have developed a technique to optically imprint complex phase patterns onto a Bose-Einstein condensate in order to create interesting topological states. This technique is analogous to sending a wave through a thin phase mask. The basic idea is to expose the condensate atoms to a short pulse of laser light with a spatially varying intensity pattern. The laser detuning is chosen such that spontaneous emission is negligible. (The phase mask can also serve as an amplitude mask by tuning closer to resonance, so that spontaneous emission is significant.) The pulse duration is sufficiently short such that the atoms do not move an appreciable distance compared to the wavelength of light during the pulse. (This Raman-Nath regime.) During the laser pulse, the AC Stark effect shifts the energy of the atoms by $U(r,t)$. Hence the effect on the atomic wavefunction is to "instantaneously" change its phase. This effect can be represented by multiplying the wavefunction by the phase factor $\exp(i\phi(r))$, where $\phi(r) = -\int U(r,t)dt/\hbar$. Since the AC Stark or light shift is proportional to the intensity of the light, any spatial intensity variation in the light field will be written onto the BEC wavefunction as a spatial variation in its phase. One example of this technique, discussed earlier, is the diffraction of atoms by an optical standing wave. A short pulse of standing wave light will imprint a sinusoidal phase onto the condensate. The subsequent evolution of the wavefunction results in wavepackets of atoms with a symmetric distribution of momenta in integer multiples of $2\hbar k$.

We used this phase imprinting technique to generate dark solitions in a BEC [23]. The condensate atoms were exposed to a pulsed, off-resonant laser beam, with a

spatial intensity profile such that only half of the BEC was illuminated. This was accomplished by blocking half of the laser beam with a razor blade and imaging this razor blade onto the condensate. Figure 8 shows the evolution of the condensate after the top half was phase imprinted with a phase step.

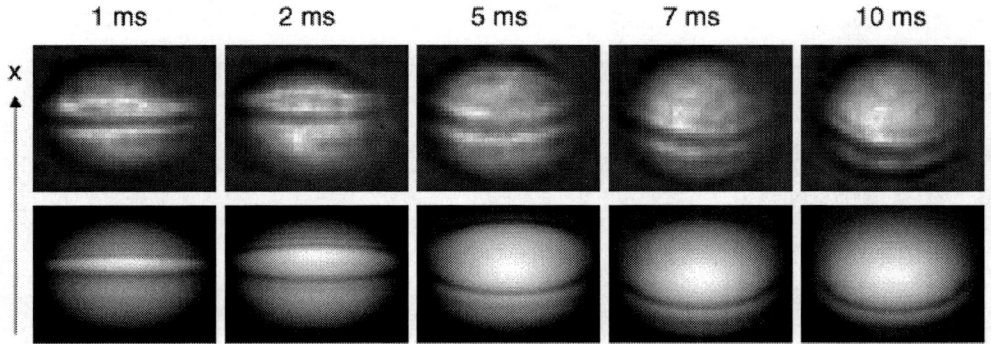

Figure 8. Experimental (upper) and theoretical (lower) images of the integrated BEC density for various times after we imprint a phase of about 1.5π on the top half of the condensate with a 1 μs pulse. The measured number of atoms in the condensate was $1.7(3)\times10^6$, and this value was used in the calculations. A positive density disturbance moved rapidly in the $+x$ direction and a dark soliton moved oppositely at significantly less than the speed of sound. Because the imaging is destructive, each image shows a different BEC. The width of the images is 70 μm.

A clear indication that the notches seen in Fig. 8 are solitons, rather than simply sound waves, is their subsonic propagation velocity. To determine this velocity, we measure the distance after propagation between the notch and the position of the imprinted phase step along the x direction. Using images taken 5 ms after the imprint, at which time the soliton has not traveled far from the BEC center, we obtain a mean soliton velocity of 1.8(4) mm/s. This speed is significantly less than the mean Bogoliubov speed of sound $v_0 = 2.8$ mm/s. From the propagation of the notch in the numerical solutions (Fig. 8, lower images) we obtain a mean soliton velocity, $v_s = 1.6$ mm/s, in agreement with the experimental value.

Atom Interferometry

Because the diffraction of the atoms by optical standing waves involves absorption and stimulated emission of photons, the process is coherent and such diffraction can be used to make mirrors and beamsplitters for atoms. At NIST, we have been applying Bragg diffraction to realize an atom interferometer with atoms from a Bose-Einstein condensate.

We've constructed a Mach-Zehnder geometry atom interferometer based on optically induced Bragg diffraction. In our Bragg interferometer, we can independently manipulate atoms in the two arms (because of their large separation) and can resolve the output ports to reveal the spatial distribution of the condensate phase. We've used this atom interferometer to determine the spatial phase profile of a BEC undergoing expansion [25], rotation [26] and soliton formation [23]. Figure 9 is illustrates the basic idea of the Bragg interferometer. An initial Bragg pulse splits the initial condensate into two states, $|A\rangle$ and $|B\rangle$, differing only in their momenta. After

they spatially separate, a phase profile (in this case a phase step for generation of a soliton) is imprinted on $|A\rangle$, while $|B\rangle$ is unaffected and serves as a phase reference. When recombined, they interfere according to their local phase difference. Where this phase difference is zero, atoms appear in port 1, and where it is π, atoms appear in port 2. Imaging the density distributions of ports 1 and 2 displays the spatially varying phase. Figure 2 shows the output of the interferometer when a phase of π was imprinted on the upper half of $|A\rangle$. The high-contrast "half-moons" are direct evidence that we did imprint a phase step of π across the condensate.

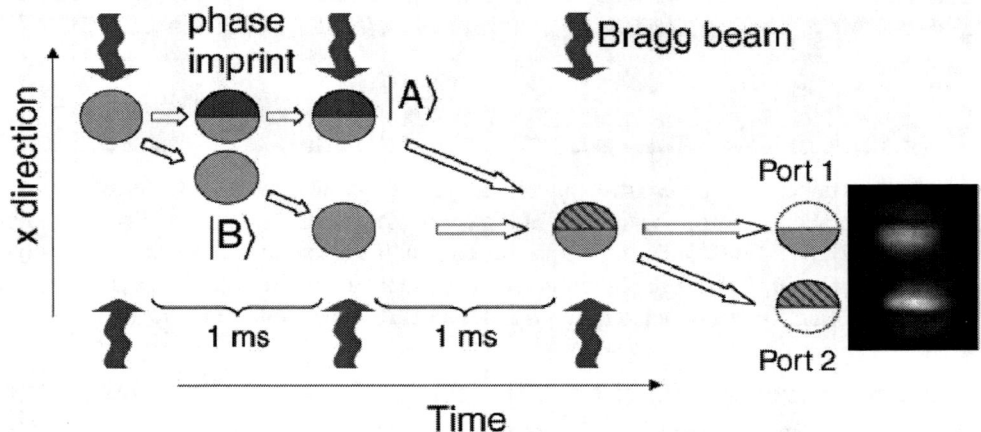

Figure 9. The Space-time diagram of the matter wave interferometer used to measure the spatial phase distribution imposed on the BEC. Three optically induced Bragg diffraction pulses formed the interferometer. The first pulse had a duration of 8 μs and coherently split the condensate into two components $|A\rangle$ and $|B\rangle$ with equal number of atoms. $|A\rangle$ remained at rest and $|B\rangle$ received two photon recoils of momentum. When they were completely separated, we exposed the top half of $|A\rangle$ to a phase imprinting pulse of π, which changed the phase distribution of $|A\rangle$ while $|B\rangle$ served as a phase reference. 1 ms after the first Bragg pulse, a second Bragg pulse of 16 μs duration brought $|B\rangle$ to rest and imparted two photon momenta to $|A\rangle$. When they overlapped again, 1 ms later, a third pulse of 8 μs duration converted their phase differences into density distributions at ports 1 and 2, which appears in the images.

Optical Manipulation For Biotechnology

The cell is arguably the basic building block of living organisms and the fundamental chemical processing plant for sustaining life. Inside the cell, complex chemical reactions typically take place in small volumes and often involve only a few, total number of molecules. Such a "nano" environment raises questions about the applicability of bulk chemical assays towards understanding and, ultimately, influencing or controlling cellular processes.

The need for studying cellular processes, evaluating cellular response and performing high-throughput reactions in the pharmaceutical industry has prompted the development of techniques for performing chemical reactions in ultra-small volumes, as well as ultra-sensitive techniques for detection of a few, if not single, molecules. In response to this need, our group is working on applying optical trapping technology for highly sensitive bioassays. In particular, we are working on using optical trapping for single biomolecule detection, for manipulation of ultra-small volume containers for

biochemical reactions, and for fabrication of nanoscale devices for biochemical applications.

Optical Tweezers

Highly focused laser beams can be used to remotely trap and manipulate dielectric objects, including cells and other biological objects, which has led to the name "optical tweezers" for such devices. Optical tweezers utilize the force (the dipole or gradient potential described earlier) resulting from the change in momentum of light transmitted by the dielectric object. When the object has an index of refraction greater than the surrounding medium, the force is such that the object will be trapped at the local maximum of the intensity of the electromagnetic field, that is, at the focus of the laser beam. The object must also be practically transparent to the light to avoid damage in the high intensity laser beam. Lasers in the near infrared are typically used for trapping.

When the wavelength of the laser light is readily absorbed, tightly focused laser beams can also used to selectively disrupt and remove material. This technique is often referred to as an "optical scalpel" and it allows one to perform microsurgery on cells and other similar sized objects. Typically, a pulsed ultraviolet (UV) laser is used for the optical scalpel. UV light is almost always absorbed by biological objects. A short pulse allows the heat generated by the absorbed light enough time to flow away and be dissipated in the surrounding medium, thereby ensuring that the optical scalpel only disrupts material localized in the focused spot.

Our apparatus is centered around a commercial inverted microscope, which we have modified to include optical tweezers and an optical scalpel [27]. We typically utilize two optical traps, which allow us to simultaneously trap two objects and move them relative to each other and to the position of the optical scalpel. The trapping light is from a continuous wave Nd:YAG laser emitting radiation at 1064 nm. The trap power can be varied up to approximately 200 mW per trap. The optical scalpel, which is at 355 nm, comes from the 3rd harmonic of a pulsed Nd:YAG laser. The pulse duration is about 5 nsec and the energy per pulse can be varied up to 5 mJoules. The focus of this beam is typically kept at a fixed position with respect to the objective lens. The apparatus also has laser induced fluorescence microscopy capabilities and an intensified CCD camera for low level light detection. A review of optical tweezers and optical scalpels, as well as details on the practical implementation of these techniques, can be found in the book by Greulich [28].

Ultra-Sensitive Bioassays

Most biological processes occur at the cellular level. Hence in the development of drugs and therapies that affect biological function, it is important to understand the effects at the cellular level. Since cellular function is often mediated by a few molecules, highly sensitive assays are needed that can detect changes at the level of a single or, at most, few molecules. There are now a number of techniques available to detect single molecules. Our research utilizes optical tweezers to measure changes in forces associated with single molecule binding. We have developed three techniques

that rely on optical tweezers as a transducer for the detection of small numbers of molecules.

An optical tweezers based immunosensor

We have developed an extremely sensitive immunoassay, which uses optical tweezers to detect antigen-antibody bonds [29]. The sensor is able to detect bio-specific adhesion due to only a few antigen-antibody binding pairs, even in the presence of non-specific binding forces. The basic principle of our device is that an antigen-coated microsphere is trapped and pulled away from an antibody coated surface using optical tweezers. The minimum amount of force applied by the optical tweezers to break the microsphere-coupled antigen to antibody bonds is measured. Detection of free antigens in solution is manifested as a reduction of this applied force due to the displacement of microsphere-coupled antigen by free antigen in the binding to the antibody.

In our experiments, 4.5 μm diameter spheres, with the antigen bovine serum albumin (BSA) covalently coupled to the surface, are placed on a silanized glass coverslip coated with the antibody anti-BSA. To demonstrate the detection of free ligands in solution, the binding force between a surface with anti-BSA and a BSA coated sphere was measured as a function of the concentration of free BSA in solution. The free BSA competes with the sphere-bound BSA for binding sites, thereby lowering the adhesion force between spheres and the surface. Our immunosensor was able to detect femtomol/L concentrations of free BSA in solution. One interesting explanation for the high sensitivity may be that the laser light used to manipulate the spheres also attracts free BSA, increasing the local concentration thereby making it easier to detect.

Optically Induced Collision Assay

We have also developed an assay that better mimics the conditions under which cellular adhesion occurs. In our new assay, which we call OPTCOL for optically controlled collisions, we collide two mesoscale particles using two independently controlled optical tweezers [30]. This assay enables precise examination of the probability of adhesion under biologically relevant conditions in which the components of the solution, the relative orientation, the impact parameter, and the relative collision velocity are all under the user's control.

To illustrate the utility of the OPTCOL assay, we studied the inhibition of viral attachment to a cell surface. The binding of influenza virus to its target cell (attachment) is the first step leading to invasion and infection. For the attachment of influenza to an erythrocyte (red blood cell) the interaction is biospecific for hemagglutinin (HA, a lectin on the surface of the virus) to sialic acid (SA, a sugar on the surface of the erythrocyte). The presence of other SA-bearing molecules can inhibit viral attachment. Polyvalent, polymeric inhibitors presenting multiple SA groups as side chains are particularly effective.

The effectiveness of inhibitors of viral adhesion to erythrocytes has traditionally been measured with techniques such as hemagglutination inhibition (HAI) assays. In

the HAI assay, the adhesion of viruses to erythrocytes forms a macroscopic gel (agglutination) that serves as an indicator of microscopic attachment. This technique is limited to ~1 nanomol/L for measuring the minimum concentration at which the inhibitor is effective. Three of the best inhibitors measured were so effective that HAI gave the lower bound of 1 nanomol/L for all of them.

In our assay, we caused a single microsphere covered with covalently bound influenza virus A (X-31) to collide with a single chicken erythrocyte, and observe whether the collision partners stick together. Dual optical tweezers independently control the positions of the erythrocyte and the microsphere, their approach velocity, and their relative orientation. The probability of adhesion is obtained by repeatedly making such measurements. The minimum concentration of inhibitor required to prevent adhesion 50% of the time was defined as the effectiveness of the inhibitor. With the OPTCOL technique, the measurements of all inhibitors with an effectiveness greater than 1 nanomol/L agreed with measurements made using the HAI assay. For those inhibitors that HAI gave a value of 1 nanomol/L, the OPTCOL assay easily distinguished their effectiveness. OPTCOL measured an effectiveness of 35 picomol/L for the best inhibitor, making it the most potent known inhibitor of attachment of influenza virus to erythrocytes ever measured. The key to the utility of OPTCOL in assaying highly effective inhibitors, where conventional assays fail, is that with OPTCOL the concentration of interactive particles can be made extremely low; much lower than the minimum concentration needed to observe hemagglutination.

Real-Time Spontaneous Dissociation Rates

OPTCOL is a highly sensitive assay for measuring equilibrium binding constants, such as the inhibition constant of potent inhibitors; however, the ability to observe and measure adhesion dynamically could provide valuable insight into the adhesion mechanism itself. Using a dual optical tweezers set-up with interferometric position detection, we can monitor, in real-time, the spontaneous (thermally driven) association and dissociation of receptors to ligand molecules immobilized on the surface of microspheres [31]. Two microspheres, one coated with a particular biomolecule and the other coated with the complementary binding pair, can be brought close to each other using the dual laser traps in which one of the traps is fixed in position and the other is mobile. The thermally driven (Brownian) motion of the microsphere in a fixed trap can be monitored directly. This thermal motion can cause the two microspheres to collide into each other. The frequency of the collision will depend on the amplitude of the Brownian motion in the trap, relative to the distance between the two microspheres. If the two microspheres collide and adhesion between the binding pairs occurs, then the two microspheres will move together. The amplitude of the Brownian motion will now reflect the motion of the two attached microspheres in the combined potential of the two laser traps. If the mobile laser trap is much stronger than the fixed laser trap, then the amplitude of the motion of the microsphere in fixed trap will be reduced. When the bond breaks and the microspheres are no longer attached to each other, the motion of the microsphere in the fixed trap will be the Brownian motion exhibited prior to attachment.

Figure 10 is a typical time record of the motion of a microsphere in a weak, fixed optical trap. The microsphere is coated with the antigen dinitro-phenol (DNP). A second microsphere, held in a strong, mobile laser trap, is brought into close proximity. The Brownian motion of the microsphere in the weak, fixed trap causes the microsphere to repeatedly collide with the second microsphere, which is coated with the DNP binding antibody IgE. During one of these collision event the DNP binds to the IgE and the amplitude of the motion decreases, reflecting the motion of the attached microspheres in the combined potential of the weak and strong laser trap.

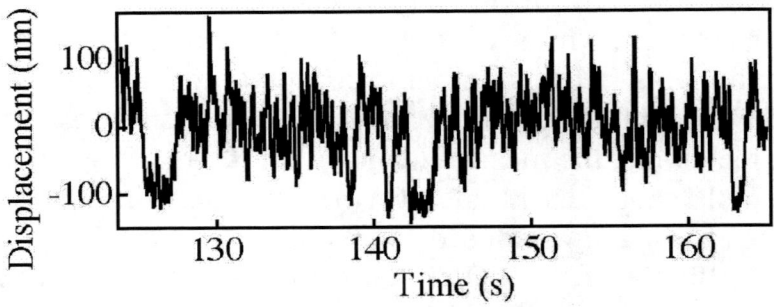

Figure 10. Typical time evolution of the signal showing successive association and dissociation events. The length of the association intervals varies stochastically.

We have sufficient sensitivity to see single antibody-antigen pair binding events. When we increase the surface density of antigens, we can observe either positive or negative cooperativity in the measured dissociation rate depending on how the antigen molecules are attached to the surface of the microspheres. That is, we observe that two microspheres, stuck together due to multiple antigen-antibody bonds, will come apart faster or slower than the rate when only one antigen-antibody bond can form due to negative and positive cooperativity between the bonds. We observe negative cooperativity in a situation where the antigen molecule is "rigidly" attached to the surface of microspheres and positive cooperativity when the antigen molecules are attached to the surface via a short, flexible tether. We speculate that rigidly attaching the molecules results in bond strain when multiple bonds are trying to form.

Ultrasmall Volume Containers

The need for performing high-throughput reactions in the pharmaceutical industry has prompted the development of ultra-small volume analytical systems such as microfluidic devices, microarrays and nanovial arrays. Nanovials offer the advantage of parallel sample analysis in a similar manner to the more common microtiter format. Microfabricated vials are generally made as indentations in a planar surface and are filled with liquid from above using ink-jet or similar technologies. The smallest vials fabricated by these methods hold between hundreds of nanoliters to hundreds of picoliters, but because the vials are not covered, evaporation is a critical issue. Without control of evaporation, droplets contained in these open vials evaporate in seconds; however, covering the vials makes it difficult to add reagents. Another disadvantage associated with nanovials fabricated in this manner is that it is difficult to

independently control individual reaction conditions (i.e., temperature and incubation time) in separate nanovials. Finally, transportation of chemicals from one nanovial to another would require the use of nanofluidic channels, a technology that is only now being developed.

Liposomes

The process of self-assembly is used widely by Nature for building complex, functional structures out of nano-scale building blocks. In particular, lipids can self-assemble into bilayer membranes that can then be formed into shapes with sizes ranging from several tens of nanometers to hundreds of microns. When the membranes assemble into liposomes (spherical structures consisting of a lipid bilayer membrane that separates the aqueous interior environment from the aqueous external environment) they are a natural choice as containers for performing biochemical reactions in ultra-small volumes.

Giant liposomes are approximately 10 μm in diameter, enclosing about a pico-liter of volume. The membrane is similar in size and composition to that of a cell and therefore giant liposomes should be a natural container for complex biochemical reactions to occur in. A number of groups are investigating the use of giant liposomes as ultra-small containers for biochemical reactions [32, 33]. The basic idea is to bring two liposomes into contact and then induce the membranes to fuse such that the contents of each liposomes mix. Our approach is to use all optical methods for manipulating the giant liposomes [27]. We use optical trapping to bring two giant liposomes into contact and then use the optical scalpel, focused at the contact point between the lipid membranes, to induce fusion of the two liposomes into one (see Figure 11).

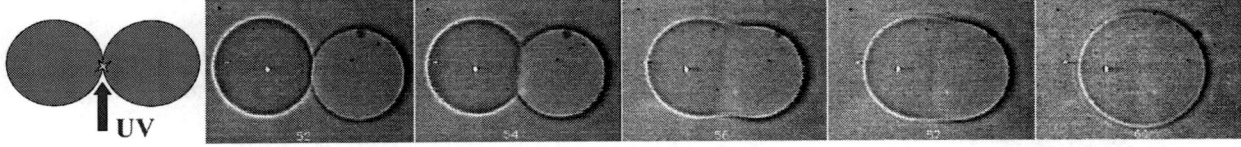

Figure 11. Images of two liposomes, brought into contact with optical tweezers and induced to fuse by the optical scalpel.

With this technique have demonstrated a controlled chemical reaction using liposomes as containers [27]. We performed the experiment using two liposomes, both approximately 10 μm in diameter. In one liposome we incorporated the chelating dye Fluo-3 in buffer. In the second liposome we incorporated Ca^{2+} in buffer. The liposomes were initially brought and held in contact with the two independent optical tweezers. The pulsed UV laser was then applied to induce fusion of the liposomes and subsequent mixing of their contents. The fluorescence of the Fluo-3 dye increased by a factor of two after fusion due to the binding of calcium ions to the dye. A similar fusion experiment performed without Ca^{2+} in the buffer showed no increase in fluorescence of the Fluo-3 dye.

Polymersomes

Polymersomes [34], which are vesicles similar to liposomes but with a self-assembled membrane composed of amphiphillic polymers, can also be used as biocompatible nano-containers. We use diblock co-polymers poly(ethyleneoxide)-poly(butadiene). They are typically longer than lipids, hence the membrane of our polymersome is typically thicker than a lipid bilayer membrane and correspondingly more robust and less prone to leakage of enclosed small molecules compared to liposomes. Similar to liposomes, polymersomes can be trapped and manipulated with optical tweezers. We have demonstrated fusion of polymersomes initiated by the pulsed UV laser. The fusion of polymersomes is typically much slower than the fusion of liposomes, often requiring several UV pulses. The slow fusion of the polymersomes is presumably due to a slower diffusion or rearrangement of the polymers due to their size and any interweaving or tangling of the polymers in the membrane.

One potentially useful feature of polymersomes is that the polymers can be cross-linked [34] to form even more stable structures. These structures can be subsequently opened up, without collapsing, using the optical scalpel. Figure 12 is a sequence of video images in which, using the pulsed UV laser, we open a hole in a polymersome that we initially cross-linked. We then trap a 4.5 μm diameter microsphere with optical tweezers and inserted it into the polymersome.

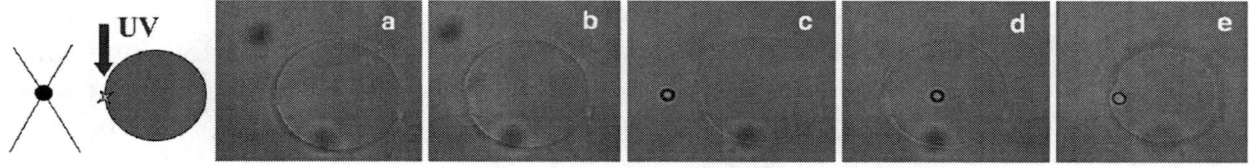

Figure 12. Video sequencing showing the opening of a hole (left side of polymersome) in a cross-linked polymersome using a pulsed UV laser, and the insertion of a 4.5 μm diameter microsphere into the polymersome using optical tweezers. The hole is localized, in that, if we move the trapped microsphere vertically (towards the reader) and then translate it to the left the microsphere runs into the polymersome wall (fig. 12e).

Hydrosomes

A number of problems associated with the incorporation of various chemicals into liposomes and polymersomes can be circumvented using a novel system, which we have invented and call hydrosomes [35]. Our hydrosome is a microscopic water droplet in a fluorocarbon background medium, such as Fluorinert from 3M. Since the index of refraction of water (1.33) is higher than that of Fluorinert (1.29), the hydrosomes are easily trapped and manipulated with optical tweezers. The solubility of water in the fluorocarbon is at the level of a few ppm, thus the water droplets are essentially stable. Further stability, especially if the water droplets are at an elevated temperature or subjected to some heating mechanism, can be achieved by using a surfactant, such as Tween-20 or Triton-100, in the water. Formation of the water droplets is easily accomplished via ultrasonic agitation, which yields droplets with

diameters in the range of 0.1 to 1 micron, or with microfluidic nozzles, which result in bigger droplets, but more mono-disperse in size. Substances that are in the water during formation of the hydrosomes are readily incorporated into the hydrosomes at the known concentration of the solution.

Fusion of hydrosomes differs from that of liposomes in two major ways. First, unlike liposomes where fusion has to be induced, hydrosomes tend to fuse spontaneously when they are brought into contact. The significance of this is that it greatly simplifies any device that might use hydrosomes in an assay (no external device such as microelectrodes or a pulsed laser beam is needed). Second, fusion of hydrosomes does not involve any rearrangement of a membrane, which is necessary for liposomes. Therefore the whole fusion process may happen faster with a corresponding increase in the mixing rate of reagents, which would greatly speed up chemical reactions and sample processing. Figure 13 is a sequence of video images showing the fusion of two hydrosomes, held in independent optical tweezers. By using two hydrosomes, one containing ladder fragments of DNA and the other containing YOYO-1 intercalating dye, we've been able to demonstrate controlled mixing and reaction.

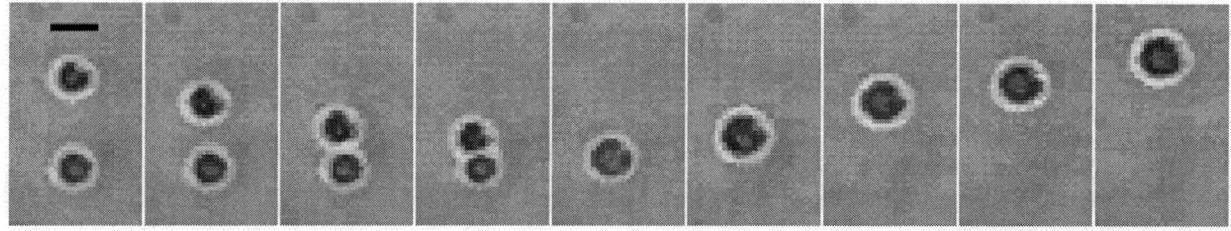

Figure 13. Series of images showing the optical manipulation of hydrosomes. Dual optical tweezers are used to grab onto two hydrosomes and bring them into contact, at which point they fuse into one hydrosome, which is then translated by a single optical trap. The scale bar in the first image indicates 1 micron.

Another application of hydrosomes, and nano-containers in general, is for single molecule studies. Single molecule spectroscopy is typically either performed transiently, where single molecules drift into and out of a focused excitation laser and the (millisecond) burst of light from the excited molecule is collected, or by attaching the single molecule of interest to a surface so that it can be interrogated for an extended period of time (typically several seconds). We have succeeded in detecting single molecules in optically trapped hydrosomes. Figure 14 is a series of time traces of the fluorescence of sulfo-Rhodamine B dye molecules in hydrosomes held stationary by the optical trap. The diameter of the hydrosome is approximately 300 nm, which is small enough to be within the confocal volume of our high numerical aperture light detection system. In addition to the detection of single dye molecules, we have also observed FRET in single DNA complexes contained in trapped hydrosomes [35].

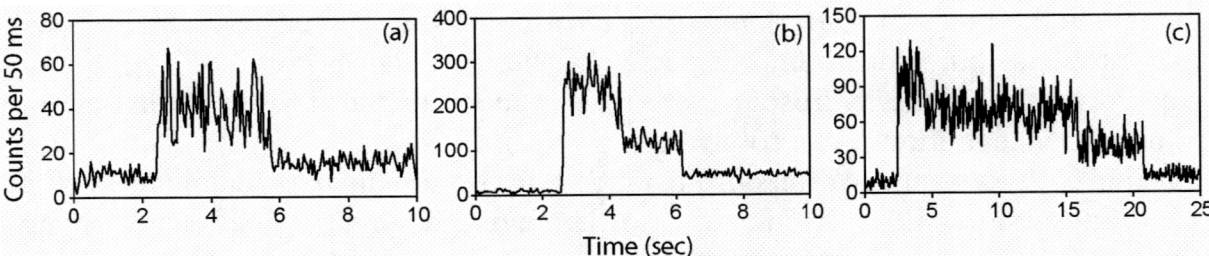

Figure 14. Three examples of single molecule detection in trapped droplets. a, b, and c illustrate the trapping and detection of 1, 2, and 3 sulforhodamine B (SRB) molecules respectively. The measurements are taken with different excitation strengths. For (a) and (c) the laser power sent into the back aperture of the microscope objective was 600 μW and for (b) the power used was 2.5 mW. The different laser powers resulted in different step sizes for photobleaching events. For these measurements a solution of 5 nmol/L SRB was used for which a 1 μm droplet is calculated to contain an average of 1.6 dye molecules.

Nanotubes

Lipid nanotubes have recently been observed between live cells and are purported to be conduits for intercellular organelle transport [36]. Pioneering experiments by Evans et al. [37] showed that a phospholipid bilayer membrane could be pulled on by micropipette techniques to form lipid nanotubes with diameters ranging from 20 to 200 nm and lengths up to 100 microns. More recent work in the group of Orwar [38] in Sweden has demonstrated that, again using micropipette techniques, lipid membranes could be stretched to form nanotubes connecting liposomes to other liposomes, thereby forming complex networks of channels and containers for processing chemical reactions. Recently, we have demonstrated that optical manipulation techniques, such as optical tweezers, could be used to stretch lipid membranes to form nanotubes. (See Figure 15.)

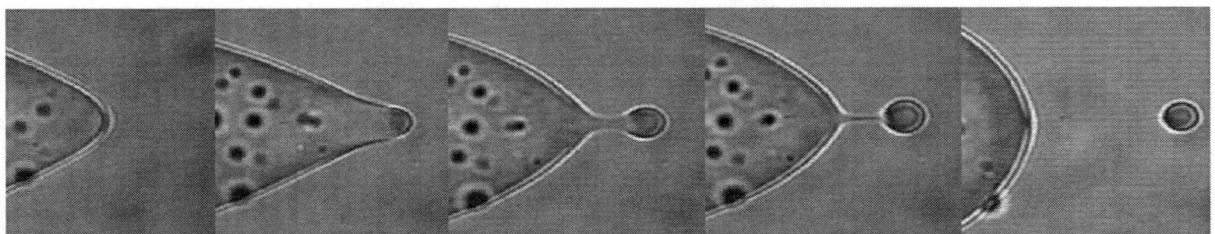

Figure 15. Sequence of video images showing the formation of a lipid nanotube by directly pulling on the membrane of a liposome with optical tweezers. The lipid nanotube in the last frame is below the resolution of our optical microscope, but is more apparent on the video as the small liposome (held by the optical trap) disappears with time as the lipids flow along the nanotube back to the initial liposome.

Although the lipid nanotubes formed are relatively stable, they are manifestations of strongly distorting lipid membranes and hence, they are not extremely robust. If the lipid nanotubes are released or the system perturbed, then the nanotubes would tend to collapse back to a structure that minimizes the energy of the system, which in this case would be the initial liposome.

Similar to liposomes, the membranes of polymersomes can be stretched to form nanotubes. This approach is particularly appealing because, as described earlier, the polymersomes are readily cross-linked. We incorporate a surfactant in the polymer membrane, which modifies the visco-elastic properties of the membrane, such that nanotubes can be readily pulled from the membrane using optical tweezers. We've been able to pull nanotubes up to a ~1 cm in length and were only limited by our translation stage. We subsequently stabilize our pulled polymer nanotubes by free radical polymerization of the hydrophobic butadiene tails [39].

The cross-linked nanotubes are extremely robust. They can be washed and removed from solution, which we did to prepare them for imaging by TEM. Figure 16a contains TEM images of the cross-linked polymer nanotubes. The left image shows a relative uniform nanotube and one that is corrugated. A number of cross-linked nanotubes in this sample were corrugated. The corrugation, which could also be seen for nanotubes in buffer by fluorescence microscopy, was only observed after cross-linking, suggesting that the cross-linking process was responsible. The image on the right-side of figure 16a is a magnified view of another pair of nanotubes. Again, one of the nanotubes has a relatively uniform profile while the other is highly corrugated. The polymer membrane is partially visible in this image and appears to have a thickness between 15 and 20 nm. With this membrane thickness the average inner diameter of the uniform nanotube is approximately 50 nm, while the corrugated nanotube's inner diameter appears to be 10 to 20 nm at the narrowest constrictions.

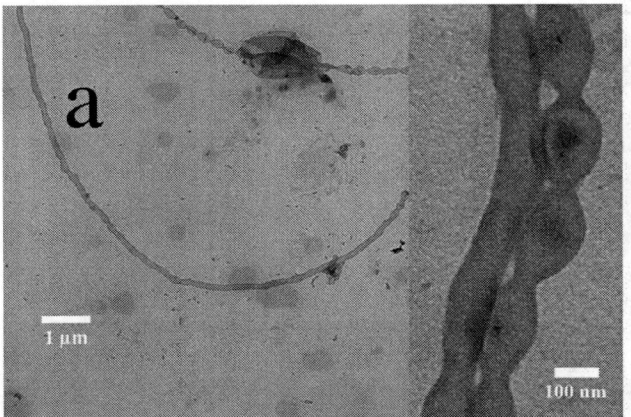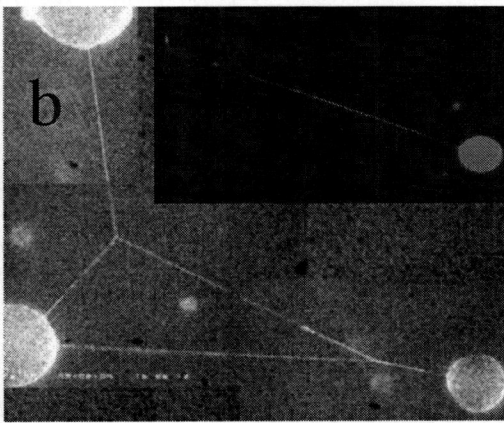

Figure 16. (a) Transmission electron microscope images of extracted and washed, cross-linked polymer nanotubes. (b) Composite image from video fluorescence microscopy of a network of polymer nanotubes and polymersomes assembled using optical tweezers. Inset image: Scanning confocal microscopy image of a nanotube pulled from a polymersome encapsulating sulfoRhodamine dye in buffer.

For applications of nanotubes as conduits for transporting biological material, it is necessary that the core of the nanotube contain water. In order to observe the water-filled core, we pulled nanotubes from polymersomes encapsulating buffer containing the fluorescent dye sulfoRhodamine B. The inset of figure 16b is a fluorescence image of a pulled polymer nanotube and the parent polymersome. The fluorescence from the

excited dye, evident in the parent polymersome, can also be seen in the pulled nanotube, indicating that buffer containing dye is in the core of the nanotube.

The polymer membranes that we pull on with optical tweezers to form nanotubes [39] appear to be as easy to manipulate as lipid membranes, especially with regard to forming network structures such as a Y-junction [38]. Figure 16b contains a composite image of a network composed of polymer nanotubes and polymersomes, imaged via a fluorescent membrane dye. Optical tweezers were used to pull nanotubes from polymersomes and then to attach them to other polymersomes. The several Y-junctions apparent in this image were formed by pulling an additional nanotube from a polymersome that already had a nanotube attached to it and the two nanotubes subsequently merged to form the junction, or by pulling a nanotube from the membrane of an existing nanotube.

ACKNOWLEDGMENTS

None of the work described in these proceedings could have taken place without the efforts and input of my colleagues. The work was primarily carried out by countless number of dedicated post-docs. Special thanks to my long time colleagues Bill Phillips, Paul Lett and Trey Porto, who are the other NIST staff members of the Laser Cooling and Trapping Group. Special thanks also to Steve Rolston, who was a past group member and a collaborator on all of the atomic physics experiments. Finally, I'm externally grateful to Rani Kishore who has been both a friend and collaborator on all of the bio-related research.

REFERENCES

1. S. Chu, *Rev. Mod. Phys.* **70**, 685-706 (1998).
2. C. N. Cohen-Tannoudji, *Rev. Mod. Phys.* **70**, 707-720 (1998).
3. W. D. Phillips, *Rev. Mod. Phys.* **70**, 721-742 (1998).
4. Feature issue on mechanical effects of light, *J. Opt. Soc. Am. B* **2**, 1706 (1985).
5. Special issue on atom optics and interferometry, *Appl. Phys. B* **54**, 321 (1992).
6. M. Kozuma *et al.*, *Phys. Rev. Lett.* **82**, 871-874 (1999).
7. E. A. Cornell and C. E. Wieman, *Rev. Mod. Phys.* **74**, 875–893 (2002).
8. W. Ketterle, *Rev. Mod. Phys.* **74**, 1131–1151 (2002).
9. *Bose-Einstein condensation in atomic gases*, Proc. of the International School of Physics "Enrico Fermi", Course CXL, edited by M. Inguscio, S. Stringari and C.E. Wieman, IOS Press, Amsterdam, 1999.
10. J.R. Anglin and W. Ketterle, *Nature* **416**, 211-218 (2002).
11. F. Dalfovo, S. Giorgini, L. P. Pitaevskii and S. Stringari, *Rev. Mod. Phys.* **71**, 463-512 (1999).
12. E. W. Hagley *et al.*, *Phys. Rev. Lett.* **83**, 3112-3115 (1999).
13. L. Deng *et al.*, *Phys. Rev. Lett.* **83**, 5407-5410 (1999).
14. Y. Ovchinnikov *et al.*, *Phys. Rev. Lett.* **83**, 284-287 (1999).
15. J. Stenger *et al.*, *Phys. Rev. Lett.* **82**, 4569-4573 (1999).
16. M.-O. Mewes *et al.*, *Phys. Rev. Lett.* **78**, 582-585 (1997).
17. E. W. Hagley *et al.*, *Science* **283**, 1706-1709 (1999).
18. B. P. Anderson and M. A. Kasevich, *Science* **282**, 1686-1689 (1998).
19. I. Bloch, T. W. Hänsch and T. Esslinger, *Phys. Rev. Lett.* **82**, 3008-3011 (1999).
20. G. Lens, P. Meystre and E. W. Wright, *Phys. Rev. Lett.* **71**, 3271-3274 (1993).
21. L. Deng *et al.*, *Nature* **398**, 218-220 (1999).
22. W. P. Reinhardt and C. W. Clark, *J. Phys. B: At. Mol. Opt. Phys.* **30**: L785-L789 (1997).
23. J. Denschlag *et al.*, *Science* **287**, 97-101 (2000).
24. S. Burger *et al.*, *Phys. Rev. Lett.* **83**, 5198-5201 (1999).
25. J.E. Simsarian *et al.*, *Phys. Rev. Lett.* **85**, 2040 (2000).

26. M. F. Andersen *et al.*, to be published in *Phys. Rev. Lett.*

27. S. Kulin, R. Kishore, K. Helmerson and L. Locascio, *Langmuir* **19**, 8206-8211 (2003).

28. K. O. Greulich, *Micromanipulation by Light in Biology and Medicine: The Laser Microbeam and Optical Tweezers*, Basel: Birkhäuser, 1999.

29. K. Helmerson, R. Kishore, W. D. Phillips, and H. H. Weetall, *Clin. Chem.* **43**, 379-383 (1997).

30. M. Mammen *et al.*, *Chemistry & Biology* **3**, 757-763 (1996).

31. S. Kulin, R. Kishore, J. B. Hubbard and K. Helmerson, *Biophys. J.* **83**, 1965-1973 (2002).

32. D. T. Chiu *et al.*, *Science* **283**, 1892-1895 (1999).

33. A. Strömberg *et al.*, *Proc. Natl. Acad. Sci.* **97**, 7-11 (2000).

34. D. E. Discher and A. Eisenberg, *Science* **297**, 967-973 (2002).

35. J. E. Reiner *et al.*, *Appl. Phys. Lett.* **89**, 013904 (2006).

36. A. Rustom, R. Saffrich, I. Markovic, P. Walther and H.-H. Gerdes, *Science* **303**, 1007-1010 (2004).

37. E. Evans, H. Bowman, A. Leung, D. Needham and D. Tirrel, *Science* **273,** 933–955 (1996).

38. M. Karlsson *et al.*, *Annu. Rev. Phys. Chem.* **55,** 613–649 (2004).

39. J. E. Reiner *et al.*, *Proc. Natl. Acad. Sci.* **103**, 1173-1177 (2006).

I. ATOMIC, MOLECULAR & CONDENSED MATTER PYSICS

Experimental Observation of the Triplet States of $^{39}K_2$ by Infrared-Infrared Double Resonance Spectroscopy

Feng Xie, Dan Li and Li Li[*]

Department Physics and Key Lab of Atomic and Molecular Nanosciences
Tsinghua University, Beijing 100084, China

Abstract. Perturbation facilitated Infrared-Infrared double resonance spectroscopy has been used to study the triplet states of K_2. The $1^3\Delta_g$, $2^3\Sigma_g^+$, and $2^3\Pi_g$ states have been observed. Resolved fluorescence spectra into the $a^3\Sigma_u^+$ and $b^3\Pi_u$ states are recorded. Perturbations between the $2^3\Pi_g$ and $4^1\Sigma_g^+$ states have been observed.

INTRODUCTION

Alkali dimers have long served as prototype molecules for the study of the spectra and structure of diatomic molecules. Triplet states are not only three times more than singlet states, but also they can provide more information which cannot be obtained from singlet states alone: spin-orbit, spin-spin, and spin-rotation interactions, as well as hyperfine structure. The ground state of alkali dimers, however, is a singlet state, $X^1\Sigma_g^+$. Transitions from singlet states to triplet states are spin-forbidden. Thus experimental study of triplet states of alkali dimers is much more difficult than the study of singlet states. Li and Field [1, 2] developed perturbation facilitated optical-optical double resonance (PFOODR) spectroscopic technique to study triplet states of alkali dimer molecules. The unique point of this PFOODR technique is to use $A^1\Sigma_u^+\sim b^3\Pi_u$ mixed levels as the intermediate *window* levels in the two-step excitation to reach high-lying triplet states. Resolved fluorescence from high-lying triplet states gives information about the low-lying $a^3\Sigma_u^+$ and $b^3\Pi_u$ states. Fig. 1 gives an energy level diagram of the PFOODR excitation and resolved fluorescence spectroscopy. With this technique, many triplet Rydberg and several doubly excited states of Li_2 and Na_2 have been observed and the $a^3\Sigma_u^+$ and $b^3\Pi_u$ states have been studied.[3,4]

The K_2 electronic states have been studied both theoretically and experimentally. Potential energy curves of 98 electronic states below the $4s+5d$ atomic limit have been calculated with high accuracy. [5,6] PFOODR spectroscopy has been applied to K_2. Recently we carried out perturbation facilitated infrared-infrared (IR-IR) double resonance excitations of K_2 and observed the $2^3\Pi_g$, $2^3\Sigma_g^+$, and $1^3\Delta_g$ states for the first time.

EXPERIMENTAL

Fig. 2 shows the experimental setup of our IR-IR double resonance excitation and resolved fluorescence spectroscopy. Potassium vapor was generated in a heatpipe oven with Ar buffer gas. Two DL 100 single mode tunable diode lasers were used as the *pump* and the *probe* lasers. The laser frequencies were measured by WA-1600 wavemeters. The *pump* laser was held fixed to excite an $A^1\Sigma_u^+\sim b^3\Pi_u$ mixed intermediate level from the ground state. The *probe* laser was scanned and transitions into the high-lying triplet states were detected by monitoring the direct or collision-induced $2^3\Pi_g \rightarrow a^3\Sigma_u^+$ yellow-green fluorescence. When the two diode laser frequencies were held fixed to excite an upper triplet level, resolved fluorescence spectra to the $a^3\Sigma_u^+$ and $b^3\Pi_u$ states were recorded by scanning a Spex 1404 monochromator.

---

[*] To whom all correspondences should be addressed. E-mail address: lili@mail.tsinghua.edu.cn.

CP888, *Modern Trends in Physics Research*,
Second International Conference on Modern Trends in Physics Research—MTPR-06,
edited by L. El Nadi
© 2007 American Institute of Physics 978-0-7354-0354-9/07/$23.00

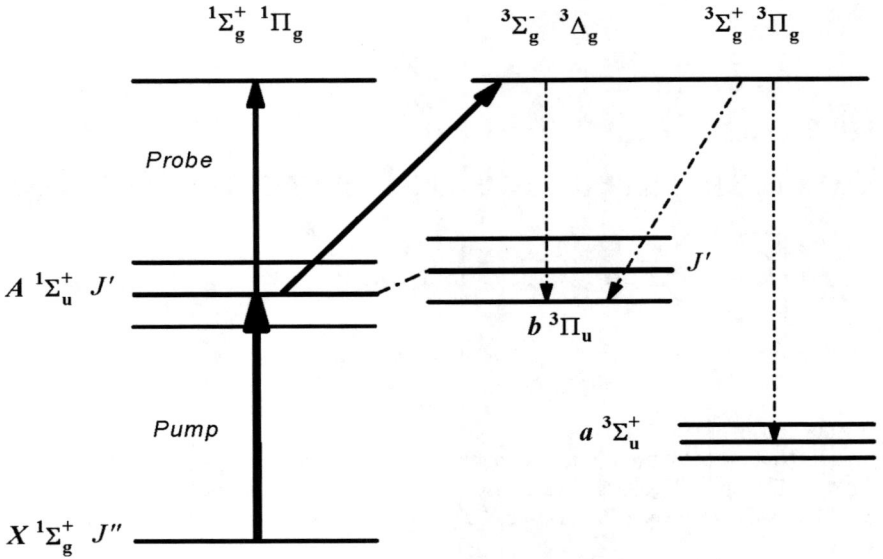

FIGURE 1. Energy level diagram of PFOODR spectroscopy. The pump laser selectively excites an A~b mixed intermediate window level, and the probe laser further excites to singlet/triplet Rydberg/doubly excited states. The $^3\Pi_g$ and $^3\Sigma_g^+$ states can fluorescence to both the $b^3\Pi_u$ and $a^3\Sigma_u^+$ states, while the $^3\Delta_g$ and $^3\Sigma_g^-$ states can fluorescence to the $b^3\Pi_u$ state.

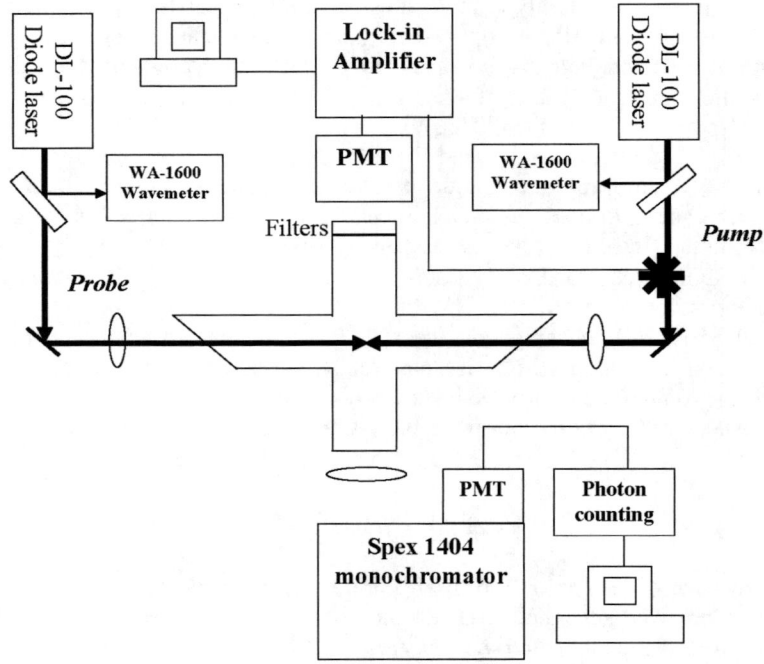

FIGURE 2. Experimental setup.

RESULTS

Ref. 7 reported the observation of the $2^3\Pi_g$ v=4-14 vibrational levels. Since then, many more data have been recorded. The complete analysis of the data will be reported in Ref. 8. Fig. 3 shows resolved fluorescence spectra

from the $2^3\Pi_{1g}$ v=16, J=54 T_{vJ}=22566.736 cm^{-1} level to the $a^3\Sigma_u^+$ state. In Fig. 3 (a), the spectrum contains lines into the bound levels of the shallow well of the $a^3\Sigma_u^+$ state (546.6 nm to 552.3 nm, not resolved in (a)), and defuse band to the repulsive wall of the $a^3\Sigma_u^+$ state. The strong peaks around 573 nm are due to quantum interference, which has been observed in the $2^3\Pi_g \rightarrow a^3\Sigma_u^+$ spectra of Li$_2$ and Na$_2$.[2,3] Fig. 3 (b) shows the bound-bound lines of the $2^3\Pi_{1g}$ v=16, J=54 $\rightarrow$ $a^3\Sigma_u^+$ transition with higher resolution. The doublets are the $2^3\Pi_{1g}$ v=16, J=54 $\rightarrow$ $a^3\Sigma_u^+$ v', N'=55 and 53 lines. The vibrational quantum numbers of the $a^3\Sigma_u^+$ state are given above the doublets. The lines into the $a^3\Sigma_u^+$ v'=3 vibrational level did not appear due to small Franck-Condon factor. Strong perturbations between the $2^3\Pi_{1g}$ and $4^1\Sigma_g^+$ states have been observed.

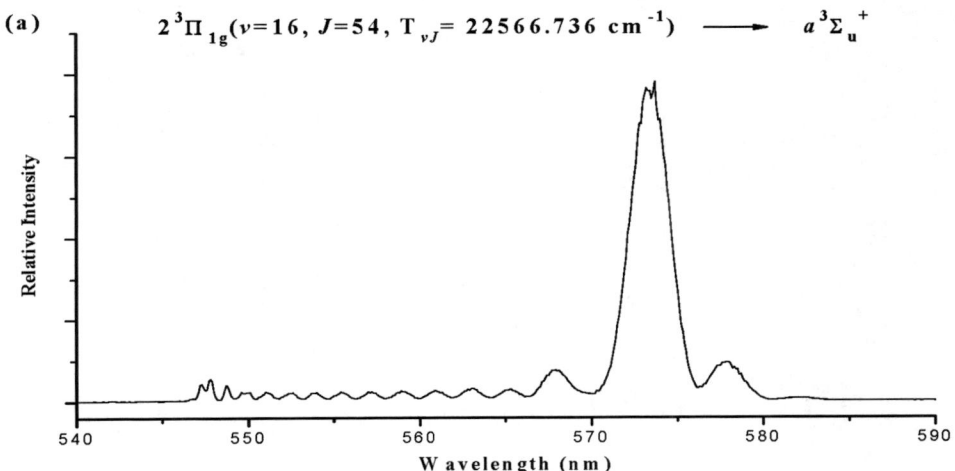

(a) $2^3\Pi_{1g}(v$=16, J=54, T$_{vJ}$= 22566.736 cm^{-1}) $\longrightarrow$ $a^3\Sigma_u^+$

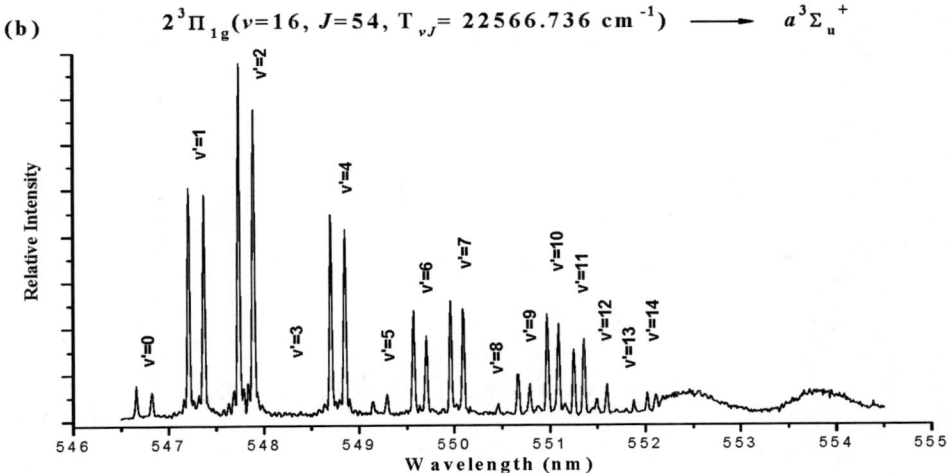

(b) $2^3\Pi_{1g}(v$=16, J=54, T$_{vJ}$= 22566.736 cm^{-1}) $\longrightarrow$ $a^3\Sigma_u^+$

FIGURE 3. (a) Resolved fluorescence spectrum from the $2^3\Pi_{1g}$ v=16, J=54, T$_{vJ}$=22566.736 cm^{-1} level to the $a^3\Sigma_u^+$ state. (b) Bound-bound lines of the $2^3\Pi_{1g}$ v=16, J=54, T$_{vJ}$=22566.736 cm^{-1} $\rightarrow$ $a^3\Sigma_u^+$ transition. The doublets are the $2^3\Pi_{1g}$ v=16, J=54 $\rightarrow$ $a^3\Sigma_u^+$ v', N'=$J\pm1$ lines. The vibrational quantum numbers of the $a^3\Sigma_u^+$ state are given above the doublets. The lines into the $a^3\Sigma_u^+$ v'=3 vibrational level did not appear due to small Franck-Condon factor.

The $1^3\Delta_g$ state dissociates to the $4s$+$3d$ atomic limit as the $2^3\Pi_g$ state. The $3d$ K atom has a spin-orbit splitting $^2D_{5/2} - ^2D_{3/2}$= -2.33 cm^{-1}, thus a small spin-orbit constant, Hund's case (a) coupling at low J, and case (b) coupling at higher J are expected for the $1^3\Delta_g$ state. Fig. 4 shows the coupling schemes at case (b) and case (a). Five groups of lines have been observed via each intermediate rotational level when J is higher as expected from Fig.4 (a), and only

$\Delta\Omega = \Delta\Lambda = 1$, $\Delta J = 0, \pm 1$ lines have been observed at the low J limit. The detailed analysis of the $1^3\Delta_g$ state will be reported elsewhere.[9]

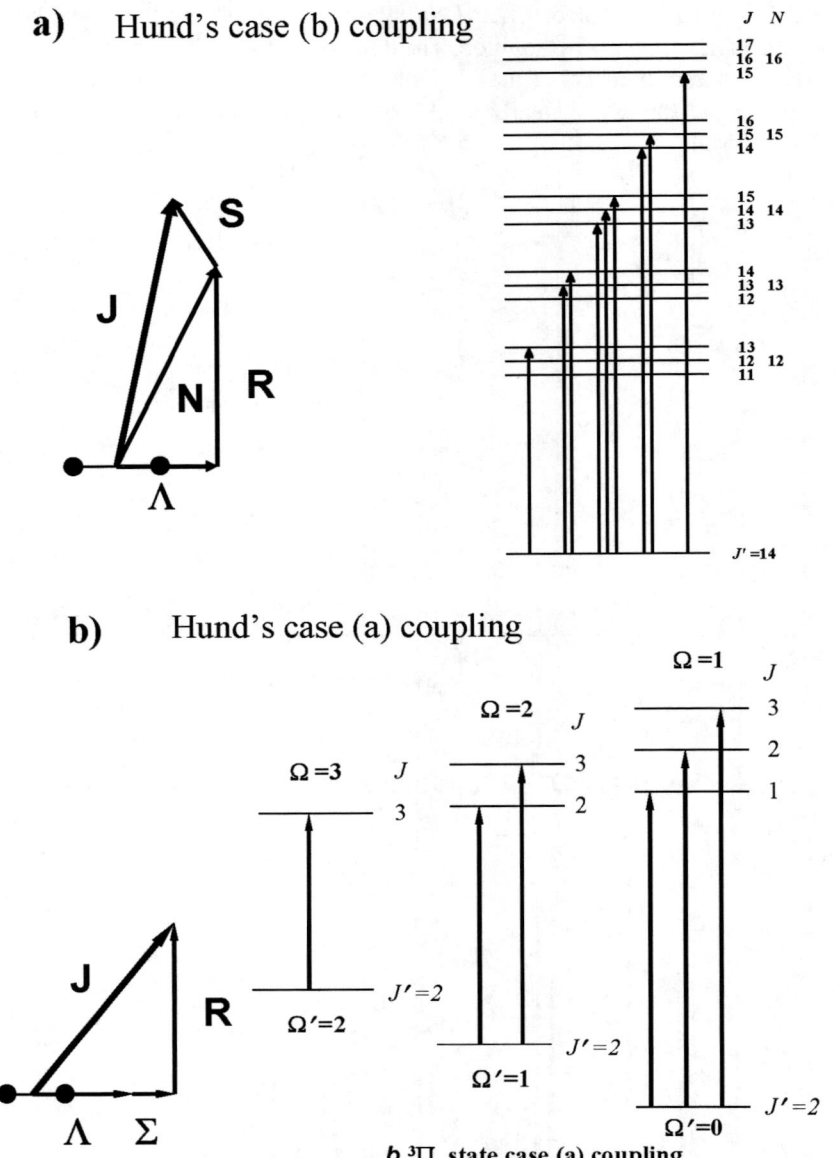

FIGURE 4. Hund's coupling schemes of the K_2 $1^3\Delta_g$ state. a) Hund's case (b), five groups of lines are expected from each intermediate $b^3\Pi_u$ rotational level. b) Hund's case (a), $\Delta\Omega = \Delta\Lambda = 1$, $\Delta J = 0, \pm 1$ lines are expected.

There are the $4^1\Sigma_g^+$ and $2^3\Sigma_g^+$ states in the same energy region as the $2^3\Pi_g$ and $1^3\Delta_g$ states.[5,6] The $4^1\Sigma_g^+$ state strongly mixed with the $2^3\Pi_{0g}$ state. The $2^3\Sigma_g^+$ levels are high vibrational levels according to *ab inito* calculation.[5,6] The results of the $4^1\Sigma_g^+$ and $2^3\Sigma_g^+$ states will be published separately.[10,11]

ACKNOWLEDGMENTS

This work was supported by NSFC (20473042) and SRFDP of China.

REFERENCES

1. Li Li and R. W. Field, *J. Phys. Chem.* **87**, 3020-3022 (1983).
2. Li Li and R. W. Field, "Continuous Wave Perturbation-Facilitated Optical-Optical Double Resonance Spectroscopy of Na_2 and Li_2" in *Molecular Spectroscopy and Dynamics by Stimulated Emission Pumping*, edited by H. L. Dai and R. W .Field, Singapore: World Scientific Co., 1995, pp. 251-277.
3. Li Li and A. M. Lyyra, *Spectrochimica Acta Part A.* **55**, 2147- 2178 (1999).
4. J. Huennekens, I. Prodan, A. Marks, L. Sibbach, and Li Li, *J. Chem. Phys.* **113**, 7384-7397 (2000).
5. S. Magnier and P. Millie, *Phys. Rev. A* **54**, 204-218 (1996).
6. S. Mangier, M. Aubert-Frecon, and A. R. Allouche, *J. Chem. Phys.* **12**, 1771-1781 (2004).
7. Y. Chu, F. Xie, D. Li, Li Li, V. B. Sovkov, V. S. Ivanov, and M. Lyyra, *J. Chem. Phys.* **122**, 074302 (2005).
8. F. Xie, D. Li, Y. Chu, Li Li, S. Magnier, V. B. Sovkov, and V. S. Ivanov, to be published.
9. F. Xie, D. Li, Li Li, R. W. Field, and S. Magnier, to be published.
10. D. Li, F. Xie, Li Li, V. B. Sovkov, and V. S. Ivanov, and S. Magnier, to be published.
11. D. Li, F. Xie, Li Li, V. B. Sovkov, and V. S. Ivanov, A. M. Lyyra, J. Heunneckens, and S. Magnier, to be published.

Electronic Band Structure and Magnetic Moment of SmCo4B

Sherif Yehia

Physics department, Faculty of science, Helwan University, Cairo, Egypt

Abstract. We present a first-principles calculation of the density of states(DOS), electronic band structure and magnetic moment of SmCo4B. The magnetic properties of SmCo4B using density functional theory(DFT) calculations are treated beyond the local density approximation (LDA) within the LDA+U scheme as implemented in the wien2k package. The calculated magnetic moment is in fair agreement with available experimental values. The features of both DOS and band structure depend on the scheme used in the calculation. The results are discussed and compared to those of SmCo5.

INTRODUCTION

The magnetic properties of RCo5 compounds and their derivatives e.g. $R_{n+1}Co_{3n+5}B_{2n}$ with n=1,2,... have always attracted much interest on both the fundamental and application levels. The compound SmCo4B is especially interesting because it has huge uniaxial magnetic anisotropy at T=4.2 K. However, both the saturation moment and the Curie temperature are about half of those of SmCo5 .Although few ab initio studies have been reported on SmCo5[1,2] no such studies were done on its derivative SmCo4B however experimental research has been done on this compound to improve its magnetic properties[e.g. 3-5]. Electronic parameters and density of states analysis were done however for other rare earths e.g. GdCo4B and YCo4B[6,7]. The RCo4B compounds crystallize in the well-known CeCo4B type structure[8] in which the Ce atoms occupy two distinct sites 1a and 1b while two and six Co atoms occupy the 2c and 6i sites respectively. The two B atoms are located in the plane which contains the 1b sites of the rare earth atoms. Since Co and Sm atoms occupy distinct sites whose neighboring atoms are different in type and number their magnetic moment and their contribution to the total DOS will be different.

In this paper we report first-principles calculation on SmCo4B using an all-electron, full-potential, relativistic linearized augmented plane wave method (FPLAPW) in the local density approximation (LDA) and the generalized gradient approximation(GGA)methods. The obtained results will be discussed in the light of available experimental data of SmCo4B and ab initio studies on SmCo5.

COMPUTATIONAL METHOD

SmCo4B crystallizes in the CeCo4B structure (space group P6/mmm,No. 191). The experimental values of a and c/a used in the calculation are 9.452 a.u. and 0.792 respectively. Our electronic structure code uses the Full-Potential Linearized Augmented Plane Wave (FPLAPW) [9] based on the Density Functional Theory (DFT) [10]. The Local Density Approximation (LDA) of Perdew and Wang [11] and the Generalized Gradient Approximation (GGA) of Perdew, Burke and Ernzerhof [12] were used for correlation and exchange potentials as implemented in the Wien2k code [13]. Both core and valence states are calculated self-consistently, the core states fully relativistically for the spherical part of the potential, and the valence states using the full potential. Local orbital extensions [14] with the converged basis of approximately 500 basis functions were used to reduce linearization errors in Sm and Co spheres. For the Brillouin zone integration we used the modified tetrahedron method for the self-consistent band structure calculations. Self-consistent calculations were performed with 48K points and the convergence was checked by varying the number of k points up to 64 in the irreducible Brillouin zone. We used the muffin tin (MT) spheres

CP888, *Modern Trends in Physics Research,*
Second International Conference on Modern Trends in Physics Research—MTPR-06,
edited by L. El Nadi
© 2007 American Institute of Physics 978-0-7354-0354-9/07/$23.00

$R_{MT}^{Sm} = 2.115$ a.u , $R_{MT}^{Co} = 2.015$ a.u , $R_{MT}^{B} = 1.8$ a.u. and the cut off energy parameters R_{Kmax} and G_{max} of 7 and 14 respectively.

Spin-Orbit (SO) interaction is an essential term for calculating the magnetocrystalline anisotropy energy and for understanding the localized nature of f electrons. The SO interaction was included using a second variational scheme [15] by taking all states below the cut off energy 1.5 Ry. The orbital moment is very much affected by spin orbit coupling and other correlation effects. The band structure and density of state of f electron materials is not easy to be understood using local density approximation (LDA) method. This difficulty is caused due to the localized nature of f electron and the subtle interplay between the spin polarization and the spin orbit interaction (SO). The spin orbit coupling and the many body correlations effects cause the orbital magnetic moment to be important. The coulomb correlation in the Sm $4f$ shell is expected to be significant factor in these systems ,the LDA+U method will be used. The LDA+U eliminate the deficiency of LDA by applying the Hubbard like interaction term for $4f$ electrons. The spin orbit coupling was included in the Hamiltonian to determine the size of the orbital magnetic moments of the two Sm atoms. In this way ,our calculation scheme corresponds to a relativistic LDA+U+SO method. The LDA+U method describes properly the orbital magnetism of solids with strongly correlated electrons. The LDA+U in the fully localized form, is appropriate for strongly localized electrons In the LDA+U scheme the U term accounts for the coulomb interaction and seems to be an essential factor in rare earth materials with $4f$ shell [9]. The two parameter needed it for carrying LDA+U method are U and J as defined for the two Sm-f shells of $SmCo_4B$ namely we have used $U_f \approx 5.2$ eV and $J_f \approx 0.75$ eV as used by Larson et at [1] for the $SmCo_5$.

RESULTS AND DISCUSSION

Fig.1 shows the density of states in the spin-up direction for $SmCo_4B$. The spin-orbit interaction is taken into account. The features of this plot is reminiscent of those of $SmCo_5$ except that Co contribution is less and in addition to the boron contribution two different Sm atoms have their different contributions to the total DOS.

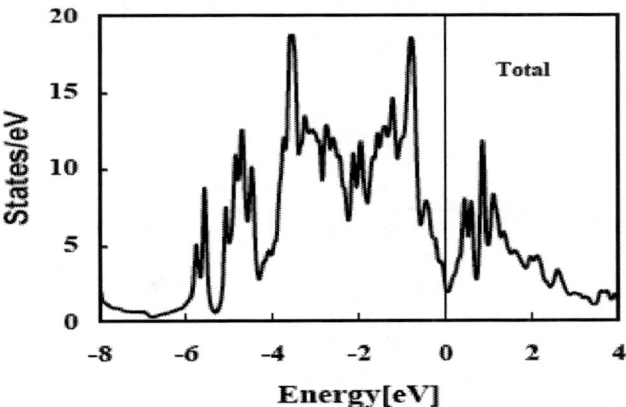

FIGURE 1. Calculated spin-up density of states (DOS) for the $SmCo_4B$ where GGA and LDA+U schemes were used with spin-orbit coupling.

Figs.(2a,2b) show in detail the local DOS curves for the Sm 1a, Sm 1b, Co-2c , Co-3g and B atoms. Several observations may be said about these plots : first, both Sm atoms contribute only few states to the DOS at Fermi energy. The DOS peaks of these two atoms are located within the same energy range but the overall DOS is different. Most of the DOS above E_f comes from the Sm atoms. Second, the 3g site of Co has the largest contribution to the DOS at E_f and third, the Boron atoms contribution to the DOS is much less than that of Co and Sm. Some of these features were also observed in $GdCo_4B$ [6].

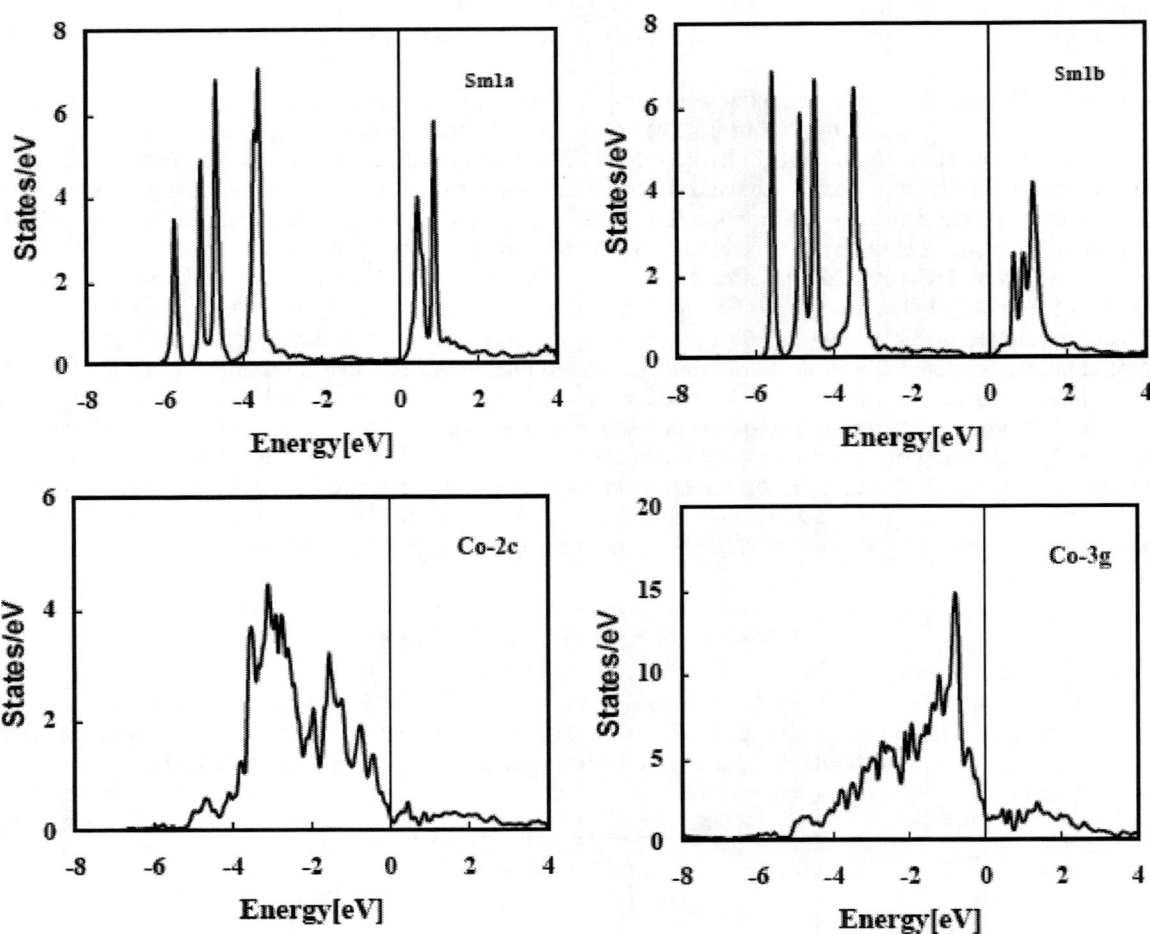

FIGURE 2a. Calculated spin-up density of states (DOS) for Sm-1a ,Sm-1b,Co-2c and Co-3g with GGA and LDA+U in SmCo$_4$B including spin-orbit coupling.

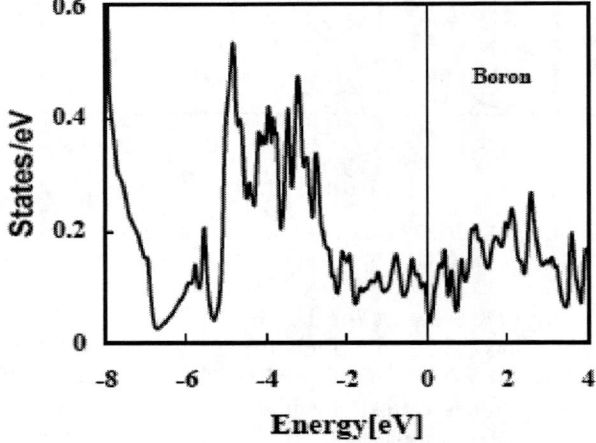

FIGURE 2b. Calculated spin-up density of states (DOS) for Boron atom with GGA and LDA+U in SmCo$_4$B including spin-orbit coupling.

In order to illustrate the role played by spin-orbit coupling on the band structure and density of states we have done calculation in the GGA, LDA+U scheme but with excluding SO interaction. Fig.3 displays the DOS of Sm 1a

in this scheme. The difference between Figs.2a and 3 is a relative shift of the DOS peaks towards lower energy and that the relative contribution of the states above E_f to the total DOS is less in the absence of spin-orbit coupling.

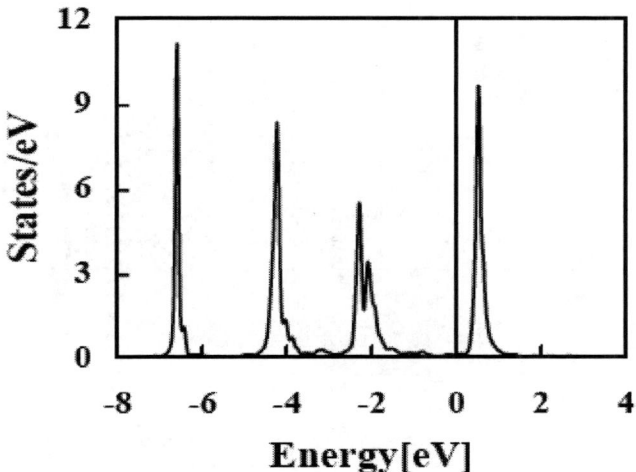

FIGURE 3. Calculated spin-up density of states (DOS) for the Sm-1a orbital with GGA and LDA+U in $SmCo_4B$ without spin-orbit coupling.

The corresponding band structure of Fig. 3 is shown in Fig. 4. The Sm 1a characters are displayed as circles and they are forming horizontal bands signifying high DOS at specific energies above and below E_f. It may be of interest here to compare the band structure of Fig.4 with our calculation of its counterpart in $SmCo_5$ (not shown). The location and number of the horizontal bands are different due to the different surroundings of the Sm atom in the two crystal structures.

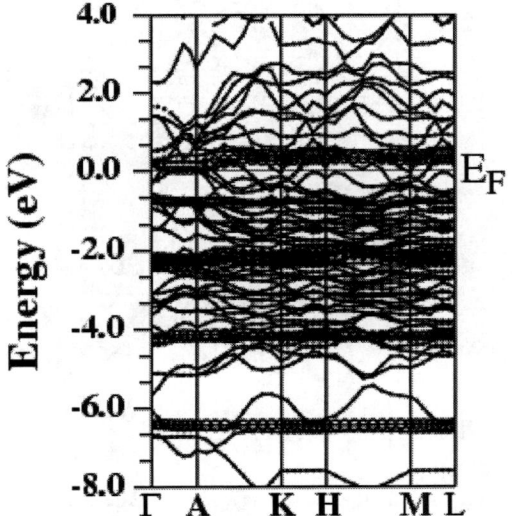

FIGURE 4. Calculated spin-up band structure for the Sm-1a orbital. GGA and LDA+U schemes were used without spin-orbit coupling. The radius of the plotting circles is proportional to Sm-1a projection of the band.

We have calculated the magnetic moment of $SmCo_4B$, in the GGA+LDA+U including spin-orbit coupling, and found the following values for the different atoms : Sm(1a)= 1.47468 , Sm(1b)= 1.65622 ,Co(2c)= 1.61715, Co(3g)= 0.67010 and B = -0.04397. The magnetic moments on the Sm atoms are the net moments after subtracting the negative orbital contribution for each. The calculated net magnetic moment of $SmCo_4B$ is 5.1 μ_b/f.u. which is in fair agreement with the experimental value 3.22 μ_b/f.u. Reported by Ido et al[3]. We may remark here that the values measured by those authors for $SmCo_4B$ and $SmCo_5$ (7 μ_b/f.u.) are rather low compared to other reported data.

Including the SO interaction allows one to calculate the band structure with the magnetization vector either parallel or perpendicular to a specific axis. The results of these calculations along [001] and [100] directions are displayed in Figs. 5 and 6 respectively.

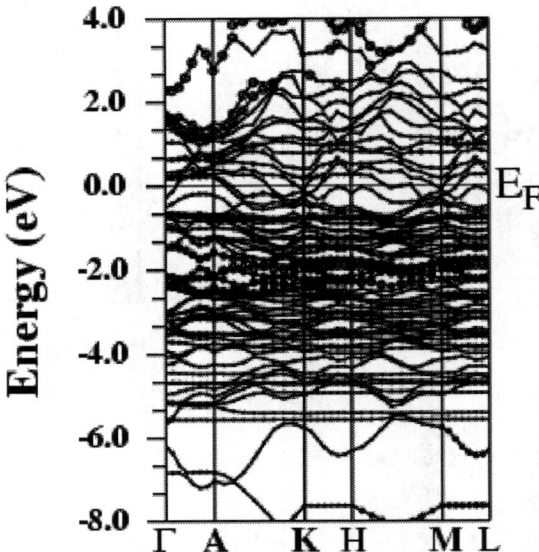

FIGURE 5. Calculated spin-up band structure for the $SmCo_4B$. GGA and DA+U schemes were used with spin-orbit coupling. The magnetization direction is along the [001] direction.

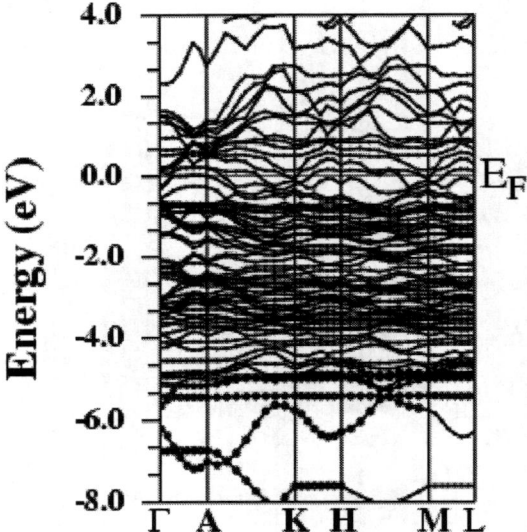

FIGURE 6. Calculated spin-up band structure for the $SmCo_4B$ .GGA and LDA+U schemes were used with spin-orbit coupling. The magnetization direction is along the [100] direction.

The Sm bands in these two figures are located at different energies and since the magnetocrystalline anisotropy, arising from spin-orbit coupling in these systems are mainly due to the f-shell of the Sm atoms it is expected that magnetic anisotropy energy will be present. We are currently studying in more detail the magnetic anisotropy in some RCo_4B to find out the reason of the huge anisotropy values in some members of this system as compared to the parent RCo_5 system.

CONCLUSIONS

We presented a ab initio calculation on the magnetic moment and electronic band structure of $SmCo_4B$ using different schemes incorporated in wein2k package. The LDA+U scheme with spin-orbit coupling seems to be adequate in calculating a magnetic moment in fair agreement with experiment . The two Sm atoms contribute differently to the total magnetic moment and to the band structure. The same is true for the 2c and 3g atoms of the cobalt sublattice.

REFERENCES

1. P. Larson, I. I. Mazin and D.A. Papaconstantpoulus, Phys. Rev. B **67**, 214405 (2003).
2. S. K. Malik, F. J. Arlinghaus and W. E. Wallace, Phys. Rev. B **16**, 1242(1977) .
3. H. Ido, K.Konno, H. Ogata, K. Sugiyama, H. Hachino, M. Date and K..Maki, J.Appl. Phys. **70**, 6128 (1991).
4. H. Ido, K. Sugiyama, H.Hachino ,M. Date, S.F. Cheng and K. Maki , Physica B **177**, 265 (1992).
5. H. Ido,O. Nashima, T. Takahashi , K. Oda and K. Sugiyama, J.Appl. Phys. **76**, 6165 (1994).
6. A. Kowalczyk, G. Chelkowska and A. Szajek, Solid State Commun. **120**, 407(2001).
7. A. Szajek, J.Magn. Magn. Mater. **185**,322(1998).
8. Y. B. Kuz,ma, N. S. Bilonizhko, N. F. Chaban and G. V. Chernyak, J. Less-Common Met **99**,L21(1984)
9. D. J. Singh, Planewaves, Pseudopotentials, and the LAPW Method (Kluwer Academic , Boston 1994)
10. W. Kohn and L. J. Sham , Phys. Rev. A **140**, 1133(1965).
11. J. P. Perdew and Y. Wang, Phys. Rev. B **45**, 13244 (1996).
12. J. P. Perdew, K. Burke and M. Ernzerhof, Phys. Rev. Lett.**77**, 3865(1996) .
13. P. Blaha, K. Schwarz, G. K. H. Madsen, D. Kvasnicka and J. Luitz, WIEN2K, An Augmented Plane Wave + Local Orbitals Program for Calculating Crystal Properties (K.Schwarz, Techn. Universitat Wien, Austria, 2001).
14. D.J. Singh, Phys. Rev. B **43**, 6388(1991).
15. D.D. Koelling and B. Harmon, J. Phys. C **10**, 3107(1977).

Slow Dissociative Collisions between He and Diatomic Molecular Ions

Feras Afaneh[1] and Lothar Schmidt[2]

[1]*Physics Department, Hashemite University, P.O. Box 150459, Zarqa 13115, Jordan.*
[2]*Institut für Kernphysik, University Frankfurt, Max-von-Laue-Str. 1, D-60438 Frankfurt, Germany.*

Abstract. We studied slow dissociative collisions between He and diatomic molecular ions using Cold Target Recoil Momentum Spectroscopy (**COLTRIMS**) technique in combination with fragment imaging technique. All final state momentum components, as well as the masses of the molecular fragments were determined. As the complete information on the kinematics is available, we are able to calculate the final state binding energy as well as the kinetic energy of the molecular fragments in the molecular center of mass system (kinetic energy release). The results show that the probability of the reaction channels is strongly dependent on the relative geometric orientation of the participants.

INTRODUCTION

Using modified versions of the **C**old **T**arget **R**ecoil **M**omentum **S**pectroscopy (**COLTRIMS**) technique together with fragment imaging, unique studies of charge transfer to aligned molecules in collision with He atoms at slow impact velocities have been performed. The results reveal the dependence of the reaction upon the relative geometric orientation of the participants. In particular we have studied charge transfer from He to the HeH^+ molecular ion forming dissociative HeH levels, to the H_2^+ molecular ion forming dissociative H_2 levels and for other molecular ions like HD^+, D_2^+.

COLTRIMS in combination with a three-dimensional fragment imaging technique allows us to measure the momentum vector of the He^+ recoil ion as well as the scattering angle and time-of-flights of both projectile fragments. Therefore, we are able to determine the final state binding energy ε with a resolution of 4 eV (FWHM). Several reaction channels involving different molecular states can be clearly separated. Within the axial recoil approximation, we measured this axis by determining the relative momentum of the molecular fragments. Thus, we were able to investigate the orientation dependence of *DEC* process. In this contribution we will present the initial results of our experiments

CP888, *Modern Trends in Physics Research,*
Second International Conference on Modern Trends in Physics Research—MTPR-06,
edited by L. El Nadi
© 2007 American Institute of Physics 978-0-7354-0354-9/07/$23.00

EXPERIMENTAL SETUP

The experiment was carried out at the ECR ion source installed in the institute for nuclear physics of the Frankfurt (M) University. Combining the well-established **COLTRIMS** reaction microscope with a three-dimensional, highly resolving fragment imaging technique, all final state momentum components of the molecular fragments were obtained [1-3]. The schematic of the experimental setup is shown in Figure 1. Briefly, a beam of 10-keV molecular ion, for example, H_2^+ extracted from ECRIS was collimated to a beam spot of less than 1 mm^2 by two pairs of collimators. The collimated beam (z axis) collided with a supersonic **He** gas jet (y axis). Following the collision, the main beam (H_2^+ ions) were deflected electro-statically and then dumped into a small Faraday cup. The time-of-flight for 10-keV H_2^+ to reach the projectile detector from the interaction region was 1 μsec, far larger than the dissociation lifetime of the H_2 states populated in the collision. The neutral fragments resulting from the dissociation of H_2 after the collision flied in straight line, through the electrostatic analyzer and finally were detected by a multi-hit position-sensitive detector. By determining the times-of-flight (**TOF**) and the impact locations of the two coincident fragments, one is able to determine the velocity vectors of both fragments and, hence, to fully reconstruct the kinematics of the collision process provided that dissociation occurs in the collision volume [4,5]. The associated **He$^+$** recoil ion was projected at right angles to the beam onto another position-sensitive detector by a weak extraction field followed by a field-free drift region. From the position and the time-of-flight of the recoil ion, measured in coincidence with projectile fragments, the recoil-ion momentum vector was determined.

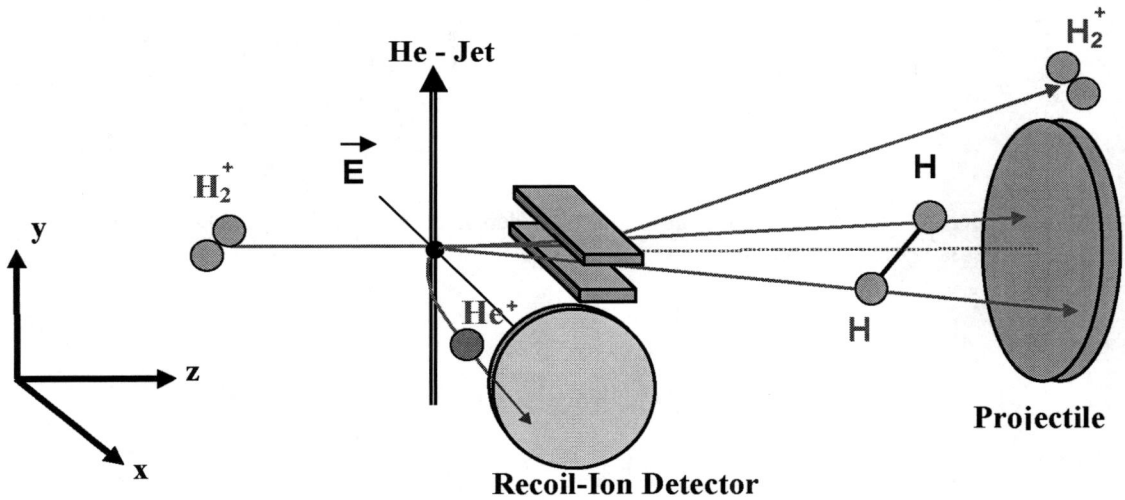

FIGURE 1. Schematic drawing of the experimental setup.

RESULTS AND DISCUSSION

Dissociative Electron Capture (*DCE*) of H_2^+ ions induced by collisions with **He** has been studied

$$H_2^+ + He \rightarrow H_2^{**} + He^+ \rightarrow H + H + He^+ \qquad (1)$$

As the complete information on the kinematics is available, we are able to calculate the final state binding energy as well as the kinetic energy of the molecular fragments in the molecular center of mass system (kinetic energy release). This allowed obtaining fragment distributions differentiated in kinetic energy of dissociation and molecular orientation for selected groups of H_2 states and for fixed collision plane and momentum transfer.

The dissociative electron capture (*DEC*) process can be viewed as a two-step reaction consisting of the formation of a doubly excited neutral-molecule state (H_2^{**}) and its subsequent dissociation into neutral atomic fragments. Characteristically, a single doubly excited state provides the coupling to the electronic continuum necessary for the capture step, but then the way toward a large number of final states is opened up as, with increasing internuclear distance, the dissociating state crosses the Rydberg series of bound potential energy curves of the neutral molecule which converge to the ionic ground state. Figure 2 shows the **Kinetic Energy Release** (*KER*) of both **H** fragments resulted from the collision as a function of the energy loss $\Delta\varepsilon$ determined from the coincident He^+ recoil-ion momentum $P_{//}$. In this figure several reaction channels involving different molecular states for the reaction can be clearly separated. Using the known potential-energy curves for H_2 and H_2^+ given in Fig. 2(a), we were able to identify the reaction channels measured in this experiment. The following reaction channels have been identified:

$$H_2^+ (1s) + He (1s^2) \longrightarrow H(1s) + H(1s) + He (1s)^+ \qquad (2)$$
$$H_2^+ (1s) + He (1s^2) \longrightarrow H(1s) + H(2l) + He (1s)^+ \qquad (3)$$
$$H_2^+ (1s) + He (1s^2) \longrightarrow H(2l) + H(2l) + He (1s)^+ \qquad (4)$$
$$H_2^+ (1s) + He (1s^2) \longrightarrow H(1s) + H(1s) + He (2l)^+ \qquad (5)$$

The molecular dynamics of these reaction channels and their dependence on impact parameter and geometric orientation will be discussed later in the coming papers.

CONCLUSIONS

In this paper we have reported a kinematically complete study of dissociative electron capture (*DEC*) of H_2^+ from **He**, in which ground and excited states of H_2 are formed. Cold target recoil ion momentum spectroscopy together with a three-dimensional, highly resolving fragment imaging technique was used to measure the momentum vector of the He^+ recoil ion as well as the scattering angle and time-of-flights of both **H** fragments. By calculating the final state binding energy as well as the kinetic energy release of the molecular fragments, the reaction channels involving different molecular states were identified.

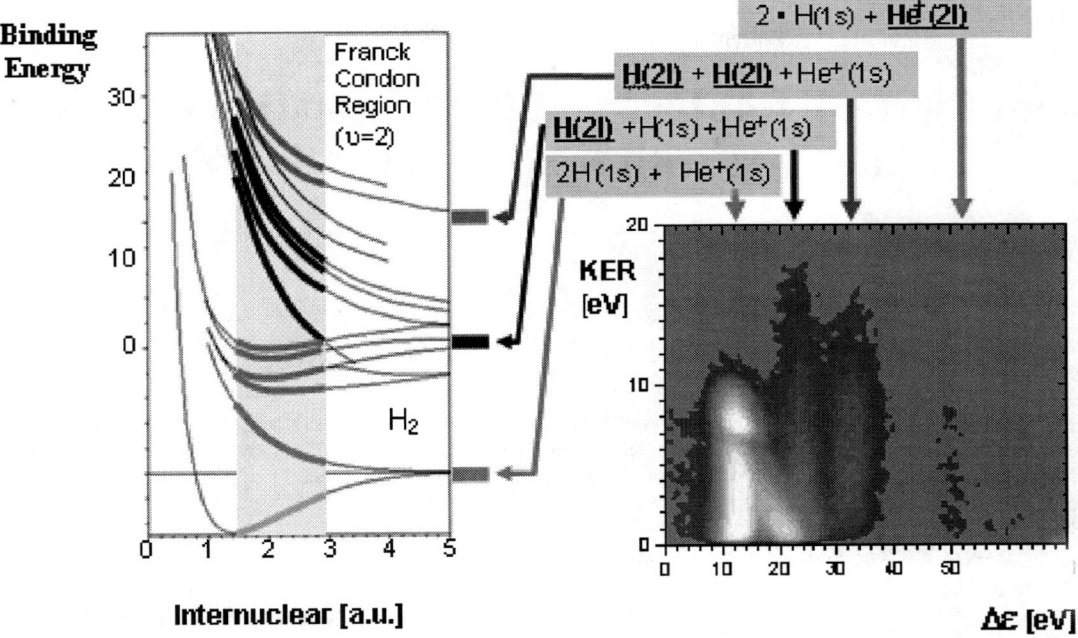

FIGURE 2. (a) **H₂** dissociative energy potential curves (b) **KER** distribution versus the change of electronic binding energy $\Delta\varepsilon$, for 10 keV $H_2^+ + He_{target} \longrightarrow H + H + He^+_{target}$. No. of reaction channels involving different molecular states have been identified

ACKNOWLEDGMENTS

We acknowledge financial support from DFG, DAAD, Frankfurt University and the Hashemite University. We would like also to thank Professor Reinhard Dörner for his support.

REFERENCES

1. Doerner, R., Mergel, V., Jagutzki, O., Spielberger, L., Ullrich, J., Moshammer, R., and Schmidt-Boecking, H., Physics Reports 330, 95-192 (2000).
2. Ullrich, J., Moshammer, R., Dorn, A., Doerner, R., Schmidt, L., and Schmidt-Boecking, H., Rep. Prog. Phys. 66, 1463–1545 (2003).
3. Afaneh, F., Doerner, R., Schmidt, L., Weber, Th., Stiebing, K.E., Jagutzki, O., and Schmidt-Boecking, H., J. Phys. B 35, L229-L235 (2002).
4. Schmidt, L., Afaneh, F., Schöffler, M., Titze, J., Jagutzki, O., Weber, Th., Stiebing, K.E., Doerner, R., and Schmidt-Boecking, H., Physica Scripta Vol. T110, 379-383 (2004).
5. Afaneh, F., Doerner, R., Schmidt, L., and Schmidt-Boecking, H., NIMB 234, 431-440 (2005).

Resonant Electron Capture and Recombination Rates for O-like Ions with K-shell Excitation.

G. Omar[*], R. Semeda and F. Shahin

[*]Physics Department, Faculty of Science, Ain Shams University, Egypt.
Physics Department, Faculty of Science, Beni-Suef University, Egypt.

Abstract. In most cases, electron–ion (e/I) collisions proceed through the capture of free electrons by positive ions. The mode of radiationless capture which leads to the formation of intermediate doubly-excited (d) states is known as resonance capture (RC) process. When d-states stabilize by emission of radiation (x-rays), dielectronic recombination (DR) is completed. The DR is an interesting process which is responsible for self-cooling of thermal astrophysical and laboratory plasma.

In the present work, the DR rates with cascade effect are computed for O-like Al^{5+}, Cl^{9+}, Ti^{14+} and Zn^{22+} ions with 1s-excitation. It is found that, the group of resonant states of the form $1s2s^22p^5n\ell$ ($\ell = 0, 1, 2$) has the highest contribution to the total DR rates for each of the previous ions. The peak values are found 1.01, 2.09, 3.12 and 3.46×10^{-14} cm^3/sec at KT = 70, 120, 210 and 400 Ry for Al^{5+}, Cl^{9+}, Ti^{14+} and Zn^{22+} respectively. This indicates that the total DR rates increase as the effective charge of the ions increases, in the same isoelectronic sequence.

INTRODUCTION

Dielectronic recombination (DR) is the most important recombination process in highly ionized plasmas. The precise knowledge of the DR rates is essential for the calculation of the ionization balance and for the analysis of the dielectronic line intensities emitted from these plasmas. DR is needed to astrophysical studies [1] and to the development of nuclear fusion plasmas. Furthermore, explanations of DR can provide knowledge about atomic structure.

Recombination of ions with free electrons can occur through several channels: radiative recombination (RR), three-body recombination (TBR) and dielectronic recombination (DR). DR process may be clarified schematically as follows:

$$e + A^{q+} \underset{\overline{A}_a}{\overset{\overline{v}_a}{\rightleftharpoons}} (A^{(q-1)+})^{**}$$
$$\text{(i)} \qquad\qquad \text{(d)}$$
$$\xrightarrow{\overline{A}_r} (A^{(q-1)+})^* + \text{x-rays}$$
$$\text{(f)}$$

Where, electrons (e) collide with ions (A^{q+}) with degree of ionization (q+). i, d and f refer to initial, intermediate and final state respectively.

In fact many theoretical methods have been developed [2-5]. For example, the DR rate coefficients for some H-like ions have been calculated by Bellanton and Hahn[6], Chen[7] as well as Karim and Bhalla [8]. DR rates for the Be and Ne were calculated by Gau et al. [9, 10] and by Hahn et al. [11, 12]. In addition, DR rates were also calculated for highly ionized Helium ions by Younger [13]. For the boron isoelectronic sequence, DR rate coefficients have been calculated for few ions, mainly using nonrelativistic wave functions in LS coupling [14, 15]. On the other hand, many experiments were performed to measure DR cross sections or rate coefficients e.g. by Belic et al. [16], Knapp et al. [17], Andersen et al. [18], Kilgus et al. [19] and Ali et al. [20]. Although, lots of theoretical and experimental works have been done, still recent works are carried out to improve the method of calculations of DR rates with final state distribution[21, 22].

Unfortunately, most of experimental works in the last decade were on simple atomic systems from He- like to Be-like ions or on closed-shell ions such as Ne-like ions. This may be attributed to their relatively simple structure in

CP888, *Modern Trends in Physics Research,*
Second International Conference on Modern Trends in Physics Research—MTPR-06,
edited by L. El Nadi
© 2007 American Institute of Physics 978-0-7354-0354-9/07/$23.00

the LS- or jj- coupling schemes. In the present work, DR rates are calculated with cascade effect for O-like ions. These ions have open shell in their ground state (i $\equiv 1s^2 2s^2 2p^4$), and 1s- excitations in their doubly excited intermediate (d) states.

The calculations of the DR rates ($\overline{\alpha}^{DR}$) are computed in the isolated resonance approximation (IRA). All the Auger and radiative rates are computed in the angular momentum average (AMA) scheme.

THEORETICAL METHOD OF CALCULATIONS

To calculate the DR rate coefficients for O-like ions with K-shell excitation the following equations are needed.

a- Auger rates in AMA scheme:
The Auger transition rate is given in the AMA scheme as follows:

$$\overline{A}_a (d \rightarrow i) = \frac{2\pi}{\hbar} \left| \langle i | V_{12} | d \rangle \right|^2 \tag{1}$$

The Auger transition probability involves two active electrons since the electron-electron interaction is a two- body operator. We denote the two electron states as

$$|d\rangle = n_a \ell_a n_b \ell_b \rightarrow n_t \ell_t k_c \ell_c = |i\rangle \tag{2}$$

Consider two separate cases depending on whether or not a and b are equivalent.

1- Nonequivalent electrons ($n_a \ell_a \neq n_b \ell_b$).
In this case the Auger transition probability is given by the following equation.

$$\overline{A}_a (d \rightarrow i) = \frac{(h_t + 1).(4\ell_a + 2 - h_a).(4\ell_b + 2 - h_b)}{(4\ell_a + 2).(4\ell_b + 2)} A_a^{(0)} (d \rightarrow i) \tag{3}$$

Where h_t, h_a and h_b are the numbers of holes in the subshells t, a and b, respectively, ℓ_a, ℓ_b are the orbital quantum numbers of the subshell a, b respectively.

2- Two quivalent electrons ($n_a \ell_a = n_b \ell_b$).
In this case, the equation (3) may be written as follows:

$$\overline{A}_a (d \rightarrow i) = \frac{(h_t + 1).(4\ell_a + 2 - h_a).(4\ell_a + 1 - h_a)}{(4\ell_a + 2).(4\ell_a + 1)} A_a^{(0)} (d \rightarrow i) \tag{4}$$

The radiationless capture probability $\overline{V}_a$ is then obtained from Auger transition rate to the ground (i) state, $\overline{A}_a (d \rightarrow i, \ell_c)$, using the formula:

$$\overline{V}_a (i \rightarrow d) = \frac{\overline{g}_d}{\overline{g}_c \overline{g}_i} \sum_{\ell_c} A_a (d \rightarrow i, \ell_c) \tag{5}$$

Where, $\overline{g}_i$, $\overline{g}_d$ and $\overline{g}_c$ are the statistical weight of the initial i-state, intermediate (d) state and the continuum (free) electron respectively.

b- Radiative rates in AMA scheme:
In the radiative emission process one electron is actively involved in the radiative transition d→f so; the radiative transition probability in angular momentum average scheme is given by:

$$\overline{A}_r (d \rightarrow f) = \frac{N_d [4\ell_{fi} + 2 - N_{fi}]}{4\ell_{fi} + 2} A_r^{(0)} \tag{6}$$

Where N_d is the number of electrons in the d-state, N_{fi} is the number of electrons in the final state initially, ℓ_{fi} is the orbital quantum number of the final state initially before transition and $A_r^{(0)}$ is the single electron transition rate which is given according to Fermi-Golden rule as:

$$A_r^{(0)} = \frac{2\pi}{\hbar} \left| \langle f | \hat{D} | d \rangle \right|^2 \rho_f \tag{7}$$

where $\hat{D}$ is the photon-electron interaction operator and ρ_f is the final state density.

c. Resonance width of d-state:

The total resonance width of d-state is given by:

$$\overline{\Gamma}(d) = \overline{\Gamma}_a(d) + \overline{\Gamma}_r(d) \tag{8}$$

Where

$$\overline{\Gamma}_a(d) = \sum_{i,\ell_c} \overline{A}_a(d \rightarrow i, \ell_c) + \sum_{j,\ell_c'} \overline{A}_a(d \rightarrow j, \ell_c') \tag{9}$$

And $\overline{\Gamma}_a(d)$: represent to the Auger width of decay of d-state to the ground (i) and excited (j) states. In addition, the radiative width $\overline{\Gamma}_r(d)$ of d-state to stable f or excited (f') final states and obtained from

$$\overline{\Gamma}_r(d) = \sum_{f,f'} \overline{A}_r(d \rightarrow f, f') \tag{10}$$

The fluorescence yield, $\overline{\omega}(d)$, is defined by

$$\overline{\omega}(d) = \frac{\overline{\Gamma}_r(d)}{\overline{\Gamma}(d)} \tag{11}$$

Equation (11) is then corrected for cascade decay processes.

The cascade processes cause a reduction of the DR rate. Its actual effect is known as cascade correction. To include this cascade effect in our calculations, the equation of $\overline{\omega}(d)$ is corrected to.

$$\overline{\omega}_{casc.}(d \rightarrow f1 \rightarrow f2 \rightarrow ...f) = \frac{\sum_f \overline{A}_r(d \rightarrow f1)\overline{\omega}(f1 \rightarrow f2)...\overline{\omega}(f2 \rightarrow f) + \sum_{f_s} \overline{A}_r(d \rightarrow f_s)}{\overline{\Gamma}(d)} \tag{12}$$

Where, the stable (f) states are reached from the d-states by successive decays (cascade decays) while the (f_s) stable states are directly obtained from d-state.

Using the equation (5) and (12) the DR rates are calculated by the following equation:

$$\overline{\alpha}^{DR}(d) = \left[\frac{4\pi Ry}{KT}\right]^{3/2} a_0^3 . \overline{V}_a(i \rightarrow d) . \overline{\omega}(d) . \exp\left[\frac{-e_c}{KT}\right] \tag{13}$$

Where: a_0 is the Bohr radius of hydrogen atom ($a_0 = 0.53 \times 10^{-8}$ cm).

The contribution of HRS is calculated by the following semi-empirical formula.

$$\sum_{n=n_c}^{\infty} \overline{\alpha}^{DR}(n) = \frac{1}{2} n_c \left[1 + \frac{1}{n_c} + \frac{1}{2n_c^2}\right] \left[\frac{n_c - 1}{n_c}\right]^3 \overline{\alpha}^{DR}(n_c - 1) \tag{14}$$

Equation (14) is applied under two conditions, first, $\overline{A}_a$ and $\overline{A}_r$ start to scale as $1/n^3$, second; all Auger channels are opened up.

RESULTS AND DISCUSSIONS

To simplify the discussion, the resonant d-states that may be formed due to the collision of these ions with free electrons are presented in Table (1). Each d-state stabilizes by allowed Auger transitions (energetically) and radiative transitions (in the dipole approx.). The final (f) states are reached from d-states by emission of radiation (x-ray). The d-state may stabilize also by Auger transitions to ground (i) state or other excited (j) states.

TABLE 1. The d-states for O-like ions with K-shell excitation. The final f-states reached from (d) by radiation while, i or j- states reached by Auger emission.

j-states	d-states	f-states $\Delta \ell = \pm 1$
$1s^2 2p^6$ $1s^2 2s 2p^5$	$d1 \equiv 1s 2s^2 2p^6$	$1s^2 2s^2 2p^5$
$1s^2 2p^5 n\ell$ $1s^2 2s 2p^4 n\ell$ $1s^2 2s 2p^5$ $1s^2 2s^2 2p^3 n\ell$	$d2 \equiv 1s 2s^2 2p^5 n\ell$ $\ell = (0, 1, 2, 3)$	$1s^2 2s^2 2p^4 n\ell$ $1s^2 2s^2 2p^5$ $1s 2s^2 2p^6$ $1s 2s^2 2p^5 n_1 \ell_1$
$1s^2 2p^4 3\ell n_a \ell_a$ $1s^2 2s 2p^3 3\ell n_a \ell_a$ $1s^2 2s 2p^4 n_a \ell_a$ $1s^2 2s 2p^4 3\ell$ $1s^2 2s^2 2p^2 3\ell n_a \ell_a$ $1s^2 2s^2 2p^3 n_a \ell_a$ $1s^2 2s 2p^3 3\ell$	$d3 \equiv 1s 2s^2 2p^4 3\ell n_a \ell_a$ $\ell = (0, 1, 2)$ $\ell_a = (0, 1, 2, 3)$	$1s^2 2s^2 2p^3 3\ell n_a \ell_a$ $1s^2 2s 2p^4 n_a \ell_a$ $1s^2 2s^2 2p^4 3\ell$ $1s^2 2s^2 2p^5 n_a \ell_a$ $1s^2 2s^2 2p^5 3\ell$ $1s 2s^2 2p^4 n_1 \ell_1 n_a \ell_a$ $1s 2s^2 2p^4 3\ell \, n_2 \ell_2$

In the following the cascade-corrected DR rates for O-like Al^{5+}, Cl^{9+}, Ti^{14+} and Zn^{22+} ions with K-shell excitation are discussed individually:

Al^{5+} ion.

The DR rates for Al^{5+} with 1s- excitation without and with cascade effect are calculated in IRA approximation using equation (12) at different thermal energies of the continuum electrons. The cascade corrected $\overline{\alpha}^{DR}$ are presented in Table (2) at three values of KT = 50, 100, and 200 Ry. The DR rates in AMA scheme are important for comparison with calculations in other coupling schemes or with experimental data in future. To compare our data in AMA schemes with the LS- coupling scheme, DR rates are calculated for dominant group of d-states in both (AMA) and LS coupling schemes and then applying the following equation:

$$\frac{\sum_{n\ell} \overline{\alpha}^{DR}(n\ell, AMA)}{\sum_{n\ell} \overline{\alpha}^{DR}(n\ell, LS)} = \frac{\overline{\alpha}_{tot}^{DR}(AMA)}{\overline{\alpha}_{tot}^{DR}(LS)} \tag{15}$$

We may predict the total $\overline{\alpha}_{tot}^{DR}$ in LS-coupling scheme. This will minimize the size of calculations. The variation of the DR rates with the continuum electron energies for Al^{5+} with cascade effect is plotted in Figure (1) for the group of d-states, $d \equiv 1s 2s^2 2p^5 n\ell$ where, $(\ell = 0, 1, 2)$. From fig. (1), it is clear also that, $\overline{\alpha}^{DR}$ changes smoothly with the energy of the continuum electron. Although, $\overline{\alpha}^{DR}$ for each group of d-states may be peaked at different values. The peak position of the total $\overline{\alpha}_{tot}^{DR}$ is not necessary coincidence with any individual groups of states. Fig. (1) shows that, the group of d-states $1s 2s^2 2p^5 np$ has the highest contribution to the total DR rates $\overline{\alpha}_{tot}^{DR}$. They are called the most dominant d-states. The reason of this dominancy is the allowed radiative transitions (2p→1s) and (np→1s) in $d \equiv 1s 2s^2 2p^5 np$. Both of these transitions have large value of $\overline{A}_r$. For example $\overline{A}_r$ (2p→1s) = 0.229 × 10^{14} sec^{-1} and $\overline{A}_r$ (3p→1s) = 0.4749 × 10^{12} sec^{-1}. This is not the situation for $1s 2s^2 2p^5 ns$ and $1s 2s^2 2p^5 nd$ groups of states, which are allowed to decay mainly by 2p→1s, ns→2p and nd→2p. Note that: ns→1s and nd→1s radiative transitions are not allowed in the dipole approx. The variation of $\overline{\alpha}_{tot}^{DR}$ with the thermal energy of the continuum

electron for the groups of d-states, $1s2s^22p^6$, $1s2s^22p^5n\ell$ and $1s2s^22p^4n_1\ell_1n_2\ell_2$ besides the total DR rates, $\overline{\alpha}_{tot}^{DR}$ are drawn in Figure (2). This will show the contribution of each group. In addition, this figure also shows that, the peak value of the cascade DR rate for the groups of d-states, $1s2s^22p^6$, $1s2s^22p^5n\ell$ and $1s2s^22p^4n_1\ell_1n_2\ell_2$ are found at 2.79×10^{-15}, 7.26×10^{-15} and 9.44×10^{-17} cm^3/sec respectively. The peak value of the total DR rate, $\overline{\alpha}_{tot}^{DR}$ is 1.01×10^{-14} cm^3/sec. The peak value of DR rates for the same groups of d-states without cascade effect were found 2.79×10^{-15}, 7.26×10^{-15} and 2.55×10^{-15} cm^3/sec respectively. The total DR rates $\overline{\alpha}_{tot}^{DR}$ peaks at 1.26×10^{-14} cm3/sec. This means that the DR rate value of the first and second group are not changed by cascade effect but the DR rate for the third groups $1s2s^22p^4n_1\ell_1n_2\ell_2$ are reduced drastically by cascade. Finally, the cascade reduction for K-shell excitation of the peak value of the total DR rates, $\overline{\alpha}_{tot}^{DR}$ is found to be about 20%.

TABLE 2: DR rates in units cm^3/sec for Al^{5+} with 1s- excitation with cascade effect at KT = 50, 100 and 200 Ry. The core $1s2s^2$ is assumed in all d-states and removed from the table for simplicity.

d-states	$\overline{\alpha}^{DR}$ cm^3/sec KT = 50 Ry	$\overline{\alpha}^{DR}$ cm^3/sec KT = 100 Ry	$\overline{\alpha}^{DR}$ cm^3/sec KT=200 Ry
$\sum2p^6$	2.59E-15	2.52E-15	1.48E-15
$\sum2p^5ns$	1.05E-15	1.09E-15	6.62E-16
$\sum2p^5np$	5.10E-15	5.31E-15	3.23E-15
$\sum2p^5nd$	3.25E-16	3.45E-16	2.11E-16
$\sum2p^5nf$	1.84E-18	1.96E-18	1.20E-18
$\sum2p^43sns$	1.04E-17	1.21E-17	7.72E-18
$\sum2p^43snp$	7.99E-18	9.16E-18	5.84E-18
$\sum2p^43snd$	2.82E-19	3.27E-19	2.10E-19
$\sum2p^43snf$	1.99E-21	2.33E-21	1.50E-21
$\sum2p^43pns$	5.41E-18	6.30E-18	4.06E-18
$\sum2p^43pnp$	4.70E-17	5.44E-17	3.48E-17
$\sum2p^43pnd$	4.51E-18	5.24E-18	3.35E-18
$\sum2p^43pnf$	2.05E-20	2.41E-20	1.56E-20
$\sum2p^43dns$	5.34E-19	6.30E-19	4.08E-19
$\sum2p^43dnp$	2.25E-18	2.67E-18	1.73E-18
$\sum2p^43dnd$	6.97E-20	8.25E-20	5.34E-20
$\sum2p^43dnf$	4.14E-21	4.93E-21	3.20E-21
$\overline{\alpha}_{Tot.}^{DR}$	**9.15E-15**	**9.364E-15**	**5.64E-15**

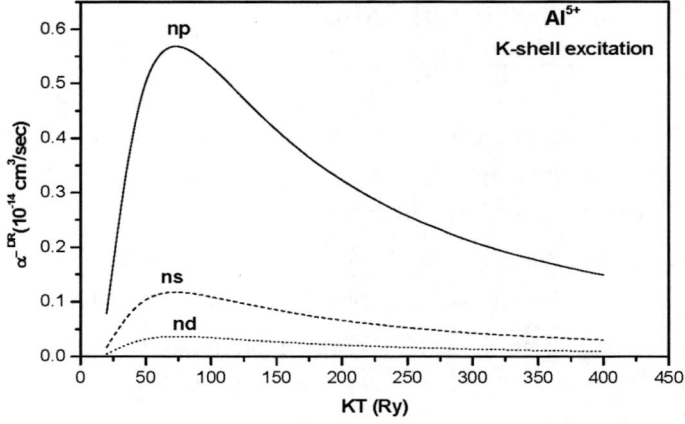

FIGURE 1: The cascade-corrected DR rates vs. KT only for the dominant subgroup d2 = $1s2s^22p^5n\ell$ ($\ell = 0,1,2$) states in Al^{5+}. $\overline{\alpha}^{DR}$ for $2p^5np > 2p^5ns > 2p^5nd$.

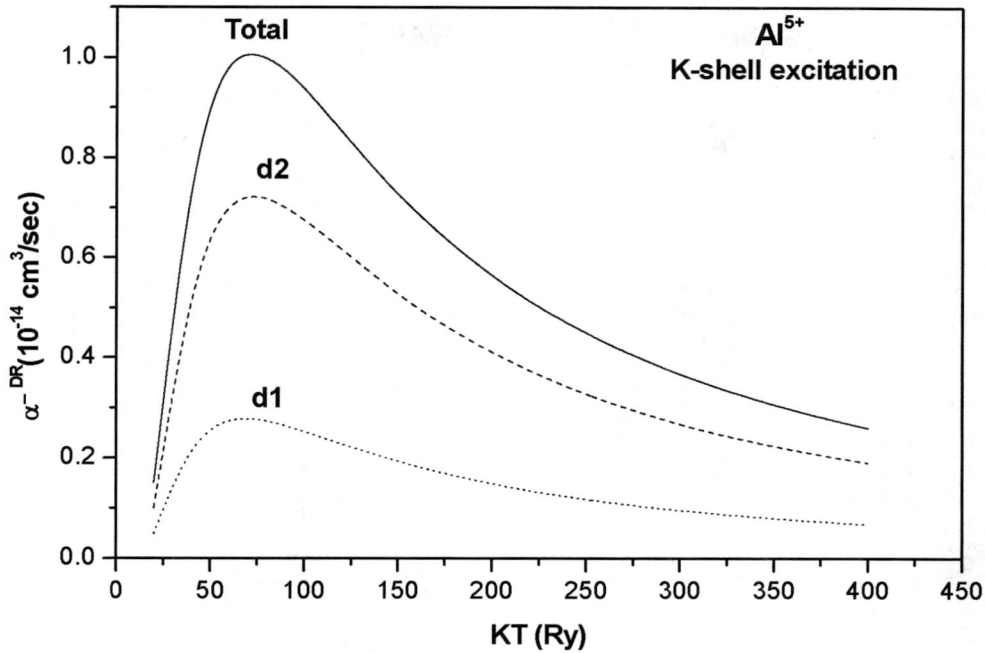

FIGURE 2: DR rates vs. KT for the groups d1= $1s2s^2 2p^6$ and d2 = $1s2s^2 2p^5 n\ell$ states as well as the total DR rates for Al^{5+}.

Cl^{9+} ion.

The cascade corrected DR rates are calculated for Cl^{9+} with K-shell excitation at KT = 60, 120, 240 Ry. These data are presented in Table (3). It is clear from the table (3) that, the values of DR rates for each group of d-states, are larger than that in the case of Al^{5+} due to the increase in the energies of the continuum electrons. The group of d-states $1s2s^2 2p^5 np$ has the highest rates relative to the other group of d-states. Figure (3) show the variation of DR rates with the thermal energy of the continuum electron for the main three groups of d-states $1s2s^2 2p^6$, $1s2s^2 2p^5 n\ell$ and $1s2s^2 2p^4 n_1 \ell_1 n_2 \ell_2$ as well as the total DR rates. It is obvious from fig. 3 that, the peak values of the DR rates for groups of d-states $1s2s^2 2p^6$, $1s2s^2 2p^5 n\ell$ and $1s2s^2 2p^4 n_1 \ell_1 n_2 \ell_2$ are $\overline{\alpha}^{DR} = 4.38 \times 10^{-15}$, 1.61×10^{-14} and 4.46×10^{-16} cm^3 /sec. the last group has small DR values so, it is disregarded in the figure. In addition, the peak value of the total DR rates before cascade effect $\overline{\alpha}_{tot}^{DR} = 2.74 \times 10^{-14}$ and after cascade effect $\overline{\alpha}_{tot}^{DR} = 2.09 \times 10^{-14}$ cm^3/sec. this means that the cascade reduction is found to be 24%. From fig. (2) and (3), we conclude that the group of the d-states, $1s2s^2 2p^5 n\ell$ is the dominant states in the case of the two ions. The total DR rates are increased with increasing the atomic number Z for the two ions. The peak value of $\overline{\alpha}_{tot}^{DR} = 1.01 \times 10^{-14}$ for Al^{5+} but it is increased to $\overline{\alpha}_{tot}^{DR} = 2.09 \times 10^{-14}$ cm^3/sec for Cl^{9+}. As the atomic number Z increases , the position of the peak of total DR rate is shifted to higher value of KT. For example $\overline{\alpha}_{tot}^{DR}$ for Al^{5+} is peaked at 70 Ry, while it peaked at KT = 120 Ry for Cl^{9+}. The cascade effect is reduced the value of DR rates for the group $1s2s^2 2p^4 n_1 \ell_1 n_2 \ell_2$ but it has no effect on other group. The cascade effect reduces the peak value of DR rates for Al^{5+} by 20% while it is increased to 24% for Cl^{9+}.

TABLE 3: DR rates (cm³/sec) for Cl⁹⁺ with K-shell excitation with cascade effect at KT = 6o, 120 and 240 Ry.

d-states	$\overline{\alpha}^{DR}$ cm³/sec KT = 60 Ry	$\overline{\alpha}^{DR}$ cm³/sec KT = 120 Ry	$\overline{\alpha}^{DR}$ cm³/sec KT = 240 Ry
$\sum 2p^6$	3.05E-15	4.36E-15	3.10E-15
$\sum 2p^5ns$	1.31E-15	2.17E-15	1.67E-15
$\sum 2p^5np$	7.24E-15	1.23E-14	9.49E-15
$\sum 2p^5nd$	9.38E-16	1.61E-15	1.26E-15
$\sum 2p^5nf$	1.26E-17	2.21E-17	1.74E-17
$\sum 2p^43sns$	2.86E-17	5.90E-17	5.03E-17
$\sum 2p^43snp$	2.37E-17	4.73E-17	3.99E-17
$\sum 2p^43snd$	1.02E-17	2.09E-17	1.78E-17
$\sum 2p^43snf$	3.08E-20	6.41E-20	5.49E-20
$\sum 2p^43pns$	9.87E-18	2.05E-17	1.75E-17
$\sum 2p^43pnp$	1.00E-16	2.03E-16	1.72E-16
$\sum 2p^43pnd$	1.67E-17	3.44E-17	2.94E-17
$\sum 2p^43pnf$	1.68E-19	3.55E-19	3.06E-19
$\sum 2p^43dns$	5.66E-18	1.20E-17	1.03E-17
$\sum 2p^43dnp$	1.75E-17	3.71E-17	3.21E-17
$\sum 2p^43dnd$	1.06E-18	2.23E-18	1.93E-18
$\sum 2p^43dnf$	7.36E-20	1.58E-19	1.37E-19
$\overline{\alpha}^{DR}_{Tot.}$	**1.28E-14**	**2.09E-14**	**1.59E-14**

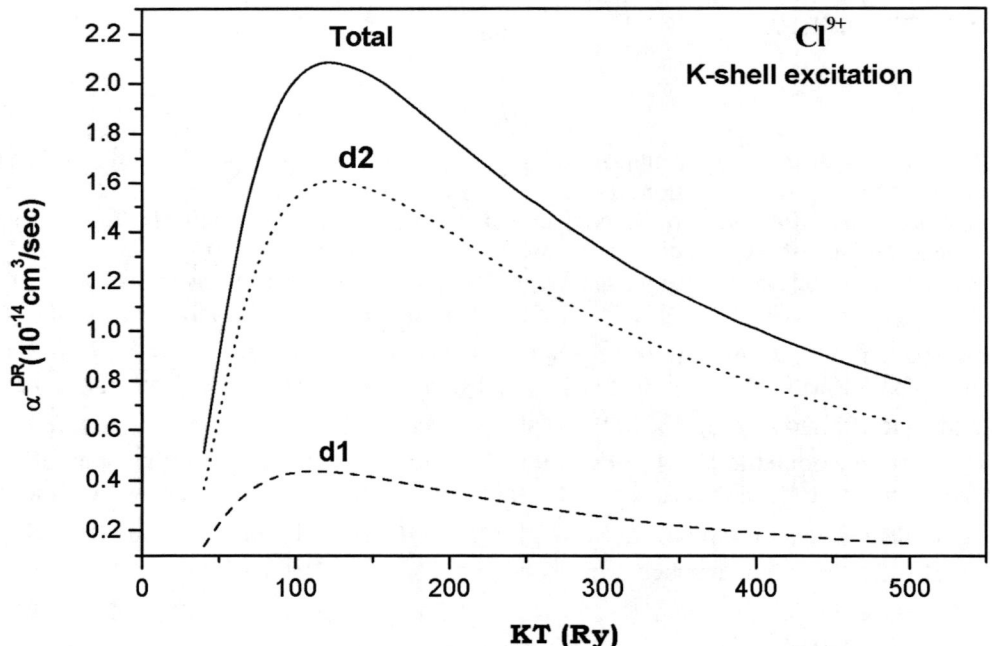

FIGURE 3: Same as fig. (2) but for Cl⁹⁺ ion.

Ti¹⁴⁺ ion.

The calculated values of cascade corrected DR rates for Ti¹⁴⁺ with K-shell excitation at KT = 110, 220, and 440 Ry are presented in Table (4). The group of d-states $1s2s^22p^5np$ has the highest DR rates as in the case of the previous two ions. The smooth behavior of $\overline{\alpha}^{DR}$ for each group of d-states as well as the total DR rate for Ti¹⁴⁺ is

shown in Figure (4). It is obvious from figure (4) that, the group $1s2s^22p^5n\ell$ has the highest peak value since, $\overline{\alpha}^{DR} = 2.43 \times 10^{-14}$. The group of d-states $d = 1s2s^22p^6$ has the peak value of DR rate, $\overline{\alpha}^{DR} = 5.79 \times 10^{-15}$ cm^3/sec. The group of $d = 1s2s^22p^4n_1\ell_1n_2\ell_2$ has the smallest value of DR rates due to the cascade processes where it has peak value $\overline{\alpha}^{DR} = 1.20 \times 10^{-14}$ cm^3/sec without cascade effect but this value reduces by cascade effects to 1.22×10^{-15} cm^3/sec. In addition, the peak value of the total DR rates is found to be $\overline{\alpha}^{DR}_{tot} = 4.18 \times 10^{-14}$ cm^3/sec without cascade effect and this value reduces to $\overline{\alpha}^{DR}_{tot} = 3.12 \times 10^{-14}$ with the cascade effect. The cascade reduces $\overline{\alpha}^{DR}_{tot.}$ by 25% for Ti^{14+}. The total DR rate is peaked at KT = 210 Ry.

TABLE 4: DR rates for Ti^{14+} with K-shell excitation with cascade effect at KT = 110, 220, and 440 Ry.

d-states	$\overline{\alpha}^{DR}$ cm^3/sec KT = 110 Ry	$\overline{\alpha}^{DR}$ cm^3/sec KT = 220 Ry	$\overline{\alpha}^{DR}$ cm^3/sec KT = 440 Ry
$\sum 2p^6$	4.61E-15	5.65E-15	3.72E-15
$\sum 2p^5ns$	2.05E-15	3.02E-15	2.18E-15
$\sum 2p^5np$	1.19E-14	1.78E-14	1.30E-14
$\sum 2p^5nd$	2.19E-15	3.31E-15	2.42E-15
$\sum 2p^5nf$	4.99E-17	7.76E-17	5.75E-17
$\sum 2p^43sns$	9.25E-17	1.72E-16	1.39E-16
$\sum 2p^43snp$	8.27E-17	1.52E-16	1.22E-16
$\sum 2p^43snd$	5.03E-17	9.28E-17	7.50E-17
$\sum 2p^43snf$	2.40E-19	4.52E-19	3.69E-19
$\sum 2p^43pns$	2.37E-17	4.43E-17	3.61E-17
$\sum 2p^43pnp$	2.21E-16	4.02E-16	3.23E-16
$\sum 2p^43pnd$	6.86E-17	1.25E-16	1.00E-16
$\sum 2p^43pnf$	8.57E-19	1.63E-18	1.33E-18
$\sum 2p^43dns$	2.49E-17	4.73E-17	3.87E-17
$\sum 2p^43dnp$	8.29E-17	1.59E-16	1.31E-16
$\sum 2p^43dnd$	8.02E-18	1.53E-17	1.25E-17
$\sum 2p^43dnf$	6.68E-19	1.29E-18	1.06E-18
$\overline{\alpha}^{DR}_{Tot.}$	**2.15E-14**	**3.11E-14**	**2.23E-14**

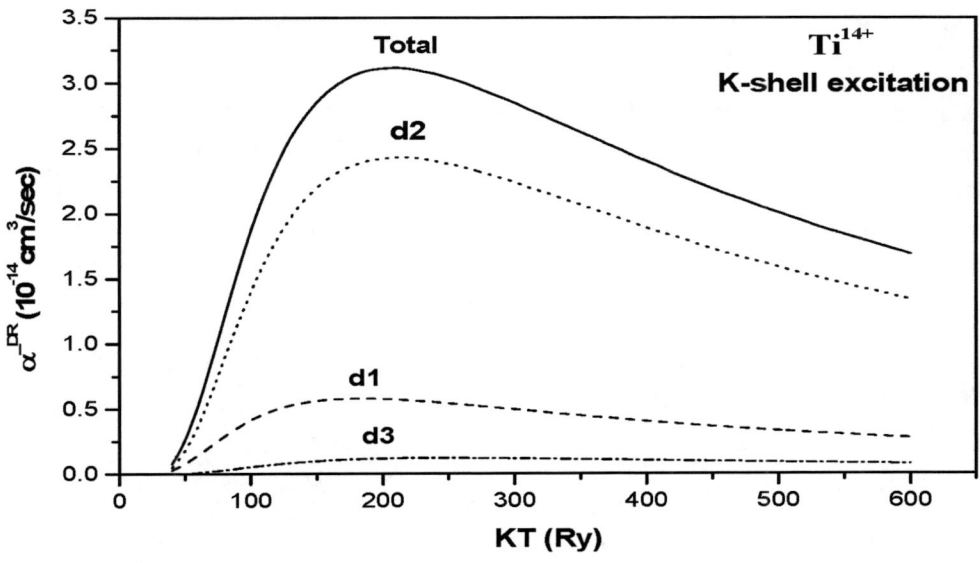

FIGURE 4: Same as Fig. (3) but, for Ti^{14+}. The contribution of d3 starts to be larger and represented on the fig.

Zn^{22+} ion.

The cascade corrected DR rates are calculated for Zn^{22+} with K-shell excitation. The contribution of high Rydberg state for this ion is calculated at n = 7. This depends on the scaling property of Auger rates and radiaive rates where, they are started to decrease as $1/n^3$. The values of DR rates for Zn^{22+} are plotted in Figure (5). Figure (5) shows that the group of d-states $1s2s^22p^5n\ell$ has the highest values of DR rates and the cascade decays of the d-states are not affected on the values of DR rates. The group of d-states $1s2s^22p^4n_1\ell_1n_2\ell_2$ has the smallest values of DR rates. The peak value of the total DR rate is larger than that of the previous ions where, $\overline{\alpha}_{tot}^{DR} = 3.46 \times 10^{-14}$ with cascade effect. The peak value is shifted also to a higher value of KT = 400 Ry. This satisfy the fact that as the atomic number increase Z the peak values of DR rates are increased and shifted toward the higher values of KT. The cascade reduction is found about 24 % as in previous ions.

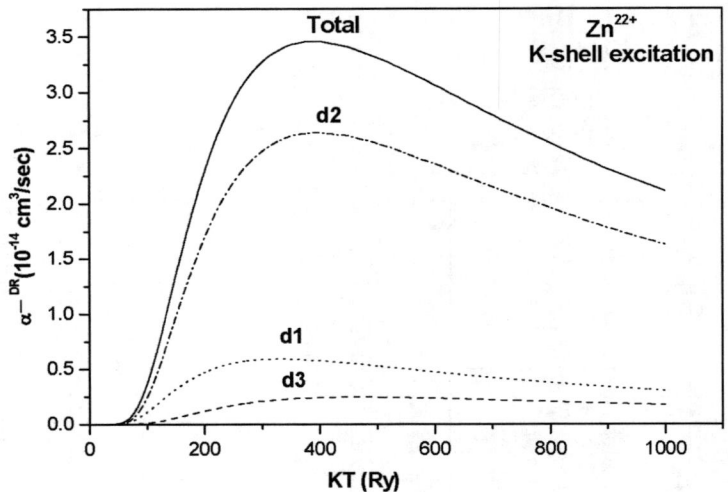

FIGURE 5: same as fig. (4) but for Zn^{22+}. The peak values and position are higher than that of the previous ions. The curves are broader than those in fig. (2, 3, 4).

Isoelectronic Trends of the DR rates.

The total cascaded DR rates for O-like Al^{5+}, Cl^{9+}, Ti^{14+} and Zn^{22+} ions are drawn together in Figure (6). This figure shows that, the peak values of $\overline{\alpha}^{DR}$ and the position of each peak (KT) increase as the atomic number (Z) of the ion increases. To study the dependence of total DR rate $\overline{\alpha}_{tot}^{DR}$ on the effective charge of the ions at its peak value. There are many formulas for the effective charge for the ions. The formula which is commonly used is $Z_{eff} = (Z_C + q)/2$ with the core charge (Z_c) of the ion and the degree of ionization (q+) of the ion. To study the dependence of $\overline{\alpha}_{tot}^{DR}$ on the effective charge the following data are presented in Table (5).

TABLE 5: The atomic number (Z), the effective charge Z_{eff}, KT at peak values and the maximum value of the total DR rates for O-like ions.

Ion	Z	Z_{eff}	KT (Ry)	$\overline{\alpha}_{tot}^{DR}$ (cm^3/sec)
Al^{5+}	13	9	70	1.01×10^{-14}
Cl^{9+}	17	13	120	2.09×10^{-14}
Ti^{14+}	22	18	210	3.12×10^{-14}
Zn^{22+}	30	26	400	3.46×10^{-14}

From table (5) the following calculations can be deduced for Al^{5+} and Cl^{9+} ions.

$$\frac{\overline{\alpha}_{tot}^{DR}(cl^{9+})}{\overline{\alpha}_{tot}^{DR}(Al^{5+})} = \frac{2.09 \times 10^{-14}}{1.01 \times 10^{-14}} = 2.07, \left[\frac{Z_{eff}(Cl^{9+})}{Z_{eff}(Al^{5+})}\right]^2 = \left[\frac{13}{9}\right]^2 = 2.09 \text{ and} \left[\frac{Z(Cl^{9+})}{Z(Al^{5+})}\right]^3 = \left[\frac{17}{13}\right]^3 = 2.2$$

This result shows that the ions with $5 \leq q \leq 9$, $\overline{\alpha}_{tot}^{DR}$ varies as Z_{eff}^2 and Z^3. We calculate the same ratios for the other ions and we conclude that, the ions with $9 \leq q \leq 14$, $\overline{\alpha}_{tot}^{DR}$ varies linearly as Z_{eff} and Z. The ions with $14 \leq q \leq 22$, $\overline{\alpha}_{tot}^{DR}$ varies as $Z_{eff}^{0.5}$ and $Z^{0.5}$.

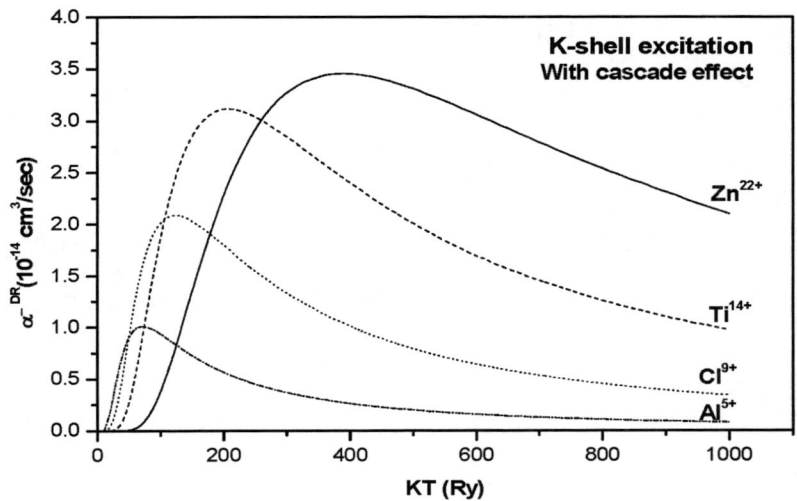

FIGURE 6: The total DR rates vs. KT for O-like ions with cascade effect.

CONCLUSIONS

The DR rates vary smoothly with the continuum (incident) electron temperature for all of Al^{5+}, Cl^{9+}, Ti^{14+} and Zn^{22+} ions. All curves of DR rates are antisymmetric around their peak value, which reflects the nature of the Maxwellian distribution for the velocities of the continuum electrons. In addition the total DR rates are found broader as Z increase in O-like ions. It is found that, the group of resonant d-states of the form $1s2s^22p^5n\ell$ ($\ell =0, 1, 2$) has the highest contribution to the total DR rates for each of the studied ions but the group of d-states $1s2s^22p^4n_1\ell_1n_2\ell_2$ have negligible DR rate because it is subjected to huge cascade decays. The cascade reduction for 1s-excitation for the peak value of the total DR rate is found to be around 24% for all ions. The peak values of DR rates increase as the effective charge (Z_{eff}) of the ions increases.

REFERENCES

1. Burgess A., Astrophys. J. **139**, 776 (1964); 141, 1588 (1965).
2. Hahn Y., Phys. Rev. **A22**, 2896 (1980).
3. McLaughlin D. J. and Hahn Y., Phys. Rev. **A29**, 712 (1984).
4. Chen M. H., Phys. Rev. **A31**, 1449 (1985).
5. Saha H. P., Phys. Rev. **A49**, 894 (1994).
6. Bellanton R. and Hahn Y., Phys. Rev. **A40**, 6913 (1989).
7. Chen M. H., UCRL- 98454 (1988).
8. Karim K.R. and Bhalla C. P., Phys. Rev. **A37**, 2599 (1988).
9. Gau J.N, Hahn Y., and Retter J.A., JQSRT **23**, 131 (1980).
10. Gau J.N, Hahn Y., and Retter J.A., JQSRT **23**, 147 (1980).

11. Hahn Y. et al., JQSRT **23**, 65 (1980).
12. Hahn Y. et al., JQSRT **24**, 505 (1980).
13. Younger S. M., JQSRT **29**, 67 (1983).
14. Ramadan H.H. and Hahn Y., Phys. Rev. **A39**, 3350 (1989).
15. Nasser I. and Hahn Y., phys. Rev. **A39**, 401 (1989).
16. Belic D.S. et al., Phys. Rev. Lett. **50**, 339 (1983).
17. Knapp D.A. et al., Phys. Lett. **62**, 2104 (1989).
18. Andersen L.H. et al., Phys. Rev. Lett. **62**, 2656 (1989).
19. Kilgus G. et al., Phys. Rev. Lett. **64**, 737 (1990).
20. Ali R., Bhalla C.P., Cocke C.L. and Stockli M., Phys. Rev. Lett. **64**, 633 (1990).
21. Omar G. and Hahn Y., Phys. Rev. E **62**, 4096 (2000).
22. Omar G. and Hahn Y., Phys. Rev. E **63**, 046407-1 (2001).

Porous Silicon Modified Photovoltaic Junctions: An Approach to High-Efficiency Solar Cells

Waheed A . Badawy

Chemistry Department, Faculty of Science, Cairo University, Giza–Egypt

E-mail: *wbadawy@chem-sci.cu.edu.eg*

Abstract: The solution of the energy problems of our universe is based on the use of the ultimate source of energy, THE SUN, as the main source of useable energy. The trials to obtain solar cells of appropriate efficiency and suitable price represent one of the main tasks of different research groups over the whole world. In this respect silicon represent the main absorber of sun light that could be converted to electricity, photovoltaic cells, or to high energy chemical products, photoelectrochemical cells. Photovoltaic and photoelectrochemical systems were prepared by the formation of a thin porous film on silicon. The porous silicon layer was formed on the top of a clean oxide free silicon wafer surface by anodic etching in $HF/H_2O/C_2H_5OH$ mixture (2:1:1). The silicon was then covered by an oxide film (tin oxide, ITO or titanium oxide. The oxide films were prepared by the spray/pyrolysis technique which enables the incorporation of foreign atoms like In, Ru or Sb in the oxide film matrix during the spray process/. The incorporation of foreign atoms improves the surface characteristics of the oxide film which leads to the improvement of the fill factor and higher solar conversion efficiency. The prepared solar cells are stable against environmental attack due to the presence of the stable oxide film. It gives relatively high short circuit currents (I_{sc}) compared to our improved silicon single crystal solar cells /6/, due to the presence of the porous silicon layer, which leads to the recorded high conversion efficiency. Although the open-circuit potential (V_{oc}) and fill factor (FF) were not affected by the thickness of the porous silicon film, the short circuit current was found to be sensitive to this thickness. An optimum thicknesss of the porous film and also the oxide layer is required to optimize the solar cell efficiency. The results represent a promising system for the application of porous silicon layers in solar energy converters. The use of porous silicon instead of silicon single crystals in solar cell fabrication and the optimization of the solar conversion efficiency will lead to the reduction of the cost as an important factor and also the increase of the solar cell efficiency making use of the large area of the porous structures.

INTRODUCTION

Porous silicon has attracted many investigators in recent years. It has been suggested that the efficient electroluminescence property of porous silicon could be accompanied by photovoltaic response [1, 2]. The porous Si layers were found to enhance the stability of n-Si photoanodes in photoelectrochemical cells [3, 4] and to increase the solar conversion efficiency in non aqueous solutions [5]. It has been shown that the electrochemical etching of Si in HF containing solutions produces porous Si layer rich in hydrogen [6]. In forming porous Si, highly textured surfaces are obtained, enhancing light trapping and its potential use as an anti-reflection coating [3]. The use of porous Si as an optimized emitter has been shown to be possible in Si solar cells, where the porous Si layer is formed onto the emitter surface by electrochemical etching [7]. The emitter consists of a porous Si layer and bulk n-Si. Encouraging results have been reported where large area solar cells were produced by converting the Si surface into porous Si by etching in controlled solutions. The use of electrochemical methods allows a good control of the morphological properties of the porous Si layers (porosity and thickness), and a two stop procedure has been developed to optimize formation conditions leading to porous Si with optimal porosity (refractive index) and thickness [8].

In this work we have tried to benefit from our experience in the improvement of the photovoltaic and photoelectrochemical characteristics with n-Si solar cells [9-22], and to use porous Si layer on the top of the Si wafer to increase the short circuit current and to achieve high conversion efficiencies. In the photovoltaic Si solar cells practical efficiencies of 10 to 14% were recorded and speculations about 25%

CP888, *Modern Trends in Physics Research*,
Second International Conference on Modern Trends in Physics Research—MTPR-06,
edited by L. El Nadi

efficient solar cells have been reported. With our improved n-Si/oxide heterojunction efficiencies up to 14% for photovoltaic cells and 9% for photoelectrochemical cells were prepared [22]. The basis of energy conversion in both cases is the same. The understanding of the theoretical aspects of the photoelectrochemical devices has increased their applications to produce clean fuels and direct conversion of solar energy into electrical energy. At a metal/metal interface, the excess charge builds up just with few tens of nanometers of the interface, whereas at a metal/semiconductor interface, the charge extended very deep into the bulk of the semiconductor. In Shottky barrier photovoltaic or photoelectrochemical cells, limitations of energy conversion are due to the decrease of the junction potential, V_j, by the ohmic drop of the series resistance, R_s, and the conductivity of the surface electrode or limitations due to the sluggish electrochemical process due to the charge transfer overpotential, η_{ch}. The value of the redox potential of the electrolyte affects the possible photopotential [23]. For n-type semiconductors, a combination of a redox electrolyte of a potential that is very close to the valance band of the semiconductor is important. One of the most serious problems of photoelectrochemical cells, and also the non-encapsulated photovoltaic cells, is the anodic decomposition of the semiconductor, which competes with the desired anodic process, leading to lower energy conversion efficiencies and deterioration of the solar cell.

In our improved cells we are using SnO_2 or TiO_2 thin films on the n-Si to inhibit the photodecomposition of the semiconductor [21, 22]. These materials are transparent to solar spectrum and function as efficient windows on the top of Si [10, 16, 24]. If Si is covered with a stable material that fulfils the requirements of transparency, conductivity and forms a stable hetero-junction with Si, a suitable photovoltaic or photoelectrochemical cell can be obtained [10, 22]. In the present paper it is aimed at the use of a thin porous Si layer on the top of the Si wafer and protect it with thin stable SnO_2 or TiO_2 film to achieve higher photocurrents and of course high solar conversion efficiency.

EXPERIMENTAL

Pure and foreign atom incorporated thin films of SnO_2 and TiO_2 were prepared conveniently by the spray/pyrolysis technique on the top of the pretreated n-Si. [14,22]. The pyrolysis reaction produces transparently clear films of an interference color depending on the film thickness. The experimental details for optical measurements, energy gap calculation, conductivity and all characteristics of the prepared oxide films are described previously [14, 15, 22].

The photovoltaic and photoelectrochemical cells were prepared by depositing an optimized oxide film on the pre-treated n-type Si. A conventional potentiostatic set-up was used to trace the current-potential characteristics of the prepared cells. The illumination was carried out using a 150 W Xenon lamp coupled with solar simulator and filter to achieve AM1 solar spectrum. The results were corrected for the series resistance as described previously [9, 25].

Pre-treatment of the Si wafers

n-Type [100]-oriented Si wafers (Wacker, Burghausen) resistivity 0.9-1.4 Ωcm were cut into 0.2 cm^2. Electrical contacts were made on the back surface with an indium-gallium eutectic. The back contacts and edges of the crystal were sealed with insulating epoxy. The crystals were etched for 10 min in H_2SO_4 : H_2O_2 (2 : 1) solution followed by 15 s etching in HF : H_2O (1 : 10). The thin homogeneous porous Si layer (PS) ~ 100 nm was formed by standard anodic etching of the Si-electrodes in HF : H_2O : C_2H_5OH (2 : 1 : 1) /26/. The thin layers are of nanoporous structure and contain columnar pores perpendicular to the samples surface. The formed layers have a thickness of ≤100nm, which is the most suitable for photovoltaic applications [27].

Details of the experimental procedures for the preparation of the photovoltaic and photoelectrochemical cells are as described elsewhere [17, 22].

RESULTS AND DISCUSSIONS

The prepared SnO_2 and TiO_2 films were optimized to fulfill the requirements of the photovoltaic and photoelectrochemical applications. The TiO_2 films have a specific conductance of 1×10^{-3} $\Omega^{-1}cm^{-1}$, whereas that of SnO_2 1×10^2 Ω^{-1} cm^{-1} [25]. The incorporation of Ru in the oxide matrix increases the film

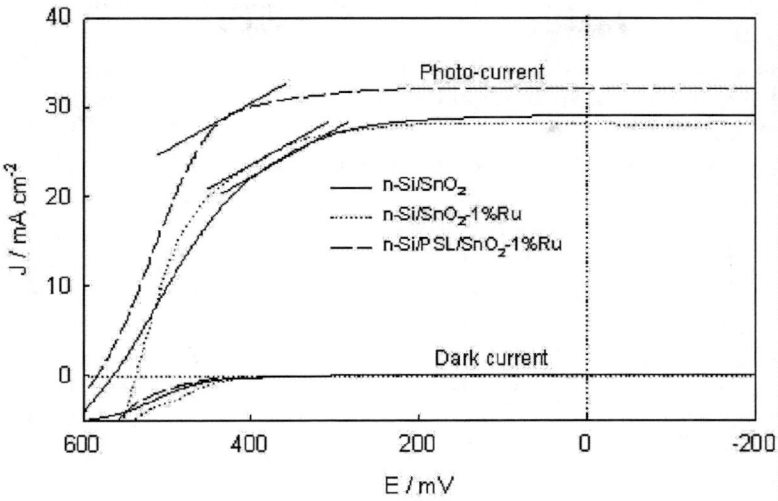

FIGURE. 1: Power characteristics of n-Si/SnO₂/Metal, n-Si/SnO₂(1%Ru)/Metal and n-Si /PSL /SnO₂ (1%Ru)/Metal photovoltaic cells under simulated solar spectrum (AM1) at 298 K.

conductivity, especially for TiO_2 film, by about three orders of magnitude. The specific conductance of the prepared films is independent of the film thickness, reflecting the high homogeneity of these films. Also, the prepared films, either pure or with foreign atom incorporation (up to 1.0%) are highly transparent to the solar spectrum > 90%. The band gap energy of the prepared film was calculated from the optical measurements and found to be 4.0 and 3.6 eV for pure SnO_2 and TiO_2, respectively, and is independent of the film thickness. Incorporation of 1% Ru in the oxide film increases the band gap. The calculated E_g for SnO_2-Ru and TiO_2-Ru was 4.10 and 3.80 eV, respectively. The band gap widening is explained by assuming that allowed states near the bottom of the conduction band of the n-type material are occupied to rather high levels and that the allowed transitions from the valence band would have correspondingly higher energies than the forbidden region [13, 15].

Photovoltaic and photoelectrochemical characteristics

To obtain an idea about the effect of the presence of porous silicon layers, the cells were prepared according to reference 22 without etching and then after the formation of the porous silicon layer on the top of the Si wafer before depositing the oxide film. The power characteristics of both n-Si/SnO₂ and n-Si/TiO₂ with and without the porous Si layer under simulated solar spectrum (100 mW/cm²) are presented in Figs. 1 and 2, respectively. The photoelectrochemical characteristics of both cells are presented in Figs. 3 and 4, respectively. The electrochemical characteristics of the pure and Ru-incorporated SnO₂ films are presented in Fig. 5.

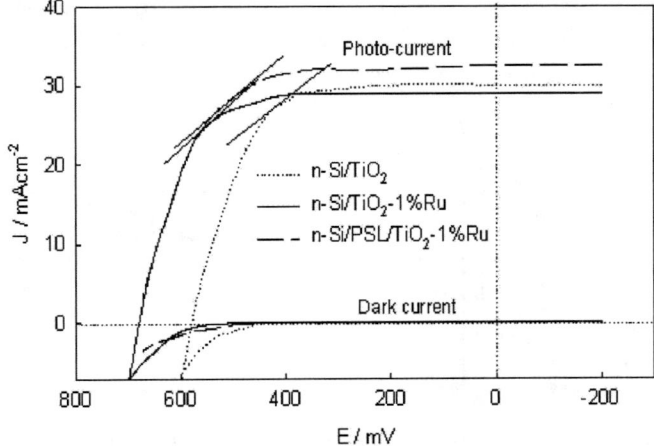

FIGURE. 2: Power characteristics of n-Si/TiO₂/Metal, n-Si/TiO₂(1%Ru)/Metal and n-Si /PSL /TiO₂ (1%Ru)/Metal photovoltaic cells under simulated solar spectrum (AM1) at 298 K.

31

In spite of the fact that the extent of foreign atoms (Ru in this case) did not exceed 1% of the whole oxide film matrix as determined by ESCA /10/, the power characteristics of the prepared cells are significantly improved [25]. The presence of the porous Si layer (PSL) did not affect the open circuit potential of the cells but a remarkable increase in the short circuit current was recorded. The increase of the short circuit current is reflected in an increase in solar conversion efficiency. The different parameters of the photovoltaic and photoelectrochemical cells are summarized in Tables 1 and 2, respectively.

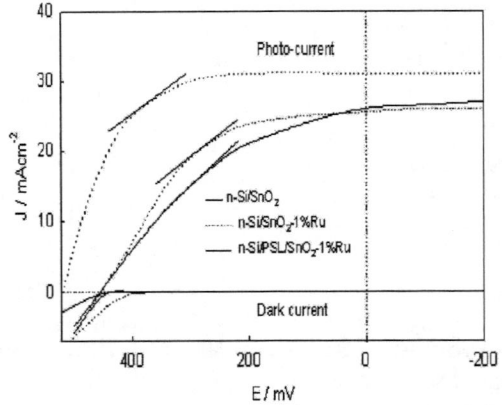

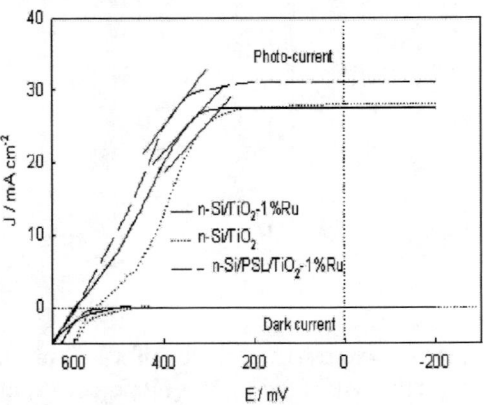

FIGURE 3. Power characteristics of n-Si/SnO$_2$/Electrolyte, n-Si/SnO$_2$(1%Ru)/ Electrolyte and n-Si /PSL/SnO$_2$ (1%Ru) / Electrolyte photoelectrochemical cells in the dark and under simulated solar spectrum (AM1) at 298 K in 0.05 M K$_3$ Fe(CN)$_6$/ 0.05 M K$_4$Fe(CN)$_6$/0.5 M KNO$_3$

FIGURE 4. Power characteristics of n-Si/TiO$_2$ /Electrolyte, n-Si/TiO$_2$(1%Ru)/ Electrolyte and n-Si/PSL/TiO$_2$ (1%Ru) photoelectrochemical cells in the dark and under simulated solar spectrum (AM1) at 298 K in 0.05 M K$_3$ Fe(CN)$_6$/ 0.05 M K$_4$Fe(CN), /0.5 M KNO$_3$

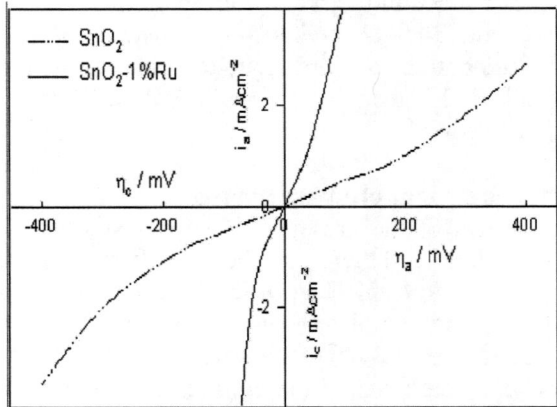

FIGURE 5. Current density–potential curves of SnO$_2$ films as disc electrodes (thickness = 100 nm) in 0.05 M K$_3$ Fe(CN)$_6$/ 0.05 M K$_4$Fe(CN)$_6$/0.5 M KNO$_3$. Scan rate=50mV/s and rotation speed =2500 rpm at 298K

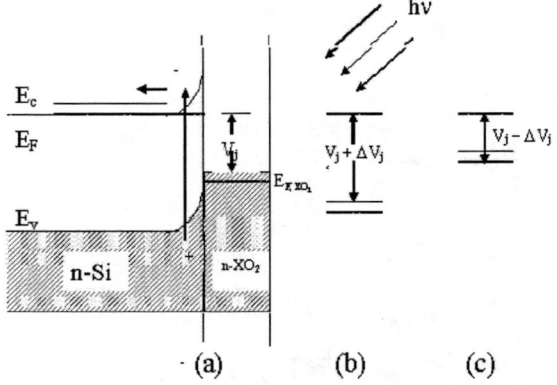

FIGURE 6. Effect of foreign atom incorporation on the junction potential, V$_j$, of the oxide film. (a) Pure oxide film.   (b) Band gap widening.   (c) Band gap shortening.

TABLE 1. Photovoltaic parameters of n-Si/oxide and n-Si/PSL/oxide solar cells under simulated solar spectrum (AMI) at 298K:

Photo voltic cell (mAcm^{-2})	J_{ph} mV	V_{oc}	FF	η%
n-Si/SnO$_2$/M	29	550	0.57	9.12
n-Si/SnO$_2$-Ru(l%)/M	28	530	0.66	9.75
n-Si/PSL/SnO$_2$/M	31.5	580	0.67	12.10
n-Si/PSL/SnO$_2$-Ru(l%)/M	31.0	580	0.71	12.76
n-Si/TiO$_2$/M	29.0	590	0.61	10.40
n-Si/TiO$_2$-Ru(1%)/M	28.0	680	0.75	14.25
n-Si/PSL/ TiO$_2$/ M	31.5	670	0.71	14.84
n-Si/TiO$_2$-Ru(1%)/M	31.0	680	0.78	16.53

TABLE 2. Photoelectrochemical parameters of n-Si/oxid and n-Si/PSL/oxide photoelectrochemical cells under simulated solar spectrum (AMI) at 298K. The electrolyte consists of 0.05 M K$_3$Fe(CN)$_6$/0.05 M K$_4$Fe(CN)$_6$/0.5 M KNO$_3$.

Photoelectrochemical cell (mAcm^{-2})	J_{ph} (mV)	V_{oc}	FF	η %
n-Si/SnO$_2$/electrolyte	26	440	0.40	4.59
n-Si/SnO$_2$-Ru(l%)/electrolyte	25	450	0.50	6.72
n-Si/PSL/SnO$_2$/Electrolyte	30.5	490	0.68	10.20
n-Si/SnO$_2$-Ru(l%)/Electrolyte	30.0	500	0.68	10.26
n-Si/TiO$_2$/ electrolyte	28.0	580	0.44	7.13
n-Si/TiO$_2$-Ru(1%)/ electrolyte	27.5	600	0.58	9.25
n-Si/PSL/TiO$_2$/ Electrolyte	31.0	600	0.59	10.92
n-Si/PSL/TiO$_2$-Ru(1%)/E	30.5	610	0.65	12.18

GENERAL DISCUSSION

The presence of porous silicon layer of thickness ≤100nm on the top of the n-Si absorber significantly improves the characteristics of the photovoltaic and photoelectrochemical cells. An increase in the fill factor, FF, which ranges between 10 and 20% for the n-Si/SnO$_2$ junctions and 13 to 25% for the n-Si/TiO$_2$ solar cells was recorded. The improvements of the fill factor and the increase of the photocurrent reflect itself in an improvement of the solar conversion efficiency. Photovoltaic cells of more than 16% efficiency could be obtained. Optimization of the preparation conditions of the PSL and control of the series resistance could improve the solar conversion efficiency more and more. The photoelectrochemical cells based on the n-Si/PSL/oxide junctions show remarkable improvement and cells with more than 12% efficiency were obtained.

The presence of foreign atoms like Ru in the oxide film window improves the characteristics of the oxide film, especially for photoelectronchemical devices. As can be seen from figure 5, the presence of 1% Ru in the SnO$_2$ matrix has increased the rate of the charge transfer process. The overpotential was decreased and the limitation of the electrochemical process was overcome. The combination of the PSL and foreign atoms incorporated oxide films is essential for the high solar conversion efficiency and the stability of the prepared solar cells. The foreign atoms incorporation in the oxide film affects the position of the Fermi level of the semiconducting oxide film. The solar conversion efficiency increases when the Fermi level of the oxide assumes lower value (cf. Fig. 6). The rate of the reaction-less one electron transfer depends on a quantum mechanical factor related to the tunnel probability of the electron and the density of electronic stats in the oxide layer energetically suitable for the redox system in the photoelectrochemical cell. Increasing the concentration of the charge carriers in the semiconducting oxide film increases the tunnel probability by decreasing the thickness of the space charge layer of the semiconductor as illustrated in Fig. 7. The limitations in the electrochemical charge transfer process lead to the deterioration of the cell quality. The increase in the rate of charge transfer at the oxide/electrolyte interface due to the incorporation of Ru in the oxide film matrix has led to a better fill factor and improved solar conversion efficiency.

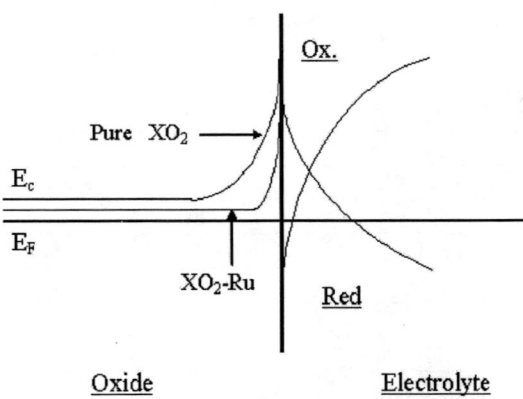

FIGURE. 7: Effect of Ru incorporation on the space charge layer of the semiconducting oxide film.

The combination of the PSL and Ru-incorporation leads to a high efficiency in both photovoltaic and photoelectrochemical devices. The system n-Si/oxide/electrolyte can be considered as a combination of photovoltaic and electrochemical characteristics which can be treated separately. The presence of a PSL in the photovoltaic part increases the photocurrent and hence a better fill factor and improved solar conversion efficiency is achieved. The presence of Ru in the oxide film improves the photovoltaic characteristics of the solid state devices and increases the rate of the charger transfer step in the electrochemical systems. The improved fill factor and the increased solar conversion efficiency of the n-Si/PSL/oxide junctions make such combinations important devices for solar energy conversion and lead to environmentally safe energy converters.

CONCLUSIONS

- Homogenous nano-structures of porous silicon were prepared conveniently on the top of silicon absorber by anodic etching in $HF/H_2O/C_2H_5OH$ mixture (2:1:1).
- The presence of a porous silicon layer significantly improves the photovoltaic and photoelectrochemical characteristics of the devices.
- The presence of 1.0% Ru as incorporation in the oxide film window improves the fill factor and increases the rate of charge transfer step at the oxide/electrolyte interface.
- The obtained devices have significantly high solar conversion efficiency which retches 16% for photovoltaic systems and 12% for photoelectrochemical devices.

ACKNOWLEDGMENT

The Alexander von Humboldt foundation (Bonn, Germany) is gratefully acknowledged for the continuous support.

REFERENCES

1. L.T. Canham; Appl. Phys. Lett. **57**,1046 (1990).
2. K-J. Kim, G.S. Kim, J.S. Hang, T-S Kang and D. Kim; Solar energy **64**, 61 (1998).
3. C. L-Clement, A. Lagoubi, M. N-Spallart, M. Rodat and R. Tenne; J. Electrochem. Soc**138**,L-69 (1991).
4. N. Koshida, H. Koyama, Y. Yamamoto and G.J. Collins; Appl. Phys. Lett. **63,** 2655 (1993).
5. D. Mao, K.J. Tim, Y.S. Tsuo and A.J. Frank, J. Phys. Chem. **99**, 3643 (1995).
6. A. Borghesi, A. Sassella, B. Pivac and L. Pavesi; Solid State Commun., **87**,1 (1993).
7. A.J. McEvoy and M. Graetzel, Solar Energy Mater. & Solar cells, **32**, 221(1994).
8. S. Streke, S. Bastide and C.L-Clement; Solar Energy Materials & Solar Cells, **58**,399 (1999).
9. W.A. Badawy, F. Decker, and K. Doblhofer, Solar Energy Materials **8**, 363 (1983).
10. F. Decker, H. Fracastero-Decker, W.A. Badawy, K. Doblhofer and H. Gerischer, J. Electrochem. Soc.**130**, 2173 (1983).

11. W.A. Badawy, K. Doblhofer, I. Eiselt, H. Greischer, S. Krause and J. Melsheimer, Electrochim. Acta **29**,1617 (1984).
12. W.A. Badawy and K. Doblhofer, Applied Physics A **35**,189 (1984).
13. W.A. Badawy, Ind. J. Technol. **24**,118 (1986).
14. W.A. Badawy and E.A. El-Taher, Thin Sold Films**158**, 277 (1988).
15. W.A. Badawy, R.S Momtaz and. E.M El-Giar, Phys. Status Solidi (a), **118**, 197 (1990).
16. W.A. Badawy, H.H. Afity and E.M. El-Giar, J. Electrochem. Soc. **137**,1592 (1990).
17. W.A. Badawy, J. Electroanal. Chem. **281**,85 (1990).
18. H.H Afify, R.S. Momtaz, W.A. Badawy. and S.A Nasser, J. Materials Sci., Materials in Electronics, **2**,40 (1991) 40.
19. W.A. Badawy, Solar Energy Materials and Solar Cells **28**,293 (1993).
20. W.A. Badawy, J. Materials Sci.**32**, 4979 (1997).
21. W.A. Badawy, Current Topics in Electrochem. **4**, 147 (1997).
22. W.A. Badawy, Solar Energy Materials and Solar Cells **71**, 281 (2002).
23. A.J. Bard and M.S. Wrighton, J. Electrochem. Soc. **124**, 1706 (1977).
24. W.A. Badawy, R.S. Momtaz, H.H. Afity and E.M El-Giar, J. Materials Science, Materials in Electronics **2**,112 (1992).
25. W.A. Badawy, in "Modern Aspects of Electrochem." **No.30**, Plenum Press, New York, pp. 187 (1996) (Chapter 2) J.O'M. Bockris, B.E. Conway, and R.E White (Eds.).
26. P. Vitanov, M. Kamenova, N. Tyutyundzhiev, M. Delibasheva, E. Goranova, M. Peneva, Thin Solid Films **297**,299 (1997).
27. P. Vitanov, M. Delibasheva, E. Goranova, M. Peneva, Solar Energy Materials and Solar Cells **61**, 213 (2000).

Optical and Electrical Properties of Sodiumborate Glass Containing V_2O_5

M.M. Eloker, H.M. Talat, N. Elkashef[*], M.F.Abd Al-Azeem

Physics Dept. Faculty of Scienc, Al Azhar University. Cairo, Egypt
[*]*Physics Dept. Faculty of Science Suez-Canal, University , Egypt.*

ABSTRACT. Glass system of the composition $(Na_2B_4O_7)_{100-x}$ $(V_2O_5)_x$, with x=0, 5,7.5,10, 12.5,15 and 20 mol% has been prepared by conventional melt-quenching technique. The optical absorption and A.c conductivity have been measured. The optical absorption reveals that the electronic transition is indirect. The exponential dependence of the absorption coefficient α (α ≈ 103-104 cm-1) as a function of the incident photon energy (hυ) suggests that the urbach role is obeyed. A.C conductivity measurements are performed in the frequency range (0.1-100 KHz), over the temperature range (300-600K).Analysis of data suggests that correlated barrier hopping (C.B.H) is the dominant conduction mechanism. The dependence of dielectric constant (έ) and dielectric loss (tanδ) on both temperature and frequency are also discussed.

INTRODUCTION

V_2O_5-based semiconducting glasses find wide application in solid-state devices, optical fibers and optical memory. Vanadate glass contain both $V^{4+}$and V^{5+} ions and conduction in such glasses is brought about by the hopping of electron / polaron from the V^{4+} sites [1].Recently ,there has been a considerable. interest in the study of their technological applications. In general transition metal ions can be used to probe the glass structure since their outer d electron orbital functions have rather broad distributions and their response to the surrounding cations are very sensitive [2]. Transition metal oxide (TMO) glasses have been frequently studied from both chemical and physical view points.

DC conductivity of vanadium oxide glasses in which V_2O_5 is associated with other glass formers such as TeO_2, P_2O_5,As_2O_3
Has been extensively studied different studies reveal that the small polaron model is most likely applies to describe the electronic transport phenomena in these materials studies on boro-vanadate glasses are very few [these glasses have their potential application as optical and electrical memory switching [3,4].Alkali borate as solid electrolytes because of their fast ionic conductions the ionic conductivity of binary lithium borate glasses have been investigated [5].In (TMO) glasses two possible mechanism can be observed .

I)   The absorption due to internal transitions between the d shell electrons. (d-d transition).

II)   The absorption due to transfer of an electron from a neibouring atom to the transition metal ion and vice versa (hopping)[6].

The present study aims to investigate the effect of V_2O_5 addition on physical properties of borate vanadate glasses containing sodium oxide.

In the present work, systematic study involves measurement of optical absorption and AC conductivity.

EXPERIMENT

Glass system of the form $(V_2O_5)_x(Na_2B_4O_7)_{100-x}$ with x=0, 5, 7.5, 10, 12.5, 15 and 20 mol% has been prepared by conventional method. The oxide mixture were melted in porcelain crucibles in an electrical furnace at temperature 1000 C for one hour, the homogenous molten were quickly casted into mould of stainless steel at room temperature. Samples were annealed at 350C for 24 hours, to get red of any internal stresses. The glass surface of all samples was made plane by polishing with five emery paper to use the measurement of both optical and electrical conductivity. The optical absorption were recorded at room temperature using (UV/Vis) spectrophotometer (perkinelmer lambad 35) in the wave

CP888, *Modern Trends in Physics Research,*
Second International Conference on Modern Trends in Physics Research—MTPR-06,
edited by L. El Nadi

length range (190-1100 nm). A.C. electrical conductivity was obtained using RLC meter (Pm6304 philips), range (50-10^5 Hz) and temperature range from (302-633 K).

RESULTS AND DISCUSSION

Optical absorption in the spectral rang (190-1100 nm) were carried out at room temperature. Glass samples of the composition $(V_2O_5)_x(Na_2B_4O_7)_{100-x}$ with x=0, 5, 7.5, 10, 12.5, 15 and 20 mol%. The transmission spectra as a function wave length are shown in Fig[1] there is no sharp absorption edge (urbach) indicating that the samples under investigation are mainly amorphous analysis of data reveals that the position of the fundamental absorption edge shifts to higher wave length (lower energy) with increasing vanadium content. Fig (2a) shows the plots of $(\alpha h\omega)^{1/2}$ against $h\omega$ and the value of Eg was determined by extra polating the linear parts of the curves $(\alpha h\omega)^{1/2}$ =0 obtained values are listed in table one.

Inspection of Fig (2b) and Table[1] show that the absorption edge is shifted from UV to visible region and the value of Eg moves toward lower energies from 3.7 to 2.9 ev with the increasing vanadium oxide from x=0 to 20 mol%. However such behavior is observed up to x=10. It starts to decrease again – the observed behavior can be accounted for the defects introduced by vanadium oxide

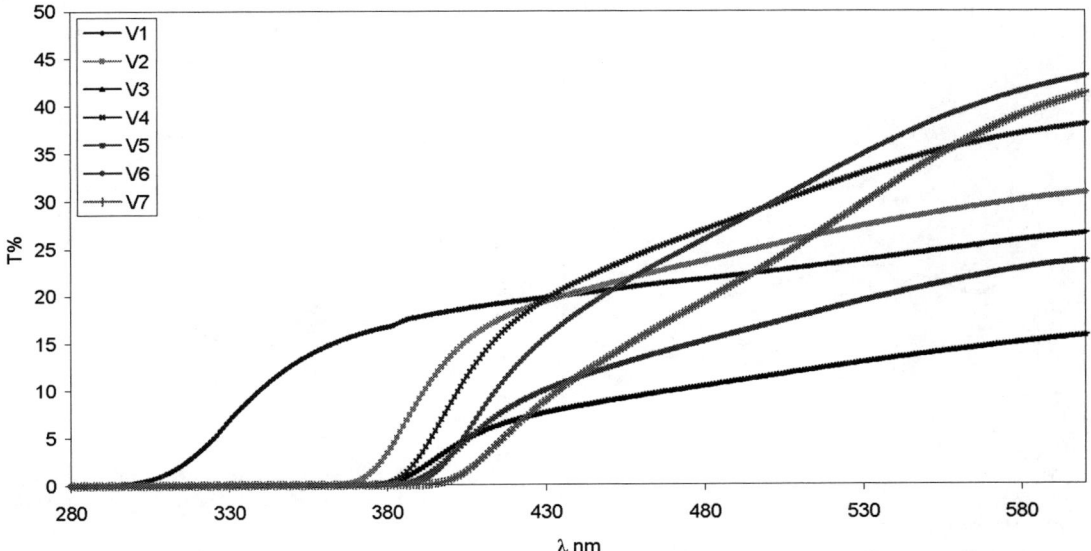

Fig [1] the relation between transimssion and wavelength for different composition

addition depend on the role of the latter. At low vanadium content x ≤ 10 which reduces the average bond energy and modifier and reduces the structural defects by decreasing the charged $(BO_4)^-$ units .i.e. reduction of non-bridging oxygen (NBO)[7]. It is generally accepted that absorption edge dependence on the oxygen bond strength in glass, which can be accounted for the increase of the v-o bonds energy in the glass network [8]. In other words at x≥10 the vanadium acts as network former, giving rise to increase the average bond length.

TABLE 1. Values of energy gap and band tail.

V_2O_5 mol%	E_{opt} (ev)	E_e (ev)
0	3.70	0.54
5	3.00	0.25
7.5	3.01	0.22
10	3.06	0.16
12.5	3.04	0.18
15	3.00	0.22
20	2.90	0.23

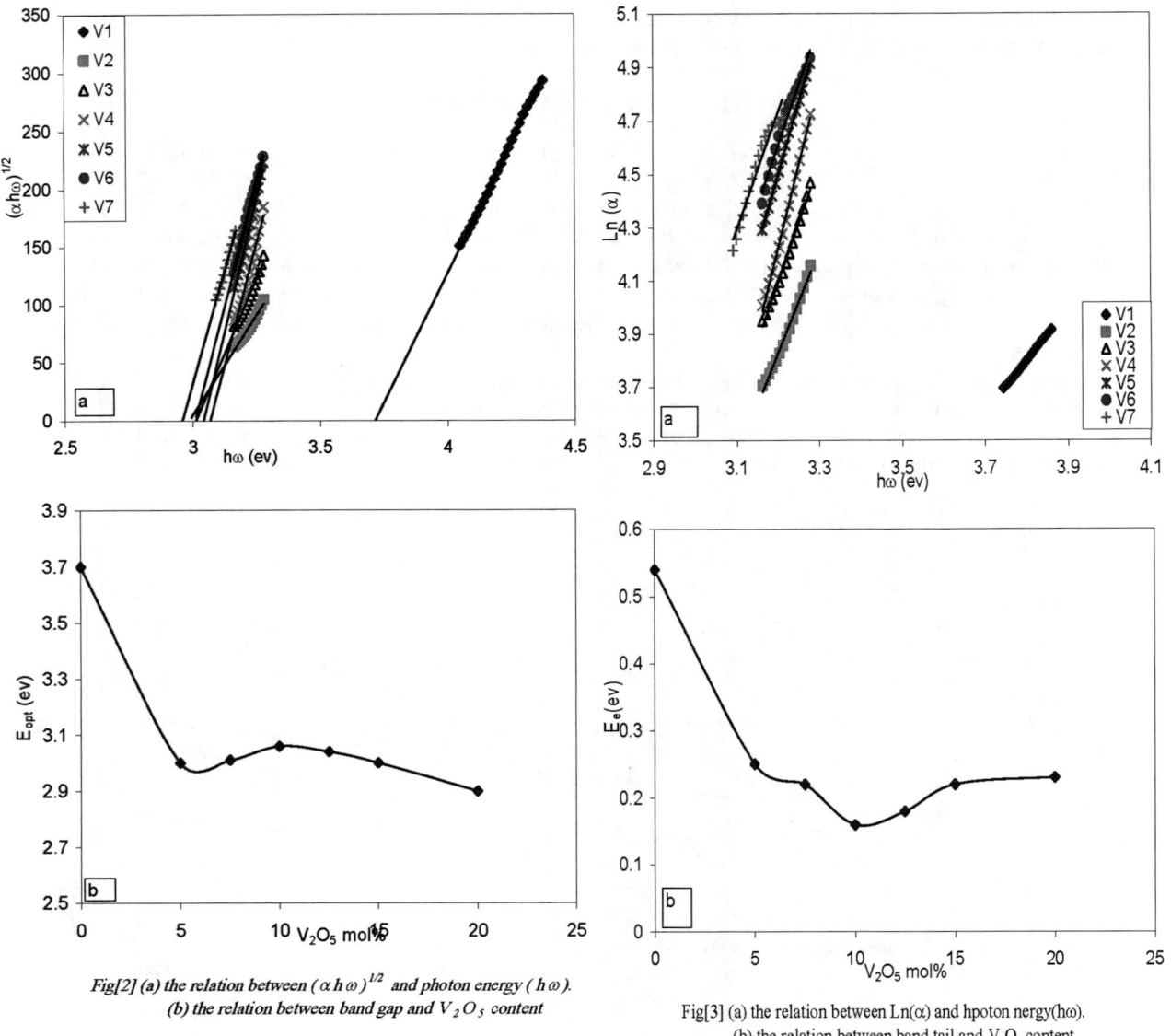

Fig[2] (a) the relation between $(\alpha h\omega)^{1/2}$ and photon energy $(h\omega)$.
(b) the relation between band gap and V_2O_5 content

Fig[3] (a) the relation between $Ln(\alpha)$ and hpoton nergy($h\omega$).
(b) the relation between band tail and V_2O_5 content.

Fig(3a) show ln α against hω for the investigated glass system the band tail width E_e is calculated from the slopes of linear region, estimated values are represented in Fig(3b) and Table[1]. It is observed that the value of Ee decreases by increasing vanadium content. In other words Ee follows opposite patterns to that of E_{opt}. The values of Ee lie between (0.46 and 0.66 ev) which is an acceptable values [9]. It is obvious that the width of the band tails of the localized states is found to depend on the concentrate of modifiers the optical absorption spectrum in Fig(4) exhibits one band characteristic of VO^{2+} ions in tetragonal symmetry for all glass samples which containing vanadium oxide. The band at (28248 cm^{-1}) was assigned $^2B_{2g}$ --- $^2B_{1g}$ transition [10]. These results are consistent with ESR observation [under to be published].

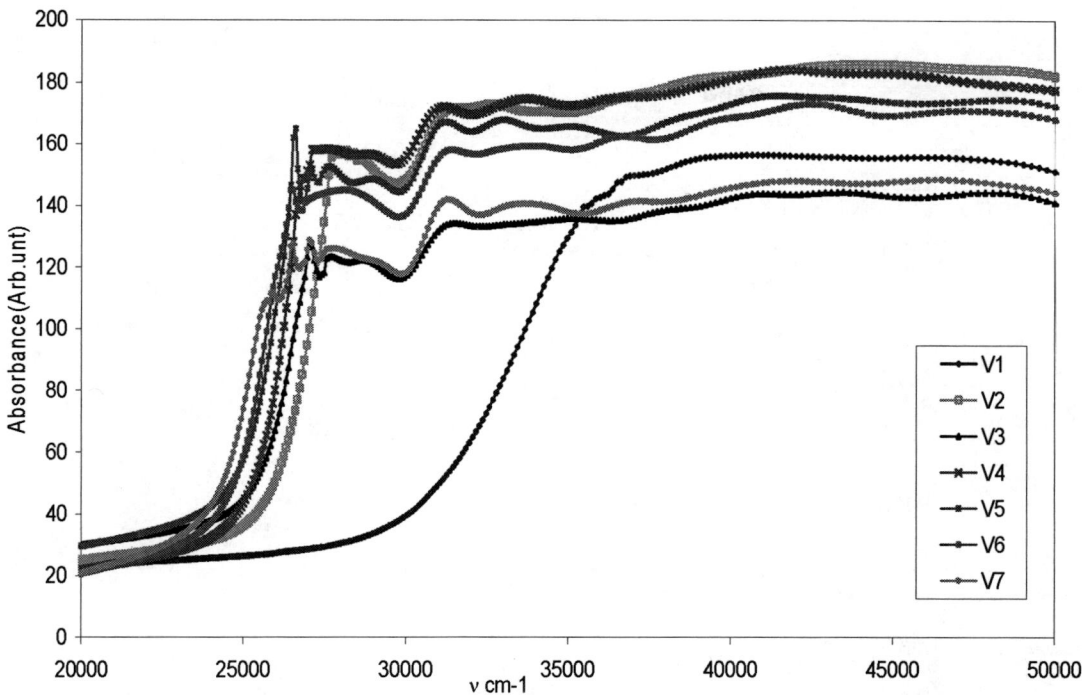

Fig[4] the relation between Absorbance and v (cm⁻¹) for different composition

A.c Conductivity

The obtained ac conductivity was found to obey the well known relation

$$\sigma(\omega) = A\,\omega^s$$

Where   is the conductivity, A is constant, w is the frequency and s is the exponent which is estimated experimentally from

$$S = \ln(\sigma)/\ln(\omega)$$

Fig[5a] show plots of ln (σ) versus ln (ω) for the investigated samples at 473 K. It is clear that the Fig [5b] show that all composition follow one pattern, where s decreases with raising temperature. Generally, the charge carries of amorphous materials are acted upon by an approximately random potential energy barriers, and $\sigma(\omega) = A\,\omega^s$, where s $\leq$ 1.
Inspection of S(T) reveal that CBH model is the dominant conduction mechanism.

Observed data reveal that the a.c. conductivity of investigated glasses tends to increase by increasing V_2O_5 content up to V_2O_5 then it starts to decrease again up to 20 mol% V_2O_5 as shown in Fig[6]. Such results are consistent with other measured properties (such as optical edge). In such case is mixed: one should expect that the conductivity is most likely ionic part (Na mobile ions) and electronic part due to hopping of electrons between V^{+5} and V^{+4}.

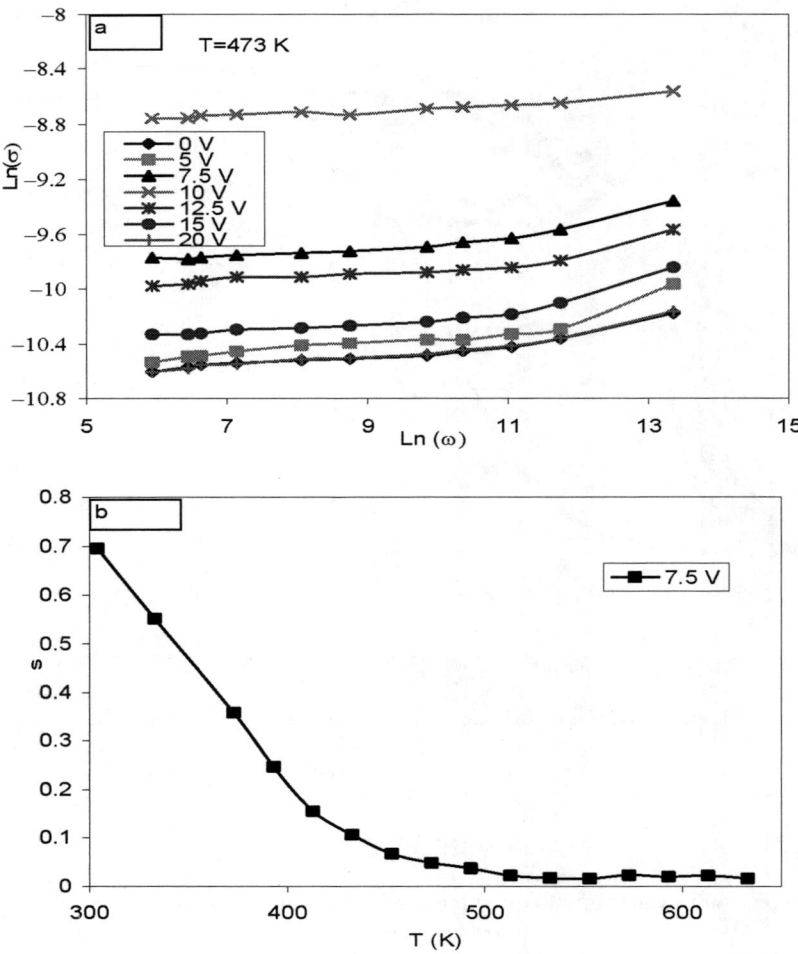

Fig [5] (a) the relation between Ln (σ) and Ln (ω) at fixed temperature for different mposition
(b) the relation between S and T(K) for x = 7.5 V_2O_5 content

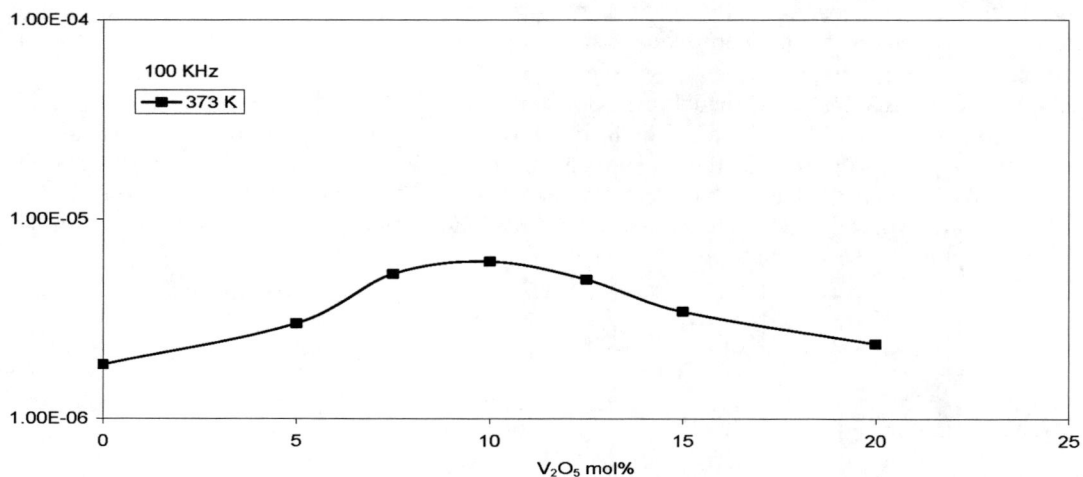

Fig [6] show the relation between conductivity and V_2O_5 content at fixed frequen and fixed temperature

At low V_2O_5 content, it is likely to that V_2O_5 acts as modifier, the latter is usually presents in from of VO^{+2}.It seems that at low V_2O_5 can concentration almost all vanadium ions are tetravalent, and hence no appreciable hopping can take place. The number of Bo_3 units dominates and allows easy paths of the charge carriers (mainly Na^+). By increasing V_2O_5 (higher than 10 mol%) the number of V^{+5} is increased and reduces the mobility of Na^+ ions. This is may be due to the fact that at higher V_2O_5 content, the latter acts as glass former [11].This in turn reduces the conductivity.

It seems that the modifying action of VO^{+2} ions, allows easy patches for the movement of the charge carries Na^+ up to 10 mol%.

REFERENCES

1. V.Sudarsan,S.k. 117 J. non-crystal solids **258**, 20 (1999).
2. R.P.Sreekanth , A.M , J.L. ,physica B **293**, 108 (2000).
3. S.Sindhu,S. Sanghi ,A. A ,V.P. ,physica B **365**, 65 (2005).
4. H.Ei Mkami ,B.Deroide . , J phy and chem. Of solides **61**, 819 (2001).
5. Y.M. Moustafa , A.K. Hassan , J non crystal solids **194**, 34 (1996).
6. C.A. Hogarth ,A.A.Hosseini , J Material science **18**, 2697 (1983).
7. M.Altaf, M. Asharf and Maria Zahid, J.Science. **14**, 253 (2003).
8. C.A. Hogarth, A.A. hosseine. J.Mate.Sci. **18**, 2697 (1983).
9. M.Pollak, Phil. Mag, **23**, 519 (1971).
10. S.Khasa, V.P.Seth, P.S.gahlot, A. Agarwal,R.M.krishna and S.K.Gupta. Physica B 334 , **347** (2003).
11. R.Balaji Rao, N.O.Gopal and N.Veeraiah, Jurnal of alloys and compounds **386**, 25 (2004).

Ab-initio Study of Structural Properties of III- Nitrides

Rashid Ahmed[1], H. Akbarzadeh[2] and Fazal-e-Aleem[1]

[1]*Centre for High Energy Physics, University of the Punjab, Lahore-54590-Pakistan*
[2]*Department of Physics, Isfahan University of Technology, 841546 Isfahan, Iran*

Abstract. We present first principles study of structural properties of Zinc-blende and Wurtzite phases of Aluminum Nitride, Gallium Nitride and Indium Nitride binary compounds. The study has been done using Full-Potential Linearized Augmented Plane Wave plus local orbitals method, within the framework of density functional theory. Results for lattice constant, bulk modulus, its pressure derivative and cohesive energy of these compounds are also compared with experimental results.

Keywords: Semiconductors; III-V Compounds; III-nitrides; WIEN2k; FP- LAPW.
PACS: 71.15.-m; 71.15. Ap; 71.15.Mb; 71.21. Nr; 71.55.Eq

INTRODUCTION

Nitride compound semiconductors in general have wide variety of applications. These nitrides such as AlN, GaN and InN materials are of current interest for their potential for optoelectronic and high power/temperature electronic devices [1-10]. Major attention in the recent past on these nitrides is due to the fact that in addition to their wide band gap (except InN) [11-17], they also have a high thermal conductivity and large bulk modulus [18-20]. Significant role of nitrogen in these compounds is in the formation of their physical properties similar to other wide band gap semiconductors such as diamond [21-22]. These properties have made them very fascinating and useful materials for different devices [23]. Therefore it is essential to understand various properties of these nitrides.

III-nitrides (AlN, GaN, and InN) usually crystallize in two phases; zincblende and wurtzite structures. The space group of zincblende structure nitrides is F $\overline{4}$3m. It contains four molecules in its unit cube and is developed on a face-centered lattice. Wurtzite structure nitrides have their space group C6mc and have two molecules in hexagonal unit cell [24]. Theoretically, III- nitrides have been studied by employing different approaches from empirical methods to first principle methods but a detail study of both phases of these nitrides is rarely found in literature. Therefore an in- depth study of structural properties of both phases of AlN, GaN and InN based on first principle should provide very useful information.

In this work, we report structural properties of III-N compounds. In our calculations we use previously calculated optimized value of internal parameter u, [17] and experimental lattice parameters. Computations have been made using Full-Potential Linearized Augmented Plane Wave ((L)APW+lo) method [25-28], within the framework of Density Functional Theory (DFT) [29-30]. For the exchange correlation potential, generalized gradient approximations (GGA) [31] has been employed. Results obtained for structural properties of AlN, GaN and InN with GGA are compared with experimental data and other theoretical work. Our results with GGA are in good agreement with earlier theoretical work.

METHODOLOGY

Calculation of electronic structure is a complex many body problem as there are about 10^{23} valence electrons in each cubic centimeter contributing to the bonding of a typical solid [32]. The exact solution of the many-body problem is almost impossible and perhaps not necessary [33]. One of the possibilities for studying such systems is by using ab-initio methods. The advantage of these methods lies in the fact that they do not require any experimental knowledge [25-26] to carry out such calculations. DFT [29-30] is an efficient and accurate first principle method for solving many-electron problem of a crystal. In this approach, many body problem of interacting system is mapped into a set of non interacting one particle system moving in an effective local potential and corresponding one particle

CP888, *Modern Trends in Physics Research,*
Second International Conference on Modern Trends in Physics Research—MTPR-06,
edited by L. El Nadi
© 2007 American Institute of Physics 978-0-7354-0354-9/07/$23.00

(electron) equations are obtained. These equations, called Kohn Sham equations [29-30], are solved iteratively till self-consistency is reached. There are several techniques to solve these equations. One of these is full potential linearized augmented plane wave plus local orbitals (L(APW+lo) method. This is one of the most accurate methods to calculate electronic structure for crystals [25-26]. In this work structural properties of both phases (Zinc-blende and Wurtzite) of AlN, GaN and InN have been calculated using FP-(L(APW+lo) method as employed in the WIEN2k code [25-26]. We calculate lattice parameters, bulk modulus, its pressure derivative and cohesive energy of both phases of each compound.

RESULTS AND DISCUSSION

Structural Parameters Calculations

Choice of a suitable exchange correlation potential is necessary for the calculations of various properties of solids. In our calculations of structural properties of III-nitrides, GGA exchange correlation potential within DFT was employed. To compute the structural properties of Aluminum, - Gallium- and Indium-nitride, we used the experimental lattice parameter and optimized value of internal parameter 'u' as given in references 24 and 17. Total energy of primitive unit cell at different volumes is calculated. By fitting the data thus obtained with Murnaghan equation of state [34], the equilibrium lattice constant, cohesive energy, bulk modulus and its pressure derivative for AlN, GaN and InN are calculated. As a sample energy–volume curve for AlN is shown in Fig.1.

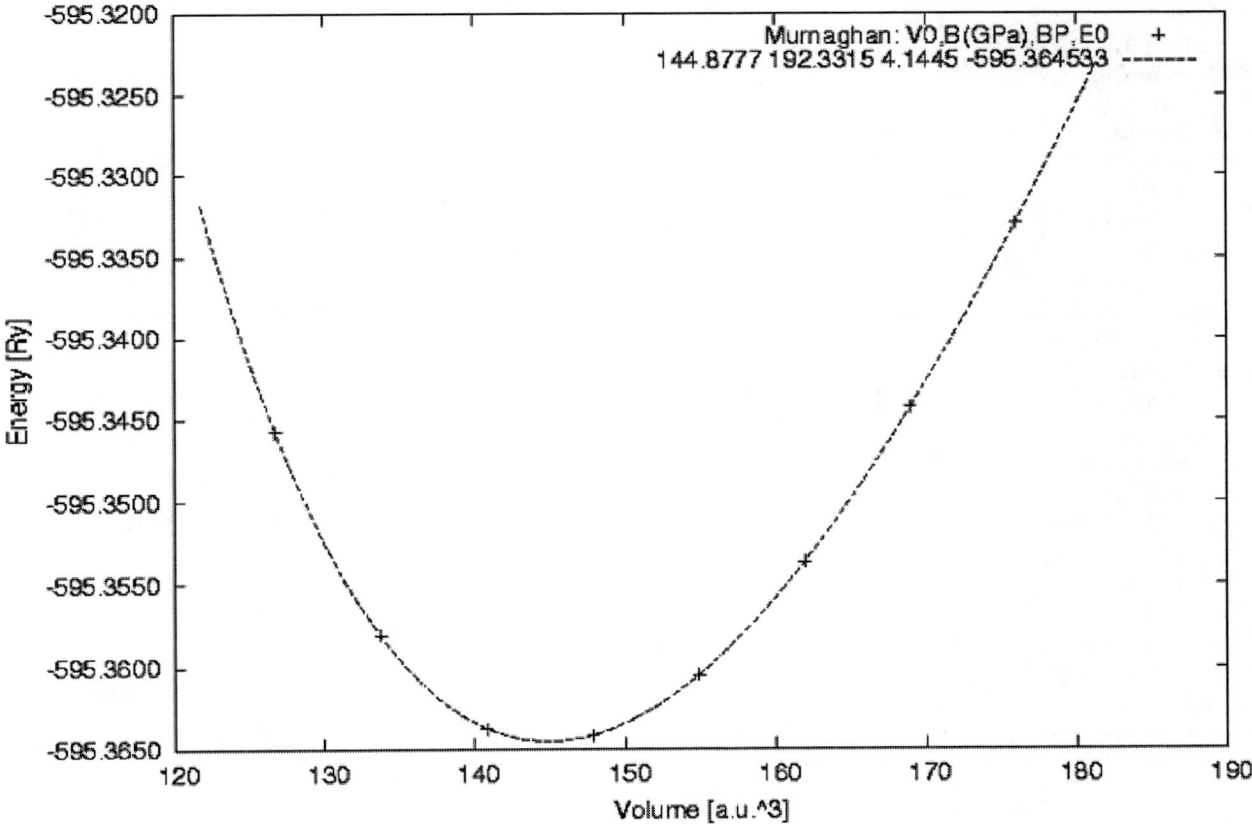

FIGURE 1. Total energy versus primitive cell volume for Zincblende AlN obtained by FP-(L(APW+lo)

The results of our computation are shown in Table 1. Comparison of the results for these nitrides with available first principle computations and experimental data has also been made. It is observed that our calculated results for

lattice parameters, bulk modulus, its pressure derivative and cohesive energy of both phases of each compound, in general, are in agreement with the experimental data and other theoretical work [35-37].

TABLE 1. Lattice Parameters (α), Bulk Modulus (*B*), it's Pressure Derivative (*B'*) and Cohesive Energy E_{coh} obtained using full potential (L(APW+lo) within GGA for AlN, GaN and InN compounds.

COMPONDS	LATTICE PARAMETERS		B(GPA)	B	E_{coh}(EV)
	α (Å)	c(Å)			
AlN (ZB)					
Present work	4.42	-	192	4.15	9.69
Calc.(GGA)	4.40 [37]	-	191 [37]	-	11.52 [37]
Calc.(PP-GGA)	4.39 [35]	-	191 [35]	3.81 [35]	13.24 [35]
Calc.(PP-GGA)	4.39 [38]	-	-	-	11.36 [35]
Calc.(GGA)	4.42 [36]	-	-	-	11.91 [38]
Experiment	4.36 [24]	-	202 [39]	-	11.52 [40]
AlN (W)					
Present work	3.23	5.25	193	4.16	11.53
Calc.(PP- GGA)	3.11 [35]	5.04 [35]	192 [35]	3.96 [35]	11.40 [35]
Calc.(PP-GGA)	3.11 [38]	4.99 [38]	-	-	12.07 [38]
Calc.(GGA)	3.14 [36]	5.03 [36]	-	-	-
Experiment	3.11 [24]	4.98 [24]	-	-	-
GaN (ZB)					
Present work	4.59	-	173	4.27	8.88
Calc.(GGA)	4.55 [37]	-	172 [37]	-	8.86 [37]
Calc.(PP-GGA)	4.59 [35]	-	156 [35]	4.25[35]	8.25 [35]
Calc.(PP- GGA)	4.54 [38]	-	-	-	9.25 [38]
Calc.(GGA)	4.56 [36]	-	-	-	-
Experiment	4.50 [24]	-	190 [39]	-	8.96 [40]
GaN (W)					
Present work	2.75	4.46	174	4.26	8.77
Calc.(PP- GGA)	3.25 [35]	5.30[35]	172 [35]	5.11[35]	8.27 [35]
Calc.(PP-GGA)	3.20 [38]	5.23[38]	-	-	9.27 [38]
Calc.(GGA)	3.16 [36]	3.22[36]	-	-	-
Experiment	3.18 [24]	5.17[24]	-	-	-
InN (ZB)					
Present work	5.06	-	121	4.71	5.96
Calc.(GGA)	5.05 [16]	-	122 [16]	-	7.35 [41]
Calc.(PP-GGA)	5.11 [35]	-	117 [35]	4.43[35]	6.86 [35]
Calc.(PP- GGA)	5.07 [38]	-	-	-	7.68 [38]
Experiment	4.98 [24]	-	137 [39]	-	7.72 [40]
InN (W)					
Present work	3.59	5.78	122	4.79	5.80
Calc.(PP-GGA)	3.61 [35]	5.88 [35]	116 [35]	7.33 [35]	6.87 [35]
Calc.(PP- GGA)	3.59 [38]	5.80 [38]	-	-	7.70 [38]
Experiment	3.53 [24]	5.69 [24]	125 [41]	-	-

ACKNOWLEDGMENTS

Part of the work was done at ICTP affiliated Center, University of Technology Isfahan, Iran. One of the authors (RA) acknowledges the financial support of the Isfahan University of Technology and University of the Punjab Lahore, Pakistan, and is deeply indebted to Dr. A. Mokhatari for his help concerning the use of WIEN2k.

REFERENCES

1. J. D. Beach, Hamda Al-Thani, S. McCray, and R. T. Collins, J. Appl. Phys. **91**, 5190 (2002).
2. H. Morko¸c, Nitride Semiconductors and Devices, (Springer, New York, 1999).
3. J. M. Van Hove, R. Hickman, J. J. Klaassen, and P. P. Chow, Appl. Phys. Lett. **70**, 2282 (1997).
4. P. M. Asbeck, E. T. Yu, S. S. Lau, G. J. Sullivan, J. Van Hove, and J.Redwing, Electron. Lett. **33**, 1230 (1997).
5. E. T. Yu, X. Z. Dang, L. S. Yu, D. Qiao, P. M. Asbeck, S. S. Lau, G. J.Sullivan, K. S. Boutros, and J. M. Redwing, Appl. Phys. Lett. **73**, 1880 (1998).
6. J. R. Mileham, S. J. Pearton, C. R. Abernathy, J. D. MacKenzie, R. J.Shul, and S. P. Kilcoyne, J. Vac. Sci. Technol. A **14**, 836(1996).
7. M. S. Minsky, M. White, and E. L. Hu, Appl. Phys. Lett. **68**, 1531(1996).
8. C. Youtsey, I. Adesida, and G. Bulman, Electron. Lett. **33**, 245 (1997).
9. S. Nakamura, Semicond. Sci. Technol. **14**, R27 (1999).
10. S. Nakamura and G. Fasol, *The Blue Laser Diode* (Springer, Berlin, 1997).
11. T. Inushima, V. V. Mamutin, V. A. Vekshin, S. V. Ivanov, T. Sakon, M. Motokawa, and S. Ohoya, J. Cryst. Growth **227/228**, 481 (2001).
12. V. Yu Davydov et al., Phys. Stat. Solidi. **B 230.** R4 (2002); **229.** R1 (2002); 234, 787 (2002).
13. J. Wu, et al., Appl. Phys. Lett. **80**, 3967 (2002).
14. J. Wu, W. Walukiewiez, K. M. Yu, J. W. Ager III, E. E. Haller, H. Lu, and J. Schaff, Appl. Phys. Lett. **80** 4741 (2002).
15. A. Yamamoto, K. Sugita, H. Takatsuka, A. Hashimoto, and V. Yu. Davy-dov, J. Crystal Growth **261,** 275 (2004).
16. Pierre Carrier and Su-Huai Wei, J. Appl. Phys. **97**, 033707 (2005).
17. Rashid Ahmed, H. Akbarzadeh and Fazal-e-Aleem, Physica B **370**, 52–60 (2005).
18. For recent reviews see, e.g., J.W. Orton, and C.T. Foxon, Rep. Prog. Phys. **61**, 1 1998; in *Gallium Nitride I*, in Semicond. Semimet. Vol. 50, edited by J. I. Pankove and T. D. Moustakas (Academic, London, 1998).
19. *Properties of Group-III Nitrides*, edited by J.H. Edgar (INSPEC,London, 1994).
20. L. E. Ramos, L. K. Teles, L. M. R. Scolfaro, J. L. P. Castineira, A. L. Rosa, and J. R. Leite, Phy. Rev. B 63, 165210 (2001).
21. A. Mokhatari, H. Akbarzadeh, Phyisca **B 324,** 305 (2002).
22. A. Mokhatari, H. Akbarzadeh, Physica **B 337,** 122-129 (2003).
23. Sadao Adachi, *Properties of Group-IV, III-V and II-VI Semiconductors*, (John Wiley & Sons, England, 2005).
24. R.W.G. Wyckoff *Crystal Structures* (2nd Edition Krieger, Malabar 1986); *Semiconductors: Data Handbook*, edited by O. Madelung (Springer, Berlin 2004).
25. K.Schwarz and P.Blaha, *Quantum Mechanical Computations at the Atomic Scale for Material Sciences.* WCCM V (Fifth World Congress on Computational Mechanics July 7-12-2002, Vienna, Austria.
26. K.Schwarz, P.Blaha, G.K.H.Madsen, *Electronic structure calculations of solids using the WIEN2k package for material sciences.* Comp. Phys. Commun. **147**, 71(2002).
27. S. Cottenier, *Density Functional Theory and the family of (L) APW-methods: a step-by-step introduction* (Instituut voor Kern-en Stralingsfysica, K.U.Leuven, Belgium), 2002, ISBN 90-807215-1-4 (to be found at http://www.wien2k.at/reg user/textbooks).
28. K. Schwarz and P. Blaha, Comput. Mat. Sci. 28, 259 (2003).
29. P. Hohenberg, W. Kohn, *Inhomogeneous electron gas*, Phys. Rev., **136B**, (1964) 864-871.
30. W. Kohn, L. S. Sham, *Self-consistent equations including exchange and correlation effects,* Phys. Rev., **140**, (1965) A1133-1138.
31. P. Perdew, K. Burke, M. Emzerhof, Phys. Rev. Lett. **77**, 3865-3868 (1996).
32. Eoin O' Reilly *Quantum Theory of solids* (Taylor and francis 2002).
33. V.V. Nemoshkalenko, and V.N. Antonov, *Computational Methods in Solids state physics* (Gordon and breach Science Publishers1998).
34. Monkhorst H J and Pack J D *Phys. Rev.* B **13**, 5188 (1976).
35. C. Stampfl and C. G. Van de Walle, Phys. Rev. B **59**, 5521-35 (1998).
36. C. Perssona,, A. Ferreira da Silvab, R. Ahujaa and B. Johansson, J. Crystal Growth **231** 397–406 (2001).
37. M. Fuchs, J. L. F. Da Silva, C. Stampfl, J. Neugebauer, and M. Scheffler, Phys. Rev. B **65**, 245212 (2002).
38. Agostino Zoroddu, Fabio Bernardini, Paolo Ruggerone and Vincenzo Fiorentini, Phys. Rev. B **64**, 045208 (2001).
39. M.E. Sherwin and T.J. Drummond, J. Appl. Phys. **69**, 8423 (1991).
40. W.A. Harrison, *Electronic Structure and the Properties of Solids,* (Dover, New York, 1989) p. 176.
41. K. Kim, W. R. Lambrecht, and B. Segall, Phys. Rev. B **53**, 16 310 (1996).

A Model for Calculating Magnetization Curves of Ferromagnetic Materials : a Review

S. H. Aly[1] , S. Yehia[2], N. El-Wazzan[1], M. Soliman[1], M. Mahgoob[1].

1. Physics Department, Faculty of Science at Damietta, Mansoura University, New Damietta, Egypt.
2. Physics Department, Faculty of Science, Helwan University, Cairo, Egypt

Abstract. We present a review on using a statistical mechanics-based model for calculating magnetic properties of hexagonal, cubic and mixed-anisotropy systems. Although our simple model is not as elaborate as, for example, monte carlo simulation it demonstrates a success in producing all the major features of magnetization curves of the ferromagnetic systems studied. Examples of these features are sequence of easy-axes of magnetization, spin re-orientation and superparamagnetism in small particles. The magnetization curves, magnetic susceptibility and orientation probability of the magnetization vector are calculated in a wide range of temperature, particle size and magnetic field. Examples of the systems we report are uniaxial ferromagnets e.g. Gd, $PrCo_5$ and $Nd_2Fe_{14}B$; cubic materials e.g. Fe and Ni and mixed-anisotropy systems, e.g. Fe on stepped Ag (001) substrates, in which a jump in the magnetization is found to depend on the ratio of the uniaxial to biaxial anisotropy present in these systems.

INTRODUCTION

Many of the magnetic materials that are of interest on both the technological and fundamental levels are among the rare-earth transition metal compounds. Extensive studies on these materials include ab initio calculations[1], experiments[2], analytical studies[3,4] and modeling [5,6]. We have reported on using the fundamentals of statistical mechanics in a simple model to calculate the dependence of magnetization and magnetic susceptibility on field and temperature for several uniaxial materials e.g. $Nd_2Fe_{14}B$, $Ho_2Fe_{14}B$ and elemental   Gd[7-9]. Our model can be well applied to systems of symmetries other than the uniaxial. The symmetry of a given crystal is embedded in the phenomenological expression of its anisotropy energy density[10]. We have extended our study to cubic systems e.g. Ni and mixed-anisotropy systems where both of cubic and uniaxial anisotropies are present. We should mention here that our model adopts the rigid model i.e. it deals with a single magnetization vector and a single set of appropriate anisotropy constants that are characteristic of a given compound as a whole regardless of its constituents. This picture is different from the sub-lattice description where each sublattice in a given compound is described by a certain saturation magnetization and a characteristic set of anisotropy constants in addition to an exchange interaction that couples the magnetizations of the two sublattices e.g in case of $PrCo_5$[11].

MODEL AND CALCULATION

The energy density E for a given system is the sum of its anisotropy E_a and zeemann energies :

$$E = E_a - \vec{H} \cdot \vec{M_s} \tag{1}$$

where $\vec{M_s}$ is the saturation magnetization, $\vec{H}$ the applied magnetic field. The exact phenomenological form of E_a depends on the crystal symmetry[10] and may involve up to five anisotropy constants[12]. Many magnetic systems of interest however may be adequately described by only two constants viz.:

$$E = k_1 \sin^2 \theta + k_2 \sin^4 \theta \tag{2}$$

CP888, *Modern Trends in Physics Research,*
Second International Conference on Modern Trends in Physics Research—MTPR-06,
edited by L. El Nadi
© 2007 American Institute of Physics 978-0-7354-0354-9/07/$23.00

For a cubic system the anisotropy energy is given by[13]:

$E = k_1 s + k_2 p + k_3 s^2 + k_4 sp$

where $s = \alpha_1^2 \alpha_2^2 + \alpha_1^2 \alpha_3^2 + \alpha_2^2 \alpha_3^2$ and $p = (\alpha_1 \alpha_2 \alpha_3)^2$

The direction cosines of the magnetization vector relative to the cubic axes

are: $\alpha_1 = \sin\varphi \cos\theta$, $\alpha_2 = \sin\varphi \sin\theta$ and $\alpha_3 = \cos\varphi$

The partition function Z, the probability P, the magnetization M, the magnetic susceptibility χ and the heat capacity C are given respectively by:

$$Z(T, H) = \int_0^{2\pi} \int_0^\pi e^{-E(T, H, \theta, \varphi)V\beta} \sin\theta \, d\theta \, d\varphi \qquad (3)$$

$$P = \frac{e^{-EV\beta}}{Z} \qquad (4)$$

$$M = \frac{1}{\beta} \frac{\partial(\ln Z)}{\partial H} \qquad (5)$$

$$\chi = \frac{\partial M}{\partial H} \qquad (6)$$

RESULTS AND DISCUSSION

Examples of the magnetization curves of $Nd_2Fe_{14}B$ at 77, 176 and 270 K are shown in Figs.1-3. The curves are calculated for field parallel to either [001],[100] or [110] directions. The temperature dependence of the saturation magnetization and five anisotropy constants of a single $Nd_2Fe_{14}B$ crystal have been reported by Bolzoni et al[12]. We have used their data to generate the magnetization curves.

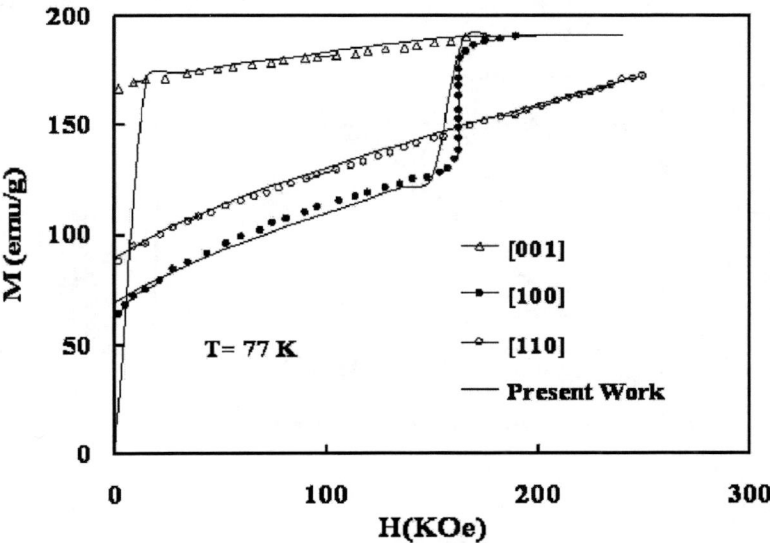

FIGURE 1. Magnetization curves for $Nd_2Fe_{14}B$ at 77K along specific crystal directions.

The agreement between our model and the experimental results are fair. It is clear that all the important features of the experimental curves are present in our simulated curves e.g. the in-plane anisotropy present at T< 176K and its disappearance at higher temperatures (Fig.3). Also the magnetization jump along[100] direction at a certain magnetic field is clear. A spin re-orientation phenomenon from easy-axis to easy-cone has been confirmed to take

place at about 150K by torque measurements[14], heat capacity[15], DSC[16], and magnetostriction and thermal expansion techniques[17]. We have confirmed this phenomenon by calculating the energy angular dependence, heat capacity[7] and probability angular distribution of the magnetization vector[8].

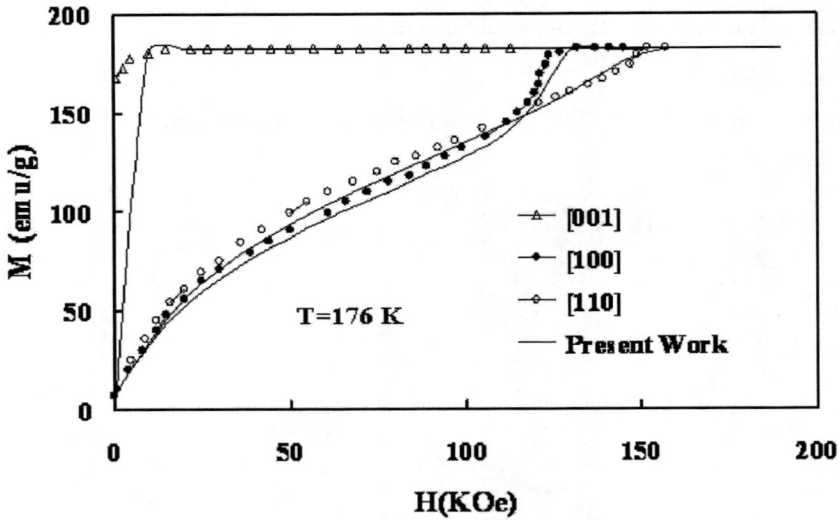

FIGURE 2. Magnetization curves for $Nd_2Fe_{14}B$ at 176 K along specific crystal directions.

In an attempt to study a model system that has been studied analytically we calculated the magnetization curves for an easy-plane system for different values of the ratio k_1/k_2 of the first to second anisotropy constants. Our simulation agree well with the analysis done by de Jesus et al[3]. Their analysis showed that the shape of the magnetization curves is sensitive to the ratio k_1/k_2 . For example for $k_1/k_2 < 0.25$ the magnetization increases smoothly from zero up to saturation although the features of the curves differ for different k_1/k_2 values. For $k_1/k_2 > 0.25$ however a totally different behavior is observed namely a sudden jump to saturation at a given critical field.

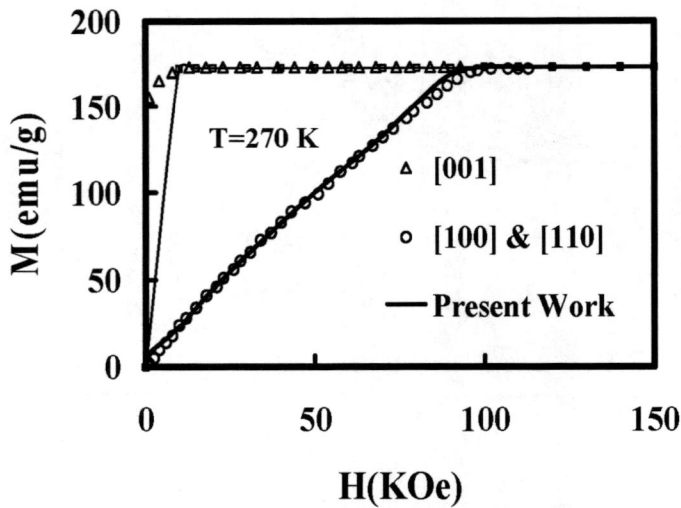

FIGURE 3. Magnetization curves for $Nd_2Fe_{14}B$ at T= 270 K along specific crystal directions.

An interesting another unaxial system is elemental Gd. It has a T_C around room temperature and its magnetization undergoes a temperature-dependent orientation off the crystallographic c-axis of this hexagonal crystal [18-20]. We concentrate our studies on the size-dependent magnetic properties in the 4-300 K range for

particles of radii in the 8-160A^0 range. The anisotropy constants of Gd have different signs but are of the same order of magnitude[21]. Fig.4 shows the isothermal magnetization at T=250 K for Gd particles in the 8-160 A^0 range with field parallel to the c-axis.

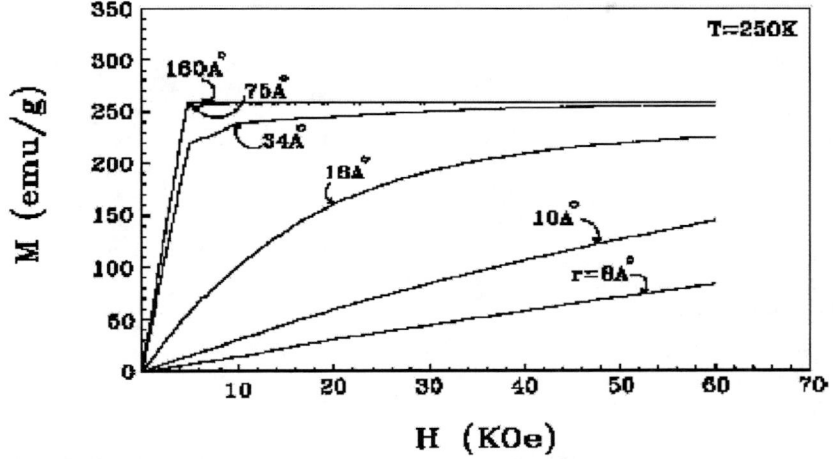

FIGURE 4. Isothermal magnetization curves at 250 K for Gd particles of different sizes .

It is evident that magnetic saturation (~260 emu/gm) is not reached for particles with radii less than about 16 A^0 in contrast to those with r>35 A^0 where saturation is achieved in small field. We have repeated this calculation at 4, 50 and 170 K to study the effect of temperature on the magnetization. At 4 K for example the magnetization showed a weak size dependence and saturation is reached for ~10 A^0 particles in only ~ 20 kOe field. An example of the temperature dependence of magnetization is shown in Fig.5.

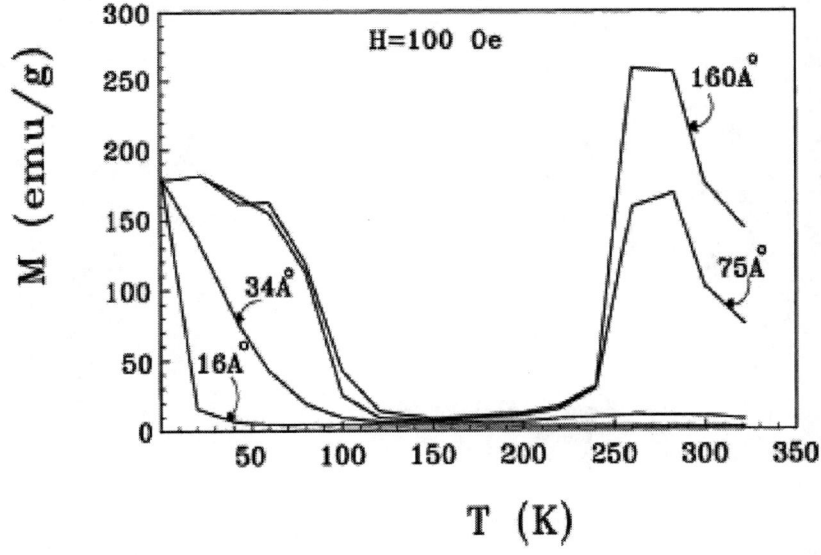

FIGURE 5. Magnetization dependence on temperature at 100 Oe for Gd particles.

An interesting feature of this figure, and of other ones calculated in different magnetic fields, is the drop in the magnetization in the temperature range 100-200K for particles with r>35 A^0 in agreement with magnetic measurements and neutron diffraction[19,20]. The magnetic susceptibility dependence on temperature for a 160 A^0 particle in a 0.1 kOe is shown in Fig.6. The two major peaks are displayed at ~100 and 250 K are associated with the spin re-orientation phenomenon in Gd. The minor peak at ~290 K is at the ferro- to paramagnetic transition.

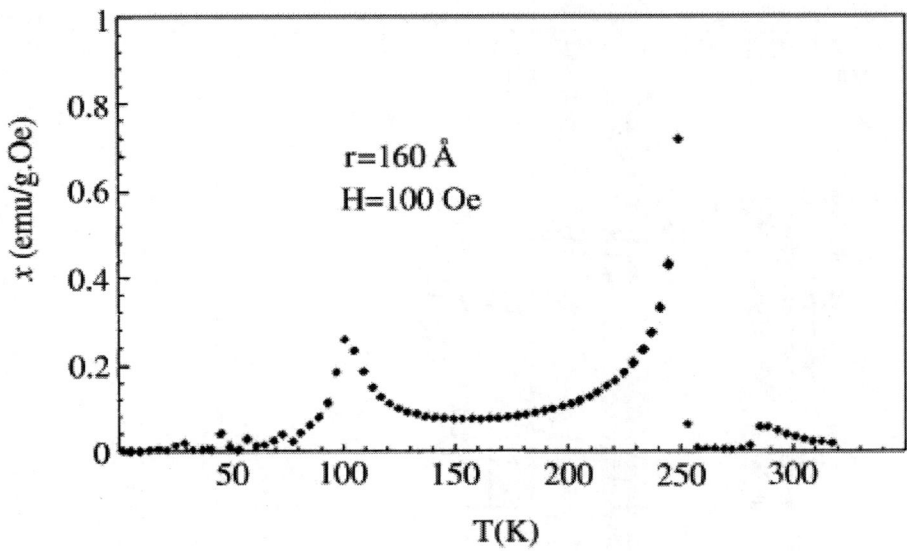

FIGURE 6. Magnetic susceptibility for a 160 A^0 Gd particle at 100 Oe as function of temperature.

The magnetic susceptibility for smaller particles showed a different behavior as superparamagnetism commences. A similar behavior has been reported in Er[22]. The dependence of magnetization on particle size at a given temperature and field is shown in Fig.7 . Magnetic saturation takes place for particles with r >35 A^0 , below this size however magnetization decreases as the particle size decreases. We have found the same trend for field in the 5-10 kOe range at temperatures from 4 to 250 K. This behavior was reported in experiments on other fine-particle systems[23].

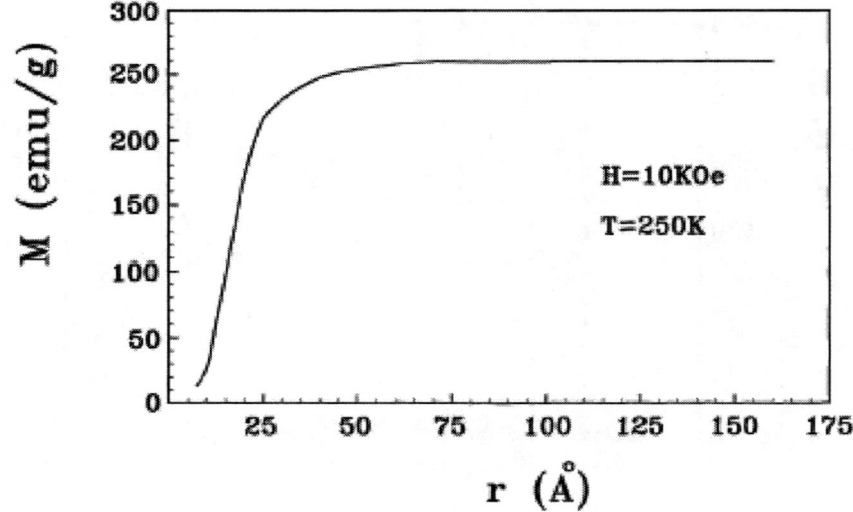

FIGURE 7. Magnetization dependence on particle size for Gd particles at 250 K In a 10 kOe field.

To confirm superparamagnetism in small Gd particles we plotted magnetization against H/T for different temperatures in a 60 kOe field. The results are displayed in Fig.8 for a 8 A^0 particle. It is well known that this phenomenon occurs in single-domain ferromagnetic particles if thermal energy is larger than the magnetic anisotropy energy barrier[24].

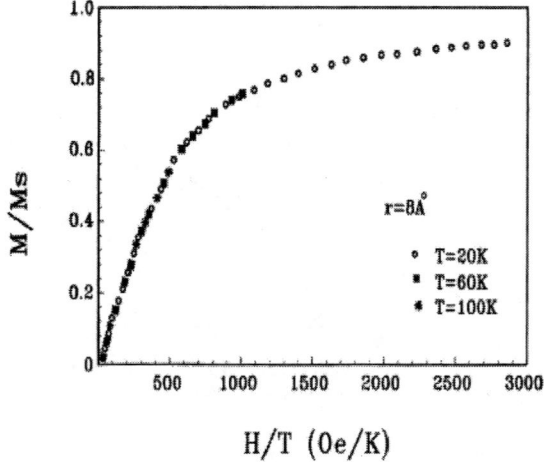

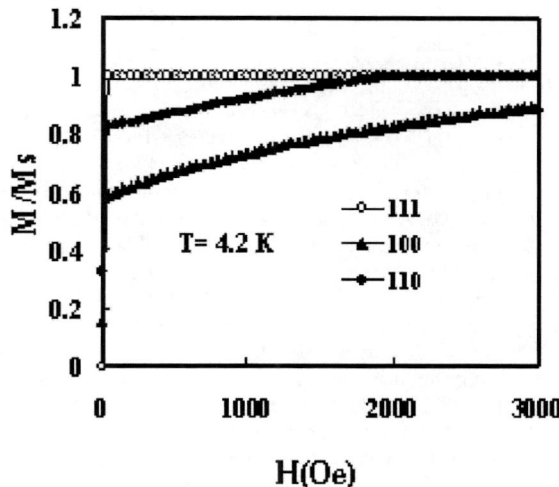

FIGURE 8. Demonstration of superparamagnetism in $8A^0$ Gd particles at different temperatures.

FIGURE 9. Magnetization curves of Ni at 4.2 K along the [100], [110] and [111] directions.

An experimental evidence of superparamagnetism is the overlap between magnetization curves measured at temperatures higher than a certain blocking temperature if they are plotted against H/T. It is clear from Fig.8 that small Gd particles meet this criterion. The same calculation have not resulted on the same behavior for larger particles. The application of our model to crystals of other symmetries has demonstrated its success to calculate magnetic properties of cubic and mixed anisotropy systems. The anisotropy constants of Fe and Ni for example were measured at 4.2 and 77 K by Tung et al[25]. In case of Ni, in contrast to Fe, the temperature dependence of these constants is strong and therefore the magnetization curves at different temperatures will reflect this strong dependence. For example the magnetization curves of Ni at 4.2 K are shown along the [100], [110] and [111] directions (Fig.9). It is evident that the [111] direction is the easy direction of magnetization and that a relatively large degree of anisotropy is present. A similar behavior is calculated at 77 K but saturation is reached at a smaller field due to the reduction of the anisotropy relative to its value at 4.2 K. As for the mixed-anisotropy systems our calculations showed that a magnetization jump to saturation is found if the biaxial anisotropy is positive and the uniaxial anisotropy is negative but are of the same order of magnitude. This takes place in easy -[100] systems in a magnetic field parallel to [010] direction. This behavior has been demonstrated experimentally in Fe films on stepped Ag(001) substrates. A detailed study of the application of our model to cubic and mixed anisotropy systems will be reported elsewhere.

CONCLUSIONS

1- Principles of classical statistical physics are used in a simple model to calculate magnetic properties in a wide range of temperature, magnetic field and particle size.

2- Model is applied to uniaxial (e.g. Gd and $Nd_2Fe_{14}B$), cubic (e.g. Ni) and mixed-anisotropy systems (e.g. ultra thin films of Co/Cu and Fe on (001) Ag substrates)

4- Interesting phenomena like spin re-orientation, in-plane anisotropy and superparamagnetism have been found in these systems

5- Our results are in fair agreement with both available experimental data and analytical studies.

REFERENCES

1. M. Richter, J. Phys. D:Appl. Phys. **31**,1017(1998)
2. J.M.D. Coy, J. Magn. Magn. Mater.**196**,1(1999)
3. J. C. O. de Jesus and W. Kleemann, J. Magn. Magn. Mater. **169**,159(1997).
4. Y. T. Millev, H. P. Oepen and J. Kirschner, Phys.Rev.B**57**,5837(1998).
5. J. Cullen and E. Cullen, J. Appl. Phys.**55**, 2426(1984)
6. Ming-hui Yu, Zhi-dong and T. Zhao, J. Magn. Magn. Mater. **195**, 327(1999)
5. S. Yehia and S. Aly, J. Magn. Magn. Mater.**212**,195(2000).
6. S. Aly and S .Yehia and M. M. Mahgoob, J. Magn. Magn. Mater.**272**,726(2004).
7. S. Aly, J. Magn. Magn. Mater.**222**,368(2000).
10. S. Chikazumi, *Physics of Ferromagnetism*,(Clarendon, Oxford 1977)
8. S. Aly and S .Yehia, J. Magn. Magn. Mater.**202**, 565 (1999).
9. F. Bolzoni, O. Moze and L. Pareti, J. Appl. Phys.**62**,615(1987).
10. E. P. Wohlfarth, *Ferromagnetic Materials*, E. P. Wohlfarth (ed.),Vol.1(North-Holland, Amsterdam 1980).
11. O. Yamada, H. Tokuhara, F. Ono, M. Sagawa and Y. Matsuura, J. Magn. Magn. Mater.**54**,585(1986) .
12. Y. Takano, Y. Sato, K. Sekizawa, J. Phys. Colloque, **C8**,561(1998).
13. C. D. Fuerst, J. F. Herbst, E. A. Alson, J. Magn. Magn. Mater.**54**,567(1986).
14. M. R. Ibarra, P. A. Algarabel, A. Alberdi, J. Bartolome and A. del Moral, J. Appl. Phys.**61**,3451(1987).
15. W. D. Corner, B. K. Tanner, J. Phys. **C9**,627 (1976).
16. H. E. Nigh, S. Legvold, F. H. Spedding, Phys. Rev. **132**,1092(1963).
17. J. W. Cable, E. O. Wollan, Phys. Rev.165,773(1968).
18. V. Mihai and J. M. M. Franse, Rev. Roum. Phys.**21**,1041(1976).
19. J. A. Cowen, B. Stolzman, R. S. Averback, H. Hahn, J. Appl. Phys.**61**,3317(1987).
20. Y.W. Du, J.Wu, H. X. Lu, T.X.Wang, Z.Q. Qiu, H. Tang and J. C. Walker, J. Appl. Phys. **61**,3314(1987).
21. C. P. Bean and J. D. Livingston, J. Appl. Phys.**30**,120S(1959).
22. C. J. Tung, I. Said and Glen E. Everett, J. Appl. Phys.**53**,2044(1982).

Relativistic Fine Structure Oscillator Strengths For Nickel Like Argon

A. Farrag

Laboratory of Lasers and New Materials
Faculty of Science, Physics Department, Cairo University

Abstract. Fine structure levels, oscillator strengths and radiative transition probabilities have been calculated for Ni-like argon. Configuration interactions and relativistic effects have been included in the calculations. The calculated energies and the transition probabilities of the ion are in good agreement with the available calculated values. The results are expected to be useful in the analysis of X-ray and EUV spectra obtained from astrophysical sources.

INTRODUCTION

There has been a continuous interest in the determination of energy levels, oscillator strengths, transition probabilities and lifetimes for ionized atoms [1,5]. These quantities are needed in the fields of astrophysics and plasma physics which usually require the knowledge of some parameters like the electron temperature and density of each charge state. The intensity of the spectral lines can be used to determine the various parameters of the plasma from the relation between the intensity of the spectral lines and the transition probabilities which is also a function of oscillator strengths. Therefore, in order to interpret the vast amount of observational data available, atomic data for energy levels, radiative rates and collision strengths are required. The experimental results for these parameters are often not available, and therefore a few theoretical attempts have been made in the past to calculate such data. However, most of the data available so far are either limited to a few transitions or lack accuracy.

The aim of the present work is to calculate the electric-dipole oscillator strengths and the transition probabilities, and to provide new values of transition probabilities for Ni XI lines which are observed in astrophysical spectra and more over it can be used to verify the previously existing data, to improve its accuracy and to interpret the available observational data.

In this paper, an approximate, rapid and practical method for obtaining radial wave functions in many-electron atoms and ions has been used. The calculations take into account the most important configuration interactions (CI) and relativistic effects. The atomic structure model used was developed with Cowan atomic structure code [6] which has Hartree-Fock and Slater parameter optimization routines. It is a semi-empirical technique which involves the adjustment of slater parameters by least squares optimization to compare the observed data with the calculated energy level.

METHOD OF CALCULATIONS

1. Transition Rates

The strength of a line is defined as the square of the reduced dipole matrix element

$$S = | \langle \psi \| \sum_{i=1}^{N} \vec{r}_i \| \psi' \rangle |^2$$

CP888, *Modern Trends in Physics Research,*
Second International Conference on Modern Trends in Physics Research—MTPR-06,
edited by L. El Nadi
© 2007 American Institute of Physics 978-0-7354-0354-9/07/$23.00

where ψ and ψ' are the physical wave functions composed of many basis states, the sum runs over all N electrons of the atom (or ion) and r_i is the radial position of the ith electron.

S is related to the radiative transition probability according to:

$$A_{u,l} = \frac{64\pi^4 e^2 a_0^2}{3hg_u} S(\Delta E)^3$$

where u and l stand for the upper and lower levels, respectively, g_u is the statistical weight of the upper level of the transition, a_0 is the Bohr radius and ΔE is the wave number of the spectral line in cm^{-1}.

The weighted oscillator strength for the transition between ψ' and ψ is defined as:

$$g_u f = \frac{2}{3}(\Delta E)S$$

The intensity of a spectral line is then given by the integral over the line of sight of the instrument and is related to the transition rates by

$$I(u,l) = \frac{1}{4\pi} \int n_e n(z,g) X(T_e, g, u) \frac{A(u,l)}{\sum A(u,k)} ds$$

Where n_e is the electron density and $n(z,g)$ is the population density of the ground level for the charge state z; $X(T_e, g, u)$ is the excitation rate coefficient at electron temperature T_e; A is the spontaneous transition probability, u and l signify the upper and lower energy levels for the transition producing the spectral line and g refer to the ground level.

That is, the line intensity of a spectral line is proportional to the spontaneous transition probability.

2. Relativistic Hartree-Fock (HFR)

The radial wave functions have been generated using the relativistic Hartree – Fock (HFR) method introduced by Cowan and Griffin 1976 [7], using the computer codes written by Cowan (1981) [6]. The relativistic corrections included in the differential equations are derived from the Pauli-approximation to the Dirac- Hartree – Fock equations, the mass-velocity and the Darwin operators are not included. All the relevant radial integrals (E$_{av}$, F^k , G^k,ζ_1 , R^k) and the reduced transition matrix elements have been computed. The Hamiltonian has been built and diagonalized within the frame work of the Slater-Condon theory.

In a further step, the radial parameters were fitted in a least square optimizing program which fits the eigen values of the Hamiltonian to the available experimental energy levels. These optimized integrals were used to compute the wavelengths and the transition rates.

RESULTS AND DISCUSSION

The calculations of the transition probabilities of the spectral lines belonging to the transition array:

$3p^6 – 3p^5\,3d$, $3p^6 - 3p^5\,4s$, and $3s\,3p^6\,3d – 3s^2\,3p^5\,3d$ of Nickel like Ar [Ni XI] ion are presented.

The calculations have used the superposition of configurations:

$3p^5\,3d + 3p^5\,4s$, for the odd parity and $3p^6$, $3p^5\,4p$, $3p^5\,4f$, for the even parity.

The interactions with more distant configurations were simulated by the inclusion of additional effective parameters such as α and β, which allows specifically for the cumulative effects of distant configurations.

The fitting procedure was applied to the $3p^6$, $3p^5 3d$, $3p^5 4s$ and $3p^5 4f$ configurations of Ni XI with the experimental energy levels taken from NIST [8]. For the $3p^5 4s$, there are three parameters (E_{av}, G^2, ζ), which are determined from the observed terms and for the $3p^5 3d$ configuration, there are six parameters (E_{av}, F^2, G^1, G^3, ζ_p, ζ_d)
Which have been determined from the observed terms: ^1P, ^3P, ^1D, ^3D, ^1F, ^3F.
As an example we have the equality ^1P, $^3P = E_{av} +7 F^2 \pm (G^1+63G^3)$, see TAS [9].

The ab initio HFR values for the Slater parameters within other configurations than those included in the fitting procedure and for the CI integrals were scaled down by a factor 0.80 as recommended by Cowan [1] to simulate CI effects, however, the ab initio values of all the Spin-orbit integrals, computed by the Blume –Watson method were used without scaling.

The present fitted parameters are shown in table 1. and the calculated HFR energy levels and the fine structure splitings for NiXI are listed in Table.2, as well as the values from NIST [8] compilation for comparison. The agreement is within few tens of cm^{-1}.
Table 3 presents the wavelengths (λ), the oscillator strengths in terms of (log gf) and the radiative transition probabilities for the electric dipole transitions. The available NIST [8] values are also presented in table 3. The NIST values are followed by a capital letter denoting the uncertainties in the atomic transition probability data which is: C for uncertainties within 25%, D for uncertainties within 50% and E for uncertainties greater than 50%. The word uncertainty is used here with the connotation "estimated extent of the deviation from the true value". The transition $3p^6 - 3p^5 3d$, which includes three electric dipole transitions, they have values compiled in NIST [8], the relative percentage is maximum 0.45% for the wavelengths, and the transition $3p^6 - 3p^5 4s$ includes two electric dipole transitions, they have also values compiled in NIST [8], the relative percentage is maximum 0.27% for the wavelengths. For the transition probabilities in both transitions our calculated values lie within the estimated true value expected by NIST uncertainties. The 36 calculated electric dipole transitions resulting from the transition array $3s 3p^6 3d - 3s^2 3p^5 3d$, have no experimental or theoretical results available in literature to compare with.

TABLE 1. Fitted parameters (in cm^{-1}) used in the HFR calculations for NiXI.

Configuration	Parameter	Fitted value
$3s3p^63d$	E_{av}	835.87
	ζ_{3d}	0.95
	$G^2(3s,3d)$	76.92
$3s^23p^53d$	E_{av}	519.92
	ζ_{3p}	15.75
	ζ_{3d}	1.95
	$F^2(3p,3d)$	128.51
	$G^1(3p,3d)$	144.66
	$G^3(3p,3d)$	98.27
$3s^23p^54s$	E_{av}	1276.14
	ζ	16.01
	$G^1(3p,4s)$	7.33
	α	9.92
	β	10.71

TABLE 2. Calculated HFR energy levels and fine structure splitting (in cm^{-1}) for NiXI

Configuration	Term	J	E_{calc}	Enist
$3p^6$	^1S	0	0	0
$3s3p^63d$	^3D	1	826755	
		2	827662	
		3	829133	
	^1D	2	858992	
$3s^23p^53d$	^3p	0	469545	469310
		1	473931	472970
		2	482981	480950
	^1p	1	673747	673960
	^3F	2	502796	504070
		3	496051	497520
		4	492808	493250
	^1F	3	543436	543220
	^3D	1	533699	534830
		2	539932	539180
		3	526822	527470
	^1D	2	530718	530830
$3s^23p^54s$	^3p	0	1290949	
		1	1292110	1292100
		2	1266928	
	^1p	1	1269996	1269900

TABLE 3. Wavelengths, Oscillator strengths, Radiative transition probabilities for electric dipole E1 transitions for NiXI

Transition	Multiplet	λ(Å)(HFR)	λ(Å)(NIST)	log gf (HFR)	A(sec^{-1}) (HFR)	A(sec^{-1}) (NIST)
$3p^6$ - $3p^53d$	$^1S_0$ - $^3P_1$	210.665	211.428	-3.339	7.05E+07	1.40E+07 E
	$^1S_0$ - $^3D_1$	187.819	186.976	-2.014	6.23E+08	4.30E+08 E
	$^1S_0$ - $^1P_1$	148.173	148.374	0.526	4.95E+11	2.34E+11 C
$3p^6$ - $3p^54s$	$^1S_0$ - $^1P_1$	78.523	78.74	-0.77	2.62E+10	6.10E+10 D
	$^1S_0$ - $^3P_1$	77.209	77.393	-0.689	2.58E+10	8.50E+10 D
$3p^53d$ - $3s3p^63d$	$^3P_0$ - $^3D_2$	280.989		-0.789	2.74E+09	
	$^3P_1$ - $^3D_1$	284.037		-0.762	4.77E+09	
	$^3P_1$ - $^3D_2$	283.306		-0.51	5.13E+09	

$^3P_1$-$^1D_2$	260.21	-2998	1.98E+07
$^3P_2$-$^3D_1$	290.501	-1.57	7.08E+08
$^3P_2$-$^3D_2$	289.737	-0.675	3.36E+09
$^3P_2$-$^3D_3$	288.508	-0.269	8.62E+09
$^3P_2$-$^1D_2$	265.625	-1.835	4.61E+08
$^1P_1$-$^3D_1$	658.47	-3.477	1.71E+06
$^1P_1$-$^3D_2$	654.558	-2.827	4.60E+06
$^1P_1$-$^1D_2$	543.169	-0.603	1.13E+09
$^3F_4$-$^3D_3$	296.682	0.141	1.50E+10
$^3F_3$-$^3D_2$	301.63	-0.003	1.46E+10
$^3F_3$-$^3D_3$	300.298	-1.642	2.41E+08
$^3F_3$-$^1D_2$	275.587	-1.324	8.32E+08
$^3F_2$-$^3D_1$	308.521	-0.18	1.54E+10
$^3F_2$-$^3D_2$	307.659	-1.673	3.00E+08
$^3F_2$-$^3D_3$	306.274	-2.534	2.97E+07
$^3F_2$-$^1D_2$	280.612	-1.22	1.02E+09
$^3D_3$-$^3D_2$	330.201	-1.251	6.86E+08
$^3D_3$-$^3D_3$	328.606	-0.249	4.97E+09
$^3D_2$-$^1D_2$	299.244	-0.413	5.76E+09
$^1D_2$-$^3D_1$	336.487	-2.14	1.42E+08
$^1D_2$-$^3D_2$	335.462	-0.541	3.41E+09
$^1D_2$-$^3D_3$	333.816	-1.287	4.41E+08
$^1D_2$-$^1D_2$	303.558	-0.434	5.30E+09
$^3D_1$-$^3D_1$	339.758	-0.589	4.96E+09
$^3D_1$-$^3D_2$	338.714	-0.845	1.66E+09
$^3D_1$-$^1D_2$	306.218	-2.516	4.34E+07
$^3D_2$-$^3D_1$	345.438	-1.389	7.60E+08
$^3D_2$-$^3D_2$	344.358	-0.732	2.08E+09
$^3D_2$-$^3D_3$	342.623	-0.796	1.30E+09
$^3D_2$-$^1D_2$	310.824	-0.513	4.24E+09

$^1F_3$-$^3D_2$	347.976	-2.399	4.38E+07
$^1F_3$-$^3D_3$	346.205	-0.434	2.92E+09
$^1F_3$-$^1D_2$	313.769	-0.216	8.24E+09

CONCLUSION

This paper presents calculations of fine structure levels, oscillator strengths and radiative decay rates for NiXI ion. There is a general agreement shown in this work between the result obtained and the values from NIST. The A values agreed to within a few percent for most of the transitions.

Accurate experimental data are urgently needed to definitely assess the reliability of the calculation discussed in the present work.

REFERENCES

1. Rice, J.E., et al.,J. Phys. B: At. Mol. Opt. Phys.**33**, 5435-5462 (2000).
2. Nahar, S. N., Astron. And Astroph. **389**, 716-728 (2002).
3. Tully, A. John,. J. Phys. B : At. Mol. Opt. Phys. **37**, 689-701 (2004).
4. Karpuskiene, R., J. Phys. B: At. Mol. Opt. Phys. **37**, 2067-2086 (2004).
5. Koc, K., J. Phys. B: At. Mol. Opt. Phys. **37**, 3821-3835 (2004)
6. Cowan, R. D., "The Theory of Atomic Structure and Spectra", University of California Press, Berkeley, California, 1981.
7. Cowan, R. D., and Griffin, D. C., J. Opt. Soc. Am. **66**, 1010 (1976).
8. Fuhr, J. R., Martin, G. A., and Wiese, W. L., "Atomic Transition Probabilities - Iron through Nickel" Journal of Physical and Chemical Reference Data, **17**, 1988, supplement No 4.
9 . Condon, E. U., and Shortley, G. H., The Theory of Atomic Spectra", Cambridge University Press, 1979.

Oscillator Strengths for Fluorine Isoelectronic Sequence

G. Omar, H. Ramadan[1] and Kh. Ammar[2]

Physics Department, Faculty of Science, Ain Shams University, Cairo, Egypt.
[1]Department of Basic Sciences, FCIS, Ain Shams University, Cairo, Egypt.
[2]Egyptian Meteorological Authority (EMA), Cairo, Egypt.

Abstract. Oscillator strengths (f-values) are calculated for five members of the F-isoelectronic sequence, namely, P^{6+}, Ar^{9+}, Sc^{12+}, Cr^{15+} and Co^{18+}. The Single Configuration Hartree-Fock program (SCHF) is used to generate the wave functions needed in the calculation of radiative transition probabilities and oscillator strengths (f-values) for singly excited states in the studied ions. Specifically, the work includes the radiative transitions of the form $n_1\ell_1 \rightarrow n_2\ell_2$. The transitions are classified into sets according to the values of ℓ_1 and ℓ_2, with n_1 and n_2 = 9, 8, ..., 3. Both ℓ_1 and ℓ_2 are limited (here) to be less than 4. The angular momentum average scheme (AMA) is utilized for all the calculations of the radiative rates A_r's, and f-values. The trends of oscillator strengths with the effective charge (Z_{eff}) and principal quantum numbers n_1 and n_2 of the upper and lower states are investigated. It is found that, the f-values decrease as $(1/n_1)^{4.5}$ for dominant transitions in low-ionized ions (P^{6+}).

INTRODUCTION

Oscillator strength (f-value) is a dimensionless quantity which has been introduced to denote the distribution of an atom or molecule to absorb or to emit radiation from a certain state. Its physical meaning can be understood through the fact that the polarizability (dipole moment induced by an electric field strength) of the atom in which each oscillator is represented with strength (f-value).

Oscillator strength is an important physical quantity in atomic and molecular spectroscopy as well as laser physics. It is related to radiative rates (A_r), radiative lifetimes (τ's) of excited states, line strengths (S's) and line intensities (I's) of spectral lines. In addition, it is related to Einstein coefficients which are playing an important role in laser research. Moreover, f-values are required in testing the accuracy of wave functions of states involved in the atomic transitions.

Oscillator strength were utilized also in the calculations of the rate coefficients of some atomic processes which are taking place in hot astrophysical and laboratory plasmas, for example dielectronic recombination [1,2]. Since last decade, many works have been done in the study of oscillator strengths [3-5] especially for ions which may lead to the production of x- rays or lasers.

In the present work, it was planned to calculate the radiative (emissive) oscillator strengths for F-like ions. Specifically, this work is concerned with the calculation of f-values for singly-excited states $(1s^2 2s^2 2p^4) n_1\ell_1$ which stabilize finally to the ground state $1s^2 2s^2 2p^5$ directly or by successive emission of radiation. Note that, the singly-excited state contains only one electron in an excited state above an outer-shell hole. Hence, no emissions of Auger-electron are permitted during its auto- stabilization. This situation is drastically different from the doubly-excited states which are often allowed to decay by Auger and/or radiative transitions.

Recently, there is a growing interest in calculations and measurements of oscillator strengths [10-19] to understand all the intricacies of f-values in various coupling schemes.

METHOD OF CALCULATIONS

The singly-excited states in ions may be formed by two methods of excitations: photoexcitation (PE) in photon-ion collisions or by direct excitation (DE) in electron-ion (e/I) collisions. Thus, for F-like ions, we may have:

(a) in PE method :

CP888, *Modern Trends in Physics Research,*
Second International Conference on Modern Trends in Physics Research—MTPR-06,
edited by L. El Nadi
© 2007 American Institute of Physics 978-0-7354-0354-9/07/$23.00

$$1s^2 2s^2 2p^5 + h\nu_1 \rightarrow (1s^2 2s^2 2p^4 n_1 \ell_1)^* \rightarrow 1s^2 2s^2 2p^4 n_2 \ell_2 + h\nu_2 \qquad (1)$$

(b) in DE method:

$$1s^2 2s^2 2p^5 + e_1 \rightarrow (1s^2 2s^2 2p^4 n_1 \ell_1)^* + e_2 \rightarrow 1s^2 2s^2 2p^4 n_2 \ell_2 + h\nu + e_2 \qquad (2)$$

where $\nu_1 > \nu_2$ and the energy of the incident electron (e_1) must be greater than that of the outgoing electron (e_2).

Now, consider for simplicity, the upper atomic excited state ($j = n_1 \ell_1$) and relaxed radiatively to the lower state ($i = n_2 \ell_2$), the oscillator strength, f ($j \rightarrow i$) may be written in its simplest form as:

$$f (j \rightarrow i) = (1/3) \Delta E . |R_D|^2 . [\ell_> / (2 \ell_j + 1)] (g_j / g_i) \qquad (3)$$

where ΔE (in Rydberg) is the energy difference between the upper j-state and lower i-state.
The radiative transition probability $A_r (j \rightarrow i)$ is simplified to

$$A_r (j \rightarrow i) = 2.6762 \times 10^9 (\Delta E)^3 . |R_D|^2 . [\ell_> / (2 \ell_j + 1)] \qquad (4)$$

Thus, the relation between f ($j \rightarrow i$) and $A_r (j \rightarrow i)$ is given by:

$$f (j \rightarrow i) = 0.125 \times 10^{-9} . [1/(\Delta E)]^2 . A_r (j \rightarrow i) . (g_j / g_i) \qquad (5)$$

In equations (3) and (4), R_D is the dipole radial matrix element:

$$R_D = \int R(n_1 \ell_1) \, r \, R(n_2 \ell_2) \, r^2 \, dr \qquad (6)$$

$R(n_1 \ell_1)$ and $R(n_2 \ell_2)$ represent the radial wave functions for the upper and lower atomic states with statistical weights g_j and g_i respectively. In the present work, these wave functions are generated using the single configuration Hartree-Fock (SCHF) program. The radiative rates, $A_r (j \rightarrow i)$, are calculated using the angular momentum average (AMA) [6, 7] scheme. Thus, the f-values obtained using eq. (5) are known as average oscillator strengths, where the transitions are defined, only through the quantum numbers n and ℓ. However, all other quantum numbers are summed over for each state. Consequently, in the AMA scheme, the two selection rules

$$\Delta \ell = \ell_j - \ell_i = \pm 1, \text{ with the parity (odd} \leftrightarrow \text{even)} \qquad (7)$$

are adequate for choosing the allowed emissive transitions between the energy states.

RESULTS AND DISCUSSION

In order to check the range of validity of the adapted AMA scheme in the calculation of the f-values, we apply this scheme on some transitions in O^{3+}, Ne^{5+} and Al^{8+} ions, where some f-values are available in other coupling schemes. Specifically, using the limited Multi-Configuration Frozen Core (MCFC) by Cohen et al (1980) [8] and the (SCHF) with LS-coupling scheme by El-Sherbini et al (1987) [9].

The present data, using AMA, are compared with that of others in Table (1). It is obvious from the table that, the present f-values are found in agreement with that calculated in other coupling schemes for some transitions. In addition, the f-values become approximately the same, in these coupling schemes, as the degree of ionization (q) of the ions increases. This may be attributed to the fact that, the wave functions are almost hydrogenic for high Rydberg states (HRS) in highly ionized ions. Thus, the relatively simple AMA scheme manifests itself as an efficient method of calculation of f-values for highly ionized ions. Thus, the adapted AMA scheme is employed extensively in the present calculations of f-values for F-like ions, especially those with at least (q = 5+). Typically, the present calculations include the radiative transitions between states with $3 \leq n_1 \leq 9$, $2 \leq n_2 \leq 9$ and ℓ_1, $\ell_2 = 0,1,2,3$ in P^{6+}, Ar^{9+}, Sc^{12+}, Cr^{15+} and Co^{18+} ions.

TABLE 1. Comparison of *f*-values (dimensionless) for O^{3+}, Ne^{5+} and Al^{8+} ions in AMA with f-values for the same ions in other coupling schemes.

Ion	Transition	present method (AMA) scheme	Ref. (9) (LS-coupling)	Ref. (8) (MCFC)
O^{3+}	$3s \rightarrow 2p$	0.0185	0.022	0.034
	$3p \rightarrow 3s$	0.682	0.669	0.615
Ne^{5+}	$3s \rightarrow 2p$	0.018	0.02	
	$3p \rightarrow 3s$	0.515	0.505	
Al^{8+}	$3s \rightarrow 2p$	0.017	0.019	
	$4s \rightarrow 2p$	0.00386	0.0036	
	$4s \rightarrow 3p$	0.0641	0.064	

To simplify the discussion, we drop the subscripts for n_1 and n_2. The (n_1) before the arrow is for the upper states. All *f*-values, here, are for emissive transitions. The absorptive *f*-values can be easily obtained from these values with some statistical factors. The emissive *f*-values for the studied F-like ions are presented and discussed in the following sections:

P^{6+} ion:

The f-values for P^{6+} are calculated and presented in Table (2). The largest value of the oscillator strengths is found at $\Delta\ell = +1$ and $\Delta n = 1$ for $\overline{f}(4f \rightarrow 3d) = 0.911$. From the table we can notice that the transitions with $\Delta\ell = +1$ have considerably large f-values in comparison with the transitions of $\Delta\ell = -1$. In addition, the transitions $n_1 d \rightarrow np$ for $\Delta n > 1$, also has large *f*-values e.g $\overline{f}(4d \rightarrow 2p) = 0.558$. It is worth noting also that the transitions (n'd $\rightarrow$ np) have considerably large f-values at $\Delta n = 3, 4$. However, for $\Delta\ell = -1$ transitions are large but still smaller than those with $\Delta\ell = 1$. The transitions which have the largest *f*-values are called dominant states. The *f*-values of these dominant states in case of P^{6+} are presented in Table (3). The dominant states have high tendency to take place than the other transitions.

Ar^{9+} ion:

It has to be noted that, the Argon is more interesting in the calculations of atomic data such as transition probabilities, oscillator strengths, and intensities of spectral lines. The reason is the availability of argon gases in laboratories. Hence, Argon ions can be easily produced.

The wave functions of the bound states, starting from Ar^{9+} with $Z_I = 9$, are traditionally easier to be generated from the Hartree-Fock program with self-consistent field method. The *f*-values for transitions with $\Delta\ell = +1$ are larger than those with $\Delta\ell = -1$.

By comparing the energy differences between the states involved in the transitions we found that, although the values of ΔE are close, the values of oscillator strengths are not for both $\Delta\ell = \pm 1$ transitions. This may be attributed to the difference in statistical weights ratios from one transition to the other. For example,

$$\frac{f(9d \rightarrow 3p)}{f(9s \rightarrow 3p)} = \frac{0.009}{0.001}, \ \frac{\Delta E(9d \rightarrow 3p)}{\Delta E(9s \rightarrow 3p)} = \frac{12.032}{11.947}, \text{ and } \frac{g_i(9d)}{g_f(9s)} = \frac{10}{2}$$

while,

$$\frac{f(7f \rightarrow 4d)}{f(7p \rightarrow 4d)} = \frac{0.074}{0.006}, \ \frac{\Delta E(7f \rightarrow 4d)}{\Delta E(7p \rightarrow 4d)} = \frac{4.442}{4.301} \text{ and } \frac{g_i(7f)}{g_f(7p)} = \frac{14}{6}.$$

In addition, the overlapping between wave functions decreases. Consequently, the *f*-values become different.

TABLE 2. The radiative oscillator strengths for P^{6+} ion. Values between parentheses are powers of ten.

	n_2 lower states	n_1 upper excited states						
		9	8	7	6	5	4	3
$n_1s \rightarrow n_2p$	2	0.15(-2)	0.23(-2)	0.36(-2)	0.63(-2)	0.13(-1)	0.33(-1)	0.16(0)
	3	0.11(-2)	0.17(-2)	0.29(-2)	0.58(-2)	0.15(-1)	0.85(-1)	
	4	0.30(-2)	0.51(-2)	1.00(-2)	0.26(-1)	0.16(0)		
	5	0.72(-2)	0.14(-1)	0.37(-1)	0.21(0)			
	6	0.18(-1)	0.48(-1)	0.27(0)				
	7	0.59(-1)	0.33(0)					
	8	0.39(0)						
$n_1p \rightarrow n_2s$	3	0.47(-2)	0.70(-2)	0.11(-1)	0.20(-1)	0.42(-1)	0.12(0)	
	4	0.80(-2)	0.13(-1)	0.22(-1)	0.45(-1)	0.12(0)		
	5	0.14(-1)	0.23(-1)	0.47(-1)	0.13(0)			
	6	0.25(-1)	0.49(-1)	0.13(0)				
	7	0.52(-1)	0.14(0)					
	8	0.14(0)						
$n_1p \rightarrow n_2d$	3	0.53(-3)	0.84(-3)	0.15(-2)	0.31(-2)	0.91(-2)	0.67(-1)	
	4	0.21(-2)	0.37(-2)	0.75(-2)	0.21(-1)	0.14(0)		
	5	0.61(-2)	0.12(-1)	0.34(-1)	0.23(0)			
	6	0.17(-1)	0.85(-1)	0.31(0)				
	7	0.61(-1)	0.40(0)					
	8	0.47(0)						
$n_1d \rightarrow n_2p$	2	0.34(-1)	0.50(-1)	0.77(-1)	0.13(0)	0.24(0)	0.56(0)	
	3	0.73(-2)	0.12(-1)	0.18(-1)	0.32(-1)	0.65(-1)	0.16(0)	
	4	0.11(-1)	0.17(-1)	0.28(-1)	0.53(-1)	0.12(0)		
	5	0.15(-1)	0.25(-1)	0.47(-1)	0.10(0)			
	6	0.24(-1)	0.43(-1)	0.93(-1)				
	7	0.41(-1)	0.86(-1)					
	8	0.74(-1)						
$n_1d \rightarrow n_2f$	4	0.35(-3)	0.61(-3)	0.13(-2)	0.36(-2)	0.22(-1)		
	5	0.16(-2)	0.33(-2)	0.92(-2)	0.53(-1)			
	6	0.59(-2)	0.16(-1)	0.90(-1)				
	7	0.24(-1)	0.130(0)					
	8	0.16(0)						
$n_1f \rightarrow n_2d$	3	0.12(-1)	0.18(-1)	0.32(-1)	0.64(-1)	0.17(0)	0.91(0)	
	4	0.24(-1)	0.39(-1)	0.74(-1)	0.18(0)	0.69(0)		
	5	0.42(-1)	0.76(-1)	0.17(0)	0.60(0)			
	6	0.76(-1)	0.17(0)	0.56(0)				
	7	0.17(0)	0.54(0)					
	8	0.51(0)						

TABLE 3. The largest f-values of all the transitions in P^{6+}.

Transition	f	$\Delta \ell$
4f →3d	0.91	
5f →4d	0.69	
6f →5d	0.60	
7f →6d	0.56	+1
4d →2p	0.56	
8f →7d	0.54	
9f →8d	0.51	
9p →8d	0.47	
8p →7d	0.40	
9s →8p	0.39	
8s →7p	0.33	-1
7p →6d	0.31	
7s →6p	0.27	
5d →2p	0.24	+1
6p →5d	0.23	
6s →5p	0.21	-1

Sc^{12+}:

The mesh size of the calculations of the dipole radial integrations is taken as Z/100 i.e. it is equal to 0.21 in the case of Sc^{12+}. The variation of oscillator strengths with Z and with the effective charge Z_{eff} are clarified and represented in figures (1) and (2) for Sc^{12+} and the rest of the other ions.

Cr^{15+}:

The *f*-values are not drastically different from those of the previous ions, because their effective charges are somewhat close.
In such cases, the energy difference between energy states in Sc^{12+}, Cr^{15+} are very small and must be calculated from the Multi- Configuration Hartree-Fock (MCHF) or Cowan's program. However, the trends of oscillator strengths will not change. It is well- known facts that, the calculations of energies from (SCHF) are good enough without loosing any physical information.

Co^{18+}:

Co^{18+} is chosen with 18 degrees of ionization (large enough) for producing better eigenfunctions from SCHF program for all configurations, involved in the emissive transitions. Above this degree of ionization, the calculations of energies and wave functions must be carried out using the relativistic HF programs. Thus the *f*-values will be more reliable.

The average oscillator strengths are calculated in the same AMA scheme to avoid errors in the fine structure oscillator strengths for the split configurations in LS coupling scheme. The *f*-values for Co^{18+} are presented in Table (4). All trends of the *f*-values in the last ion Co^{18+} are consistent with the previous ions.

Trends of oscillator strengths

The trends of oscillator strengths with the effective charge Z_{eff}, and the principal quantum numbers n_1, n_2 (the upper and lower levels respectively) are investigated. The dominant transitions (with large *f*-values) are determined. This dominancy is discussed in view of tables (2) and (3).

TABLE 4. The radiative oscillator strengths in Co^{18+}.

	n_2 lower states	n_1 upper excited states						
		9	8	7	6	5	4	3
$n_1s \rightarrow n_2p$	2	0.12(-2)	0.17(-2)	0.28(-2)	0.48(-2)	0.94(-2)	0.24(-1)	0.12(0)
	3	0.86(-3)	0.13(-2)	0.23(-2)	0.44(-2)	0.11(-1)	0.55(-1)	
	4	0.23(-2)	0.39(-2)	0.75(-2)	0.19(-1)	0.92(-1)		
	5	0.56(-2)	0.11(-1)	0.27(-1)	0.13(0)			
	6	0.14(-1)	0.35(-1)	0.17(0)				
	7	0.43(-1)	0.21(0)					
	8	0.25(0)						
$n_1p \rightarrow n_2s$	3	0.81(-2)	0.12(-1)	0.20(-1)	0.37(-1)	0.82(-1)	0.28(0)	
	4	0.14(-1)	0.23(-1)	0.42(-1)	0.91(-1)	0.31(0)		
	5	0.26(-1)	0.46(-1)	0.10(0)	0.34(0)			
	6	0.51(-1)	0.11(0)	0.37(0)				
	7	0.12(0)	0.40(0)					
	8	0.43(0)						
$n_1p \rightarrow n_2d$	3	0.30(-3)	0.48(-3)	0.83(-3)	0.17(-2)	0.47(-2)	0.27(-1)	
	4	0.13(-2)	0.22(-2)	0.44(-2)	0.17(-1)	0.64(-1)		
	5	0.38(-2)	0.75(-2)	0.20(-1)	0.11(0)			
	6	0.11(-1)	0.28(-1)	0.15(0)				
	7	0.37(-1)	0.19(0)					
	8	0.24(0)						
$n_1d \rightarrow n_2p$	2	0.31(-1)	0.46(-1)	0.73(-1)	0.13(0)	0.25(0)	0.63(0)	
	3	0.11(-1)	0.16(-1)	0.27(-1)	0.50(-1)	0.11(0)	0.41(0)	
	4	0.18(-1)	0.28(-1)	0.51(-1)	0.11(0)	0.37(0)		
	5	0.29(-1)	0.52(-1)	0.11(0)	0.36(0)			
	6	0.53(-1)	0.11(0)	0.36(0)				
	7	0.12(0)	0.37(0)					
	8	0.38(0)						
$n_1d \rightarrow n_2f$	4	0.26(-3)	0.46(-3)	0.95(-3)	0.27(-2)	0.16(-1)		
	5	0.13(-2)	0.26(-2)	0.71(-2)	0.39(-1)			
	6	0.47(-2)	0.13(-1)	0.67(-1)				
	7	0.19(-1)	0.99(-1)					
	8	0.13(0)						
$n_1f \rightarrow n_2d$	3	0.11(-1)	0.17(-1)	0.30(-1)	0.61(-1)	0.17(0)	0.95(0)	
	4	0.23(-1)	0.39(-1)	0.74(-1)	0.18(0)	0.77(0)		
	5	0.43(-1)	0.79(-1)	0.18(0)	0.70(0)			
	6	0.81(-1)	0.18(0)	0.67(0)				
	7	0.19(0)	0.66(0)					
	8	0.66(0)						
$n_1f \rightarrow n_2g$	5	0.20(-3)	0.43(-3)	0.13(-2)	0.81(-2)			
	6	0.13(-2)	0.37(-2)	0.22(-1)				
	7	0.71(-2)	0.39(-1)					
	8	0.59(-1)						

Z-dependence of f-values

Consider the transitions $n_1f \rightarrow n_2d$ and $n_1d \rightarrow n_2p$ where, their corresponding emissive oscillator strengths are the largest all over the whole transitions. These transitions are considered as samples for the

Z-dependence. The dominant transitions are: 4f → 3d, 5f → 4d, 6f → 5d, 4d → 2p and 5d → 2p which have $\Delta n \neq 0$. The f-values are drawn versus the atomic number (Z) and effective charge (Z_{eff}) in Figures (1) and (2) for the above transitions, respectively. The atomic number may be written as, $Z = N + Z_I$, where, N is the number of electrons that exist in the ion, Z_I is the degree of ionizations, respectively. The f-values are drawn versus the atomic number for (6p → 3s, 5p → 3s, and 4p → 3s) transitions in figure (1), the effective charge is taken as $Z_{eff} = \dfrac{Z_I + Z}{2}$. The f-values are drawn as a function of Z_{eff} in figure (2)

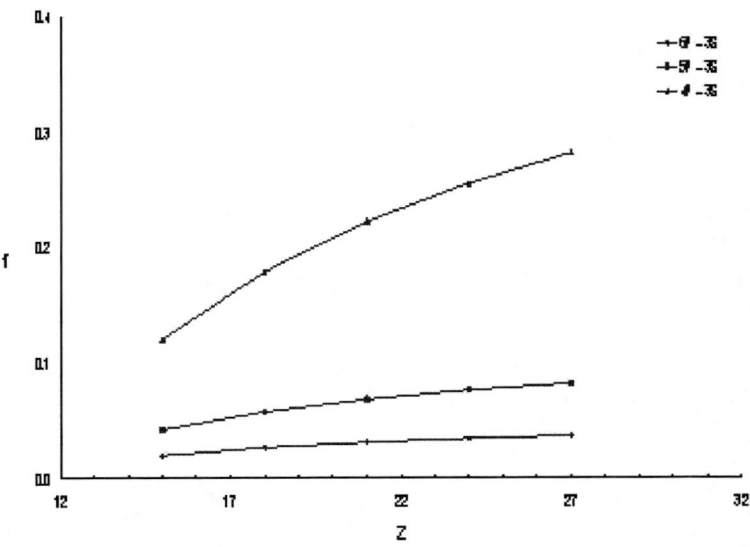

FIGURE 1. f versus Z for the transitions group 6p → 3s, 5p → 3s, and 4p → 3s

From the figures, it is clear that the trends of f-values are behaving similarly for both Z and Z_{eff}. The dependence on Z is clarified with the example

$$\frac{P^{6+}f(6p \to 3s)}{Ar^{9+}f(6p \to 3s)} = \frac{0.02}{0.03} = 0.66 \qquad \text{while} \qquad \frac{P^{6+}(Z)}{Ar^{9+}(Z)} = \frac{15}{18} = 0.83$$

This ratio changes for highly-ionized ions

$$\frac{Cr^{15+}f(6p \to 3s)}{Co^{18+}f(6p \to 3s)} = \frac{0.03}{0.04} = 0.75 \qquad \text{and} \qquad \frac{Cr^{15}(Z)}{Co^{18+}(Z)} = \frac{24}{27} = 0.9$$

As for the relation between the oscillator strengths and Z_{eff}, it may be clarified as follows:

$$\frac{P^{6+}f(6p \to 3s)}{Ar^{9+}f(6p \to 3s)} = \frac{0.02}{0.03} = 0.66 \qquad \text{while} \qquad \frac{P^{6+}(Z_{eff})}{Ar^{9+}(Z_{eff})} = \frac{10.5}{13.5} = 0.78$$

This ratio changes for highly-ionized ions

$$\frac{Cr^{15+}f(6p \to 3s)}{Co^{18+}f(6p \to 3s)} = \frac{0.03}{0.04} = 0.75 \qquad \text{and} \qquad \frac{Cr^{15}(Z_{eff})}{Co^{18+}(Z_{eff})} = \frac{19.5}{22.5} = 0.87$$

The relation between f and both Z and Z_{eff} is almost linear, especially for transitions with high n_1.

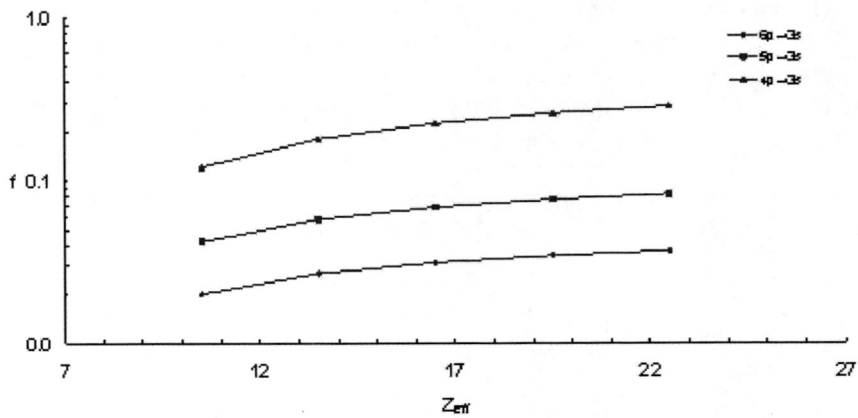

FIGURE 2. f versus Z_{eff} for the transitions group 6p → 3s, 5p → 3s, and 4p → 3s.

n_1-dependence of f-values

Figure (3) shows that the f-values decreases as $1/n_1^{4.5}$, where n_1 is the principal quantum number of the upper states. This is supposed to be the case for dominant transitions (n_1d → 2p) for low-ionized ions such as P^{6+}.

This may be the case for the other transition sets or the other ions with higher degrees of ionization such as Cr^{15+}, or Co^{18+}. The power of n_1 as a trend of f-values, is approximately 4.5. In other words, f-values are scaled as $1/n_1^{4.5}$. For example,

$$\frac{f(4d \rightarrow 2p)}{f(5d \rightarrow 2p)} = \frac{0.56}{0.24} = 2.33 \text{ and } \frac{n_1 (5d \rightarrow 2p)}{n_1(4d \rightarrow 2p)} = \frac{5}{4} = 1.25$$

Thus, it is concluded that, the f-values decreases roughly as $1/n_1^{4.5}$ for all ions in the F-isoelectronic sequence.

It has to be noted that, the trends of oscillator strengths are obtained with average oscillator strengths, which are calculated in the present work. These trends may not be exact for the f-values in other coupling schemes adapted with other Hartree-Fock versions. Figure (3) represents the fact that, the oscillator strengths for the transitions set n_1d → 2p are not quite different. This is one of the disadvantages of the calculations of energies from (SCHF) for atomic structure problems. However, for atomic collision problems, the energies and wave functions from (SCHF) may be good enough for quantitative studies.

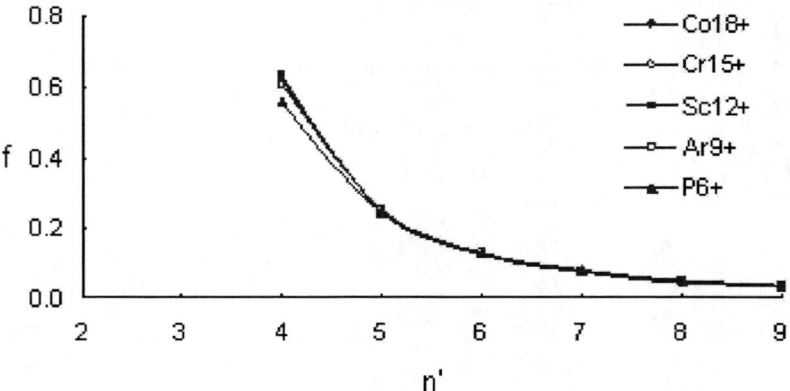

FIGURE 3. f versus n' (principal quantum number of the upper states) for the group transitions n'd → 2p

n₂-dependence of *f*-values

The principal quantum number n_2 of the lower states in case of $8d \rightarrow n_2p$ is drawn versus *f*-values for all ions in Figure (4). The *f*-values drops suddenly at $n_2 = 2$ to $n_2 = 3$, and then increase smoothly. This increase is not as rapid as the fall from $n = 2$ to $n = 3$ which is evident if we compare the values of f at $n = 5$ with $n = 2$. For all the five F-like ions of the F isoelectronic sequence under study, it is found that the *f*-values of the transitions $8d \rightarrow 2p$ are somewhat equal in all ions. As Z increases it is noticed that the differences in the *f*-values (for the same n_2) become negligible. This may be attributed to the inaccurate energies obtained from the SCHF. Thus, it is recommended for transitions in highly- ionized ions, the energies must be obtained using the MCHF or Cowan programs. Unfortunately, this model of MCHF is not running at Ain Shams Theoretical Group. In future, theses *f*-values may be calculated and improved.

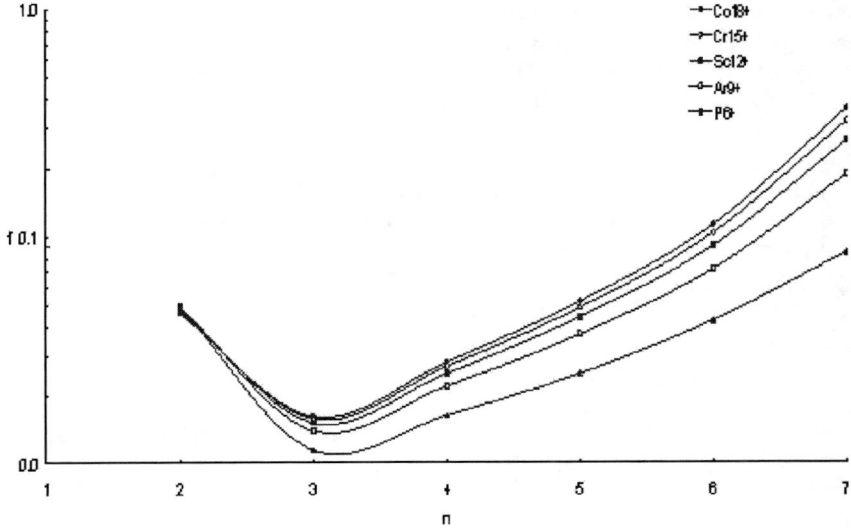

FIGURE 4. f versus n (principal quantum number of the lower levels) for the transitions $8d \rightarrow np$

CONCLUSION

Oscillator strengths are calculated in the AMA scheme for P^{6+}, Ar^{9+}, Sc^{12+}, Cr^{15+}, and Co^{18+}, with the seven sets of emissive transitions $n_1s \rightarrow n2p$, $n_1p \rightarrow n_2s$, $n_1p \rightarrow n_2d$, $n_1d \rightarrow n_2p$, $n_1d \rightarrow n_2f$, $n_1f \rightarrow n_2d$, and $n_1f \rightarrow n_2g$. The trends of the oscillator 1strengths with the atomic number (Z), the effective charge (Z_{eff}), n_1 and n_2 (principal quantum numbers of the higher and lower states respectively) are investigated.

It should be taken into consideration that the results of *f*-vlues are only average since we use Angular Momentum Average (AMA). The calculations of energies from (SCHF) for atomic structure problems are not very reliable. However, for atomic collision problems, the energies and wave functions from (SCHF) may be good enough for quantitative studies

The results may be summarized in the following points:

- The transitions $n_1f \rightarrow n_2d$ and $n_1d \rightarrow n_2p$ with $\Delta\ell = 1$ have the largest *f*-values of all the transitions.
- The dominant transitions are found to have the forms: $4f \rightarrow 3d$, $5f \rightarrow 4d$, $6f \rightarrow 5d$, $4d \rightarrow 2p$ and $5d \rightarrow 2p$ with $\Delta n \neq 0$.
- It is also found that, although the values of ΔE are close, the values of oscillator strengths are not for both $\Delta\ell = \pm1$ transitions. This is may be attributed to the difference in statistical weights ratios of upper
  and lower levels.
- The oscillator strengths for the transitions set $n_1d \rightarrow 2p$ are not quite different.

- For all the five F-like ions under study, the f-values of the transitions $8d \rightarrow 2p$ are somewhat equal in all

  ions.
- The f-values decreases as $(1/n_1)^{4.5}$ where n_1 is the principal quantum number of the upper states. This is supposed to be the case for dominant transitions ($n_1 d \rightarrow 2p$) in low-ionized ions such as P^{6+}. The power of n_1 as a trend of f-values is approximately 4.5. In other words, f-values are scaled as $(1/n_1)^{4.5}$.
- The overlapping between wave functions decreases as the atomic number Z increases in F-like ions.
- The trends of f-values are found to behave similarly for both Z and Z_{eff}. The relation between f and both Z and Z_{eff} is almost linear.

REFERENCES

1. Burgess A., Astrophys. J, **139**, 776 (1964).
2. Burgess A., Astrophys. J, **141**, 1588 (1965).
3. Zilitis V. A, Opt. Spec. **72**, 47, 801 (1992).
4. Hibbert A. et al, Atom. Nucl. data tab. **53**, 23 (1993).
5. Tiwary S. N. et al, Nuovo Cimento **15**, 77 (1993).
6. Gau J., and Hahn Y., JQSRT **23**, 121 (1980).
7. Gau J., and Hahn Y., and Ritter J. A., JQSRT **23**, 147 (1980).
8. Cohen M and Nabon J., J, Phys. B: At. Mol. Phys. **13**, 4325 (1980).
9. El- Sherbini Th., et al Ann. der Phys., **44**, 412 (1987).
10. Guet C. et al , Phys. Lett. A **143**, 384 (1990).
11. Fawcett B., At. Nucl. Data tables, **47**, 319 (1991)
12. Vasilyev A. et al., phys.Rev A **58**, 732 (1998).
13. Torodoir X et al., Eur. Phys. J. **D6**, 3 (1999).
14. Kato D. et al Physica Scripta. **T92**, 126 (2001).
15. Gorczyca T.W, Badnell N.R., Chen M.H., et al, the Astrophys. J. **592**, 636 (2003).
16. Allam S., Farag A., Refaie A., and El-Sherbini Th, Arab J. of Nucl. Sci. Appl., Spec.Issue, **32**, 89 (2003).
17. Zohny E., Allam S., El Sherbini Th., Arab J. Nucl. Sci. and Appl., Spec. **36**, 425 (2003).
18. Ramadan H. and Elkilany S. BPL **12**, 46 (2004).
19. Younis W., Allam S., and El Sherbiuni, Arab J. of Nucl. Sci. and Appl. Speci. **38**, 9 (2005).

Electron Impact Excitation and Ionization Rate Coefficients and Electron Densities Calculations for Lithium-like Ion Si XII and S XIV

A. I. Refaie

Laboratory of Lasers and New Materials, Physics Department, Faculty of science, Cairo university, Giza, Egypt.
E- Mail : amal1_ibrahim1@yahoo.com

Abstract. Absolute excitation and ionization rate coefficients have been evaluated for arbitrary excited states at certain electron temperatures kT_e and their corresponding electron densities N_e of the Lithium-like ions Si XII and S XIV. The populations of the chosen excited levels are calculated namely, for the doublet state of the Li-like ions. The calculations have been carried out by using the coupled rate simultaneous equations in which, the monopole and quadruple transitions have been introduced in the calculations in addition to the dipole transitions.

A theoretical population model has been developed to study the influence of the different processes that might contribute to the population of the different levels at the plasma parameters. The population densities of these different levels were then derived from these rate coefficients. For the most levels the theoretical excitation rate coefficients were found to be in good agreement with the available theoretical and experimental data. The other calculations were found to be in fair agreement with available ones in literature.

INTRODUCTION

Atomic processes are of great importance from astrophysical and controlled thermonuclear fusion research point of views. Especially the low energy recombination cross sections and rate coefficients are required for understanding the fusion and astrophysical plasmas[1]. The demand for reliable rate coefficients for highly ionized multi-electron atoms by electron collisions has considerably increased during the last decade and can only be met by theory. However, cross-checking the different theoretical approaches by experimental results are still necessary.

Rate coefficients for ionization, recombination and excitation of highly charged multi-electron atoms by electron collisions are needed in many areas of physics. Now a further step towards the study of the heavier elements Silicon and Sulfur have been made[2-4]. Certain recombinations play significant role in astrophysics because they are the dominant electron-ion recombination processes for most ions in low density, photo-ionized and electron-ionized cosmic plasmas.

A detailed knowledge of the different rate coefficients is needed in the study of the high temperature plasmas [5-7], the ionization balance, thermal structure and line emission of cosmic plasmas in astrophysics and in controlled thermo-nuclear fusion research. Only if these and the transition probabilities are known the radiation emitted by a non-equilibrium plasma can be quantitatively described by a theoretical population model. For electron – induced ionization, recombination, excitation and de-excitation, mainly from excited atomic states, a detailed analysis is presented for the dependence of the rate coefficients on the electron energy, the temperature and the atomic parameters.

More recently the use of heavy-ion accelerators and electron coolers of ion storage rings has greatly advanced the experiment[8]. Rate coefficients for various ions have been measured at electron energies from threshold to hundreds of electron volts (eV) [9,10].

The ion Si XII which is studied in this work is the heaviest ion for which it has been possible to prove that rapid recombination scheme does work to develop a VUV or soft X-ray laser. A series of calculations of atomic data of various Sulfur ions were motivated mainly by the recent observations of ultraviolet emission lines from the plasma torus for determining the composition and nature of the interstellar medium.

In the present work, the absolute electron impact excitation, ionization, recombination and de-excitation rate coefficients of about 13 configurations for the doublet state are considered for lithium-like ions Si XII and S XIV.

CP888, *Modern Trends in Physics Research,*
Second International Conference on Modern Trends in Physics Research—MTPR-06,
edited by L. El Nadi
© 2007 American Institute of Physics 978-0-7354-0354-9/07/$23.00

Oscillator strengths for allowed and forbidden transitions including relativistic effects in Berit-Pauli approximation were calculated by using Cowan code and also by using the available results. These oscillator strengths are used in the calculations of the rate coefficients.

The level populations are calculated as functions of the electron density and the plasma temperature for the configurations $1s^2 2s$, $1s^2 2p$, $1s^2 3s$, $1s^2 3p$, $1s^2 3d$, $1s^2 4s$, $1s^2 4p$, $1s^2 4d$, $1s^2 4f$, $1s^2 5s$, $1s^2 5p$, $1s^2 5d$ and $1s^2 5f$ for each ion.

THEORY

Rate Coefficients

We consider the electronic $|p\rangle \longrightarrow |n\rangle$ and $|p\rangle \longrightarrow |i\rangle$ transitions in an atom, where p and n are the (effective) principal quantum numbers of initial and final states $|p\rangle$ and $|n\rangle$, and $|i\rangle$ denotes the ion ground state [7]. The following notation is used : E_{pi} and $E_{pn} = E_n - E_p$ are the ionization, excitation (for $E_n > E_p$) and de-excitation ($E_n < E_p$) energies, E_e and E are the incident electron and the energy transfer to the atom respectively.

1. Ionization rate

The ionization rate Coefficient is given by [11]:

$$K_{pi} = \frac{9.56 \times 10^{-6} (kT_e)^{-1.5} \exp(-\varepsilon_{pi})}{\varepsilon_{pi}^{2.33} + 4.38 \varepsilon_{pi}^{1.72} + 1.32 \varepsilon_{pi}} \quad cm^3 s^{-1} \tag{1}$$

Where, ε_{pi} is the energy transfer ($\varepsilon_{pi} = E_{pi} / kT_e$) and kT_e is expressed in eV.

2. Recombination rate

The recombination rate coefficient is given by:

$$K_{ip} = \frac{3.17 \times 10^{-27} (kT_e)^{-3} \left(g_p / g_i \right)}{\varepsilon_{pi}^{2.33} + 4.38 \varepsilon_{pi}^{1.72} + 1.32 \varepsilon_{pi}} \quad cm^6 s^{-1} \tag{2}$$

Where g_p and g_i are the statistical weights of level $|p\rangle$ and of the ion ground state $|i\rangle$.

3. Excitation rate

An empirical formula which represents the numerical rate coefficients for excitation rate with energy transfer ε_{pn} is:

$$k_{pn} = \frac{1.6 \times 10^{-7} (kT_e)^{0.5} \exp(-\varepsilon_{pn})}{kT_e + \Gamma_{pn}} \times \left[A_{pn} \ln\left(\frac{0.3 kT_e}{R} + \Delta_{pn} \right) + B_{pn} \right] \quad cm^3 S^{-1} \tag{3}$$

Where, kT_e is in eV , R (Rydberg energy) in eV and $\varepsilon_{pn} = E_{pn} / kT_e$.

$$\Gamma_{pn} = R \ln\left(1 + \frac{p^3 kT_e}{R} \right) \left[3 + 11 \left(\frac{s}{p} \right)^2 \right] \times \left(6 + 1.6ns + \frac{0.3}{s^2} + 0.8 \frac{n^{1.5}}{s^{0.5}} |s - 0.6| \right)^{-1}$$

$$R = 13.595, \quad p = z_{eff} \times \sqrt{R / E_{pi}}, \quad n = z_{eff} \times \sqrt{R / E_{ni}}, \quad s = |n - p|$$

$$A_{pn} = \left(2R / E_{pn} \right) f_{pn} \tag{4}$$

f_{pn} being the absorption oscillator strength.

$$\Delta_{pn} = \exp\left(-\frac{B_{pn}}{A_{pn}}\right) + \frac{0.06s^2}{np^2}$$

where $B_{pn} = \dfrac{4R^2}{n^3}\left(\dfrac{1}{E_{pn}^2} + \dfrac{4E_{pi}}{3E_{pn}^3} + b_p\dfrac{E_{pi}^2}{E_{pn}^4}\right)$ and $b_p = \dfrac{1.4\ln p}{p} - \dfrac{0.7}{p} - \dfrac{0.51}{p^2} + \dfrac{1.16}{p^3} - \dfrac{0.55}{p^4}$

4. De-excitation rate

The de-excitation rate is given by:

$$k_{pn} = \frac{1.6\times10^{-7}(kT_e)^{0.5}\,g_n/g_p}{kT_e + \Gamma_{np}} \times \left[A_{np}\,\ln\left(\frac{0.3kT_e}{R} + \Delta_{np}\right) + B_{np}\right] \tag{5}$$

Where Γ_{np} and Δ_{np} are obtained from Γ_{pn} and Δ_{pn} by interchanging p and n. g_p and g_n are the statistical weights of level $|\,p\rangle$ and $|\,n\rangle$.

Population Densities

Level population can be calculated by solving the steady-state rate equations[11]

$$N_j\left[\sum_{i<j} A_{ji} + N_e\left(\sum_{i>j} C_{ji}^d + \sum_{i>j} C_{ji}^e\right)\right] =$$

$$N_e\left(\sum_{i<j} N_i C_{ij}^e + \sum_{i>j} N_i C_{ij}^d\right) + \sum_{i>j} N_i A_{ij} \tag{6}$$

where N_j is the population of level j, A_{ji} is the spontaneous decay rate from level j to level i (transition probabilities), C_{ji}^e is the electron collisional excitation rate coefficient, C_{ji}^d is the collisional de-excitation rate coefficient.

The population of the jth level is obtained from the identity

$$N_j = \left(\frac{N_j}{N_I}\right)\left(\frac{N_I}{N_T}\right)\left(\frac{N_T}{N_e}\right)N_e \tag{7}$$

where N_I is total number density of all the levels of the ion under consideration, and N_T is the total number density of all ionization stages. Since the populations calculated from Equation (6) are normalized such that

$$\sum_{j=1}^{n}\left(\frac{N_j}{N_I}\right) = 1 \tag{8}$$

where n is the number of all the levels of the ion under consideration, the quantity actually obtained from Equation (6) is the fractional or reduced population N_j/N_I. We calculated the population densities and rate coefficients by using the method given by Vriens[11] and by Feldman[13-15].

RESULTS AND DISCUSSION

Excitation, Ionization and Recombination Rate Coefficients

For the evaluation of electron impact ionization, excitation, de-excitation and recombination rates, for excited atomic states in Si XII and S XIV, the rate coefficients have been evaluated by using an empirical formula which is published by Vriens[12]. The calculations were carried out by using a computer program (CRMO code)[16]. The calculations include all forbidden and allowed transitions

that are necessary for the calculations. Therefore, in addition to the dipole transitions we have introduced the monopole and quadrupole transitions in the calculations.

The rate coefficients are determined for $2s^2 2s$ ($^2S_{1/2}$), $2s^2 2p$ ($^2P_{1/2}$), $2s^2 3s$ ($^2S_{1/2}$), $2s^2 3p$($^2P_{1/2}$), $2s^2 3d$ ($^2D_{3/2}$), $2s^2 4s$ ($^2S_{1/2}$), $2s^2 4p$ ($^2P_{1/2}$), $2s^2 4d$ ($^2D_{3/2}$), $2s^2 4f$ ($^2F_{5/2}$), $2s^2 5s$ ($^2S_{1/2}$), $2s^2 5p$ ($^2P_{1/2}$), $2s^2 5d$ ($^2D_{3/2}$) and $2s^2 5f$ ($^2F_{5/2}$) excited atomic states in Si XII and S XIV. The calculations were carried out at different electron temperatures (in eV).

The ionization rates for all the levels are drawn versus the electron temperature (kT_e) as shown in figures (1-2) for Si XII and S XIV ions. The curves show the usual behavior for ionization trend.

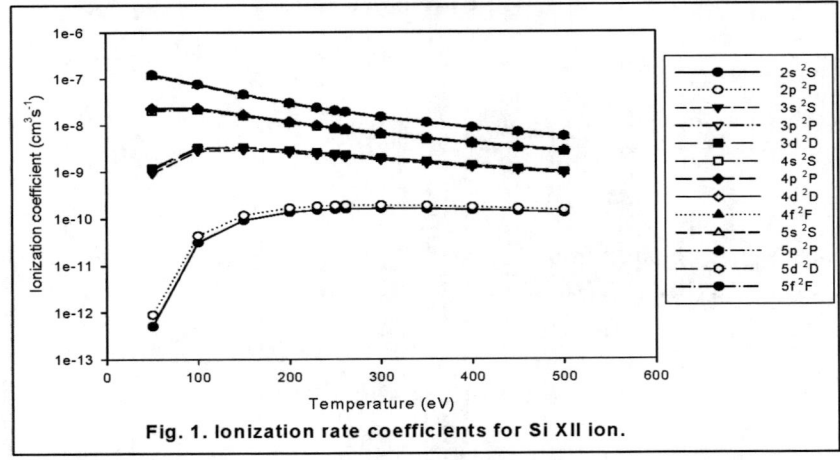

Fig. 1. Ionization rate coefficients for Si XII ion.

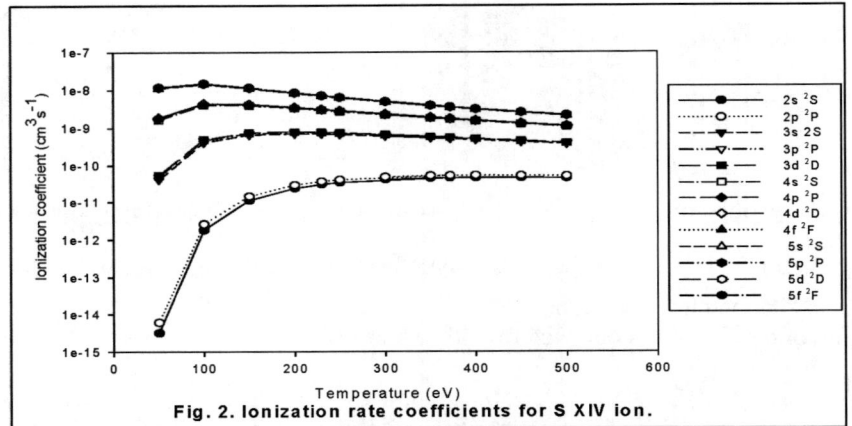

Fig. 2. Ionization rate coefficients for S XIV ion.

The ionization rates for the levels ns (^2S) and np (^2P), where n = 2-5 are drawn versus the electron temperature (kT_e) as shown in figures (3-4) for Si XII and S XIV ions.

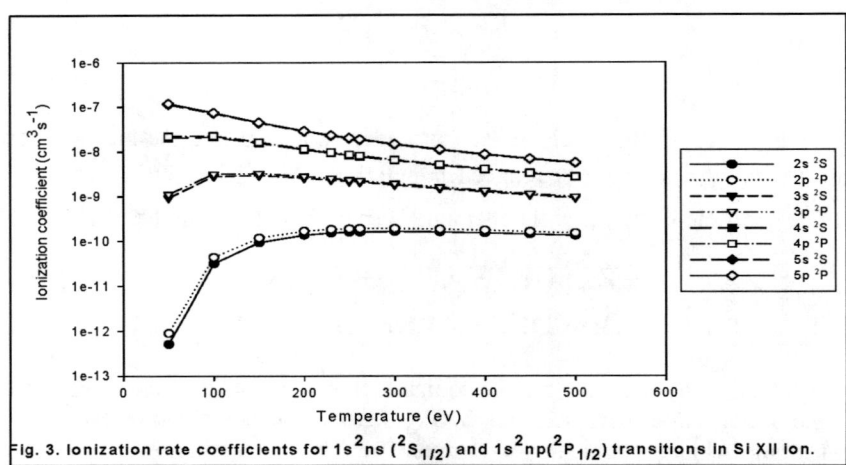

Fig. 3. Ionization rate coefficients for $1s^2$ns ($^2S_{1/2}$) and $1s^2$np($^2P_{1/2}$) transitions in Si XII ion.

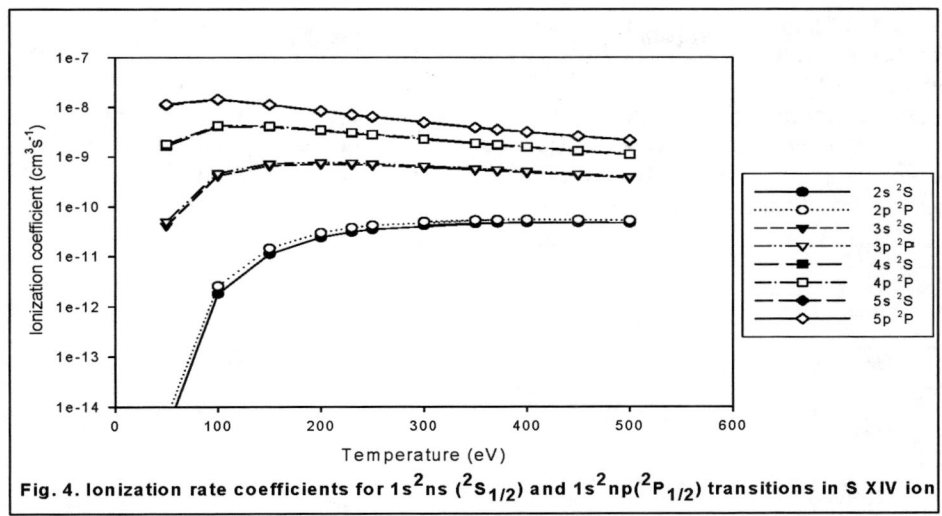

Fig. 4. Ionization rate coefficients for $1s^2ns$ ($^2S_{1/2}$) and $1s^2np$($^2P_{1/2}$) transitions in S XIV ion

Figures (3-4) show that the curves are divided into groups according to the principle quantum number "n", that is as n decreases the coupling between the electrons and the nucleus getting larger. When n =2 and by increasing the temperature the behavior of the energy states shows saturations. This phenomenon might be referred to the filment of the electrons in these ionization states by increasing the temperature. When n increases, the ionization rate coefficients are decreasing by increasing the temperature because they are far from the nucleus.

The excitation rate coefficients are drawn versus the electron temperature kT_e in (in eV) as shown in figures (5-6) for some forbidden and allowed transitions for Lithium-like Si XII and S XIV ions.

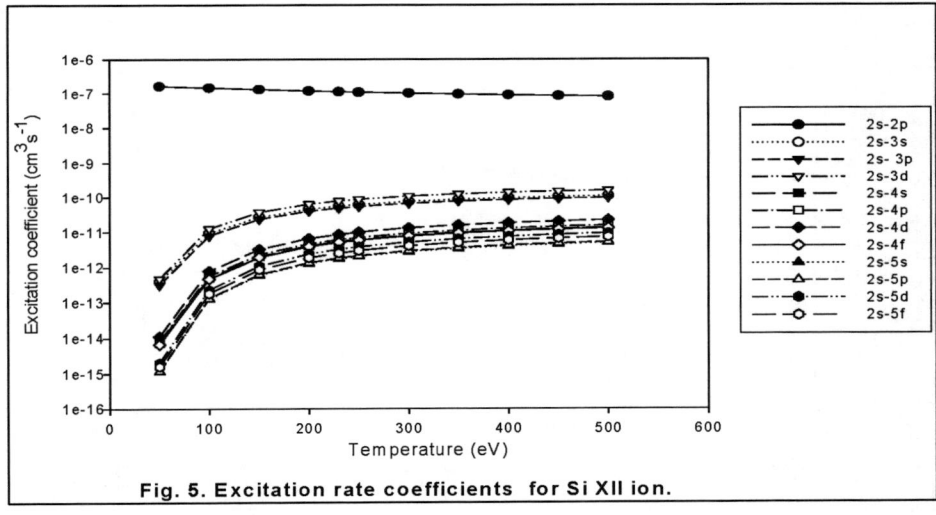

Fig. 5. Excitation rate coefficients for Si XII ion.

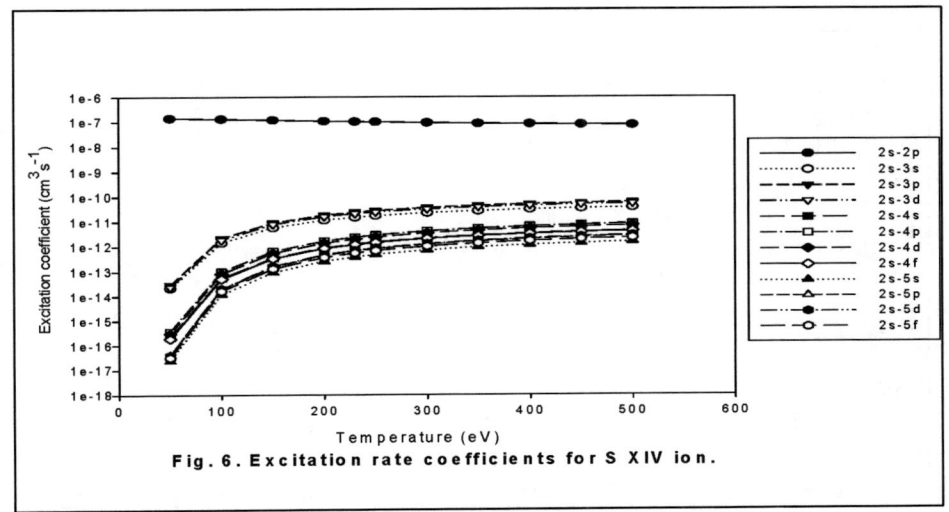

Fig. 6. Excitation rate coefficients for S XIV ion.

A detailed study of both radiative and forbidden transitions showed a comparable behavior with respect to both the theoretical and experimental work as shown in figures (7-13) in case of Si XII ion for $\Delta n=2$ and 3. For the most levels the theoretical excitation rate coefficients of this work were found to be in good agreement with the available theoretical and experimental data as shown in figures (8-10 and 12). The rate coefficients that were calculated by using the Coulomb-Born Oppenheimer approximation[17,19] are comparable with this work. The other calculations[18] were found to be in fair agreement with those available in literature.

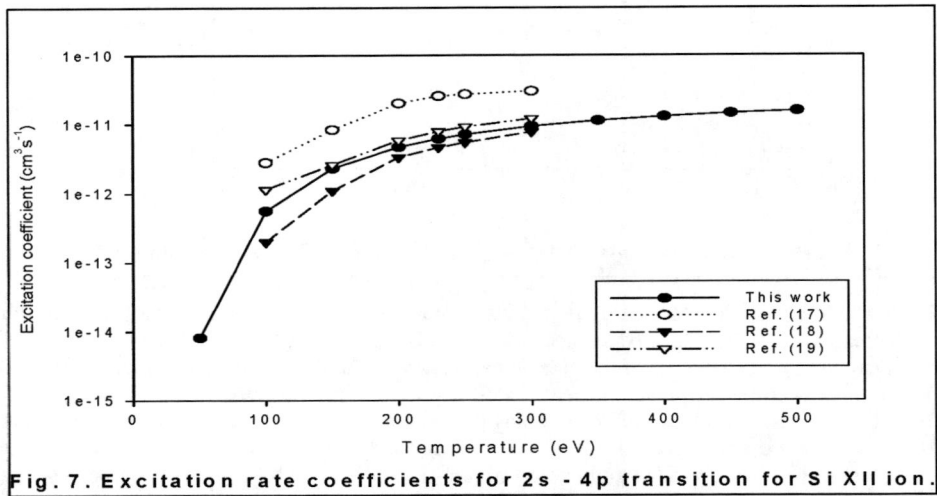

Fig.7. Excitation rate coefficients for 2s - 4p transition for Si XII ion.

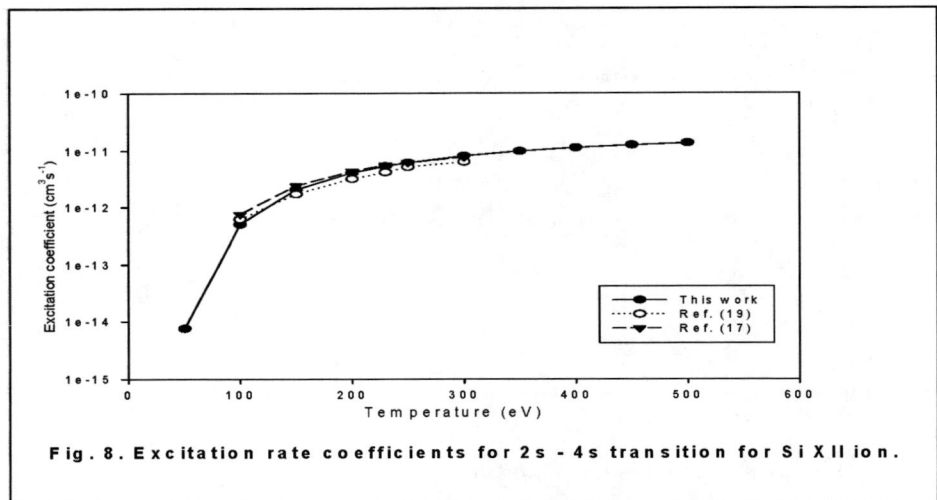

Fig.8. Excitation rate coefficients for 2s - 4s transition for Si XII ion.

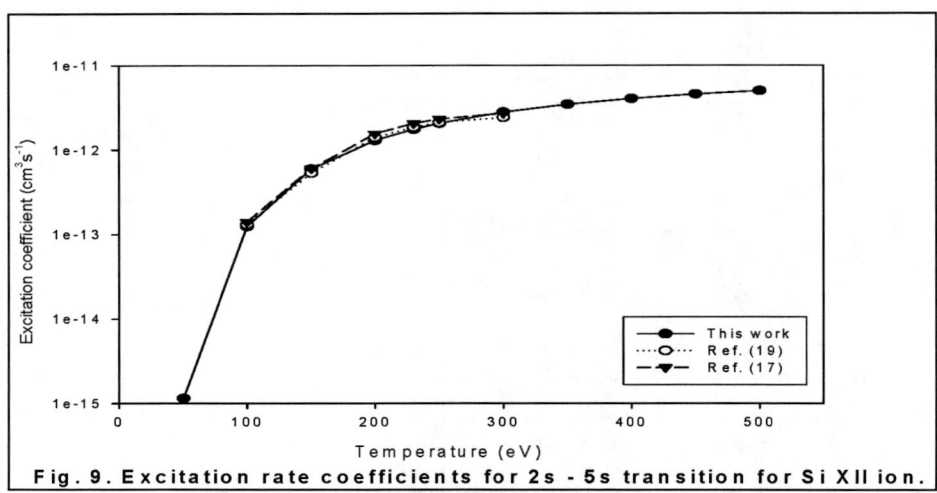

Fig.9. Excitation rate coefficients for 2s - 5s transition for Si XII ion.

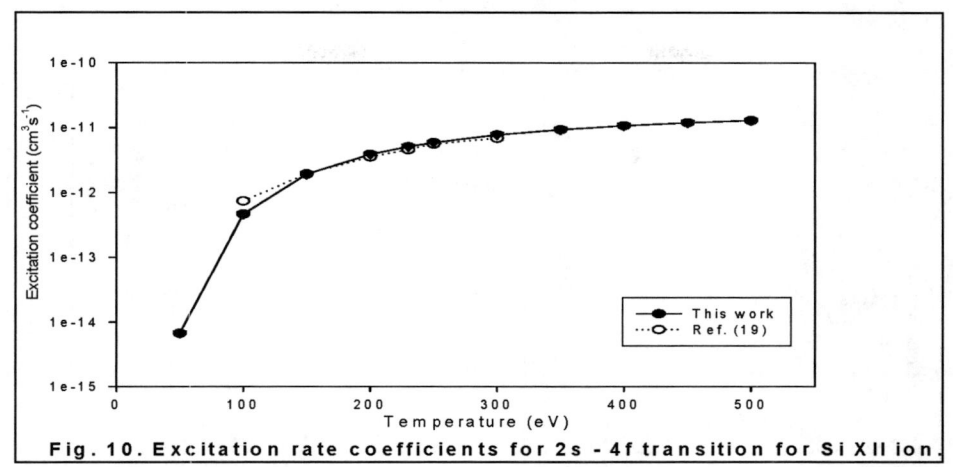

Fig. 10. Excitation rate coefficients for 2s - 4f transition for Si XII ion.

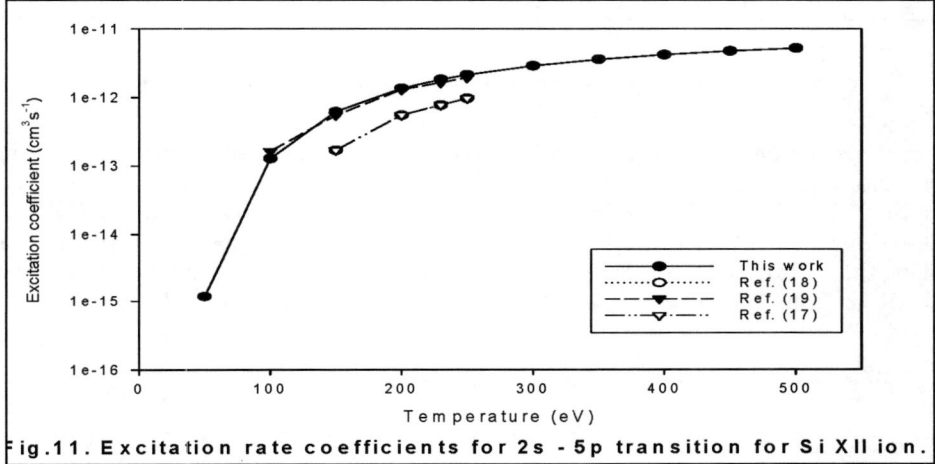

Fig.11. Excitation rate coefficients for 2s - 5p transition for Si XII ion.

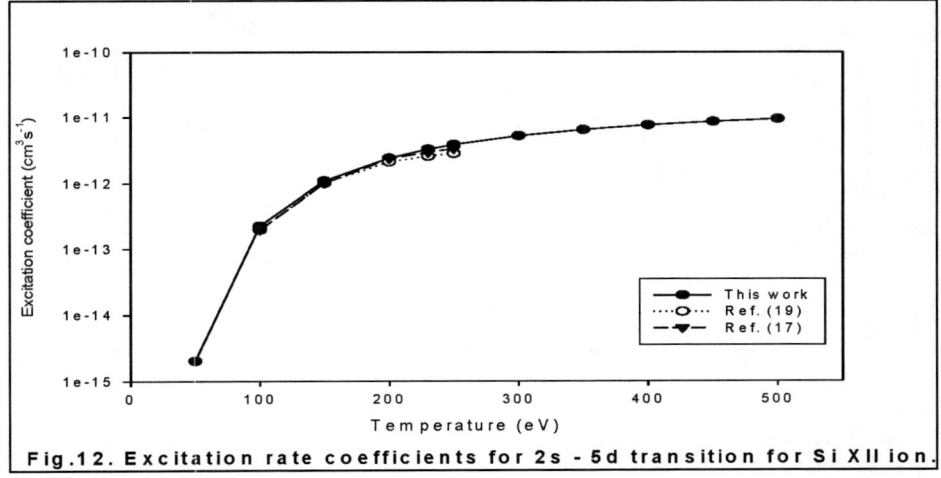

Fig.12. Excitation rate coefficients for 2s - 5d transition for Si XII ion.

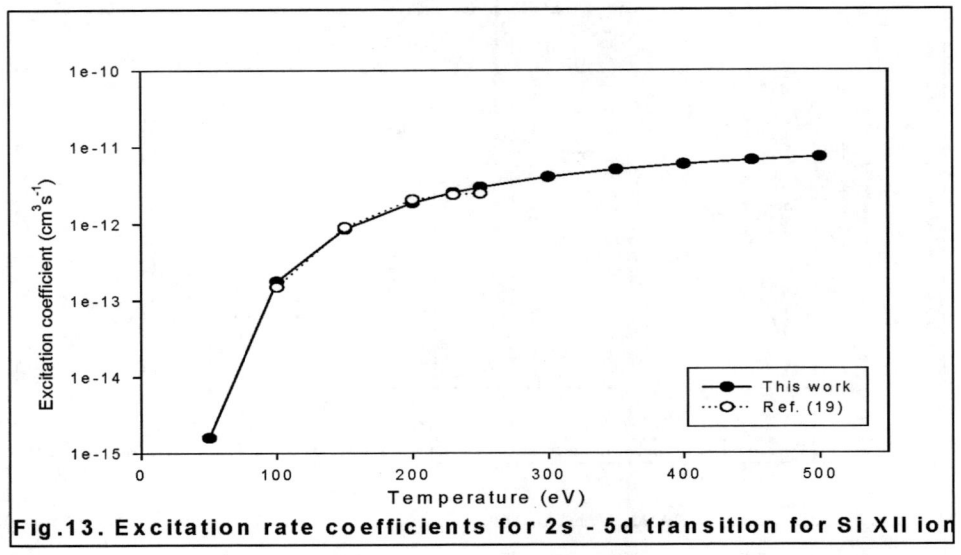

Fig.13. Excitation rate coefficients for 2s - 5d transition for Si XII ion

3.2. The Population Density of the Exited Levels

The level populations N_j are calculated by solving the 13 coupled rate equations (6) belonging to the configurations $2s^22s$ ($^2S_{1/2}$), $2s^22p$ ($^2P_{1/2}$), $2s^23s$ ($^2S_{1/2}$), $2s^23p$($^2P_{1/2}$), $2s^23d$ ($^2D_{3/2}$), $2s^24s$ ($^2S_{1/2}$), $2s^24p$ ($^2P_{1/2}$), $2s^24d$ ($^2D_{3/2}$), $2s^24f$ ($^2F_{5/2}$), $2s^25s$ ($^2S_{1/2}$), $2s^25p$ ($^2P_{1/2}$), $2s^25d$ ($^2D_{3/2}$) and $2s^25f$ ($^2F_{5/2}$) excited atomic states in Si XII and S XIV ions. Fractional populations are obtained by the aid of equations (7) and (8) using the CRMO code[16]. Calculations are performed at electron temperatures 230 eV in the case of Si XII ion[2] and 371.37 eV (1/2 the ionization potential) in case of S XIV ion[13].

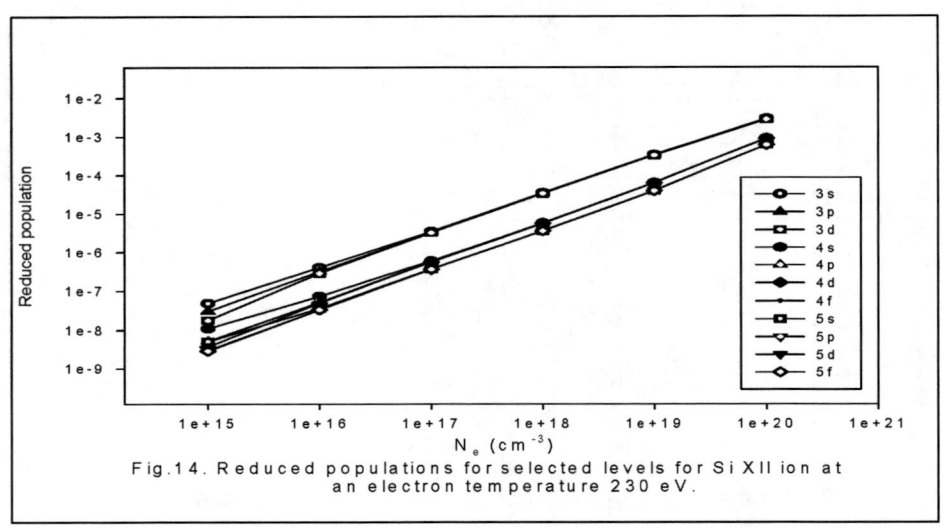

Fig.14. Reduced populations for selected levels for Si XII ion at an electron temperature 230 eV.

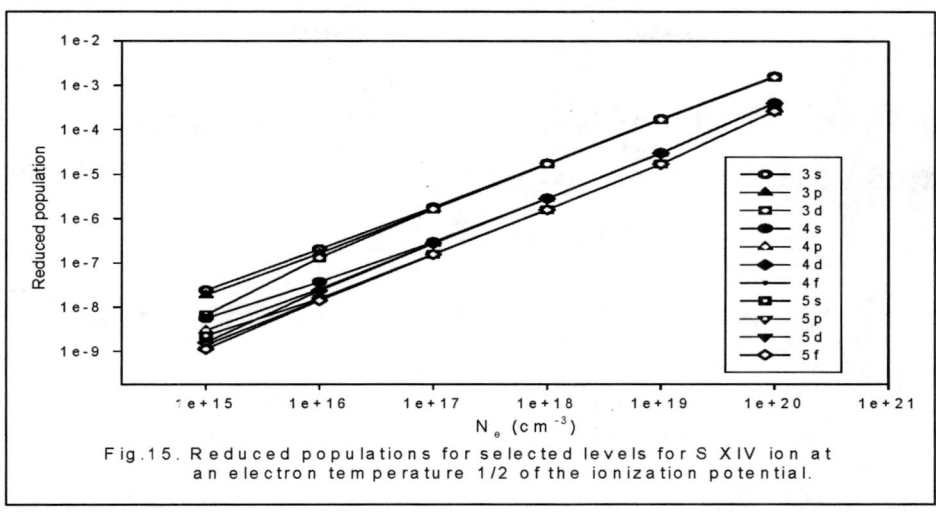

Fig.15. Reduced populations for selected levels for S XIV ion at an electron temperature 1/2 of the ionization potential.

The calculated reduced populations (fractional level populations per unit statistical weight N_j/g_jN_i) as a function of electron densities ($N_e = 10^{15}$-10^{20} cm^{-3}) are drawn in figures (14-15) for Si XII and S XIV ions respectively. The behavior of level populations of the Si XII and S XIV ions can be explained as follows at low electron densities the reduced populations increase. This is due to the increase in the collisional excitation rates with density[13-15]. At high electron densities ($N_e \geq 10^{18}$-10^{20} cm^{-3}) where the collisional excitation rates exceed the radiative decay rates, the reduced populations are independent of electron density and are approximately equal.

CONCLUSION

The ionization, recombination and excitation rates are calculated for 13 levels up to n = 5 in Si XII and S XIV ions. The rate coefficients were comparable with the available data.
The population density for each level is calculated and hence the reduced populations were determined.

REFERENCES

1. G. F. Gribakin, S. Sahoo and V. Dzuba, Physics. **29**, (2004).
2. R. Kőnig, K. H. Kolk and H.-J. Kunze, Physica Scripta.**48**, (1993).
3. R. Kőnig, K. H. Kolk and H.-J. Kunze, Physica Scripta.**53**, (1996).
4. S. N. Nahar, Astro. Phys. **10** (2002).
5. T. Fujimoto, J. Phys. Soc. Jpn. **47**, 273 (1979).
6. L. Vriens, J. Appl. Phys. **49**, 3814 (1978).
7. J. Stevefelt, J. Boulmer, and J. F. Delpech, Phys. Rev. A **12**, 1246 (1975).
8. G. Kilgus , Phys. Rev. Lett. **64**, 737 (1990)
9. D. R. DeWitt, Phys. Rev. A **50**, 1257 (1994); D. R. DeWitt, J. Phys. B **28**, L147 (1995).
10. O. Uwira, Hyperfine Interact. **108**, 149 (1997); S. Mannervik, Phys. Rev. Lett. **81**, 313 (1998).
11. W. C. Martin, R. Zalubas and A. Musgrove, J. Chem.Ref. Data **19**, 821 (1990).
12. L. Vriens, Phys. Rev. A. **22**, 940 (1980).
13. U. Feldman, J. F. Seely and A. K. Bahatia, J. App. Phys. **56**, 9 (1984).
14. U. Feldman, A. K. Bahatia and S. Suckewer, J. App. Phys. **54**, 5 (1983).
15. U. Feldman, J. F. Seely and A. K. Bahatia, J. App. Phys. **58**, 11 (1985).
16. S. H. Allam, "CRMO-Collisional Radiative Model" computer code, private communication.
17. R. Mewe, Astron. Astrophys. 20, 215 (1972); O. Bely, Proc. Phys. Soc. **88**,587 (1966).
18. J. Davis, P. C. Kepple and M. J. Blaha, J. Quant. Spectr. Rad. Trans. **18**, 535 (1977).
19. R. E. H. Clark, D. H. Sampson and S. J. Goett, Astrophys. J. Suppl. Ser. **49**, 545 (1982); R. E. H. Clark and S. J. Goett, Astrophys. J. Suppl. Ser. **45**, 603 (1981).

A Mean-Field Study of the Magnetic and Magneto-Thermal Properties of Selected   Amorphous R_x-Fe_{1-x} and Crystalline $RFe_{10}V_2$ Systems.

S. H. Aly[1] , S. Yehia[2], G. El-Alfi[1], and M.Hammam[2].

1. Physics Department, Faculty of Science at Damietta, Mansoura University, New Damietta, Egypt.
2. Physics Department, Faculty of Science, Helwan University, Cairo, Egypt

Abstract. We present a mean-field theory study on the magnetization, magnetic specific heat and entropy for a selected amorphous R-Fe and crystalline $RFe_{10}V_2$ alloys. The systems we report are a-Gd_xFe_{1-x} and a- $Er_x Fe_{1-x}$ where the atomic concentration x lies in the range 0.42 to 0.60 and 0.085 to 0.20 for these two systems respectively. These systems allow for a systematic study on the dependence of the magnetic and magneto-thermal properties on the rare-earth concentration within a given series. On the other hand the fixed-composition $RFe_{10}V_2$ system, where R=Y, Lu, Nd, Gd, Tb, Ho, Er and Tm, enables a detailed study of the dependence of the aforementioned properties on the R species. A two-sublattice magnetic system is devised for each of the studied alloys where inter- and intra- sublattice exchange interactions are taken into account. The temperature dependencies of the sublattice magnetizations and their temperature derivatives are used in calculating the magnetic specific heats and the associated entropies either in zero field or in the presence of an external field. In particular, both of the specific heat and its associated entropy, at constant field, consist of four terms: one for each magnetic sublattice and two for inter-sublattice contributions. For the $RFe_{10}V_2$ system, where R is magnetic, we found that the Fe-Fe contribution to the zero-field specific heat and entropy, although not the same for all R, is the largest compared to the R-R and R-Fe contributions. We calculated the largest zero-field S_{max} for Ho and Dy ($\sim$ 1.2 J/K.cm^3) and the smallest for Y and Lu ($\sim$ 0.9 J/K.cm^3). Furthermore we found that heavy rare earths have higher S_{max} compared to the light ones for example ~1.1 J/K.cm^3 for R=Nd. In all cases our calculation fairly agree with the equation: S_{max}=Nk_B ln (2J+1). Our calculations showed that the specific heat anomaly and S_{max} of the amorphous alloys increase with increasing the Fe content. For the Gd system they lay in the range 0.15-0.22 and 0.12-0.14 J/g. K respectively. For the Er-system the corresponding ranges are 0.19-0.30 and 0.18-0.20 J/g. K respectively. The effect of the external field on the specific heat is to reduce it below T_c and increase it above T_c relative to its zero-field value. Its effect on the entropy, however, is to reduce it at temperatures close to and above T_c relative to its zero-field value as expected.

INTRODUCTION

Thermomagnetic properties of amorphous and crystalline RT alloys have attracted much interest [1-3]. In particular magnetization dependence on temperature has been calculated in the MF analyses[1,2]. Experiments on deducing the magnetic contribution to the specific heat and hence the magnetic entropy have been reported for different RT systems[e.g. 4] but no such studies were reported for the systems we report in the present work except for the measurements done on thin films of the Gd-Fe amorphous alloy by Hellman et al[5]. In 1987 the magnetic properties of the then novel $RFe_{10}V_2$ where R=Y,Lu, Nd, Gd, Tb, Dy, Ho, Er and Tm compounds were reported by de Boer et al [6]. Examples of studies on this system are: magnetic anisotropy [7], spin re-orientation [8] and mean field calculation [9]. As far as we know there are no MF analysis of the magnetic specific heat and entropy of the systems we report so we setup our purpose to calculate these quantities from the calculated magnetization in a temperature range up to their Curie temperatures. The study of the fixed-composition $RFe_{10}V_2$ enables a systematic study of the role of different rare-earth species in the magnetic properties. In addition it provides information on the Fe sublattice when R is non-magnetic e.g. Y and Lu.

CP888, *Modern Trends in Physics Research*,
Second International Conference on Modern Trends in Physics Research—MTPR-06,
edited by L. El Nadi
© 2007 American Institute of Physics 978-0-7354-0354-9/07/$23.00

MODEL AND CALCULATION

The exchange energy between two magnetic moments m and n is given by

$$e_{mn} = -2\Im_{mn}\, j_m\, j_n \tag{1}$$

In the mean-field approach however the effect of all the magnetic moments on a given central moment m is

$$e_m = -\Im_{mn}\, j_m\, \langle j_n \rangle \tag{2}$$

Therefore for Z nearest neighbors in a two-sublattice system the energy will be

$$E_m = -Z_m\, j_m \sum_{n=1}^{2} x_n \Im_{mn} \langle j_n \rangle \tag{3}$$

where xn is the concentration of the nth sublattice.

The angular momentum average at a given temperature T is given by the well-known Brillouin function[10] viz:

$$\langle j_m \rangle = \left(J_m + \frac{1}{2}\right)\coth\left[\frac{Z_m(2J_m+1)\sum_{n=1}^{2}x_n\Im_{mn}\langle j_n\rangle}{2kT}\right] - \frac{1}{2}\coth\left[\frac{Z_m\sum_{n=1}^{2}x_n\Im_{mn}\langle j_n\rangle}{2kT}\right] \tag{4}$$

The magnetization of the sublattice n is

$$M_n = Nx_n \mu_b g_n \langle j_n \rangle \tag{5}$$

To calculate the magnetic specific heat we start by the definition of the energy density

$$U = -\frac{1}{2}\left(\lambda_{11}M_1^2 + \lambda_{22}M_2^2 + \left(\lambda_{12} + \lambda_{21}\right)M_1 M_2\right) \tag{6}$$

Where the exchange constant is

$$\lambda_{ij} = \frac{Z_i x_j \Im_{ij}}{N_j \mu_b^2 g_i g_j} \tag{7}$$

The specific heat will depend on the magnetizations and their temperature dependences in the following manner:

$$C_v(T) = -\lambda_{11}M_1\frac{dM_1}{dT} - \lambda_{22}M_2\frac{dM_2}{dT} + \frac{1}{2}\left(\lambda_{12} + \lambda_{21}\right)\left(M_1\frac{dM_2}{dT} + M_2\frac{dM_1}{dT}\right), \tag{8}$$

i.e. $C_v(T) = C_{11} + C_{12} + C_{21} + C_{22}$ \hfill (9)

where the four terms corresponding to those in Eq.8 The magnetic entropy at a temperature T is given by:

$$S(T) = \int_0^T \frac{C_v(T)}{T}dT \tag{10}$$

RESULTS AND DISCUSSION

I) The RFe₁₀V₂ System

a) Magnetization

We have calculated the sublattice and total magnetizations of the members of this system as function of temperature up to ~700K which is above their Curie temperatures. In case of Y and Lu the net magnetization is that of Fe only while for other R's it is resultant of the sublattice magnetizations. Examples of these plots are shown in Figs.1 and 2 for R=Nd and Ho respectively.

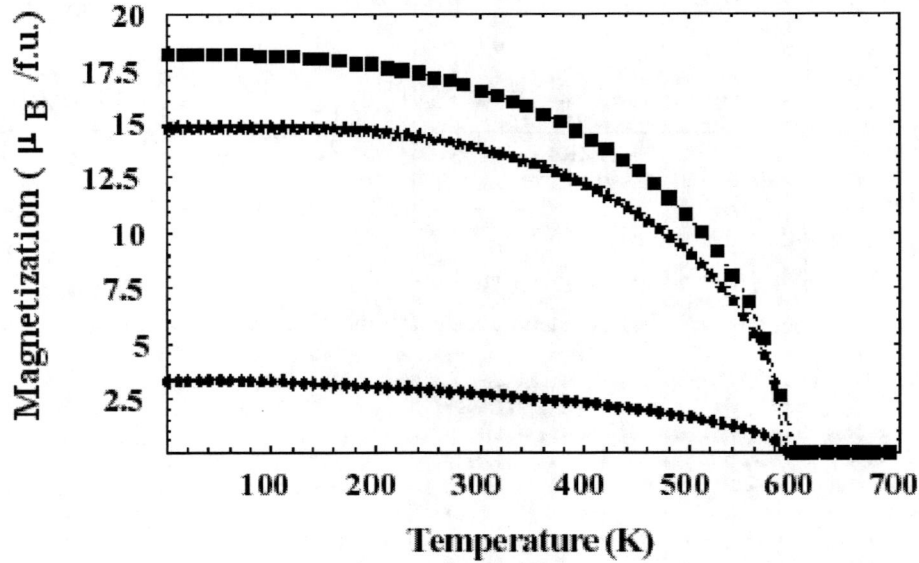

FIGURE 1. Calculated sublattice (♦ Nd,* Fe) and net magnetizations(■) vs. temperature for NdFe₁₀V₂ in zero field

The coupling between Nd and Fe is ferromagnetic. Their magnetizations add up in the entire temperature range with maximum magnetic moment of ~17.7 μ_b/f.u at very low temperatures. The coupling in case of Ho is evidently ferrimagnetic with only ~6 μ_b /f.u. It may be seen also from these two figures that Fe magnetization is different in these two compounds. We have performed these calculations for the other R species and found out that the magnetic order is ferromagnetic for Y, Lu and Nd and ferrimagnetic for the others. The calculated magnetization-temperature curves are reminiscent of ferrimagnetic systems [10].

The Fe sublattice dominates the R sublattice and therefore no compensation points were found in these ferrimagnetic systems. Our results are in good agreement with the measurements of de Boer et al [6]. The zero-temperature calculated magnetization, calculated and measured   Curie temperatures are shown in Table 1. The highest T_c is that for R=Gd in which the Fe-Fe coupling is the strongest in the series. It is well known that the T-T interaction has the   strongest effect on T_c for the R-T compounds [11]. This trend is consistent with the   variation of de Gennes factor $(g-1)^2 J(J+1)$ of the R species[12]. The calculated saturation magnetizations are in agreement with the available experimental data. For example Chuang et al [13] have reported a value of 18.5 μ_b/f.u for R=Nd. For R=Dy the corresponding value is 6.73 μ_b/f.u at 5K with Fe moment of 1.67 μ_b /atom.

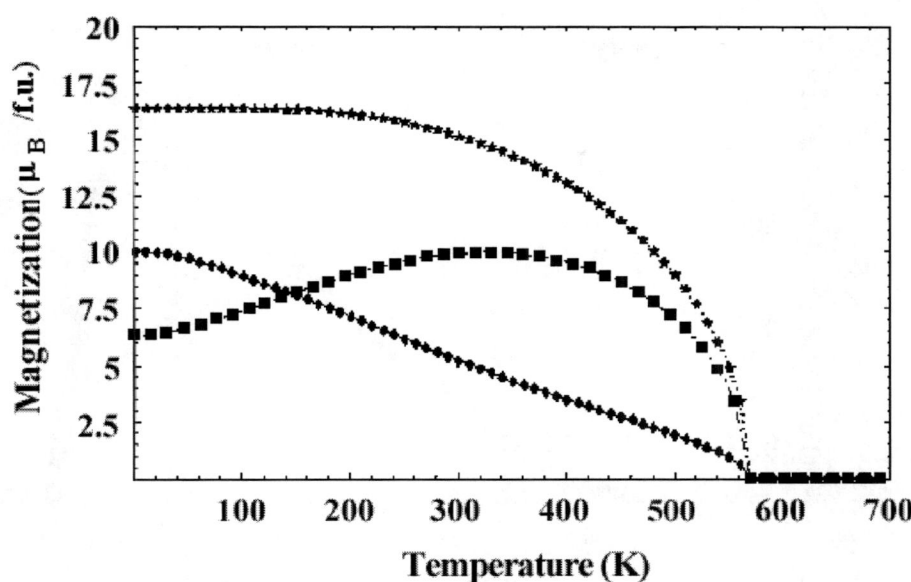

FIGURE 2. Calculated sublattice (♦ Ho,* Fe) and net magnetizations (■) vs. temperature for $HoFe_{10}V_2$ in zero field.

TABLE 1. Curie Temperatures and Magnetizations of $RFe_{10}V_2$ Compounds

RE	Calculated T_c (K)	Measured T_c (K)	Magnetization M_T (0) (μ_B/f.u.)
Y	533	532	16.1
Lu	487	483	15.6
Nd	566	570	18.1
Gd	619	616	8.9
Tb	564	570	7.9
Dy	538	540	6.67
Ho	524	525	6.3
Er	507	505	8.4
Tm	495	496	11.2

b) Magnetic Specific Heat and Entropy

An example of the magnetic specific heat and its field and temperature dependences is shown in Fig.3 for $NdFe_{10}V_2$ system. The mean features of this figure are: a) the sharp drop in the zero-field specific heat at T_c. This is typical of MF calculation which is different from the lambda -type experimental drop[14,15] ; b)The specific heat below the transition in the presence of an applied magnetic field is less than the zero- field value just at the transition temperature. The opposite trend however occurs above T_c, and c) the peak is smeared out at the transition upon applying a magnetic field. These features are typical to ferromagnetic second-order phase transitions. MF analysis predicts a discontinuity in the specific heat at T_c equals to $5NkJ(J+1)/(J^2+(J+1)^2)$ where N is the concentration of the magnetic species.

The parameters used in calculating the magnetization i.e. the angular momenta, densities and atomic numbers are of course those used in calculating the jump in the zero-field specific heat. The quantity $5J(J+1)/(J^2+(J+1)^2)$ saturates to 2.5 for J>>1. For rare earths J is between 3.5 and 8 and therefore this quantity is quite close to 2.5 across the series.

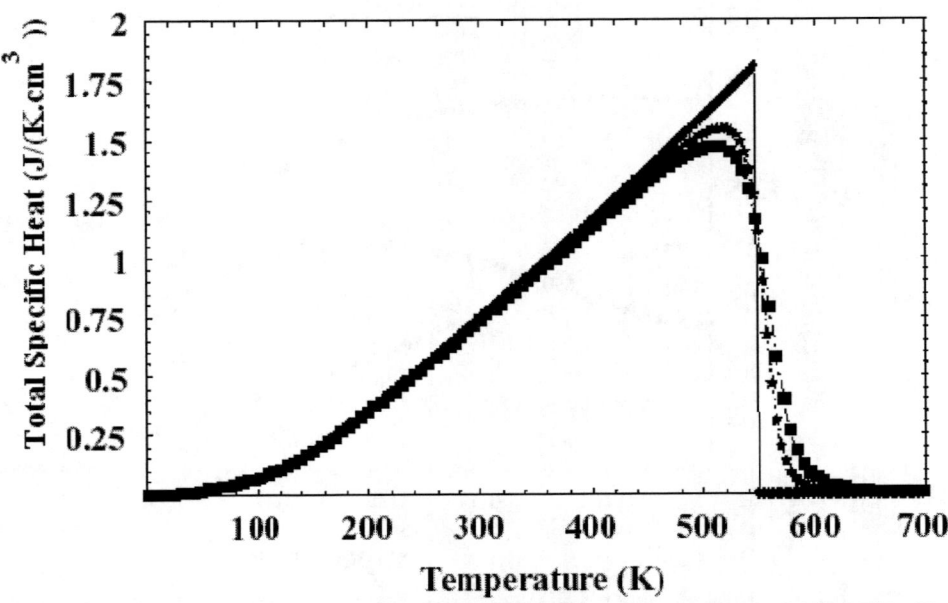

FIGURE 3. Magnetic specific heat in H=0 (♦), 40 (*) and 80 kOe (■) as a function of temperature for the $NdFe_{10}V_2$ compound.

In addition we calculated the Fe-Fe, Nd-Nd and the intersublattice contributions to the zero-field specific heat of the Nd-compound and found out that the Fe contribution is the highest while the other three interactions contribute only ~16% to the total magnetic specific heat. We have obtained similar results for the other R-species. The associated entropies for H=0 and 80 kOe for R= Nd and Ho are shown in Figs. 4,5 respectively. It is clear that the zero-field entropy saturates to different values in the two compounds due to the different angular moments of the rare earths. The saturation in the entropy takes place close to T_c as the transition from ferro (ferri) magnetic order to paramagnetism (disorder) commences.

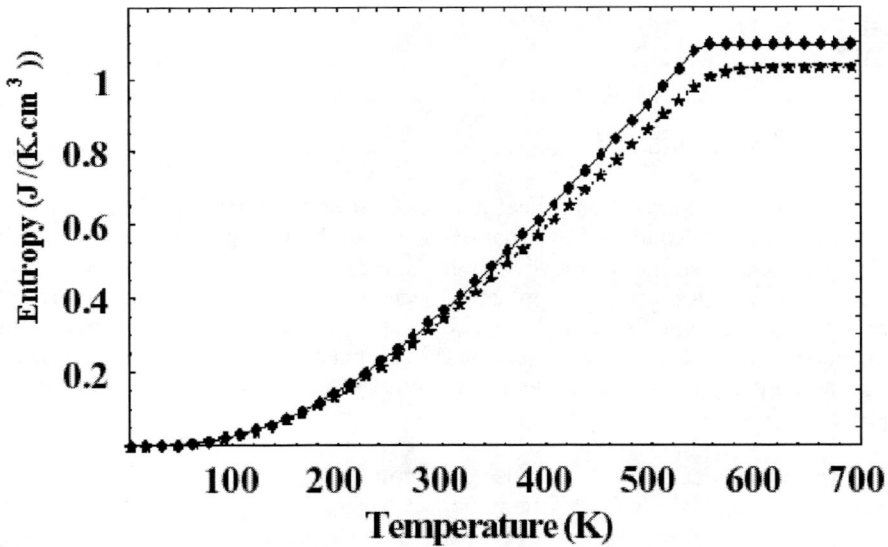

FIGURE 4. Magnetic entropy for the $NdFe_{10}V_2$ compound as a function of temperature for H= 0 (♦) and 80 kOe (*).

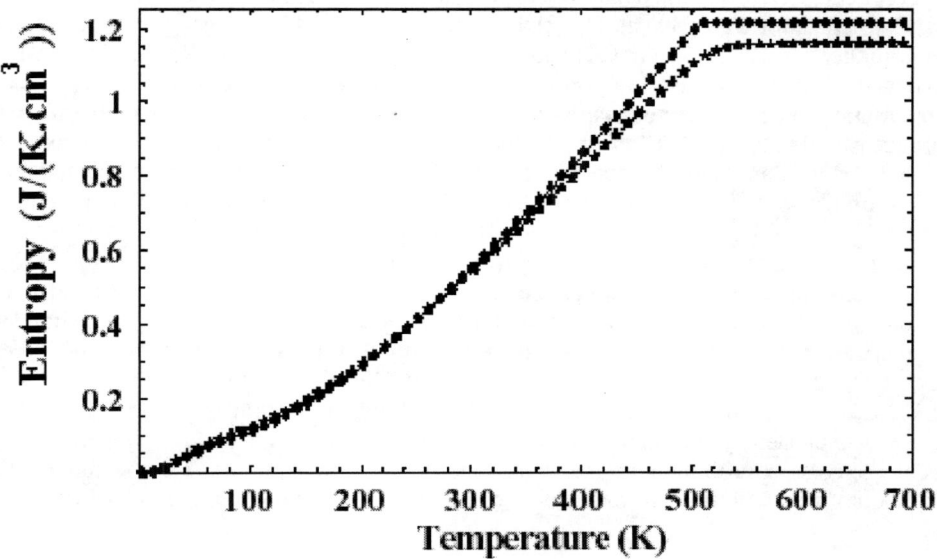

FIGURE 5. Magnetic entropy for the $HoFe_{10}V_2$ compound as a function of temperature for H= 0 (♦) and 80 kOe (·)

The entropy is reduced in the presence of a magnetic field as expected in particular at high temperatures. Again the Fe-Fe contribution to the total entropy exceeds the contribution of the other interactions. The study of the entropy in this series leads to the following observations: 1) compounds with non-magnetic R are of the smallest entropies (~0.9 J/K.cm^3); 2) members with light R have smaller S_{max} than those of heavier ones and 3) the maximum entropy is achieved in Ho (~1.2 J/K.cm^3). The variation of the entropy with R is proportional to the logarithm of the multiplicity i.e. ln J (J+1). This factor is ~2.8 and 2.3 for Ho and Nd respectively for example.

II) The R-Fe Amorphous System

Fig.6 displays our mean field calculation of the temperature dependence of the zero-field magnetization for a-Er_x-Fe_{1-x} alloys with x in the 8.5-28 range..

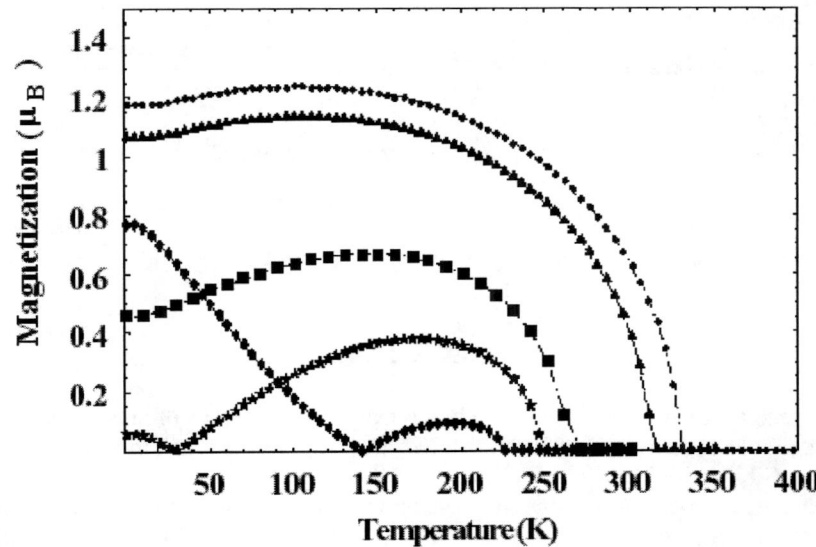

FIGURE 6. The temperature dependence of the magnetization, in zero field, for a-Er_x Fe_{1-x} alloy .(♦ x = 0.28, · 0.215, ■ 0.165, ▲ 0.10, ● 0.085)

These curves reflect a ferrimagnetic behavior [10] e.g. they display compensation points, for x = 0.215 and 0.28 , at which the magnetization vanishes at a temperature below T_c. The compensation takes place if the Er-magnetization is higher than that of Fe at 0K, and since the Er-Er interaction is weaker than Fe-Fe interaction one would expect that the former decrease more rapidly with temperature than the latter and this behavior results in a magnetic compensation. The temperature variation of the magnetization curves and their 0K values are in good agreement with measured values[16]. The Curie temperature of the alloys increases with increasing their Fe content. The zero-field specific heat and entropy increase as well with increasing the Fe- content of the alloys. The effect of the magnetic field on the specific heat and entropy is similar to that described in the previous section. The results of our calculation of the thermomagnetic properties of Gd-Fe amorphous system can be summarized as follows: a) The Fe-rich alloys have higher Curie temperatures and lower spontaneous magnetization compared to the Fe-poor alloys; b) The magnetization-temperature plots exhibit no maxima in the temperature range studied but rather decrease rapidly to zero magnetization at T_c. This is a feature of a ferrimagnetic two-sublattice systems in which the sublattice with larger magnetization (Gd in the present case) is more easily disturbed as the temperature increases as compared to the lattice with smaller magnetization (e.g. Fe)[10]; c) The temperature at which the entropy saturation commences increases with increasing Fe content; d) The entropy maximum is higher for alloys with higher Fe-content and e) the heat capacity jump increases with increasing the Fe-content and for a given composition the Fe-Fe sublattice has the largest contribution to the discontinuity as shown in Fig. 7 for example. For more details concerning the model used in this work and the calculation of the magnetic properties of these systems one may consult Ref.16

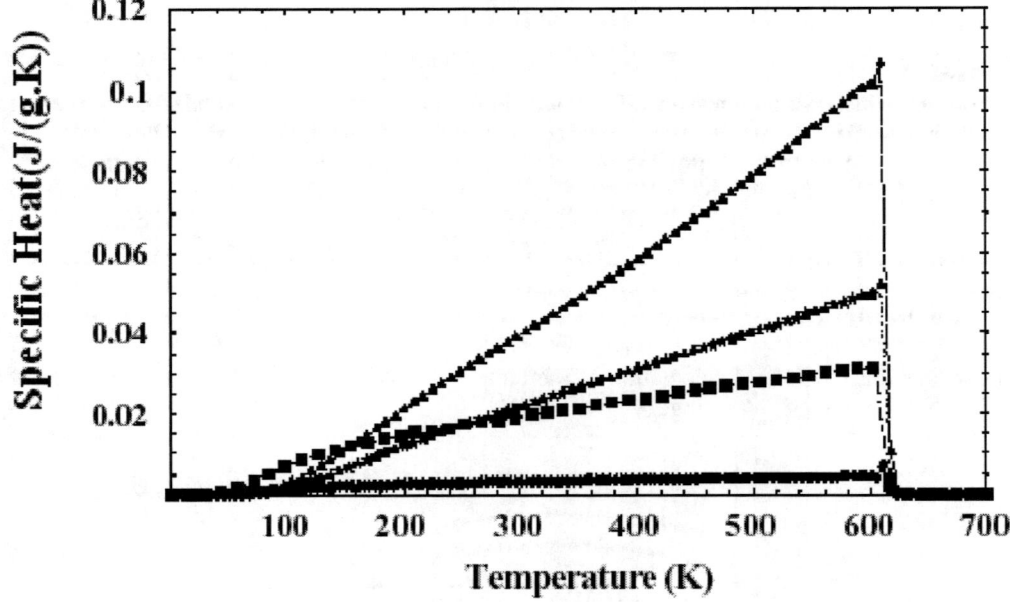

FIGURE 7. Temperature dependence of the subnetwork and inter-subnetwork contributions to the zero-field specific heat for an amorphous $Gd_{0.42}Fe_{0.58}$ alloy : ♦ C_{11}, ▲ C_{22} , ∗ C_{12} , ■ C_{21}, (see text)

CONCLUSIONS

1- We have used mean-field analysis to study the magnetic properties of crystalline $RFe_{10}V_2$ and R-Fe amorphous systems where R=Y, Lu, Nd, Gd, Tb, Ho, Er and Tm in the former system and Er and Gd in the latter.

2- The temperature dependence of the magnetization up to T_c showed that the magnetic order in the crystalline alloys with R= Y, Lu and Nd is ferromagnetic but ferrimagnetic in the others. Compensation points were found in the ferromagnetic Er-Fe alloys with x=0.215 and 0.28 only.

3- The magnetic specific heat in the systems studied show a temperature and field dependence

that is characteristic of a second order phase transition.

4- The magnetic entropy increased with temperature up to T_c above which it saturate to a value proportional to the logarithm of the angular momentum multiplicity. The magnetic field reduces the entropy in particular at temperatures above T_c .

5- The largest entropies in the $RFe_{10}V_2$ system are found for R=Ho and Dy

6- The Fe-Fe sublattice has the largest contribution to the magnetic entropy and to the discontinuity in the specific heat at T_c.

7- Our calculation is in fair agreement with available experimental data.

REFERENCES

1. Z.-S. Liu, Mater. Lett.**58**, 3111(2004)
2. Z.-S. Liu, M. Divis and V. Sechovsky, Physica B **367**, 48(2005)
3. S. H. Aly, J.Magn.Magn.Mater. **232**,168(2001)
4. J. H. Hagmusa , E. Biik, F. R. de Boer and K. H. J. Buschow , J. Alloys. Comp. **198**, 77 (2000).
5. F.Hellman, E. N. Abarra and A. L. Shapiro, Phys.Rev. B 58, 5672(1998)
6. F. R. de Boer, Huang Ying-Kai, D. B. de Mooij and K.H. J. Buschow, J. Less- Common. Met. **135**, 199 (1987).
7. W. R. Zhao and J. Y. Ping, J. Mater. Science&Tech.**14,** 447 (1998).
8. Z. Drzazga, A. Winiarska, D. Echert, M. Wolt. J. Szade and K. H. Muller, J. Magn. Magn Mater. **182**, 225(1998).
9. Huang Rui-Wang, Zhang Zhong-Wa, Hu Bin , Kang Jian-Dong, Zhang Zhi-Dong, X. K. Sun and Y.C.Chuang, J. Magn. Magn. Mater. **119**, 180(1993)
10. S. Chikazumi, "*Physics of Ferromagnetism*", 2 nd ed. (Clarendonn Press, Oxford 1997).
11. E. P. Wohlfarth *"Ferromagnetic Materials"*, ed (E. P. Wohlfarth),Vol.1, Ch.4 (North. Holland publishing company, Amsterdam 1980).
12. S.Khnelevskyi and P.Mohn, J.Phys.Condens.Matter.**12**, 9453 (2000).
13. Y. C. Chuang, D. Zhang, T. Zhao, Z. D. Zhang, W. Liu, X. G. Zhao, X. K. Sun and F. R. de Boer, J.Alloys. Comp. **221**, 60 (1995).
14. F. Luis, P. Infante , J. Bartolome , R. Burriel , C. Pique , R. Ibarra and K. H. J. Buschow , J. Magn. Magn. Mater.**140**, 1045 (1995).
15. H. Wada, S. Tomekawa and M. Shiga, J. Magn. Magn. Mater. **196**, 689 (1999)
16. G. El-Alfi, " A Study of the Magnetic and Magneto-Thermal Properties of some Amorphous and Crystalline Compounds", M.Sc. Thesis, Mansoura University, 2003.

Ionization Waves in ZNS

F.M.Abou El-Ela

Department of Physics, Faculty of Girls, Ain Shams University,
Heliopolis, Cairo, Egypt

Absract. Ionization waves, analogous to positive column of gaseous plasma, are shown to occur in ZnS. This is demonstrated with a model in which a single type of carrier (electron) impact ionizes a deep level trap in ZnS. The no-local nature of impact ionization is described using lucky drift theory.

Ionization waves were discovered by using our numerical simulation model and were seen to arise as a consequence of the non-local nature of the impact ionization process. This is contrast to the prediction of the local-impact ionization theory. Under certain conditions space-charge striations are produced and a NDR occurs for some film thicknesses

INTRODUCTION

The importance of non-local effects in impact ionization has been long recognized, particularly in connection with breakdown in gases, where Crook's dark space and glow striation provides classic examples. The impact ionization at point z in an electric field $F(z)$ is a consequence of electron drifting down the potential gradient upstream of the point z, say z_0 picking up an energy E given by

$$E = \int_{z_0}^{z} e \ F(z) dz \tag{1}$$

where $E > E_I$, E_I the threshold energy (as determined from conservation of momentum and energy). The ionization rate, is therefore a function of the carrier density $n(z)$, and the average field, $\overline{F}(z, z_0)$ Where carrier density and field vary with distance non-local effects may become of crucial importance

The impact ionization rates in Semiconductors were calculated numerically by Baraff [1] with assumption of local field, uniform carrier density and parabolic energy band. The dark space effect on Baraff's theory was included by Okuto and Crowell [2] .A Monte Carlo simulation was used by Shichijo and Hess [3] to include the full band structure in the impact ionization calculation. The lucky-drift approach was offered as a simple method to calculate the impact ionization (Ridley [4]; Burt [5]), and recently this method was modified to include soft threshold (low probability of impact ionization at threshold com pared with that for relaxing energy) (Ridley [6] ; Marsland [7]). Non local effects in junction were discussed using lucky-drift theory by Childs [8].

In the present paper we use the soft-threshold lucky-drift approach to include non-local effects and dark space. Applied to thin film of ZnS, the model shows space-charge anomalies, and under certain circumstances a spatial oscillation of the field occurs, which leads to current—controlled negative differential resistance (NDR) for some film thicknesses.

MODEL SYSTEM AND THE SOFT LUCKY-DRIFT FORMULA

In our model system we have considered a thin film of ZnS where there is a high concentration ($10^{18} \ cm^{-3}$) of both deep level and shallow level defects, with the deep-level half filled at a lattice temperature of 300 K. A current of electrons is injected at the cathode by tunnelling. The barrier height at the cathode is assumed to be 1.0 e.V and the deep-level trap is 3 e.V below the C.B while the shallow

CP888, *Modern Trends in Physics Research,*
Second International Conference on Modern Trends in Physics Research—MTPR-06,
edited by L. El Nadi
© 2007 American Institute of Physics 978-0-7354-0354-9/07/$23.00

level is 0.15 e.V below the C.B., and we have assumed that holes are not injected at all .The impact ionization of the trap is modeled by soft threshold lucky-drift theory. Non-local effect and dark space are incorporated in a straight forward manner (Ridley and El-Ela, [9]). Due to lack of information concerning, mean free path and how soft the thresholds are, we have had to rely on extrapolation from the better known III-V S.C, which agreed reasonably well with our transport simulation and with existing experiments (Thompson and Allen [10], F.M. El—Ela [11]) .

The dark-space width z_d is defined by .

$$z_d = \frac{E_I}{e\overline{F}} \tag{2}$$

where $\overline{F}$ is defined as the average field, the impact ionization coefficient (α_t) zero for $z < z_d$, then increases as E_{max} increases, where

$$E_{max} = e\overline{F}z \qquad z > z_d \tag{3}$$

The rates of impact ionization in "local" theory were determined by the local field and drift velocity. The "non-local" model takes into account that the impacting particle achieves the necessary energy not instantaneously but over a distance as it drifts down the potential gradient. The energy is determined by the average field, but retain the local relation between field and drift velocity. This approximation will be valid over a distance of order of the energy relaxation length.

In order to include soft threshold energy effect, we considered the lucky drift approach. If $P(E)$ is the probability of an electron starting from zero energy and reaching energy E then:

$$P(E) = \exp\left(-\int_0^t \frac{dt}{\tau_i(E)}\right) = \exp\left(-\int_0^E \frac{dE'}{e\varepsilon v \tau_i}\right) \quad (\varepsilon > 0) \tag{4}$$

where v is the velocity (group velocity for ballistic carriers and drift velocity for lucky-drift carriers)

In this study, we have used the Keldysh ionization relaxation rate [12], to be:

$$\tau_i^{-1} = W_{ph}(E_I) P \left(\frac{E - E_I}{E_I}\right)^2 \qquad E > E_I \tag{5}$$

Where $W_{ph}(E_I)$ represents the scattering rate at threshold, usually determined by phonons, and in which P is a numerical factor. It measures the hardness of the threshold.

The probability for impact ionizing becomes

$$P_I = \int_{E_I}^\infty P(E) \exp\left[-\int_{E_I}^E \frac{dE'}{e\varepsilon v \tau_i}\right] \frac{dE}{e\varepsilon v \tau_i} \tag{6}$$

The ionization coefficient is then written as:

$$\alpha = \int_{E_I}^\infty \frac{e\varepsilon}{E} P(E) \exp\left[-\int_{E_I}^E \frac{dE'}{e\varepsilon v \tau_i}\right] \frac{dE}{e\varepsilon v \tau_i} \tag{7}$$

The total ionization coefficient consists of a ballistic contribution and a lucky-drift contribution. Equation (7) depends on a factor p , which is defined by

$$p = \frac{\tau_E}{\tau_i} \tag{8}$$

Equation (8) can be written as

$$p = p_0 \sqrt{\frac{E}{E_I}} \left(\frac{E - E_I}{E_I}\right)^2 \tag{9}$$

A better parameter to measure hardness p_0, becomes known as Keldysh factor, defined by:

$$p_0 = W_{ph}(E_I)\tau_E(E_I)P = \frac{E_I(2n(\omega)+1)}{\hbar\omega}P \qquad (10)$$

where $\tau_E(E_I)$ is equal to the energy-relaxation time at threshold energy (E_I) and $n(\omega)$ is Bose Einstein number at the optical phonons angular frequency ω.

The rate of change of trap concentration, neglecting thermal generation and Auger-trapping, is given by (Fig. 1)

$$\frac{dn_t}{dt} = c_t n(N_t - n_t) - e_t n_t n' \qquad (11)$$

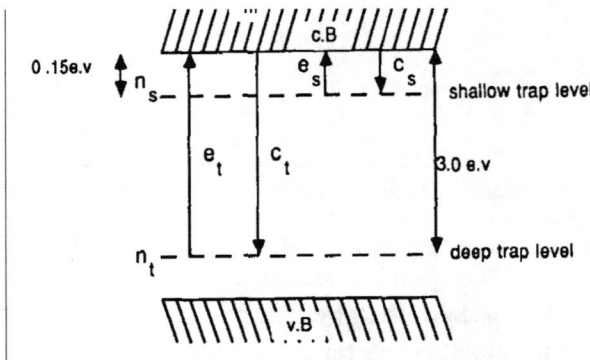

FIGURE (1). Emission and capture processes involving deep and shallow traps in ZnS

we define n' as the cathode electron density, n is the local density, e_t is volume ionization coefficient related to the ionization rate (α_t) via

$$e_t = \frac{v_d}{n_t}\alpha_t \qquad (12)$$

(α_t) is given by soft threshold lucky-drift theory, c_t is volume capture coefficient, v_d is the drift velocity and (N_t - n_t) represent the vacant trap concentration.

We used an empirical drift velocity formula which roughly fit the result of Monte Carlo simulation for ZnS [10,12] given by

$$v_d = v_s[1 - (1 - 3.25 \times 10^{-5} F)\ \exp(-7.5 \times 10^{-6})] \qquad (13)$$

where $v_s = 5 \times 10^6$ cm s^{-1}, F is the local field

The current density, J is given by our choice of the cathode field, hence

$$n' = \frac{J}{eV_s} \qquad (14)$$

Current continuity demands that for all z

$$n = \frac{J}{eV_d} \qquad (15)$$

The variation of field in one dimension given by

$$\frac{dF}{dZ} = \frac{-e}{\varepsilon}(N_e - N_{e0}) \qquad (16)$$

where $N_e = n + n_t + n_s$, $N_{e0} = n_{t0}$

N_{e0} is defined as the electron density required for neutrality (and we assume $n_s = 0$)

For given current the procedure of the calculation was to obtain (α_t) from lucky- drift theory at a given z, calculate (n_t) using the value at $z - \Delta z$,to obtain N_e and F , calculate the local and average field, obtain n from current continuity, repeat after advancing z to $z + \Delta z$.

RESULTS AND DISCUSSION

The material parameters of ZnS used in our calculation are shown in table 1, which represent the best known values presently available. Let us start first with only a deep level trap, which is half occupied by elec trons to begin with.   Fig.(2) shows the local and average field profiles, when the concentration of trap is 10^{18} cm^{-3} and the current density is 1 A cm^{-2}. In the cathode dark space capture of injected electrons causes the field to rise. When impact ionization begins electrons are ejected and the field drops sharply. When the field drops and ionization weakens, capture can build up a large nega tive charge which causes a rapid increase of field to the point where ionization can again become significant. As a consequence, the field profile exhibits spa tial oscillations (field striations). The variations of impact ionization coef ficient for the deep level, and the trapped electron density inside the sample are shown in fig.(3) and fig.(4) respectively, for the same current. The corresponding variation of free electron density is shown in fig.(5) for different current densities .

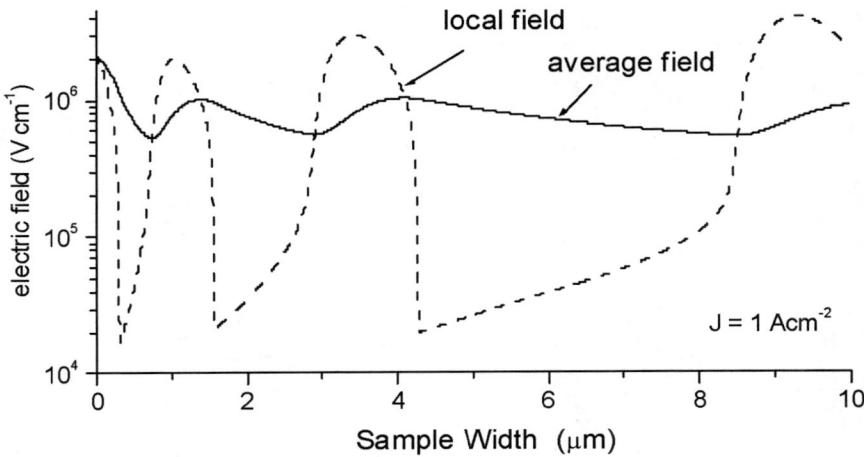

FIGURE (2). Calculated local field and average field profiles in the "non local" model.

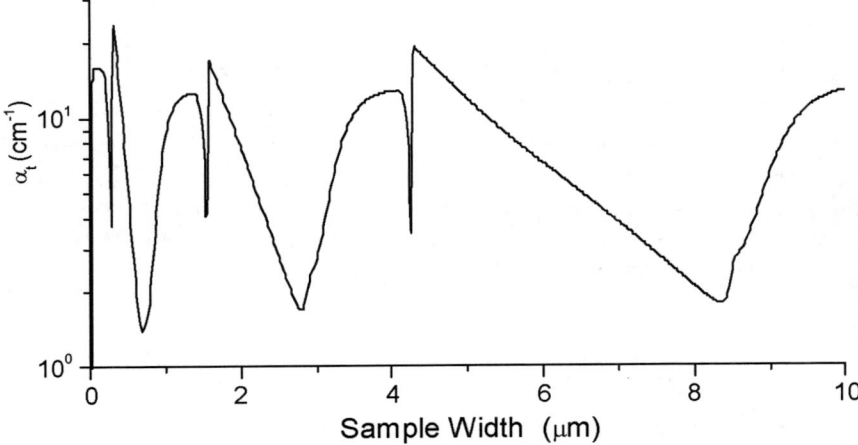

FIGURE (3). Variation of ionization coefficient with distance, as in fig. 2.

89

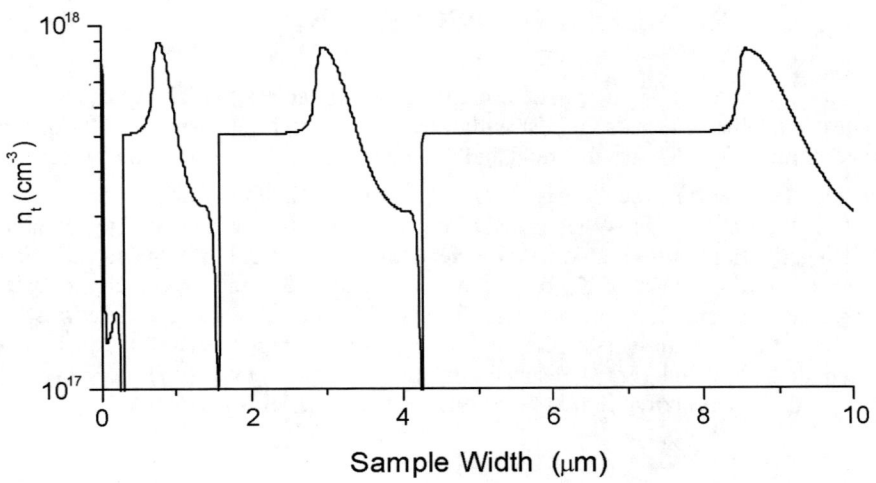

FIGURE (4). Variation of of trapped electron density n_t with distance, as in fig. 2.

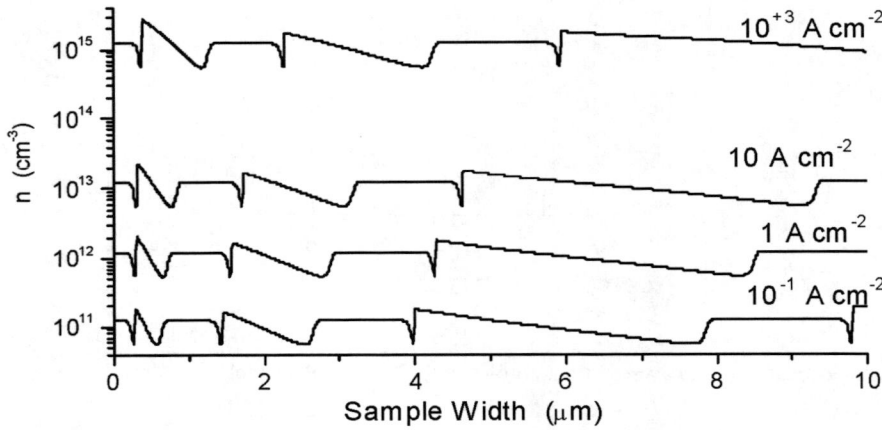

FIGURE (5). Variation of of free- electron density profile n with distance, as in fig. 2.

Figure (6) shows a comparison between the average field in the non-local model and the local field in the local model, where there is $10^{18} cm^{-3}$ trap half filled for a current density of $1 \, A \, cm^{-2}$. The main features are the field striations in the non-local model, which become more and more widely separated with distance from the cathode. In the local model, on the other hand, the field merely drops to a value sufficient to sustain the ionization required to support the steady current

TABLE1. Paramterters for ZnS Model

$\varepsilon_s / \varepsilon_0$	$\hbar\omega$	$\lambda(0)$	E_I	E_s	$c_t = c_s$
8.32	33 meV	64 A°	3.0 eV	0.15 eV	$10^{-10} cm^3 s^{-1}$

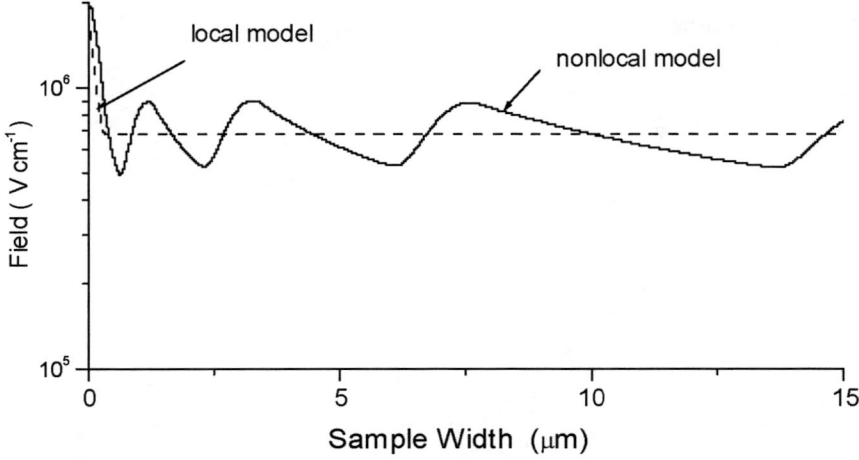

FIGURE (6). Variation of of local field, in the "local" model and the average field in the "non-local" model with distance, as in fig. 2.

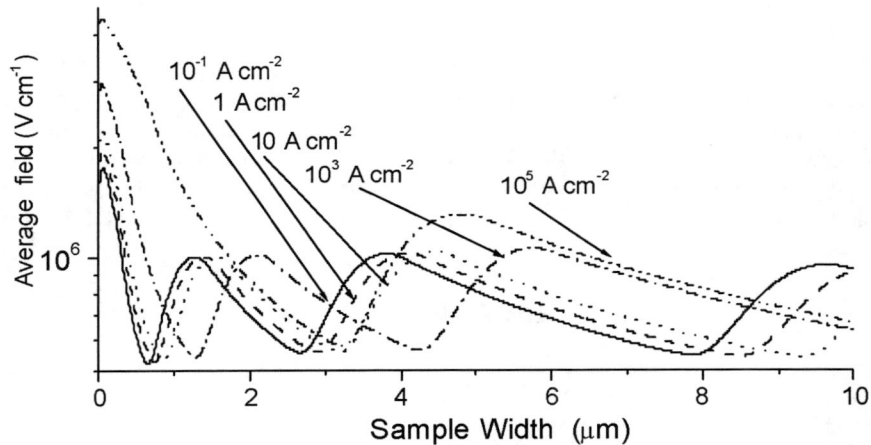

FIGURE (7). Variation of the average field vesus distance, for current densitits as shown in figure

A major result is the appearance of NDR for certain film thicknesses. This represents a novel NDR mechanism which requires the presence only of a single partially filled trap, and so it should be widely observable. To test if the NDR survives with introduction of a second trap species, we ran the model with addition on a shallow trap, 0.15 e.v below the conduction band, initially empty. In fig.(7) we present the variation of the average field profile versus sample length for a number of current densities. The phase of the oscillations is a function of current density, trapped-electron concentration and the initial occupation of the deep trap. It was found that the NDR still existed in the two-trap model as shown in fig.(8). The system exhibits a positive resistance for thickness 1.5μ, while for other thickness e.g $0.9 - 1.2\mu$ the system exhibits negative differential resistance .

We should emphasize that we are only demonstrating qualitative phenomena in these model. The actual voltage, thickness, current, etc. depend upon the magnitude of parameters chosen, such as the barrier height, capture and ionization cross-section, and the drift velocity as a function of fields, which we known only very imperfectly.

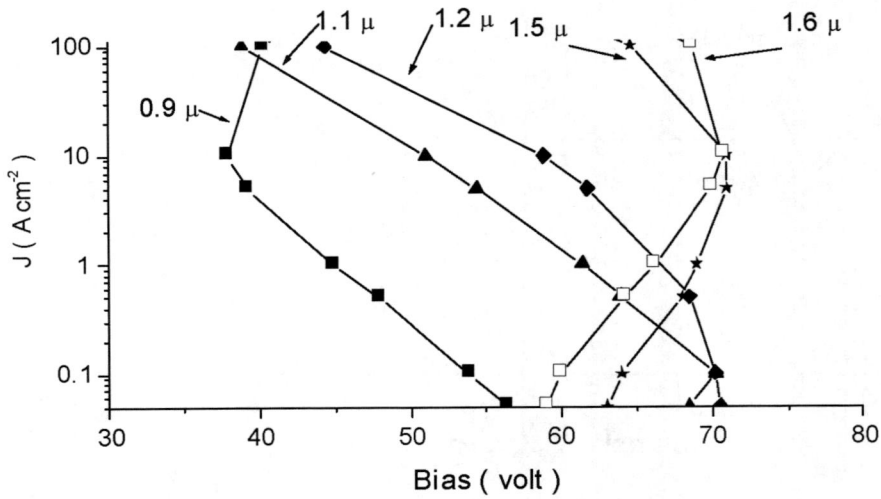

FIGURE (8). Current –voltage characterstics for film thicknesses as shown in the figure.

CONCLUSION

We summarize our main result as follows; for deep level concentration of 10^{18} cm^{-3} and more, non-local impact ionization effects reproduce the dark space and striations familiar in gas discharges. The spatial oscillations of the fields means that I-V characteristic is dependent on film thickness. NDR appears only for certain thicknesses and not for others. These effect depends only on the presence of a partially occupied trap. Introduction of an empty shallow trap does not materially affect these phenomena .

REFERENCES

1. G. A. Baraff, Phys. Rev. **128**, 2507(1962).
2. Y. Okuto and C.R Crowell phys. Rev. **102**,369(1974).
3. H. Shichijo and K.Hess, Phys. Rev. B **23**, 8, 4197(1981).
4. B. K. Ridley, J. Phys. C.Solid St. Phys.**16**, 3373(1983), B. K. Ridley, J. Phys.C.Solid St. Phys.**16**, 4733(1983).
5. M. G. Burt and S. Mckenzie, Physica B **134**, 247(1985).
6. B. K. Ridley, Semicond. Sci. Tecknol. **2**, 116(1987).
7. J. S. Marsland, Solid State Electron **30**, 125(1987).
8. P.A. Childs. J. Phys. C.Solid St. Phys. **20**, L243 (1987).
9. B. K Ridley and F. M. Abou El –Ela, J. Phys. Condens. Matter **1**, 7021(1989).
10. T.D. Thompson and J.W. Allen , J. Phys. C.Solid St. Phys., **20**, L 4GS. (1987) .
11. F.M.Abou El-Ela, Pramana – J. Phys. **63**, 1089 (2004), F.M.Abou El-Ela, First Conference of Modern Trends in Physics Research MTPR-04, American Institute of Physic (AIP) conference proceedings, **748**,134 (2004).
12. L. V. Keldysh, Sov Phys. JEPT **10**,509(1960).

Effect of k.p Band Structure Models on the Electron Lucky Drift in Zinc-Blende-Type Semiconductor

Samy. S. Montasser [a], F. M. Abou El-Ela [b] and Taroob. A. Abdel-Baset [c]

[a] *Department of Physics, Faculty of Science, Cairo University*
[b] *Department of Physics, Faculty of Girls, Ain Shams University*
[c] *Department of Physics, Faculty of Science, Fayoum University*

Abstract. We have studied the various k.p models including Kane's model, the 8x8 matrix model and 14x14 matrix models. We include the nonparabolic terms in the energy formula using each of these models where we took into account the fourth and sixth order terms in the energy–wavevector expansion. We then calculate the electron impact ionization threshold energy in terms of these nonparabolic parameters for some of zinc-blende-type semiconductor. Using the impact ionization energy, we extract the lucky drift ionization parameters. Our calculation shows that the 14x14 matrix model gives the closest result to that of the ab initio calculation of impact ionization threshold energy and that of Monte Carlo empirical pesudpotintial calculation. The advantage of our work is that by using simple derivation and computation, we manage to get a closer result to that obtained by heavy computational resources.

INTRODUCTION

The performances of avalanche photodiodes depend critically on the ratio of the electron and hole ionization coefficients. Carrier multiplication through impact ionization is the basic mechanism in operation of avalanche photodiodes. The impact ionization of semiconductors offers a mechanism for carrier multiplication, which can be used to amplify small signals, and is an essential mechanism in the operation of many semiconductor devices.

In the impact-ionization, free carriers (electrons and holes) are accelerated by a high electric field until they gain sufficient energy to promote an electron from the valance band into the conduction band. The electric fields required to observe impact ionization depend on the band gap of the material[1]. The first simple theoretical calculation of the impact ionization coefficient was done by, Wolff[2] where he applied the gas discharge theory to solve Boltzman equation. On the other hand, Shockley[3] explained that few electrons are lucky to avoid several collisions and reach the ionization critical energy. The time independent Boltzmann equation has been solved by Baraff[4], his numerical results agreed with the Shockley model at low field while converged to the Wolff results at high field. The lucky drift model of impact ionization, was first proposed by Ridley[5] and was developed by Burt and McKenzie[6,7]. Mont Carlo simulations of electron impact ionization coefficient in silicon were presented by Tang and Hess[8], Brennan and Hess[9] and Sano et al[10]. A fully *ab initio* calculation of impact ionization rates in GaAs within the density functional theory framework, using a screened-exchange formalism and the highly precise all-electron full-potential linearized augmented plane wave method are presented by Asahi et al[11].

The threshold energy of impact ionization depend on the nonparabolicity of semiconductors In most of the above mentioned survey the nonparabolicity was either

CP888, *Modern Trends in Physics Research,*
Second International Conference on Modern Trends in Physics Research—MTPR-06,
edited by L. El Nadi
© 2007 American Institute of Physics 978-0-7354-0354-9/07/$23.00

taken into account through Kane's model or through a full band numerical models. Up to our knowledge, non of the previous studies took the nonparabolicity through the more advanced models which include 8x8 or 14x14 matrices or higher.

LUCKY DRIFT WITH A SOFT THRESHOLD

A soft threshold can be incorporated into lucky drift theory in a straightforward manner[5-7]. The total probability of impact ionization is

$$P_I = \int_{E_I}^{\infty} P(E) \exp\left[-\int_{E_I}^{E} \frac{dE'}{e\varepsilon v \tau_i} \right] \frac{dE}{e\varepsilon v \tau_i} \tag{1}$$

where v is the appropriate velocity , and $\tau_i = W_I^{-1}$,

The ionization coefficient is then

$$\alpha = \int_{E_I}^{\infty} \frac{e\varepsilon}{E} P(E) \exp\left[-\int_{E_I}^{E_i} \frac{dE'}{e\varepsilon v \tau_i} \right] \frac{dE}{e\varepsilon v \tau_i} \tag{2}$$

The total ionization coefficient consists of a ballistic contribution and a lucky-drift contribution in the ballistic contribution. $v = v_g$, where v_g is the group velocity, and hence

$$e \varepsilon v \tau_i = \frac{e\varepsilon \lambda}{r P} \cdot \frac{E}{E_I} \tag{3}$$

While in lucky drift $v = v_d$, where v_d is the drift velocity, and hence

$$e \varepsilon v \tau_i = \frac{(e\varepsilon \lambda)^2}{2 r P E_I} \tag{4}$$

Where ε is the electric field, λ is the mean free path between scattering events, and

$$r = \frac{\hbar\omega}{E_I(2n(\omega)+1)} \qquad , \quad P = \frac{\tau_E}{\tau_i} \tag{5}$$

P(E) for ballistic flight and lucky drift are given by

$$P = p_0 \sqrt{\frac{E}{E_I}} \left(\frac{E - E_I}{E_I}\right)^2 \tag{6}$$

A better parameter to measure hardness p_0, becomes known as Keldysh factor, defined by:

$$p_0 = W_{ph}(E_I)\tau_E(E_I)P = \frac{E_I(2n(w)+1)}{\hbar w}P \qquad (7)$$

where $\tau_E(E_I)$ is equal to the energy-relaxation time at threshold energy (E_I). And $n(\omega)$ is the Bose – Einstein number and $\hbar\omega$ is the phonon energy, and W_I is the energy dependence of impact ionization scattering rate and given by :

$$W_I = W_{ph}(E_I)P\left(\frac{E-E_I}{E_I}\right)^2 \qquad E > E_I \qquad (8)$$

Where E_I is the threshold energy determined by momentum and energy conservation, and $W_{ph}(E_I)$ represents the scattering rate at threshold, usually determined by phonons, and P is a numerical factor which measures the hardness of the threshold.

A fit of soft threshold lucky drift theory to experiment has been made using the least–square algorithm E04FDF, for the cases of electrons in GaAs, InP, InAs and InSb.

THE NONPARAPOLICTY EFFECT ON THE THRESHOLD ENERGY

The threshold energy required for impact ionization depends on the band structure of semiconductors[12]. For the case of parabolic conduction bands, it can be written as

$$E_i = E_P = E_g\left(\frac{1+2\mu}{1+\mu}\right) \qquad (9)$$

where E_g is the energy gap, and $\mu = \dfrac{m_c^*}{m_v^*}$

where m_c^* and m_v^* are the effective mass of the conduction and valance bands respectivly

The nonparabolicity was taken into account up to the fourth order by Ridley[13]. He introduced the relation

$$E_{th1} = E_i = E_P\left[1+\left(\frac{\alpha\mu}{1+\mu}\right)E_P\right] \qquad (10)$$

where he used the energy nonparabolicity formula,

$$\frac{\hbar^2 k^2}{2m_c^*} = E(1+\alpha E) \qquad (11)$$

we have extended equation (11) to include the sixth order term to be

$$\frac{\hbar^2 k^2}{2m_c^*} = E(1+\alpha E + \beta E^2) \qquad (12)$$

where α and β are related to the coefficients of the energy-wavevectors expansion

$$E = a_2 k^2 + a_4 k^4 + a_6 k^6 \tag{13}$$

by

$$\alpha = -\frac{a_4}{a_2^2} \tag{14}$$

$$\beta = 2\alpha^2 - \frac{a_6}{a_2^3} \tag{15}$$

We have derived a formula for E_i in terms of the nonparabolic coefficients α and β which have the form,

$$E_{th2} = E_i = E_P \left[1 + \left(\frac{\alpha\mu}{1+\mu} \right) E_P + \left(\frac{\mu}{1+\mu} \beta + 2 \left(\frac{\alpha\mu}{1+\mu} \right)^2 \right) E_P^2 \right] \tag{16}$$

The values of α and β depend on the model of the derivation. We will now discuss these models.

1. Kane's model

Kane[14] has derived an expression for the energy of the conduction band where he assumed that the spin orbit splitting $\Delta \gg E_g$. In terms of his results we have

$$\alpha = \frac{1}{E_g} \left(1 - \frac{m_c^*}{m_o} \right)^2 \tag{17}$$

$$\beta = -\frac{2}{E_g^2} \frac{m_c^*}{m_o} \left(1 - \frac{m_c^*}{m_o} \right)^3 \tag{18}$$

where m_o is the free electron mass.
In the more general case, equation (17) and (18) become

$$\alpha = \frac{1}{E_g} (1 - M_S)^2 \left(1 - \frac{(E_g * \Delta)}{3*(E_g + (2/3)\Delta)(E_g + \Delta)} \right) \tag{19}$$

$$\beta = -\frac{2}{E_g^2} M_S (1 - M_S)^3 \left(1 - \frac{2\Delta E_g}{(\Delta + E_g)(2\Delta + 3E_g)} + \frac{\Delta^2 E_g}{(\Delta + E_g)(2\Delta + 3E_g)M_S} \frac{\Delta + E_g(1 + M_S)}{(\Delta + E_g)(2\Delta + 3E_g)} \right) \tag{20}$$

where $M_S = \dfrac{m_c^*}{m_o}$ \hfill (21)

2. The 8x8 and 14x14 models

In the 8x8 model the interaction between the lower conduction band Γ_{6c}, the upper valence bands Γ_{8v} and the split off band Γ_{7v} were treated exactly while the interaction between these bands and the higher

conduction bands Γ_{8c} and Γ_{7c} were treated by perturbation theory[15]. The 14x14 model takes all the seven bands into consideration in an exact way [16].

RESULTS

In the previous section, we study the different between the three band structure models to calculate the threshold energy for the electron lucky drift of impact ionization in zinc-blend materials. Table (I) shows the comparison between these model in case of GaAs, InP, GaSb, InSb.

TABLE 1. The calculated of the threshold energy for some zinc-blende-type semiconductors.

	Kane's		Kane's delta		8x8		14x14	
	E_{th1}	E_{th2}	E_{th1}	E_{th2}	E_{th1}	E_{th2}	E_{th1}	E_{th2}
GaAs	1.78289	1.79948	1.77272	1.78608	1.79743	1.89705	1.79935	1.84705
GaSb	0.867651	0.875137	0.8618	0.862918	0.862802	0.867112	0.863512	0.868822
InAs	0.39271	0.393893	0.3907	0.390756	0.391774	0.392827	0.391788	0.392557
InP	1.62981	1.63607	1.62626	1.63204	1.66602	1.72595	1.66623	1.7265

Where E_{th1} and E_{th2} are given by equation (10) and (16)
From these different values of the threshold energy we can give the fitting parameter to study the impact ionization coefficients.

Tables II demonstrate the effect of the different values of the threshold energy on the values of the optimum mean free path $\lambda(0)$ and the optimum keldysh factor P_o

TABLE II. The effect of the different values of the threshold energy on the values of the optimum mean free path $\lambda(0)$ and the optimum keldysh factor P_o for some zinc-blende-type semiconductors.

a-GaAs

	E_I (eV)		$\lambda(0)$ $A°$		P_o	
	E_{th1}	E_{th2}	E_{th1}	E_{th2}	E_{th1}	E_{th2}
Kane	1.78289	1.79948	109.72	110.6	3.639	3.712
Kane (Δ)	1.77272	1.78608	109.45	109.8	3.595	3.653
8x8	1.79873	1.84553	110.14	111.37	3.71	3.92
14x14	1.79935	1.84705	110.16	111.40	3.71	3.93

b-InP

	E_I (eV)		$\lambda(0)$ $A°$		P_o	
	E_{th1}	E_{th2}	E_{th1}	E_{th2}	E_{th1}	E_{th2}
Kane	1.6298	1.63607	74.98	75.114	1.1375	1.144
Kane (Δ)	1.62626	1.63204	74.898	75.03	1.1337	1.1399
8x8	1.66602	1.72595	75.77	77.05	1.1758	1.241
14x14	1.66623	1.7265	75.77	77.06	1.176	1.242

c-GaSb

	E_I (eV)		$\lambda(0)$ $A°$		P_o	
	E_{th1}	E_{th2}	E_{th1}	E_{th2}	E_{th1}	E_{th2}
Kane	0.86765	0.87514	174.11	174.69	463.36	499.05
Kane (Δ)	0.8618	0.86292	173.67	173.76	437.33	441.91
8x8	0.86280	0.86711	173.73	174.08	442.88	460.19
14x14	0.86351	0.86882	173.79	174.21	445.29	468.24

d-InAs

	E_I (eV)		$\lambda(0)$ $A°$		P_o	
	E_{th1}	E_{th2}	E_{th1}	E_{th2}	E_{th1}	E_{th2}
Kane	0.3927	0.394	640.39	641.1	2.76	2.79
Kane (Δ)	0.3907	0.39075	640.67	640.67	2.74	2.74
8x8	0.392	0.3928	639.87	640.36	2.75	2.76
14x14	0.392	0.3926	639.87	640.4	2.75	2.758

Lucky drift ionization rate has the advantages of great simplicity and less intensive computations compared to detailed Monte Carlo simulation. The experimental data, in case of electron in GaAs, InP, GaSb and InAs have been used to fit the soft lucky drift of impact ionization in semiconductors.

The experimental data of GaAs have been taken from the work of Bulman et al[17], for InP from Umebu et al[18] , Kao and Crowell[19] and Cook et al[20], for GaSb from Hildebrand et al[21], and for InAs from Brennan et al[9].

Figure (1) shows the fitting of α and the experimental values for impact ionization coefficients as a function of reciprocal electric field (ε^{-1}).

a-GaAs

b-InP

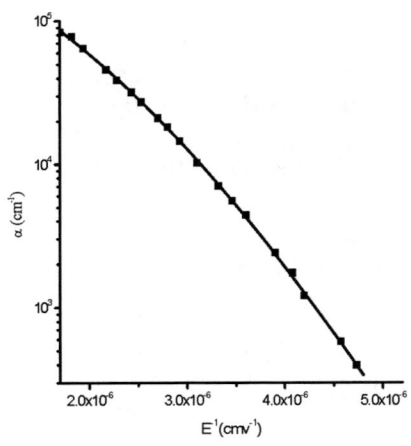

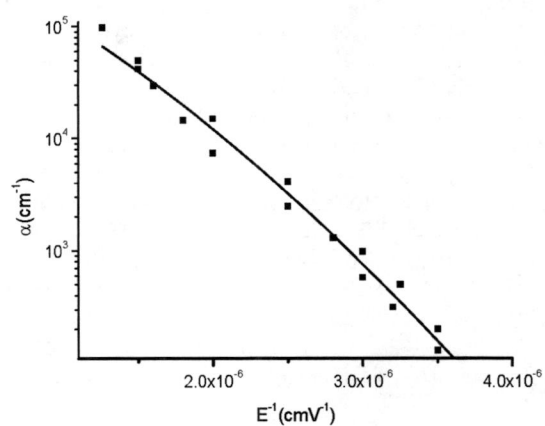

c- GaSb

d-InAs

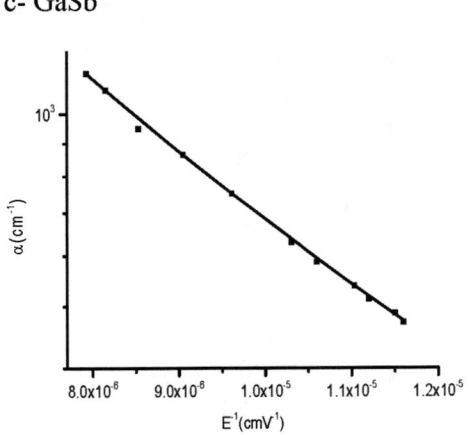

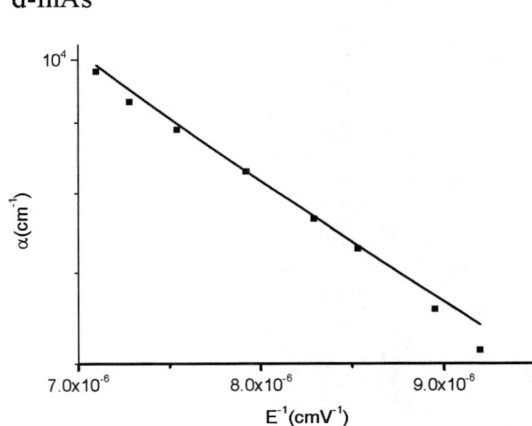

FIGURE 1. Calculated (α) and experimental values for impact ionization coefficients as function of reciprocal electric field (ε^{-1}) for different semiconductors materials. Experimental: electron ■ and The solid curve represents our theoretical prediction.

CONCLUSION

Our calculation shows that the 14x14 matrix model gives the closest result to that of the ab initio calculation of impact ionization threshold energy [11] and that of Monte Carlo empirical pesudpotintial calculation. A fit of soft threshold lucky drift theory using a generalized Keldysh formula to experimental data has been made using the least squares algorithm. It is useful to note that generalized Keldysh formula becomes appropriate for realistic bands and reflects the density of states of energy bands.

REFERENCES

1. F. Capasso, in Semiconductors and Semimetals, edited by R.K Willards and A.C Beer, Acdemic, New York, 22D, 1985, p.1
2. P.A.Wolff, Phys. Rev. **95**, 1415 (1954).
3. W. Shockley, Solid State Electron **2**, 35 (1961).
4. G.A. Baraff, Phys. Rev. **128**, 2507 (1962).
5. B.K. Ridley, J. Phys. C.Solid St. Phys.**16**, 3373 (1983), B.K.Ridley, J. Phys. C.Solid St. Phys.**16**, 4733 (1983).

6. S. Mckenzie and M.G. Burt, J.Phys.C.Solid St. Phys.**19**, 1959 (1986).

7. M.G. Burt and S. Mckenzie, Physica B **134**, 247 (1985).

8. J.Y. Tang and Karl Hess , J.Appl.phys.**54**, 9 (1983).

9. Kevin Brennan and Karl Hess , Phys. Rev. B **29**, 5581–5590 (1984).

10. Nobuyuki Sano, Takahiro Aoki , and Akira Yoshii , Appl.Phys.Lett. **55**, 14 ,2 (1989).

11. S. Picozzi[*] R. Asahi ,C.B. Geller, A. Continenza ,A. J. Freeman ; Phys. Rev. B **65**,   113206 (2002).

12. B.K.Ridley " Quantum processes in semiconductors " third edition, Oxferd science publications (1993).

13. B.K.Ridley, J. Appl. Phys.**48**, 754 (1977).

14. Kane, E. O. in "Semiconductors and Semimetals" vol. 1, physics of III-V compounds eds. Willardson , R.K. and Beer, A,c. (Aca-demic press, New York 1966) P.75

15. H.R. Trebin, U. Rossler, and R. Ranvaud, Phys. Rev. B **20**,686 (1979).

16. H. Mayer and U. Rossler, Phys. Rev. B **44**, 9048 (1991).

17. G.E. Bulman, V.M.Robbins and G. E. Stillman, IEEE Trans. Electron. Dev. ED-32, 1985 2454.

18. J. Umebu, A.N.M.H. Choudhury and P.N Robson, Appl Phys.Lett. **36**, 302 (1980).

19. C.W.Kao and C.R Crowell, Solid State Electron **23**, 881 (1980).

20. L.W. Cook ,G.E. Bulman and G.E.Stillman, Appl. Phys. Lett. **40**, 589 (1982).

21. Hildebrand, 0., W. Kuebart, and M. H. Pilkuhn, *Appi. Phys. Lett.* **37**, 9 , 801-803 (1980) .

Monte Carlo Simulation of Electron Transport in $Ga_{0.47}In_{0.53}As$

F.M. Abou El-Ela, N.T.Mokhtar

Department of Physics, Faculty of Girls, Ain Shams University, Cairo, Egypt

Absract. Monte Carlo simulation of electron transport in $Ga_{0.53}In_{0.47}As$ has been performed for three valley conduction band model. Scattering Sources include polar optical phonons, non–polar intervalley phonons, non-polar acoustic phonons, charged impurity and random potential alloy. A negative differential mobility is observed at field of about 5.0×10^5 v/m. $Ga_{0.53}In_{0.47}As$ exhibits higher low-field drift mobility and a higher peak electron drift velocity than both GaAs and InP.

INTRODUCTION

Electron transport in III—V semiconductors has been the focus of considerable investigation over the past few years. The determination of the dependence of the electron drift velocity on the applied electric field has been a central goal of such studies [1-4]. Many III—V semiconductors have been found to exhibit negative differential mobility, in which the drift velocity achieves a peak at a certain critical field, beyond which further increase in the applied electric field leads to a reduction in the corresponding drift velocity. The negative differential mobility exhibited by these semiconductors is commonly attributed to intervalley transitions between the lower and upper valleys; the upper valley electrons are heavier and thus slower than their lower valley counterparts.

The electron transport properties of semiconductors have been studied extensively for many years. GaAs, InP, and $Ga_{0.53}In_{0.47}As$ have received a great deal of attention because they have demonstrated great potential for many device applications, including GaAs metal- semiconductor field effect transistors (MESFETs), InP Gunn diodes, and heterojunction transistors such as modulation doped field-effect-transistors (MODFETs) [5-6]. The excellent electrical characteristics of these materials exploit the small Γ valley effective mass and relatively large Γ to L valley separation for example, 0.079 m_0 and 0.61 eV for InP; 0.037 m_0 and 0.55 eV for $Ga_{0.53}In_{0.47}As$, where m_0 is electron rest mass. Very large low-field mobility and peak velocity at 300 K with doping concentration $N_d = 10^{16}$ cm^{-3} have been reported by Hauser et al [7]. Therefore, there has been a considerable interest in $Ga_{0.53}In_{0.47}As$ as ternary semiconductors material in modern electronic microwave devices and opto-electronic applications, where it is possible to tune their physical properties such as lattice constant and band structure. Calculations of velocity-field relation in $Ga_{0.53}In_{0.47}As$ based on Monte Carlo method was reported by Littlejohn et al. [8,11,12].

The aim of the present study is to perform a numerical calculation of steady state of electron transport in $Ga_{0.53}In_{0.47}As$. Ensemble Monte Carlo simulation is used for various doping concentration and lattice temperature. The velocity field relation, the electron heating and the intervalley population are investigated with direct connection to the change of lattice temperature and doping concentration.

CP888, *Modern Trends in Physics Research,*
Second International Conference on Modern Trends in Physics Research—MTPR-06,
edited by L. El Nadi
© 2007 American Institute of Physics 978-0-7354-0354-9/07/$23.00

MONTE CARLO TECHNIQUE

The Monte Carlo method allows the Boltzmann transport equation to be solved using a statistical numerical approach. The transport history of one or more carriers is simulated subject to the action of external forces. The forces on the particles consist of applied field and scattering mechanisms. The technique of the Monte Carlo method lies in the generation sequences of random numbers with specified distribution probabilities used to describe quantities such as scattering events that determine the time between successive collisions of carriers [9-11].

The electron transport in semiconductor requires material parameters, knowledge of energy band structure, lattice temperature and the definition of the applied electric field. The process of simulating the electron motion involves a number of computational steps to determine the duration of free flight, the scattering process and the choice of state after scattering.

The state of an electron is specified by its wave vector, k, which is related to the momentum p, and the electron energy ε in non-parabolic conduction bands by

$$p = \hbar k \tag{1}$$

$$\varepsilon \, (1 + \alpha \, \varepsilon) = \frac{\hbar^2 k^2}{2m^*} \tag{2}$$

Where α is the band non-parabolicity given by,

$$\alpha = \frac{1}{E_g} \, (1 - \frac{m^*}{m_0})^2 \tag{3}$$

where E_g is the energy gap between the conduction band and the valence band, m^* is the electron effective mass in the material and m_0 is the free electron mass.

The motion of the electron in an electric field E is given by

$$k. = k_0 + \frac{qE \, \tau}{\hbar} \tag{4}$$

where q is the electron charge, τ is the free flight time, and k_0 is defined as the wave vector at time $t = 0$.

SCATTERING PROCESSES

The most important scattering sources that determine electron transition in the $Ga_{0.53}In_{0.47}As$ at $300\,K$ are:

1-polar optical phonon scattering
2-acoustic phonon scattering
3-equivalent intervalley phonon scattering
4-non-equivalent intervalley phonon scattering
5-random alloy scattering
6- impurity scattering
7- piezoelectric scattering

The most important phonon scattering rates were taken from references [9-11] . We also included the random alloy scattering according to reference [8,11] which contain band edge derivative instead of the band independent potential well parameter.

The following alloy scattering rate is included :

$$W_{al}(k) = \frac{3\sqrt{2\pi}}{8\hbar^4 N_{al}} \left[\chi(1-\chi)\right] \left[\Delta E\right]^2 \left(m^*\right)^{3/2} \gamma^{1/2}(E)(1+2\alpha E) \qquad (5)$$

where $\gamma(E) = E(1 + \alpha E)$ represents nonparabolic conduction bands and ΔE is the electron affinity difference between two end binaries. N_{al} is the density of alloying sites and χ is the mole fraction of the binary AC in the ternary alloys $A_\chi B_{1-\chi} C$.

Three non-parabolic valley ($\Gamma - L - X$) band structure for bulk $Ga_{0.53}In_{0.47}As$ was considered. The model consists of three types of valley: a central valley at Γ point, four equivalent valleys at L points separated by a large energy gap $E_{gL} = E_g^L - E_g^\Gamma$ from the center of Γ valley, and three equivalent valleys at X point with a large energy gap $E_{gX} = E_g^X - E_g^\Gamma$ above the Γ valley. E_g^Γ, E_g^L and E_g^X are respectively the energy band gap of the Γ, L and X valleys.

The parameters appearing in the phonon scattering rates are chosen with the range described in references [4,8,11].

Figure (1) shows the energy dependence of random alloy scattering rates for $Ga_{0.53}In_{0.47}As$ in the Γ, L and X valleys at 300 K. Equation (5) have been used for these calculations.

Figure (2) depicts the computed scattering rates for phonons, charged impurity, piezoelectric and random alloy in the L valley at 300 K. It is clear that charged impurity scattering rate weakens towards high electron energies, as does piezoelectric scattering rate. Neither of these mechanisms is expected to be a serious source of scattering of hot electrons at 300 K; Also, figure (2) shows how small acoustic phonon scattering rate is. Indeed, the principal source of scattering depends on optical-type phonons either through polar interaction or through intervalley deformation potential mechanism. This is also true for both L-valley and X-valley. A comparison between ionized impurity scattering of the Brooks Herring (B.H) model [12] and the Ridley model (R) [13] showed a large reduction of the scattering rate in the (R) model compared to the (B.H) model, especially at small electron energy at the impurity concentration of 10^{21} m^{-3}.

It is noted that high energy electrons in the upper valley interact with phonons mainly through the deformation potential, since the polar interaction falls off with increasing energy. The scattering rate associated with non-polar optical phonons scattering or intervalley phonons scattering has simple property that is proportional to the density of states.

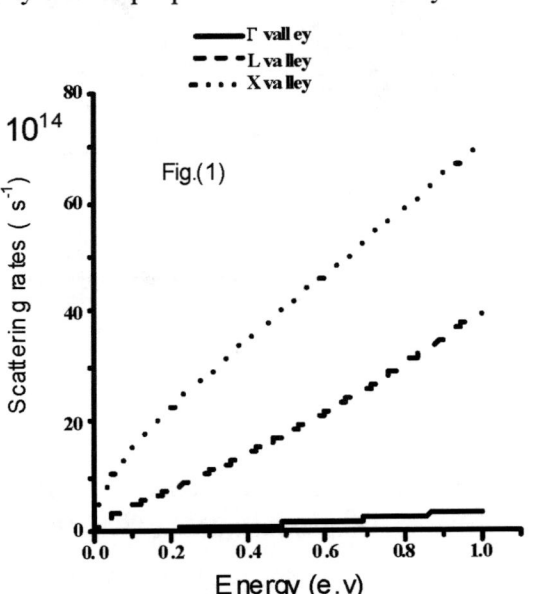

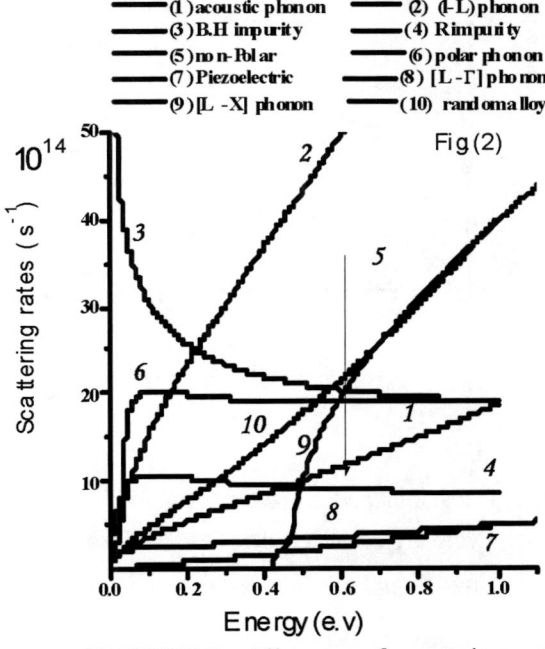

FIGURE 1. Alloy scattering rate in the Γ, L and X valleys at 300 K in $Ga_{0.53}In_{0.47}As$.

FIGURE 2. All types of scattering rate in $Ga_{0.53}In_{0.47}As$ in the L valley at 300K.

RESULTS AND DISCUSSION

GaInAs has been widely regarded as a promising material for high-speed device applications. Its high electron mobility, small effective mass and large intervalley energy separation suggest that it can provide electron velocities significantly higher than those of GaAs and InP. These matrials have non-linear behavior of the velocity field characteristic at fields up to the Gunn Effect threshold. The effects of doping and temperature on the electron velocity in this range of fields are important in device modeling and in understanding electron transport in such semiconductors.

Figure (3) shows our Monte carlo simulation of the way the drift velocity varies with electric field in Γ, and L valley, in addition to the total average in all valleys. The main features are an ohmic region at low fields, followed by a negative differential resistance (NDR) and then tend toward saturation. The drift velocity calculated to be around 10^5 m s^{-1} at lattice temperature 300 K. The NDR occurs as a consequence to intervally electron transfer mainly to L valley. It can be seen that the upper valley, L starts to contribute to the total average velocity at electric fields (equivalence to peak valley field) approximately 5×10^5 V m^{-1}.

Figure (4) shows the calculated total average drift velocity for electrons as a function of electric field strength for various lattice temperatures. It is seen that the peak velocity and the low field mobility rise as a result of temperature cooling; in fact, this is due to the low phonon occupation number. Also it can be seen that a slight dependence of lattice temperatures on the threshold field occurs for onset of negative differential mobility. The threshold field varies from 3×10^5 V m^{-1} at 77 K to 5×10^5 V m^{-1} at 500K. The average energy variation versus electric field is presented in Fig. (5). The mean energy contribution is due to Γ valley and L valley, in addition to the total average electron energy.

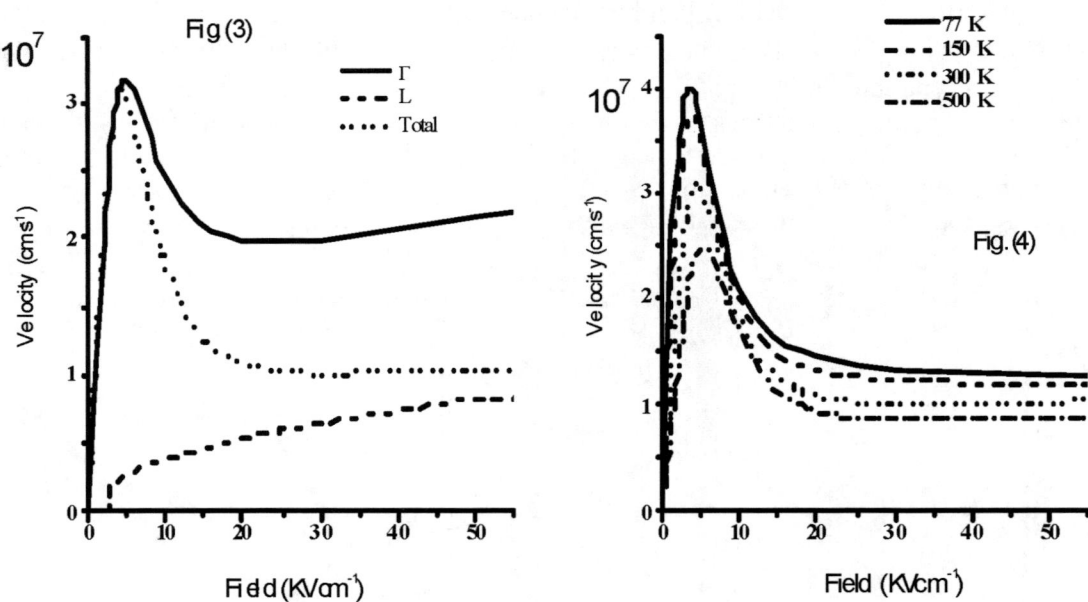

FIGURE 3. The total and average electron drift velocity versus electric field at 300 K in the Γ and L valley of Ga$_{0.53}$In$_{0.47}$As .

FIGURE 4. The total electron drifts velocity versus electric field in Ga$_{0.53}$In$_{0.47}$As at 77 K, 150 K, 300 K and 500 K.

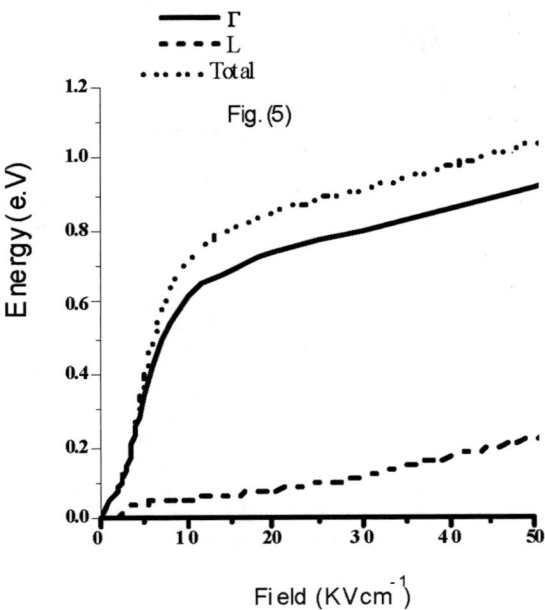

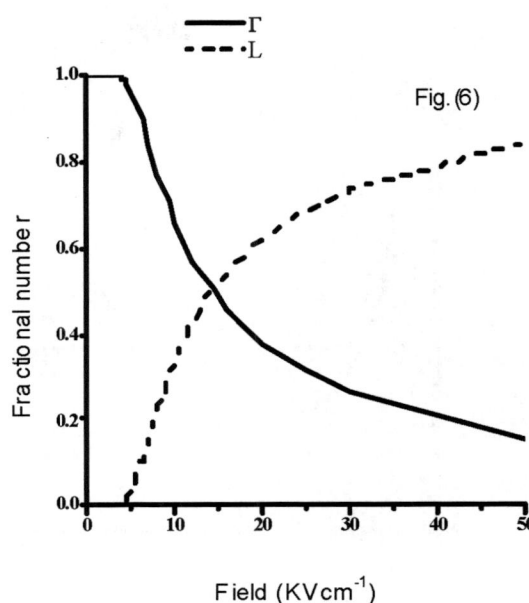

FIGURE 5. The total and average electron energy versus electric field at 300 K in the Γ and L valley of $Ga_{0.53}In_{0.47}As$.

FIGURE 6. The fractional electron number versus electric field at 300 K in the Γ and L valley of $Ga_{0.53}In_{0.47}As$.

Variation of the electron fractional number ($\frac{n_\Gamma}{N}, \frac{n_L}{N}$) in Γ and L valley against electric field is presented in Fig. (6). Where n_Γ and n_L are numbers of electrons in Γ and L valleys, respectively. N is the number of electrons in all valley. It is shown that electrons start to populate the upper valley at electric fields equivalence to peak valley field ($\approx 5 \times 10^5$ V m^{-1}).

Figures (7-8) show the average electron velocity and the average electron energy over all valleys versus time , for different applied electric fields and at the lattice temperatures of 300K. The electrons are first accelerated by the external electric field. As the electron energy increases, the chance of scattering from intervalley and optical phonon mechanisms also rises; consequently, their drift velocity decreases with increasing the electric field. Overshoot velocity occurs only for fields above 5×10^5 V m^{-1}. It is clear from the above figures, that the momentum relaxation time over which electron velocity overshoot occurs and becomes shorter as the field gradually rises. Indeed, this is due to the fact that the electron energy relaxation time becomes smaller as the electric field increases, since electrons gain energies that enable them to transfer faster to L valley

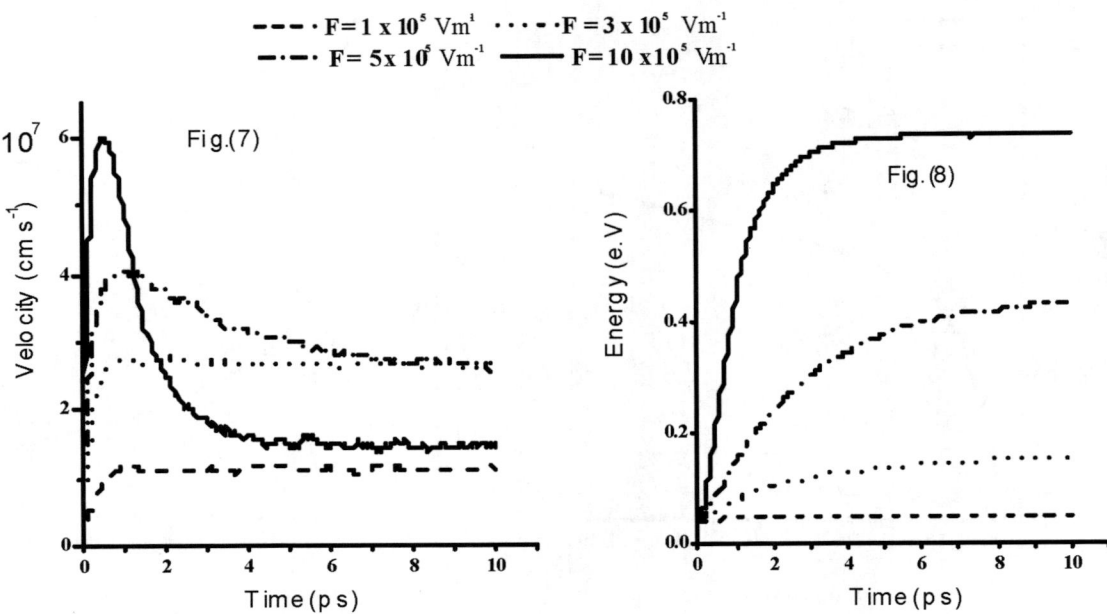

FIGURE 7. The total electron drift velocity versus time at 300 K in $Ga_{0.53}In_{0.47}$ As

FIGURE 8. The total electron energy versus time at 300 K in $Ga_{0.53}In_{0.47}$ As .

The distribution function in Γ valley at 300 K is also shown in Fig. (9) where the electric field varies in the range of $(2-8)\times10^5$ V m^{-1}. There is a kink in energy distribution that appears at electron energy corresponding to phonon energy ($\approx$ 0.031 e.V). Also, the influence of intervally scattering appears at electron energy corresponding to 0.58 eV, showing a peak at electric field equivalence to 8×10^5 V m^{-1}. The distribution function in the Γ valley of $Ga_{0.53}In_{0.47}$ As in three-dimensional **k**-space is presented in Fig.(10) for fields varying in the range of $(1-10)\times10^5$ V m^{-1} at lattice temperature of 300 K. These figures show that the Maxwellian distribution at 1×10^5 V m^{-1} becomes non-Maxwellian at higher electric fields due to electron transfer from the Γ valley to the L valley.

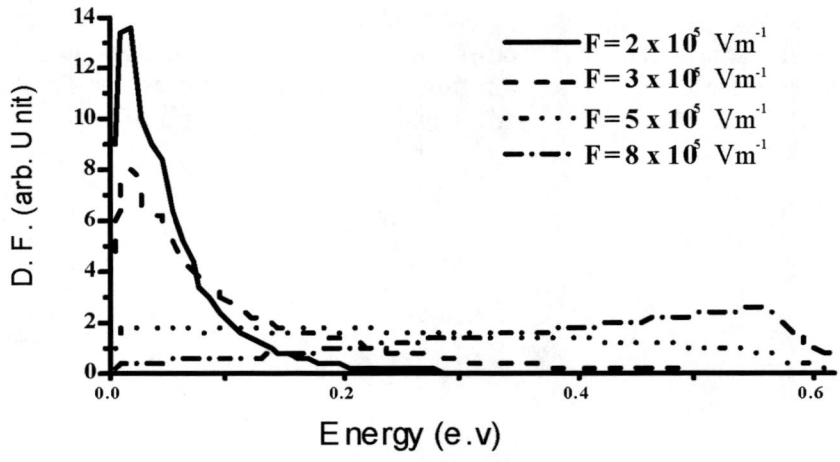

FIGURE 9. The number of Distribution Function. against electron energy at 300 K in the Γ valley of $Ga_{0.53}In_{0.47}$ As .

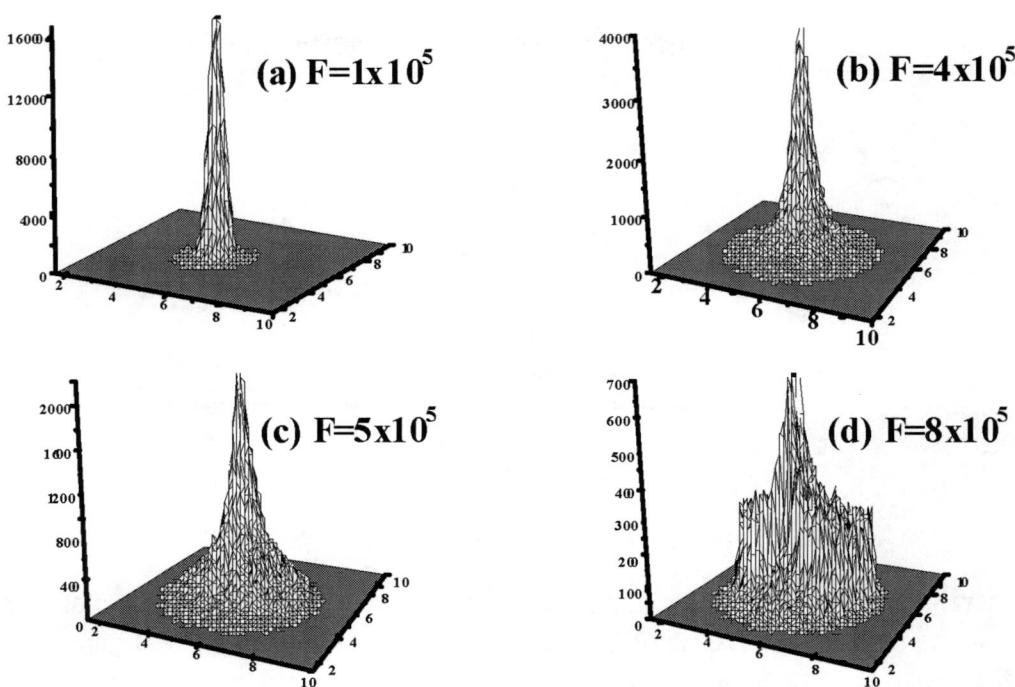

FIGURE 10-a,b,c,d. The distribution function in k_x and k_r for different Fields, for electrons in the Γ valley of $Ga_{0.53}In_{0.47}As$ at 300 K, where $k_r = \left(k_y^2 + k_z^2\right)^{1/2}$.

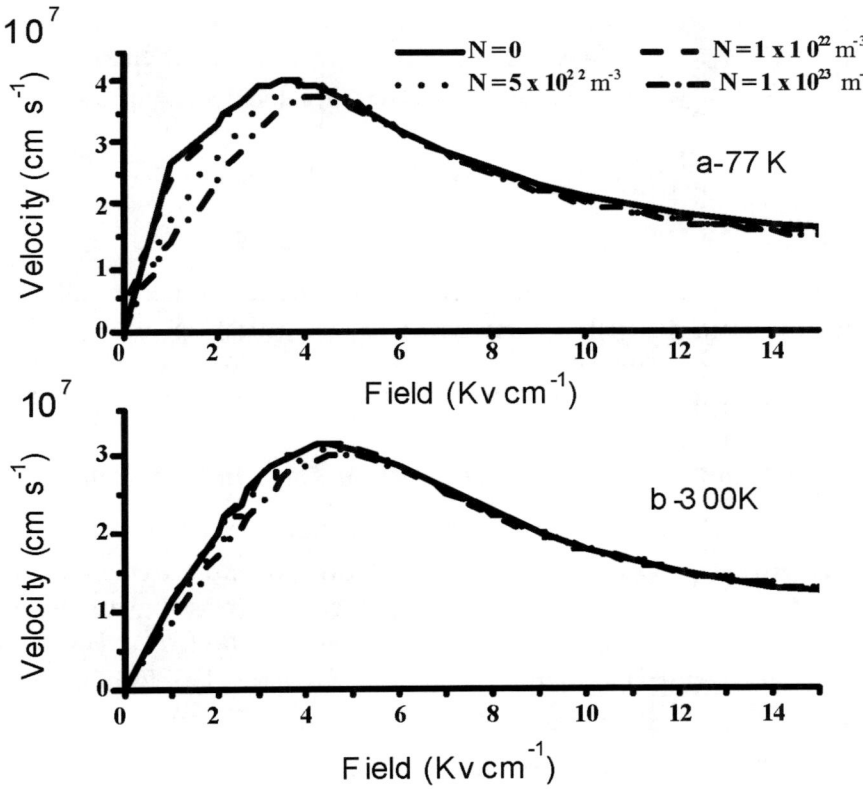

FIGURE 11-a-b. The total electron drifts velocity versus electric field at 77 K and 300 K respectively in GaAs, InP and $Ga_{0.53}In_{0.47}As$ at different impurity concentrations

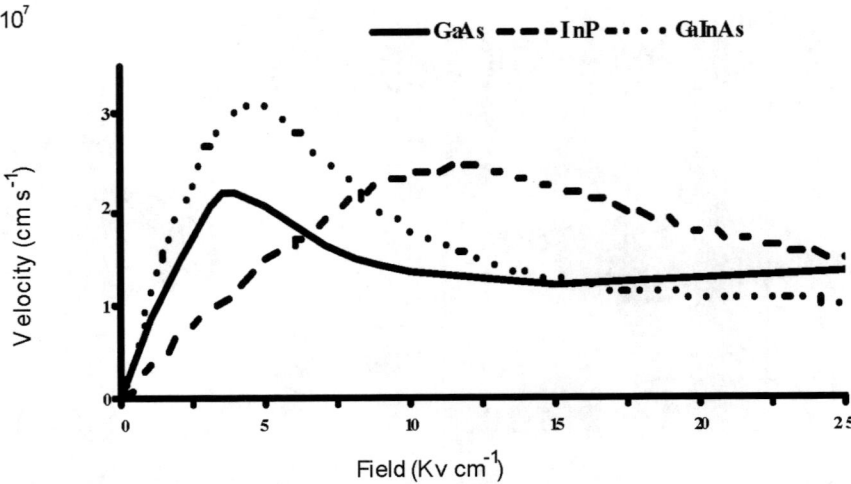

FIGURE 12. The total electron drift velocity versus electric field at 300 K in GaAs, InP and $Ga_{0.53}In_{0.47}As$.

Figures (11a) and (11b) show the total electron drift velocity versus electric field at 77 K and 300 K respectively in $Ga_{0.53}In_{0.47}As$ at different impurity concentrations. The figures show that the impurity scattering is important at low lattice temperature only. In materials with high carrier concentration, ionized impurity scattering decreases the low-field mobility and thus tends to reduce this curvature. The threshold velocity decreases with increasing carrier concentration. This behavior has an effect on the performance of high-speed devices fabricated with heavily doped materials. At room temperature, this scattering mechanism in $Ga_{0.53}In_{0.47}As$ has little effect on thecharacteristics unless the carrier concentration is greater than 10^{23} m^{-3} .

Figure (12) illustrates the total electron drift velocity versus electric field at 300 K in GaAs, InP and $Ga_{0.53}In_{0.47}As$. $Ga_{0.53}In_{0.47}As$ has high electron mobility, small effective mass and large intervalley energy separation. This suggests that can provide electron velocities significantly higher than those of GaAs and InP.

CONCLUSION

In conclusion, we have calculated the transport properties of $Ga_{0.53}In_{0.47}As$ up to field 5×10^5 V m^{-1}. At that field up 80 % of electrons have been transferred to the upper valley. The electron kinetic energy (measured from the valley minimum) is quite modest up to the mentioned electric field. It should be safe to work with known many-valley conduction band structures and conventional transport theory. It is necessary to review the situation at very high electric fields ($>5 \times 10^5$ V m^{-1}) and to enter new elements.

The threshold field for negative differential resistance (NDR) in $Ga_{0.53}In_{0.47}As$ changed over the lattice temperature range 77K-500K. The threshold field varies from 3×10^5 V/cm at 77K to 5×10^5 V/cm at 500K. The absence of phonon absorption at 77 K explains the extremely high low-field mobility at that temperature. Also, the average electron energy increases with the lattice temperature cooling as a consequence of the lower phonon emission scattering rate. The impurity scattering is important at low lattice temperature only. $Ga_{0.53}In_{0.47}As$ provides electron velocities significantly higher than those of GaAs and InP due to high electron mobility, small effective mass and large intervalley energy separation.

REFERENCES

1. W. Fawcett, A. D. Boardman and S. Swain, J. Phys. Chem. Solids **31**, 1963 (1970).
2. J. Xu, M. Shur, Phys. Rev. B **36**, 1352 (1987).

3. S. K. O'Leary, B.E. Foutz, M.S. Shur, U.V. Bhapkar and L.F. Eastman, J. Appl. Phys. **83**, 826 (1998).

4. K.Y. Choo and D.S.Ong , J. Appl. Phys. **96**, 5649 (2004).

5. M. Shur. Ga,4s Dcrices and Circuits , 1986, Plenum., New York.

6 C. Chao, P. M. Smith, K. H. Dub, J. M. Ballingall, C. F. Lester, B R. Lee, and A. A. Sabra, IEEE Electron Device Meeting. 410 (1987).

7 .J. R. Hauser, T. H. Glisson. and M. A. Littlejohn, Solid-State Electron., **22**, 487 (1979).

8. M. A. Littlejohn, J. R. Hauser, and T. H. Glisson, AppI. Phys. Lett. **30**, 242 (1977).
   M. A. Littlejohn, J. R. Hauser, T. H. Glisson, D. K. Ferry, and J. W. Harrison, Solid-State Electron. **21**, 107 (1978).

9. C. Jacobani and P.Lugli, The Monte Carlo method for Semiconductors Devices Simulation, Computation Microelectronics, Vol.2, , 1889, Springer-press, Wien, New York .

10. F. M. Abou El- Ela, PH.D Thesis, Essex University, England (1989).

11. T. P. pearsall, GaInAsP Alloy Semiconductors, Johon Wiley & sons Ltd.,(1982).

12. Baynas, A.Krotkus., A.Stalnionis, A.T.Gorelionok, N.M.Shmidt, J.A.Tellefsen, *Appl Phys.A*, **51**, 357 (1990).

13. H. Brooks, C. Herring, Phys. Rev. **83**, 879(1951).

14. B. K.Ridley, J. Phys. C **10**, 1589 (1977).

Preparation and Characterization of the Porous (TiO₂) Oxide Films of Nanostructure for Biological and Medical Applications

Sahar A.Fadl-Allah, Rabab M.El.Sherief and Waheed A.Badawy

Chemistry Department, Faculty of Science, University of Cairo
corresponding author: Wbadawy@chem-sci.cu.edu.eg

Abstract In this paper, galvanostatically and potentiostatically formed surface oxide film on titanium in H_2O_2 free and H_2O_2 containing H_2SO_4 solutions were investigated. Conventional electrochemical techniques and electrochemical impedance spectroscopy (EIS) measurements beside the scanning electron microscope (SEM) were used. In absence of H_2O_2, the impedance response indicated a stable thin oxide film which depends on the mode of anodization of the metal. However, the introduction of H_2O_2 into the solution resulted in significant changes in the film characteristics, which were reflected in the EIS results. The film characteristics were found to be affected by the mode of oxide film growth and polarization time. The H_2O_2 addition to the solution has led to a significant decrease in the corrosion resistance of the passive film. The electrochemical and the use of equivalent circuit models have led to the understanding of the film characteristics under different conditions.

INTRODUCTION

Titanium and its alloys represent an important category of materials that have been used for many years in different industrial applications as implant material. This is partly due to the stable passive oxide films that they possess; therefore they are applied for corrosion protection. In general, passive films are competing with other techniques of surface protection, like phosphating, electro deposition of paints and others [1]. In high-tech systems, especially in micro and nanotechnology, passivation is competing with other micro structuring techniques, like PVD. The superior corrosion resistance of titanium and its alloys in comparison to stainless steels has been widely reported [2-4]. Here, titanium oxide thin films with nanoporous structures are desirable for many applications due to their large surface area and high reactivity for some applications. It is well known that Ti and its alloys are always used in medical applications [5]. The good biocompatibility and osteoinegrability of titanium are due to the presence of a very thin and adherent film, which is spontaneously formed on the metal surface. This so called passive condition of titanium, when placed in a physiological environment, allows a good corrosion resistance. On the other hand, the presence of the oxide layer on the surface plays an important role for the favorable tissue response to titanium implants [5]. The biocompatibility is determined by the chemical processes occurring at the interface between prosthesis and organic tissue which in the case of titanium, consists properly of TiO_2 layer [5, 6]. It is possible to increase the range of biomaterial applications by depositing a thicker layer of TiO_2 on Titanium surface [7] or by covering it with another biocompatible material e.g. hydroxyapatite $Ca_{10}(PO_4)_6(OH)_2$ [8]. Titanium oxide surfaces show a wide range of structural and chemical properties, depending on their preparation and handling. Hence, the Ti biocompatibility could be improved through oxide layer growth by several surface treatments, for example, heat treatment, sol-gel technique, or anodic oxidation (AO) [9, 10]. The application of anodic oxidation to the surface preparation of specificied surfaces of titanium alloys is a recent procedure that has been proposed to improve the wear resistance and the adhesive properties of these materials [11]. The improvement of these materials was extended to their biocompatibility [12, 13]. The biological performance of titanium oxide films seems to be related to the morphology and pore structure of those films [14]. So the approach of anodic oxidation to build porous titanium oxide film of controllable pore size, good uniformity, and conformability over large areas at low cost seems to be a great challenge which can be verified [15].

Various electrochemical technique, including ac-impedance spectroscopy, photo-electrochemistry, scanning electron microscopy, and ellipsometric studies have demonstrated the changes in the growth rate and film properties with the anodizing conditions [16-25]. During the last decades , various oxide

CP888, *Modern Trends in Physics Research,*
Second International Conference on Modern Trends in Physics Research—MTPR-06,
edited by L. El Nadi
© 2007 American Institute of Physics 978-0-7354-0354-9/07/$23.00

films on metal surfaces have been characterized by electrochemical impedance spectroscopy (EIS), e.g. anodic oxides on Aluminum [26, 27], zirconium [28] as well as passive films on titanium [29-31].

In this paper we are aiming at the optimization of the fabrication of nano-structured TiO_2 films on Ti for biomedical applications. In this respect, Ti was anodized in H_2O_2 free and H_2O_2 containing H_2SO_4 solutions. The formed oxide films were investigated by the conventional electrochemical techniques and electrochemical impedance spectroscopy (EIS). A comparison was made between the oxide films formed galvanostatically and those formed potentiostatically under the same conditions. The formed porous films were examined by the scanning electron microscope (SEM).

EXPERIMENTAL

The substrate material used was commercially pure Ti (cp-Ti). The exposed circular surface area of the investigated material was $1cm^2$. Prior to immersion in the electrolyte, the electrodes were abraded using successively grades emery papers down to 2000 girt, then rubbed with a soft cloth until they acquired a mirror bright surface and washed with triple distilled water. The electrochemical set-up, electrochemical cell and methodology are as previously reported [32, 33].

To investigate the oxide film preparation, two sets of experiments were performed in parallel. In One set, passive films of titanium were formed galvanostatically at (10-100) mA cm^{-2} in H_2O_2 free and H_2O_2 containing 0.5 M H_2SO_4 for 30 min. After anodization, the electrode impedance was followed as a function of time. All potentials were measured against and referred to a $Hg/Hg_2SO_4/Na_2SO_4$ (sat), reference electrode. In order to reduce phase shift errors at high frequencies a low impedance reference system was established connecting a platinum probe in parallel to the reference electrode by a 10 μF capacitance. The Haber-Luggin capillary of the reference electrode and the platinum probe were adjusted at nearly the same distance to the working electrode surface. In the second set of experiments, the passive films were formed potentiostatically at (1-10 V) in H_2O_2 free and H_2O_2 containing 0.5 M H_2SO_4 for 30 min. The anodization experiments were carried out using a dc supplier (1- 50 V; 1- 10 A) working in a two – electrode cell with a large surface platinum counter electrode. After oxide film formation the electrode impedance was followed as a function of time. All measurements were carried out in naturally aerated solutions at a constant room temperature of 303 ± 2 K.

Electrochemical impedance spectroscopy (EIS) experiments were performed with the IM6d.AMOS system (Zahner Elektrik GMBH & Co., Kronach, Garmany). The input signal was usually 10 mV peak to peak in the frequency domain 100 mHz - 100 KHz. Before each experiment, the working electrode was immersed in the test solution until a steady – state potential was reached and then the impedance data were recorded. The data were fitted to a theoretical model using a complex non-linear least squares (CNLS) circuit fitting software [34]. Scanning electron micrographs (SEM) were performed using a JEOL-840 Electron probe micro analyzer.

RESULTS AND DISCUSSION

3.1. Oxide film formation and characterization

In this series of experiments the pure titanium electrodes were anodized galvanostatically and potentiostatically as described in the experimental section. The anodization process was carried out in naturally aerated H_2O_2 free 0.5 M H_2SO_4 solutions. The characterization of the formed oxide films under different conditions is carried out using the electrochemical impedance spectroscopic technique (EIS). EIS is a non-destructive sensitive technique that enables the detection of any changes taking place at the electrode/electrolyte interface. All experimental data presented as Bode plots, which are always recommended as standard impedance plots, The Bode plot format enables equal presentation of all impedance data and the presence of the phase angel θ, as a sensitive parameter for any surface changes, gives direct information about the electrode surface [35, 36].

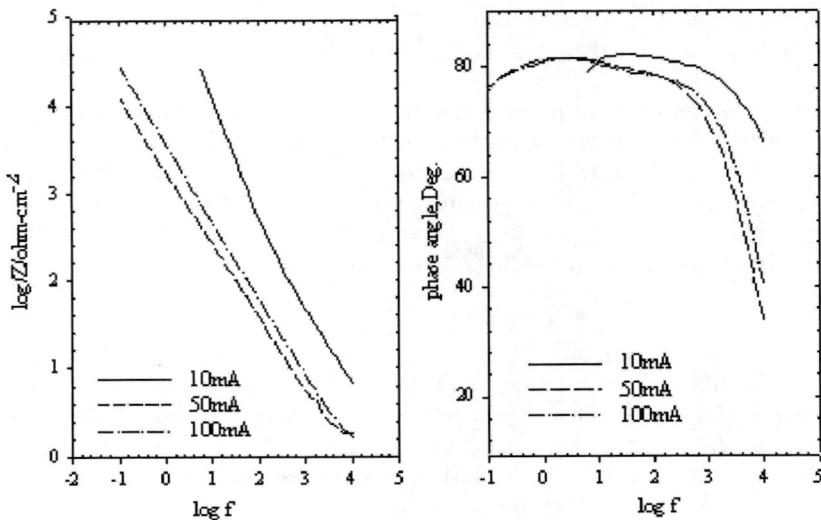

FIGURE 1. Bode plots, modulus and phase of oxide film formed galvanostatically for 30 min at different current density in 0.5 M H$_2$SO$_4$ after 120 min exposure.

Fig.1 presents the Bode impedance plots of galvanostatically formed oxide films on Ti for 30 min in 0.5 M H$_2$SO$_4$ at different current densities ranging between 10 and 100 mA cm^{-2}. The impedance data were recorded after 2h of electrode immersion in the anodizing solution. The obtained impedance spectra represent the typical behavior of passivated titanium [30, 37, 38]. For comparison, the impedance data of potentiostatically passivated Ti in the anodizing medium and under the same conditions are presented in Fig.2. According to the anodization potential, the passive film acquires an interference color, starting from light gold, deep gold, light blue, dark blue to violet, which is in good agreement with the previously reported data [9].

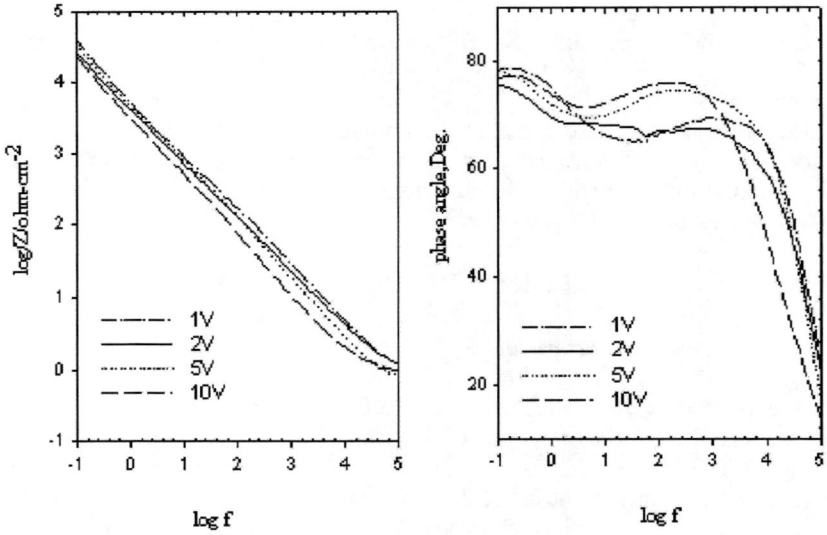

FIGURE 2. Bode plots, modulus and phase of oxide film formed potentiostatically for 30 min at different cell voltage in 0.5 M H$_2$SO$_4$ after 120 min exposure.

The general features of the impedance plots are consistent with a passive film behavior, which shows a phase angle approaching 90°C over a wide range of frequency (10^{-1}-10^{5}Hz) [9,30,37-39].
As can be seen from the phase shift vs log f presentation, in both Fig.1 and especially, Fig.2, the electrochemical impedance responses display two time constants. This may be taken as evidence, that the passive film formed on Ti, which may always described as a continuous single layer of TiO$_2$, is more likely to consist of two layer, a dense inner layer and a porous outer layers [40-42].

TABLE 1: Equivalent circuit parameters for T different time of immersion in 0.5 M H_2SO_4 at 25°C after different current density.

CD, mACm^{-2}	Time / min	R_p / kΩcm^2	R_b / MΩcm^2	C_p / μFcm^{-2}	C_b / μFcm^{-2}
10	5	0.5	3.9	6.4	0.76
	120	192.3	$0.8*10^3$	23.6	$3.2*10^{-3}$
50	5	$1.1*10^3$	$5.5*10^3$	14.8	0.012
	120	$0.65*10^3$	$3.5*10^3$	20.8	0.012
100	5	$17.7*10^{-3}$	$6.7*10^3$	9.6	6.8
	120	$6.7*10^{-3}$	$3.2*10^3$	15.2	5.6

The experimental impedance data were fitted to theoretical data according to proposed equivalent circuits. The best fitting was obtained using the equivalent circuit presented in Fig.3a, where a duplex nature of the passive film was proposed. The results of data fitting according to this model for both the galvanostatically and potentiostatically formed oxide films are presented in Fig.4a and Fig.4b, respectively.

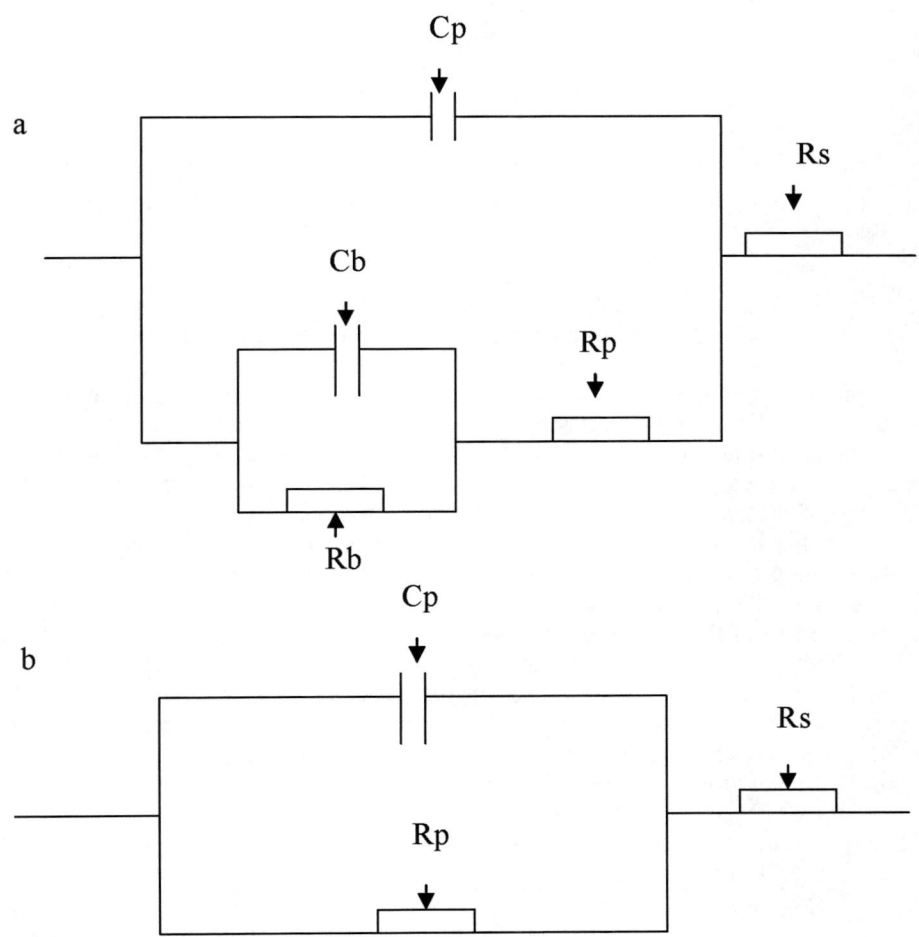

FIGURE 3. Equivalent circuits used for fitting the oxide film formed on titanium under different conditions of formation.

As can be seen from the results of data fitting, there is an excellent agreement between the experimental data and the proposed model, especially for oxide films formed potentiostatically (Fig.4b). The slight deviation observed in Fig.4a can be attributed to the fact that the oxide films formed galvanostatically are less homogeneous than those formed potentiostatically [43].

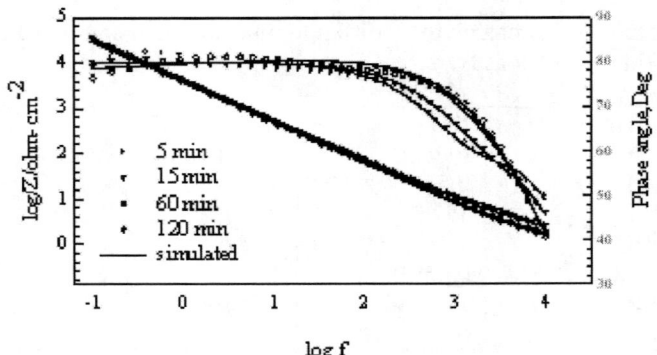

FIGURE 4 (a)   Impedance data recorded for oxide film formed galvanostatically at 100mAcm^{-2} in 0.5 M H$_2$SO$_4$ for different time of immersion, —— simulated data.

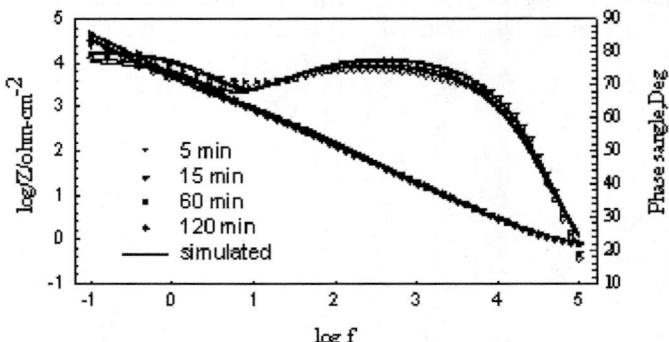

FIGURE 4 (b)   Impedance data recorded for oxide film formed potentiostatically at 5V in 0.5 M H$_2$SO$_4$ for different time of immersion, —— simulated data.

The proposed equivalent circuit model consists of solution resistance, R_s, in series to two parallel combinations, R_b and C_b representing the barrier film resistance and capacitance, and then R_p and C_p, representing the porous film resistance and capacitance, respectively the values of the equivalent circuit parameters for the galvanostatically formed TiO$_2$ films at different current densities are presented in Table 1. It is clear that the values of R_p are relatively high and those of C_b are low. It is also worth to mention that the value of C_b decreases and that of R_p increases with the immersion time in the electrolyte. The decrease of C_b and the increase of R_p may be attributed to a slow growth of the oxide film, indicating a long term stability of the passive film formed on Ti in H$_2$O$_2$ free H$_2$SO$_4$ solutions [29]. The decrease of R_p with the increase of the formation current density leads to the conclusion that the use of high current density leads to the formation of less resistive open porous structure. On the other hand the recorded increase of the porous film capacitance, C_p, with the increase of the immersion time in the electrolyte indicates a porous film thinning with time. These results are in good agreement with the XPS investigation and independent optical measurements that concludes a small total thickness with a very thin outer porous layer [9].

TABLE 2:   Equivalent circuit parameters for Ti-pure after 120 min of immersion in 0.5 M H$_2$SO$_4$ at 25^{0}C after different cell-Voltage

Cell-voltage	R_p/ kΩcm^2	R_b/ MΩcm^2	C_p/ μFcm^{-2}	C_b/ μFcm^{-2}
1	1.25	37*10^4	5.2	5.2
2	0.6	217*10^3	6.8	2.2
5	4.1	18.1*10^3	8.8	5.0
10	10.0	17.8*10^3	14.8	10.8

The results of data fitting for oxide films formed potentiostatically are presented in Table 2. It is clear from this table that the barrier film resistance, R_b, is relatively high and decreases with the increase of the formation voltage.

The mode of oxide film growth and the mechanism of film formation depend on the anodization technique. Fig.5 shows a clear comparison between oxide films growth potentiostatically (Fig. 5a) and that formed galvanostatically (Fig. 5b). In both cases, a capacitive behavior of the passive film over a

wide frequency range is recorded. In the case of films grown galvanostatically only a single time constant controls the film formation kinetics, which implies that the compact inner layer is dominating.

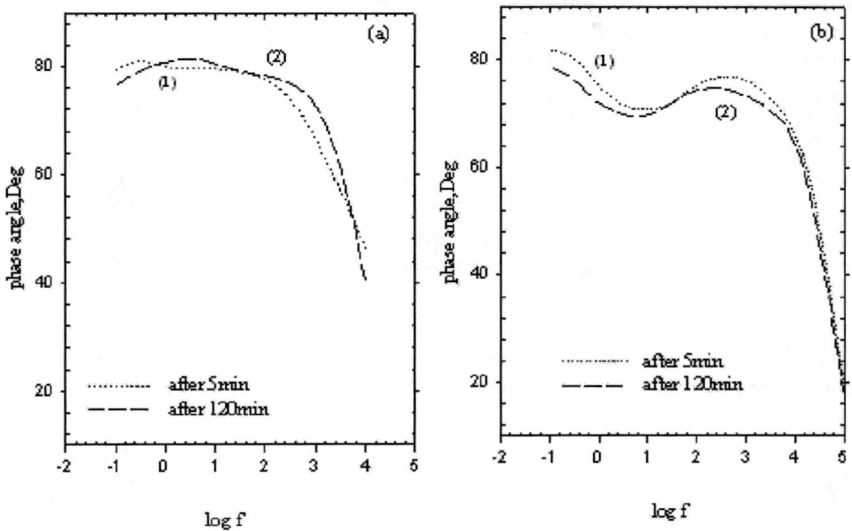

FIGURE .5 Electrochemical impedance response of oxide film formed (a) potentiostatically at 5V, (b) galvanostatically at 100mAcm^{-2} on cp-Ti in 0.5 M H$_2$SO$_4$ after (1) 5 min and (2) 120 min time of exposure.

On the other hand, the films formed potentiostatically show a splitting in log f vs phase diagram (Fig.5a) which means that the kinetics are controlled by two time constants. This means that both the inner compact layer and the outer porous layer are contributing to the film growth kinetics.

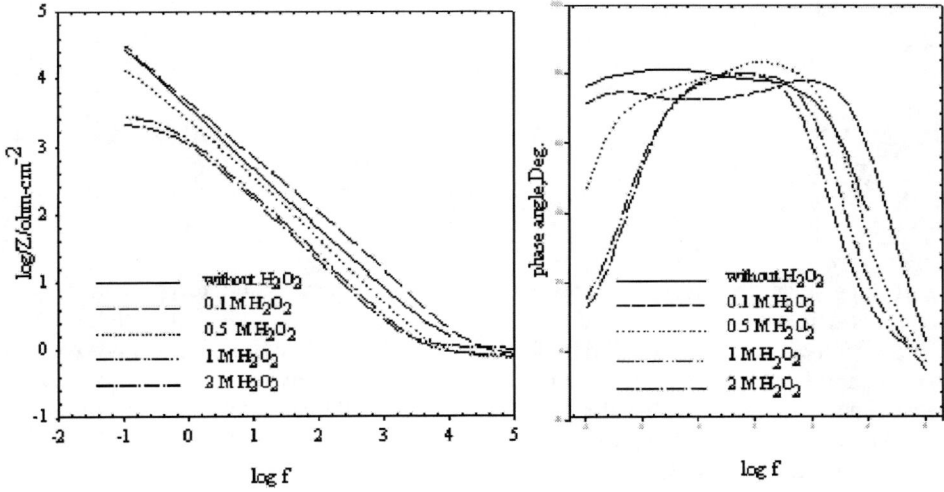

FIGURE 6 Bode plots, modulus and phase of oxide film formed galvanostatically for 30 min at 100mAcm^{-2} in 0.5 M H$_2$SO$_4$ with different H$_2$O$_2$ concentration after 120 min exposure.

3.2 Effect of H$_2$O$_2$ on the passive film characterization

In this series of experiments the effect of H$_2$O$_2$ on the passive film formation on Ti was investigated. In this respect, the films are formed under the same conditions described in section 3.1 but the electrolyte was provided by different concentrations of H$_2$O$_2$ ranging between 0.1 and 2 M. The impedance spectra of the films formed galvanostatically at 100mAcm^{-2} are presented in Fig.6 and those of film formed

115

potentiostatically at 5V are presented in Fig.7. Comparison between the data presented in Figs 6 and 7 and those presented in Figs 1 and 2 indicates that the presence of H$_2$O$_2$ changes the passive film properties. From the data presented in both Fig. 6 and Fig. 7, it is clear that the presence of relatively low concentration of H$_2$O$_2$ ($\leq$ 0.1M) in the passive film formation medium leads to the presence of two phase axima that indicates the presence of the two layers as described before [42].

In these cases, The impedance data of these measurements were fitted also to the equivalent circuit model presented in Fig.3a and the results of data fitting are quite satisfactory indicating that the passive film consists of a dense inner layer of high corrosion resistance and an outer porous layer of relatively low resistance. The whole film thickness is small as inferred from the values of the capacitance.

The increase of H$_2$O$_2$ concentration (>0.5M) in the anodizing solution leads to impedance characteristic with a single phase maximum which is characteristic for corroding systems governed by the simple Randles equivalent circuit presented in Fig.3b [30,38]. In this model R$_p$ represents the charge transfer (corrosion) resistance of the passive film, C$_p$ its capacitance and R$_s$ is the ohmic drop in the electrolyte [29, 33].

TABLE 3: Equivalent circuit parameters for cp Ti after different time of immersion in 0.5 M H$_2$SO$_4$ with different H$_2$O$_2$ conc. at 25°C at 100mACm^{-2}

[H$_2$O$_2$]	Time / min	R$_p$ / kΩcm^2	R$_b$ / MΩcm^2	C$_p$ / μFcm^{-2}	C$_b$ / μFcm^{-2}
	5	17.7*10^{-3}	6.7*10^3	9.6	6.8
	15	12.8*10^{-3}	7.0*10^3	10.8	6.4
0 M	60	8*10^{-3}	4.8*10^3	14.0	6.0
	120	6.7*10^{-3}	3.2*10^3	15.2	5.6
	5	1.55	20.5*10^3	5.8	5.2
	15	1.15	25.0*10^3	6.8	4.4
0.1 M	60	0.43	18.8*10^3	8.8	3.5
	120	0.28	14.6*10^3	10.4	3.3
	5	1.7*10-3	68*10^3	5.6	12.8
	15	1.2*10-3	0.5*10^3	8.8	12.0
0.5 M	60	2.7*10-3	90*10^3	17.6	9.6
	120	2.2*10-3	29.3	24.8	10.0
	5	1.5	29.7	20.0	4.8
	15	0.25	9.6	27.6	4.8
1 M	60	3.7	-	42.6	-
	120	2.9	-	54.0	-

The results of impedance data fitting are presented in Table 3. These results reveal that the increase of H$_2$O$_2$ concentration in the formation medium above 0.5 M leads to a decrease of the barrier film resistance R$_b$, which means that the barrier film becomes defective and starts to dissolve. Since the oxide film formation is governed by dissolution /oxidation mechanism [42], the presence of H$_2$O$_2$ leads to an increased rate of dissolution and the oxide film becomes thinner as reflected from the increased C$_b$ values. It should be mentioned that the effect of H$_2$O$_2$ is the same, whether the oxidation process is carried out galvanostatically or potentiostatically (cf. Figs. 6&7).

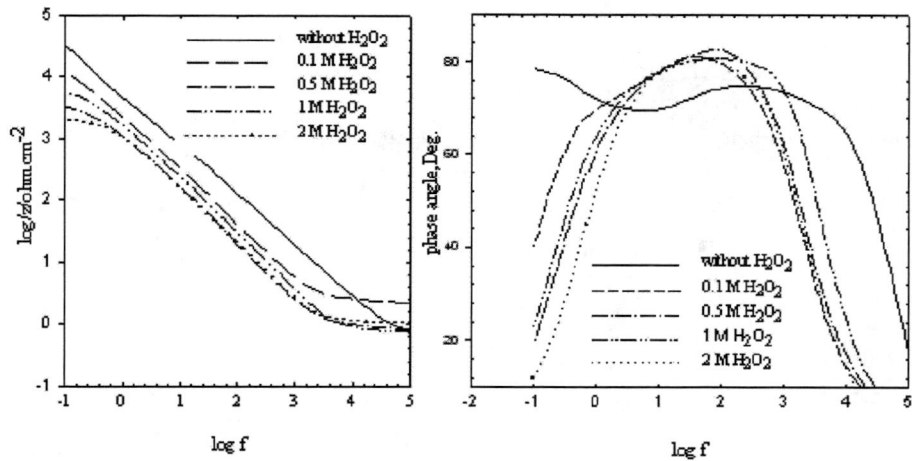

FIGURE 7. Bode plots, modulus and phase of oxide film formed potentiostatically for 30 min at 5 in 0.5 M H$_2$SO$_4$ with different H$_2$O$_2$ concentration after 120 min exposure.

The effect of the time of electrode immersion in the electrolyte before the impedance measurements on the equivalent circuit parameters is also investigated. The results are presented in Fig.8 a, b. It is clear that the films formed in H$_2$O$_2$ free 0.5 M have always constant R$_p$ and C$_b$ values.

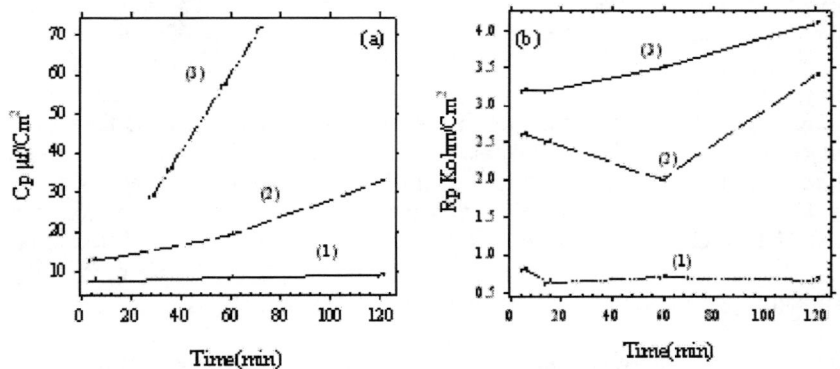

FIGURE 8. Equivalent circuit parameters (a) Capacitance (b) Resistance of porous layer on cp-Ti after different time of immersion in (1) 0.5 M H$_2$SO$_4$ (2) 0.5 M H$_2$SO$_4$+0.1 M H$_2$O$_2$ (3) 0.5 M H$_2$SO$_4$ + 0.5 M H$_2$O$_2$

In the presence of H$_2$O$_2$ and especially at concentrations ≥ 0.5 M, an increase in C$_p$ is recorded, which means that the passive film is subjected to continuous dissolution. It is well known that valve metals like Ti form barrier layers in aqueous solutions even without anodization. The characteristics of this barrier layer depends essentially on the electrolyte composition and anodization conditions if any [30,37], Under special conditions, an immediate increase in the local conductivity occurs and a passivity breakdown takes place [30,44].

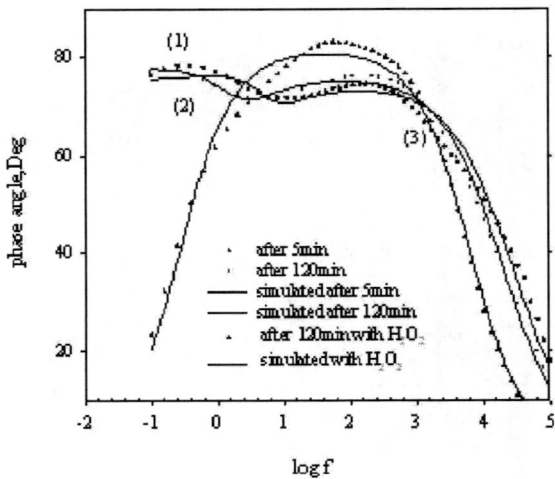

FIGURE 9. Electrochemical impedance response of oxide film formed potentiostatically at 10V on cp-Ti in H_2SO_4 after (1) 5 min (2) 120 min (3) 120 min in presence of 0.1 M H_2O_2.

The passivity breakdown is depending mainly on the electrolyte composition, its concentration, and to less extent on the current density, temperature and roughness factors [44].

To emphasize the effect of anodization potential in the potentiostatic formation mode, some oxide films are formed in H_2O_2 free H_2SO_4 solutions at 10V under the same conditions. The impedance data of the formed oxide films were fitted to the theoretical models presented in Fig.3.

As can be seen from Fig.9 the addition of small concentration of H_2O_2 leads to the disappearance of one of the two phase maxima, and only a single phase maximum is recorded. Also, the data fitting according to the theoretical models presented in Fig.3 show a deviation compared to the data fitting of oxide films formed at 5V, especially in the presence of H_2O_2. The deviation of the experimental impedance values obtained with the passive films formed at 10V from the values of the equivalent circuits which are in excellent agreement with the experimental values obtained for those films formed at 5V may lead to the conclusion that the increase of the formation voltage leads to a less homogeneous film with more defective nature. Such a conclusion was reported previously and confirmed experimentally [30, 43]. The experimental results presented in this paper represent the preliminary investigations of Ti/TiO$_2$/electrolyte system. A detailed study of the effects of different parameters like electrolyte composition, pH, temperature and possible implants is now in progress. These investigations are always complimented by surface analytical studies.

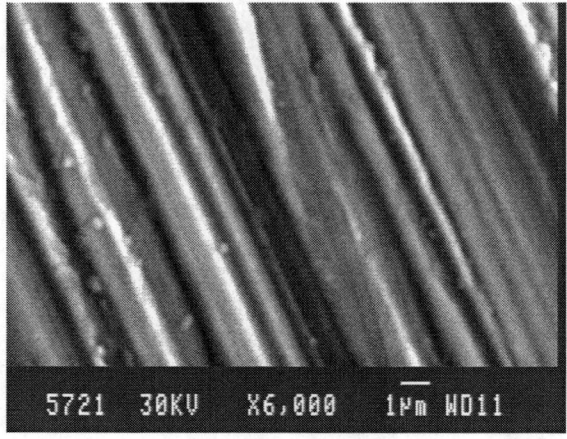

FIGURE 10. Scanning electron microscope for non-anodized cp-Ti surface.

3.3 Scanning Electron Microscope Investigations:

To have an insite on the morphology of the metal surface and the morphology changes before and after anodization and also the effect of the presence of H_2O_2 in the electrolyte, the electrode surface was investigated by the SEM. The results of these investigations are presented in Figs.10, 11&12.

Fig.10 presents the SE micrograph of the mechanically polished Ti surface. This micrograph represents the general morphology of mechanically polished polycrystalline metallic surface [45]. Fig.11 shows the SE micrographs of the potentiostatically anodized surface at 5V in 0.5 M H_2SO_4 after immersion in H_2O_2 free 0.5 M H_2SO_4 for 30 min.(Fig.11a), after immersion in 0.1 M H_2O_2 containing 0.5 M H_2SO_4 for the same time (Fig.11b) and after 30 min immersion in 1.0 M H_2O_2 containing 0.5 M H_2SO_4 solution (Fig.11c), inspection of Figs.11a,b,andc, shows that the surface treated in H_2O_2 free H_2SO_4 solutions is covered with a thin homogeneous compact film of uneven nanocrystals of TiO_2[46].

The presence of such a homogeneous film is also confirmed by the electrochemical characteristics presented in Fig.4.

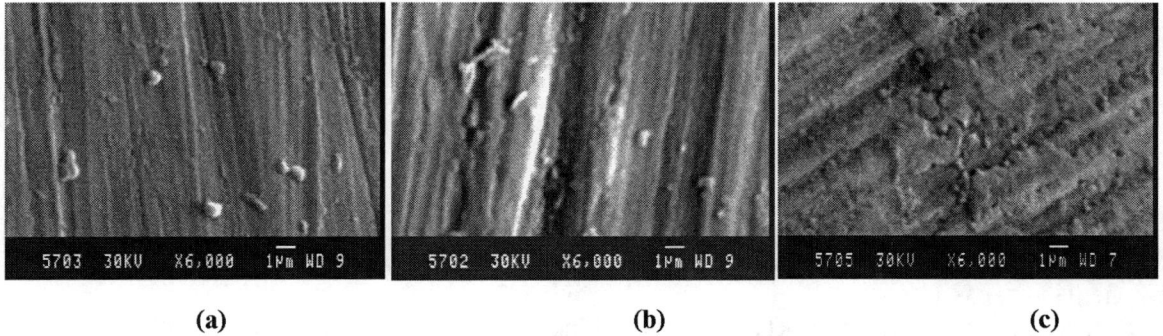

(a)   (b)   (c)

FIGURE 11. Scanning Electron microscope of titanium samples anodized potentiostatically at 5V in (a) 0.5 M H_2SO_4 for 30 min (b) 0.5M H_2SO_4 + 0.1 M H_2O_2 (c) 0.5M H_2SO_4 + 1 M H_4 after 120 min exposure.

The presence of a low concentration of H_2O_2 (≤ 0.1 M) leads to the formation of dissolution patterns in the nano range which gives the nano-porous structure assigned to the outer layer (cf.Fig.11b). The increase of the concentration of H_2O_2 (>0.5 M) leads to a continuous dissolution of the passive film and a complete damage of the compact film is recorded (cf.Fig.11c). The damage of this passive film is reflected in the passivity breakdown observed electrochemically [44]. The inhomogeneity assigned to passive films formed at 10V was also confirmed by SEM. Fig.12 shows the scanning electron micrograph of the film formed at 10V after 30min immersion in 0.5 M H_2SO_4. Comparison of this micrograph with that of the film formed at 5V under the same conditions (Fig.11a) shows clearly that the use of higher formation voltage leads to heterogeneous passive films with microstructures as an outer layer which covers a dense compact film with several crades. The results of SEM confirm the conclusions mode from the experimental results of EIS.

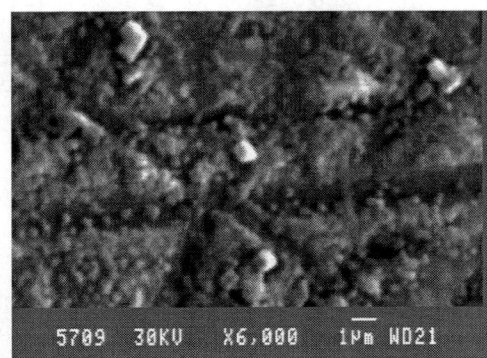

FIGURE 12. Scanning Electron microscope of titanium sample anodized potentiostatically at 10 V in 0.5 M H_2SO_4 for 30min.

CONCLUSION

-Anodic oxide films on Ti are consisting of two layers, an inner compact and an outer porous layer.
-Anodic oxide films formed galvanostatically on Ti are controlled by the compact layer, whereas those formed potentiostatically are controlled by both layers.
-Potentiostatically formed films at higher voltages ($\geq 10V$) are less homogeneous with more defective nature.
-The presence of low concentrations of H_2O_2 (≤ 0.1 M) leads to the formation of a nano-porous oxide film.
-Increase of H_2O_2 concentration leads to a dissolution of the passive film.

REFERENCE

1- J.W. Schulze, M.M. Lohrengel, .Electrochim Acta **45**, 2499(2000).

2- R.J. Solar, in: A. Syrett, A. Acharya (Eds.), Corrosion and Degradation of Implant Materials, First International Symposium, American Society for Testing and Materials, Philadelphia, **PA**, 259(1979).

3- E.J.sutow, S.R.Pollack, in: D.F.Williams(Ed.), Biocomatibilty of clinical Implant Materials, CRC Press, Boca Raton, FL., 1981, p.45.

4- D.F Williams, in: D.F. Williams (Ed.) Biocompatibility of Clinical Implant Materials, CRC Press, Boca Raton, FL., 1981, P.99.

5- B.D.Ratner, J.Biomed.Res.**27**, 37(993).

6- E.A.Vogler, J.Electronspectrosc.Relat.Phenom **81**, 237(1996).

7- J. Lausmaa, J. Electronspectrosc. Relat.Phenom.**81**, 343(1996).

8- A.Montenero, F.Ferrari, M. Cesori, G.Gnappi, E.Salvioli, L.Mattogno and S.Kaciulis, J. Mater.Sci.**35**,1(2000).

9- D.Velten, V.Biehl, F.Aubertinetal, Journal of Biomedical Materials research **59**, 18(2002).

10- T-Y.Xiong,X-Y.Cui, H-M.Kimetal, Key Engineering Materials **254**, 375(2004).

11-M.E.Sibert, J.Electrochem.Soc.**25**, 65(1983).

12- T.Kokubo, Acta Mater.**46**, 2519(1998).

13- Y-T.Sul, Biomaterials, **23**, 3893(2003).

14- Y-T.Sul, C.B. Johansson, Y.Jeong et al., Clinical Oral Implants research, **13**, 252(2002).

15- T.Oishi,T.Matsubara, and A.Katagiri, Electrochemistry (in Japanese) **68**, 106(2000).

16- K.Azumi,M.Seo, Corros.Sci.**43**,(533)2001.

17- D.J.BlackWood,L.M.Peter, Electrochim Acta **34**, 1505(1989).

18- D.J. Blackwood, R.Greef, L.M.Peter, Electrochim Acta **34**, 120(1989).

19- C.da Fonseca, M.G. Ferreira, M.da Cunha Belo, Electrchim Acta **39**, 2197(1994)

20- F.Di Quarto, S. Pizza, C.Sunseri, Electrochim Acta **38**, 29(1993).

21-M.Kozlowski,W.H.Smyrl,L.Atanasoska,R.Atanasoski, Electrochim. Acta **34**, 1763(1989).

22- J. Marsh, D.Gorse, Electrochim. Acta **43**, 659(1998).

23- T.Ohtsuka, N.Nomura, Corros.Sci **39**, 1253(1997).

24- T.Ohtsuka, T.Otsuki, Corros.Sci **40**, 951(1998)

25-E.M.Oiverira, C.E.B.Marino, S.R.Biaggio,R.C.Rocha- FillO, ElectrochemCommun **2,** 245(2000).

26- F. Mansfeld, Analysis and Interpretation of EIS data for metals and alloys,Chapter, Technical Report **26**, Solartron-Schlumberger 1993.

27-J.l.Dawson, G.E.Thompson and M.B.H.Ahmadun, Electrochemical impedance: Analysis and Interpretation STP **1188**, (EdibyJ.R.Scully,   D.C. Silver man And M.W. Kendig), (P.255) American
society for testing and Materials, Philadelphia 1993.

28-J.A.Bardwell and M.C.H. McKubre, Electrochim Acta **36**, 314(1991).

29-T.P.Cheng, J.T.Lee and W.T.Tsai, Electrochim Acta **36**, 2069(1991).

30- W.A.Badawy, S.S.Elegamy and Kh.M.Ismail, Br.Corros.J. **28**, 133(1993).

31- D.G.Kolman and J.R. Scully, J.Electrochem.Soc.**141**(2633)1994.

32- F.M.Al-Kharafi and W.A.Badawy, Electrochim.Acta **40**, 1811(1995).

33-W.A.Badawy,F.M.Al.Kharafi and A.S.El-Azab Corros.Sci.**41**, 709(1999).

34-J.R. Macdonald, Complex Nonlnear least squares Immitance Fitting Program, University of North Carolina, Chape/ Hill,NC, Version 3.02.

35- J.R. Macdonald (ED), Impedance spectroscopy John wiley & sons, New york,Chapter **4**, 1978.

36- F.M.Alkharafi and W.A.Badawy, Electrochim.Acta **42**,579(1997).

37-A.Felske,W.A.Badawy,W.j.Plieth;J.Electrochem.soc.,**137**, 1804(1990).

38-W.A.Badawy and Kh.M.Ismail; Electrochimica Acta **38**, 233(1993).

39-I-Garcia and J.J De Damborenea,Corros.Sci **40**,1411(1998).

40-N.D.Tomashov, G.P.chernova, Yu.S.Ruscol andG.A.Ayuyan. Electrochim. Acta **19,** 159(1974)

41- M.M.Hefny, A.A.Mazhar and M.S.ElBasiouny, Br.Corrosion.J **17**, 38(1982).

42- M.S. El Basiouny and A.A.Mazhar, corrosion-NACE **38**, 237(1982)

43- J.PAN,D.THIERRYand .Leygraf, Electrochim Acta **41**, 1143(1996).

44-N.Sato, G.Okamoto,in: J.O.M.Bockr (Ed),ComperehensiveTreatise of Electrochemistry **4**, 193(1981).

45- A.s.Mogoda, Y.H.Ahmed and W.A.Badawy,J.Materials and Corrosion **55**, 499(2004).

46-AiKaterini G.Mantzila, Mamas I.Prodromidis, Electrochimi Acta **51**, 3537(2006).

Does Ti^{+4} Ratio Improve the Physical Properties of Cd$_x$Co$_{1-x+t}$Ti$_t$Fe$_{2-2t}$O$_4$?

M.A.Ahmed, S.F.Mansour* and S.I.El-Dek

Materials Science lab. (1), Physics department, Faculty of Science , Cairo University , Giza, Egypt.
* Physics Department, Faculty of science, Zagazig University, Zagazig, Egypt.

Abstract. The microstructure and magnetic properties of Ti substituted CoCd ferrites of the general formula Cd$_x$Co$_{1-x+t}$Ti$_t$Fe$_{2-2t}$O$_4$, x = 0.20, 0.00≤t≤0.25 have been reported. The ferrite samples were prepared by standard double sintering ceramic technique and structural analysis was carried out using X-ray diffraction. The spinel structure is confirmed for all concentration. Some physical properties (such as lattice parameter, density and porosity) have been also calculated. Temperature and magnetic field dependence of susceptibility is illustrated for all Ti contents. Both experimental and theoretical values of the effective magnetic moment were increased with increasing Ti content. The Curie temperature increases with the addition of Ti up to t = 0.15 after which it decreases. The effect of mechanical pressure on the dc resistivity at room temperature enhances the use of the samples in some applications.

INTRODUCTION

Ferromagnetic cubic spinel, or namely ferrites possess properties of both magnetic materials and insulators and they are important in many technological applications. The spinel structure allows the introduction of different metallic ions which can change the magnetic and electrical properties considerably [1]. Deny and Ghose [2] reported the synthesis and characterization of nanocrystalline Co$_{0.2}$Zn$_{0.8}$Fe$_2$O$_4$. Magnetic interaction in Co-Zn Fe-O system [3] and spectral analysis of Co$_{0.2}$Zn$_{0.4}$-Fe$_2$O$_4$ with different soaking time were reported elsewhere [4]. Co-ferrite has high cubic magnetocrystalline anisotropy and reasonable saturation magnetization and is promising for the use in isotropic permanent magnets, magnetic recording and magnetic fluids. Nanosized particles of ferrite play a significant role in many medical applications either as diagnostic tool or as treatment for many tumors. Ferrofluids are dispersions of magnetic nanosized particles stabilized by coating these particles with surfactant material. These magnetic fluids have multiple interesting properties which offer new medical and technical applications. Furthermore, magnetic fluids based on cobalt-ferrite nanoparticles surfaces coated with oleic acid have been used to treat tumors in dogs [5]. There are several applications in development using cobalt ferrite particles e.g. the magnetocrystalline anisotropy of such particles for magnetic inkjet printing which allows the production of magnetic readable structures [6].

Co-ferrite crystallizes in partially inverse spinel structure with cation distribution as (Co$^{2+}_x$Fe$^{3+}_{1-x}$) [Co$^{2+}_{1-x}$Fe$^{3+}_{1+x}$]O$_4$ where x depends on thermal history and preparation condition [7-9]; with a Curie temperature T$_C$ around 520°C and a relative large magnetic hysteresis.

M.A.Ahmed et al. [10-12], prepared and studied the electric and magnetic properties of Co-Zn and LiCo ferrite doped with different rare earth elements

The goal of our study is to prepare and investigate the physical characterizations of Ti substituted Co-Cd ferrite. Also, we aimed to probe the sensing properties of the prepared samples by applying external mechanical pressure

EXPERIMENTAL TECHNIQUES

Samples of the formula Cd$_x$Co$_{1-x+t}$Ti$_t$Fe$_{2-2t}$O$_4$; x=0.2., 0.00≤t≤0.25 were prepared by the conventional solid-state reaction from analar grade form oxide (BHD). Stoichiometric amounts were good mixed and grinded using agate mortar for three hours, transferred to agate ball mill for another three hours. The samples were pressed into pellets form using uniaxial mess of pressure 5x10^8 N/m^2. Presintering was carried out in air at 850°C for 6 hours with a heating rate of 2°C/min and then cooled with the same rate as that of heating down to room temperature, regrinded again, sieved and pressed into discs of diameter 1cm and thickness of = 1.5 mm and fired finally at 1150°C for another 10 hours in air with a heating rate of 2°C/min. The prepared samples were checked by X-rays diffraction using Diano corporation of target (λ=1.5418 Å) to assure the complete reaction. The dc magnetic susceptibility measurements were carried out using Faraday's method from room temperature up to

CP888, *Modern Trends in Physics Research,*
Second International Conference on Modern Trends in Physics Research—MTPR-06,
edited by L. El Nadi

about 800K at five different values of magnetic field intensity ranging from 1733 to 3800Oe. The point of maximum gradient at which the sample was inserted was obtained at each field from the calibration curves and a very small amount of fine powdered sample was inserted in a cylindrical glass tube. For the electrical properties measurements, the two surfaces of each pellet were good polished, coated with silver paste and then checked for good conduction. The effect of mechanical pressure on the dc resistivity was carried out using a homemade apparatus where the pressure was calculated from P =mg/A

RESULTS AND DISCUSSION

The X-ray diffratograms of $Cd_xCo_{1-x+t}Ti_tFe_{2-2t}O_4$; x=0.20, $0.00 \leq t \leq 0.25$ reveal a single phase spinel cubic structure and homogeneity as shown in Fig. (1). The main reflection planes (220), (311), (222), (400), (422), (511), (440), (620) and (533) of the spinel structure appear in all patterns except the plane

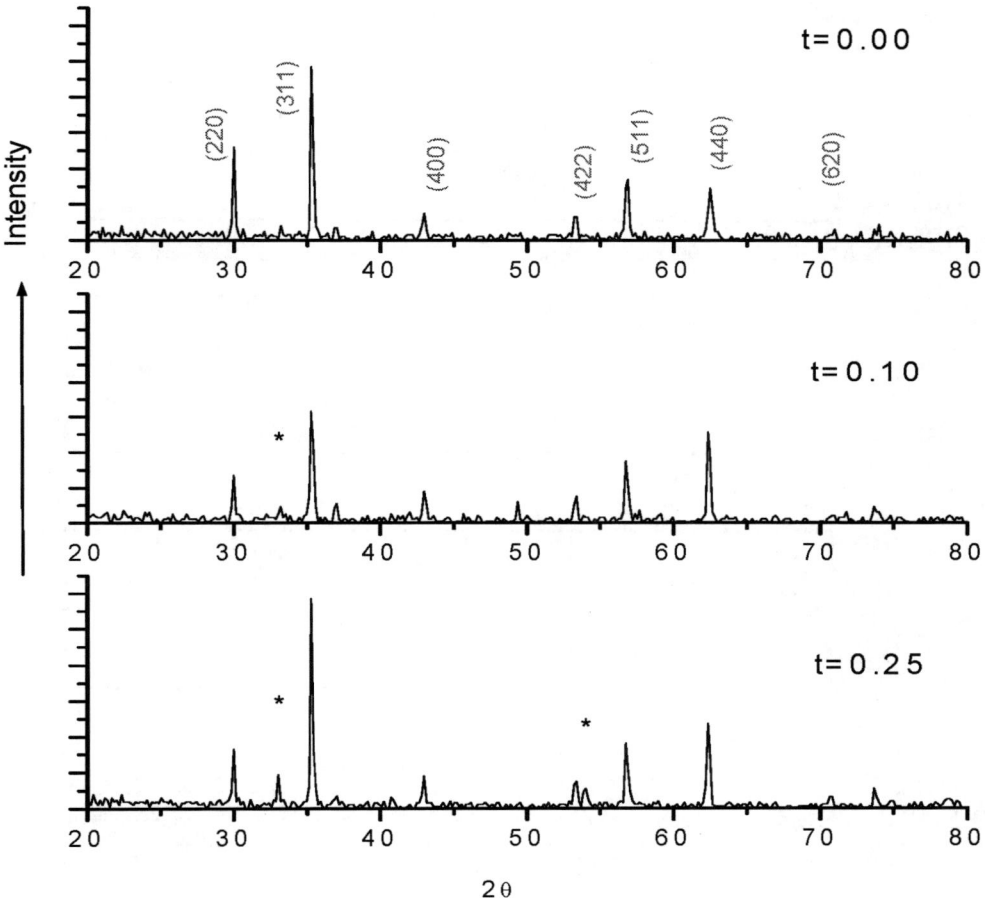

FIGURE 1. X-ray diffraction patterns for $Cd_xCo_{1-x+1}Ti_tFe_{2-2t}O_4$, x = 0.20, $0.00 \leq t \leq 0.25$. * $CoTiO_3$

(111) which was disappeared in all samples.

The precise values of the lattice parameter (a) were calculated using the extrapolation function F(θ) [13]. The lattice parameter was calculated theoretically using the following equation [14]

$$a_{th} = \frac{8}{3\sqrt{3}} \left[(r_A + R_o) + \sqrt{3}(r_B + R_o) \right]$$ (1)

where R_o is the radius of oxygen ion (0.132 nm), r_A and r_B are the ionic radii of the tetrahedral (A-site) and octahedral (B-site) respectively. In order to calculate r_A and r_B it is necessary to suggest a cation distribution as follows

$(Cd_{0.2}Fe_{1-y-t})^A[Co_{0.8+t}Ti_tFe_{1+y-t}]^B$ for t≤0.10, y=0.03   (2)

$(Cd_{0.2}Co_zFe_{1-y-t})^A[Co_{0.8+t-z}Ti_tFe_{1+y-t}]^B$ or t≥0.15, y=0.03, 0.03≤z≤0.05   (3)

123

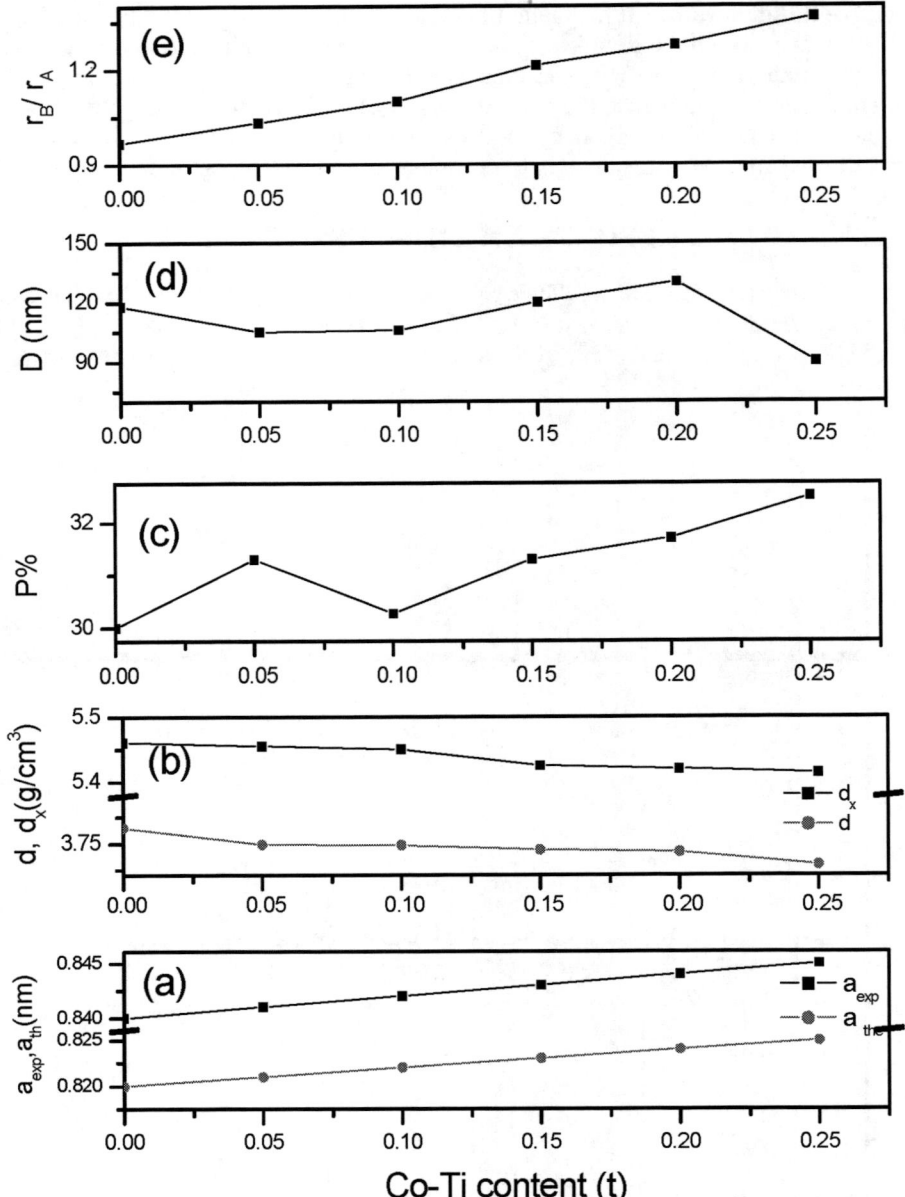

FIGURE (2:a-e) Variation of the experimental and theoretical lattice parameter a_{exp}, a_{th}, X-ray density d_x, experimental density, porosity, grain size D (nm) and octahedral and tetrahedral ionic radii r_B /r_A with the composition (t) respectively.

and $r_A = 0.2 r_{Cd}^{+2} + (1-y-t) r_{Fe}^{+3}$ (4.a)

$r_B = 0.5 [(0.8+t) r_{Co}^{+2} + t r_{Ti}^{+4} + (1+y-t) r_{Fe}^{+3}]$ for $t \leq 0.10$ (4.b)

Where r_{Cd}^{+2}, r_{Fe}^{+3}, r_{Co}^{+2} and r_{Ti}^{+4} are the ionic radii of Cd^{+2}, Fe^{+3}, Co^{+2} and Ti^{+4} respectively. The theoretical and experimental values of the lattice parameter (a_{th} and a_{exp}) are plotted against CoTi content (t) in Fig. (2.a). From this figure, one can see that (a_{th} and a_{exp}) increase slightly with increasing the composition (t). This increment may be due to substitutional occupancy since the ionic radii of Co^{+2}, Fe^{+3}, Ti^{+4} in octahedral sites are (0.074 , 0.64 and 0.060 nm) respectively. The difference between a_{exp} and a_{th} can be attributed to the presence of some divalent iron ions Fe^{+2}. The increase of Ti^{+4} (t) and Co (0.8+t) ions content consequently leads to the creation of Fe^{+2} in the spinel lattice.

$Co^{+2} \longleftrightarrow Co^{+3} + e$

$Fe^{+3} + e \longleftrightarrow Fe^{+2}$

From X-ray data and except for x = 0.0 there exists a secondary phase indexed as Co titanate (CoTiO$_3$) which is increased in intensity with t. The variation of the apparent density (d) with composition as well as X-ray density (d$_x$) is represented in Fig. (2.b). It is clear that both densities decrease slightly with increasing (t) i.e. the apparent density reflects the same general behavior of the theoretical density. This can be ascribed to the atomic weight and density of Co^{+2} (58.93, 8.96 gm cm^{-3}) and Ti^{+4} (47.9, 4.51 gm cm^{-3}) respectively which are lower than those of Fe^{+3} (55.847, 7.86 gm cm^{-3}). The percentage of porosity was calculated using the relation P = 100 [1-d/d$_x$] where (d) is the apparent density and (d$_x$) is the X-ray density. Fig. (2c) shows that when the porosity increases, the density becomes lower. The porosity of ferrite samples results from two sources intragranular porosity P$_{intra}$ and intergranular porosity P$_{inter}$. Thus, the total porosity P is the sum of two types [15] that is P = P$_{intra}$ + P$_{inter}$. It was reported by Rezelescu et al. [16] that as the grain size increase the intergranular porosity increase. Consequently, the slight increase in the grain size with (t) points towards the small contribution of the intergranular porosity and the pronounced effect of the intragranular one.

Figure (2.d) illustrates the dependence of crystallite size (D in nm) on (t) as calculated from Scherrer formula [17]. It is observed that D increases slightly with (t) reaching maximum value at t = 0.20 and then decreases again. This can be interpreted as an increase of the lattice parameter with (t). Moreover the appearance of secondary phase CoTiO$_3$ as indexed from XRD on the grain boundaries impedes the grain growth. This secondary phase produced on the surface of ferrite grains during the sintering process induces lattice distortions in the internal grain region. These lattice distortions correspond to a lattice contraction might be attributed to the compression induced by the discrepancy in the thermal expansion coefficient between bulk and intergranular material or to the lattice mismatch between grain and grain boundary phase. Since at t = 0.25, this secondary phase was more pronounced then it leads to the decrease of the grain size. From Fig. (2.e), it is obvious that the ratio between the octahedral and tetrahedral site radii as calculated from eq. (4.a, b) increases with increasing t

Figure (3: a, b) illustrates the relation between the molar magnetic susceptibility and the absolute temperature as a function of the magnetic field intensity for the samples Cd$_{0.2}$Co$_{0.8+t}$Ti$_t$Fe$_{2-2t}$O$_4$; (t= 0.0 and 0.25). The general trend of the data is the same where the susceptibility increases slightly reaching a hump at ≈ 600 K for the sample with t = 0.0 which varies in position and intensity depending on the magnetic field intensity. This means that the magnetic ordering increases slightly achieving maximum at this hump and then decreases sharply reaching the Curie temperature at about 780K which is in good agreement with the previously reported for substituted Co ferrite [18]. After the Curie temperature, the sample behaves as a typical paramagnetic material. The values of χ_M decrease with increasing the

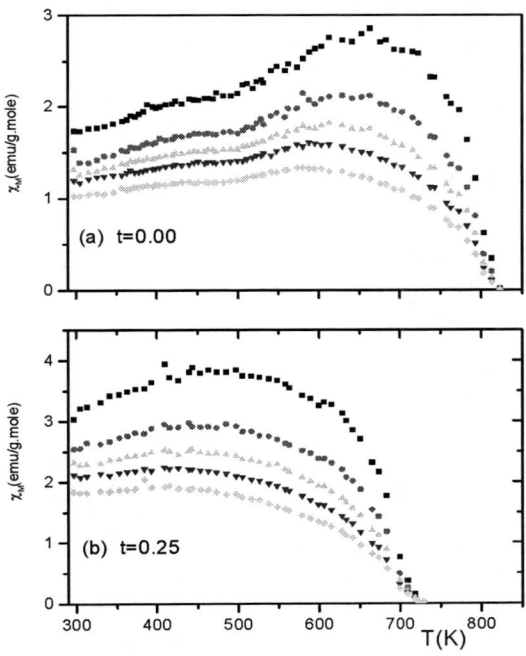

FIGURE (3: a, b): Temperature dependence of the molar magnetic susceptibility χ_M as a function of magnetic field intensity for the samples with t = 0.0 and 0.25.

magnetic field intensity due to the saturation of the magnetic moment.

The magnetic susceptibility value increases by the introduction of Co and Ti on the expense of Fe^{3+} ions with small amount of t = 0.05 as in Fig. (4. a). Further increase in t decreases χ_M and achieving a minimum value at t = 0.15 and then increases again to the higher point at t = 0.25. This trend can be interpreted on the basis of the dependence of the exchange interaction on the cation distribution. In other words, in the samples with $0.0 \leq t \leq 0.10$ Co^{2+} ions reside on the octahedral site, thereby decreasing the magnetization of the B site (M_B). Increasing t, i.e. $0.15 \leq t \leq 0.25$ a small number of Co^{2+} ions occupy the tetrahedral sites, this number increases slightly with increasing Co-Ti content. Consequently a weakening of J_{BB} with respect to J_{AA} will take place which is enhanced by the increase of the ratio (r_B/r_A) with increasing t as in Fig. (2.e).

According to Néel model, the Curie temperature for ferrites is related to number of Fe ions on both sites their intersublattice distance and bond angles. The increase in T_C with increasing (CoTi) substitution, Fig.(4.b), is mainly attributed to the increase in the strength of J_{AB} with increasing t. When Co-Ti (t) content is increased assumed that Co^{2+} ions are in high spin; it caused the increase of AB interaction and then T_C increased until t = 0.15 and decreased for t $\leq$ 0.2 as a consequence of the migration of some Co^{2+} ions from the B to the A site. This explains the peculiarity that occurred at t=0.15.

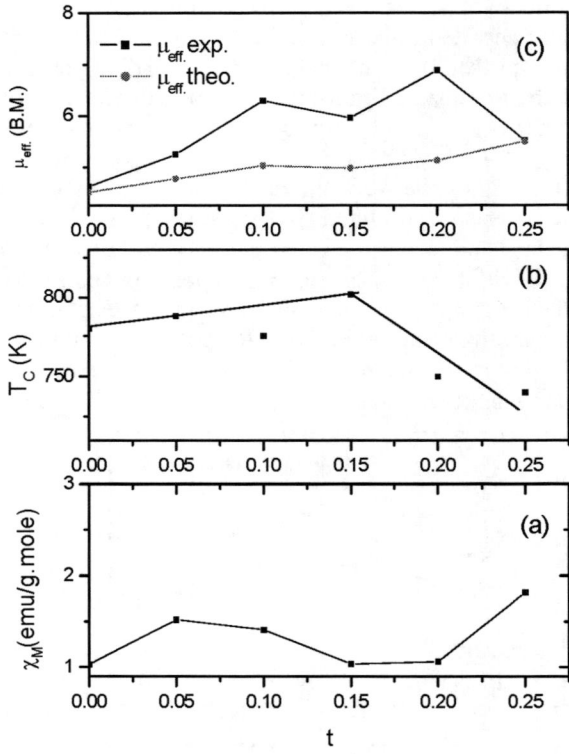

FIGURE. (4: a-c) Dependence of the room temperature magnetic susceptibility χ_M, Curie temperature, experimental and theoretical magnetic moment respectively at 3000Oe on the composition (t).

The variation of the effective magnetic moment with Co-Ti content is clarified in Fig. (4.c). The theoretical magnetic moment was calculated from the postulated cation distribution mentioned before assuming that Co^{2+} ions are always in high spin state configuration ($t_{2g}^5 e_g^2$) on both B and A sites, the values are reported in table (1). The calculated values of the effective magnetic moment increase with increasing t both experimentally and theoretically except the anomaly that appears at t = 0.15. The difference between the experimental and theoretical values may be due to the existence of Fe^{2+} ions generated from the well known Verway mechanism and the probability of the valence exchange between Co^{2+} and Co^{3+} ions.

TABLE 1. Values of the magnetic constants for the samples $Cd_{0.2}Co_{0.8+t}Ti_tFe_{2-2t}O_4$; $(0.0 \leq t \leq 0.25)$

t	$\mu_{exp.}$(B.M.)	$\mu_{theor.}$(B.M.)	C (emu/g.mole)K	θ (K)
0.00	4.62	4.51	2.67	760
0.05	5.25	4.77	3.45	748
0.10	6.30	5.05	4.90	719
0.15	5.96	4.97	4.45	752
0.20	6.86	5.13	5.92	687
0.25	5.52	5.49	3.82	670

CONCLUSION

1. The ferrimagnetic sample $Cd_xCo_{1-x+t}Ti_tFe_{2-2t}O_4$; x=0.20, $0.00 \leq t \leq 0.25$ reveal a single phase spinel cubic structure and homogeneity.
2. The experimental and theoretical lattice parameter increase slightly with increasing Co-Ti content.
3. The values of magnetic susceptibility χ_M, Curie temperature T_C and magnetic moment give peculiar behavior at t = 0.15.

REFERENCES

1. T.T.Ahmed, I.Z.Rohman, M.A.Rahman, J. Mat Process Technol **153-154**, 797 (2004).
2. S.Dey and J.Ghose, Mater. Res. Bull. **38**, 1653 (2003).
3. Misbah, UI Islam, M.U.Rana, T. Abbas, Mater Chem Phys.**57**, 190 (1998).
4. O.M.Hemeda ,M.I.Abd El-Ati, Materials Lett. **51**, 42 (2001).
5. R.Chandrasekhar et al , Journal of Imaging Technology **13**, 55 (1987).
6. M.Sincai D.Ganga D.Bica L.Vekas, J. Mag.Mag.Mat. **225**, 235 (2001).
7. K.Haneda A.H.Morrish J.Appl.Phys **63**, 8 (1988).
8. P.N. Vasambekar C.B. Kolekar A.S. Vaingankar J. Mater Chem. Phys **601**, 282 (1999).
9. M.Rajendran, R.C. Pullar, A.K. Bhattacharya, D.Das, S.N.Chintalapudi, C.K.Majumdar J. Mag.Mag.Mat **232**, 71 (2001).
10. S.A.Mazen and M.A. Ahmed, Phys. Stat Sol (a) **73**, K307 (1982).
11. M.A.Mousa, A.M.Summan, M.A.Ahmed and A.M.Badawy Thermochimica Acta **144**, 45 (1989).
12. M.A. Ahmed and Samiha T.Bishay, Journal of Physics and Chemistry of Solids **64**, 769 (2003).
13. E.C.Snelling, "Soft Ferrites Properties and Applications" (London, life books lid, (1969).
14. S.A.Mazen, M.H. Abdallah, B.A.Sabrah and H.A.M.H. Hashem Phys. Status Solidi a., **134**, 263 (1992).
15. W.D.Kigery, H.K. Bowen and D.R.Uhlmann, Introduction of ceramics John Wiley and Sons, 1975.
16. N Rezelscu, E. Rezelscu, C.Pasnicu, and M.L.Craus, J.Phys. Cond. Matter **6**, 5707 (1994).
17. B.D.Cullity , Elements of X-ray diffraction, Addisson-Wesley INC, 1978.
18. J.Smit and H.P.J.Wijin, Ferrites, 1959.

Preparation and characterization of Ni-Ferrite Nano Particles

M.M. El-Okr, Ayser S. Al-Aloosy,[*] R.M. El-Okr[**], M.A. Salem,
M.Ashoush

*college of health studies, Kuwait.
** NCR, Nasr city, Cairo.

Abstract: Mixed ferrite system of com position $Ni_x Fe_{3-x} O_4$ Where (x= 1,1.25,1.5,2) (inverse spinel) has been prepared in nano form. Ni and Fe nitrates are mixed and shacked thoughrly, then NaOH has been used as a precipitating agent. Obtained samples were washed by the ionized water and ethanol. Finally dried at $70^{\circ}C$ for one day the ingots have been annealed at different temperatures . XRD and ME spectra have been measured. It was found that the particle size increases by increasing annealing temperature. All samples annealed at $T \leq 650^{\circ}C$ show superparamagnetic effects, Samples annealed at 900,$1000^{\circ}C$ exhibit magnetic hyperfine interaction reflecting high magnetic moments.

INTRODUCTION

Over the past decade, nanomaterials have been the subject of enormous interest. These materials, notable for their extremely small feature size, have the potential for wide – ranging industrial, biomedical, and electronic applications. Nanomaterials can be metals, ceramics, polymeric materials, or composite materials [1]. Their defining characteristic is a very small feature size in the range of 1-100 nm. The nanoworld lays midway between the scale of atomic and quantum phenomena, and scale of bulk materials. As the size reduces into the nanometer range, the materials exhibit peculiar and interesting mechanical and physical properties [1]. Nanomaterials can be classified into nanocrystalline materials and nanoparticles. The former are polycrystalline bulk materials with grain sizes in the nanometer range (< 100nm), while the latter refers to ultrafine dispersive particles with diameters below 100nm. Nanoparticles are generally considered as the building blocks of bulk nanocrystalline materials. It has been found that a combination of covalent and non-covalent synthesis is the best way for the preparation of materials of finite dimension in the nano-size range [2].

Huang et al. [3] reported that nanocrystalline copper has a much higher resistivety than bulk copper. They attributed this effect to the grain-boundary enhanced scattering of electrons. Also, when the crystallites of materials are reduced to nanometer scale, there is an increase in the role of interfacial defects: grain boundary, triple junction, and elastically distorted layers [4].

Synthesis and characterization of ferrite nanocrystalline materials have been shown to be very promising because of their potential fields of applications [5-8].

Xiong Gong et al. [9] synthesized and studied Co Cr FeO_4 nanocrystals. They found that this composition exhibit a supermagnetic transition at temperature of about 249 K. Compared to the bulk form, nanoscaled spinel zinc ferrite (Zn $Fe_2 O_4$) shows unusual magnetic properties, which has attracted much attention[10,11].

Nickel ferrite, Ni $Fe_2 O_4$, is an inverse spinel in which the tetrahedral sites (A), are occupied by Fe2+ ions and the octahedral sites (b) by Fe2+ and Ni2+ ions [12]. This material is largely used in electric devices and in catalysis. Thus, different convential and non-convental methods for obtaining Ni-ferrite in the form of nano-structured powders are current subject [13, 14]. Nano ferrite system can exhibit ferro or ferrimagnetic ordering beside their magnetic moments. In the present work, we investigated the synthesis of the nano-structured Ni Fe_2O_4 powders by the co-precipitation method, followed of annealing by temperatures between $350^{\circ}C$ and $1000^{\circ}C$. The structural and magnetic properties of the obtained samples were characterized by X-ray diffraction, transmission electron microscope (TEM), and Mossbauer spectroscopy.

CP888, *Modern Trends in Physics Research*,
Second International Conference on Modern Trends in Physics Research—MTPR-06,
edited by L. El Nadi
© 2007 American Institute of Physics 978-0-7354-0354-9/07/$23.00

EXPERIMENTAL

1. Sample preparation

Ni-ferrite powder was synthesized by dissolving Fe and Ni nitrates in deionized water in the required mole proportion according to the formula $Ni_x Fe_{3-x} O_4$ (x=0,0.5,1,1.5 and 2). After vigorous stirring for one hour, the precipitate was filtered, washed by deionized water and ethanol about three times. After drying for 24 hour at 70°c, ultra thin powder was obtained. Samples with different average particle size were prepared by annealing the precipitate at temperatures between 300oc and 600oc for 2h under ordinary atmosphere. A well-crystallized reference sample was prepared by heat treatment at 1100°c . The co-precipitation synthesis of nickel ferrite can be described by the following chemical reaction:

$$1Ni (NO_3)_2.6H_2O+2Fe(NO_3)_3.9H_2O+8NaOH \rightarrow NiFe_2O_4+8NaNO_3+28H_2O$$

2. CHARECTERIZATION OF THE SAMPLES

X-ray diffraction (XRD) patterns were recorded in the range $5° < 2\theta < 80°$ in a diffractometer using Cu Kα radiation (λ= 1.5418 A).
Transmission electron microscope (TEM) was performed under acceleration voltage of 200 kV.
Mossbauer spectra were recorded with the spectrometer of electro-mechanical type with constant acceleration. A source 57Fe in rhodium matrix, held at room temperature was used. AFe foil was used for calibration. The fitting of Mossbauer data was carried out by using the least square fitting program considering Lorentz shape function. The isomer shift values reported in the present work are computed with respect to α-Fe at room temperature.

RESULTS AND DISSCUSSION

XRD:

Representative XRD patterns of prepared samples are shown in fig (1). Broad well defined peaks are observed of samples annealed at 650, 900, and1000°C. However samples annealed at T < 650°C show only humps. Obtained XRD data were used to roughly estimate the particle size. The results follow the expected pattern where the particle size increases by increasing temperature (table 1) all the prepared samples lie in the nano range.

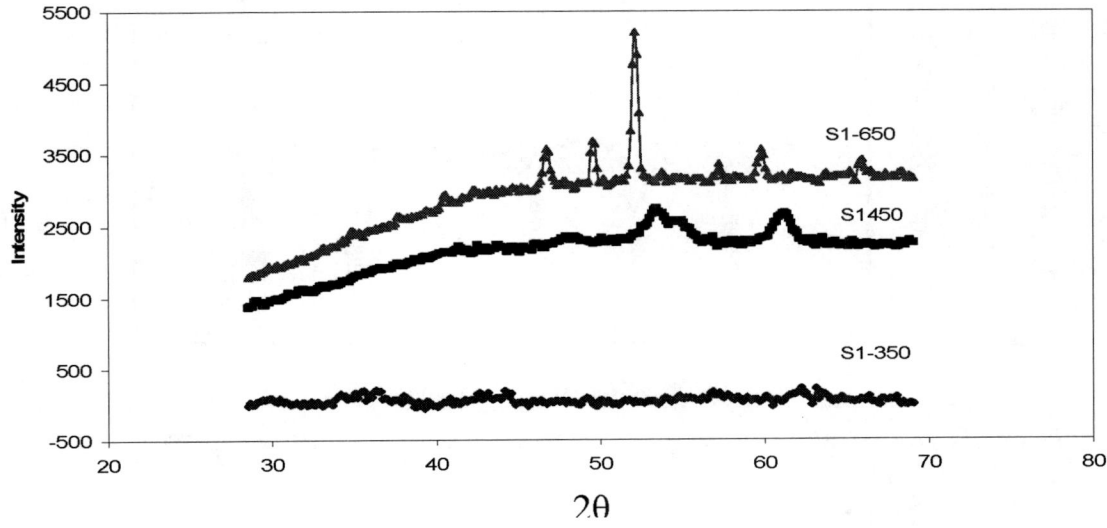

FIGURE 1. XRD of sample with x=1which annealed at 350, 450, 650° C

129

TABLE 1. Represented the particle size of the different samples estimated by XRD.

x T	1	1.25	1.5	2
R.T	7	9	11	8
350	10	11	12	10
450	12	13	14	12
650	12	14	15	14
900	16	16	16	14
1000	16	16	18	16

ME SPECTRA

ME spectra of prepared samples were obtained at R.T. All the as prepared ferrite nanoparticles exhibit strong superparamagnetism. This confirms the small particle size. However, no observed difference in the position of the peak by changing compositions. [Fig2] shows representative spectra of some as prepared samples. Moreover, for samples annealed up to 650oc the estimated isomer shift I.S. was found to be relatively large indicating law electron density at nucleus. Moreover, I.S. of such samples is almost particle size independent. However, samples annealed at 900 and 1000oc exhibit reduction in I.S. and appearance of magnetic hyperfine interaction (Obtained values of ME parameters are listed in table 2).

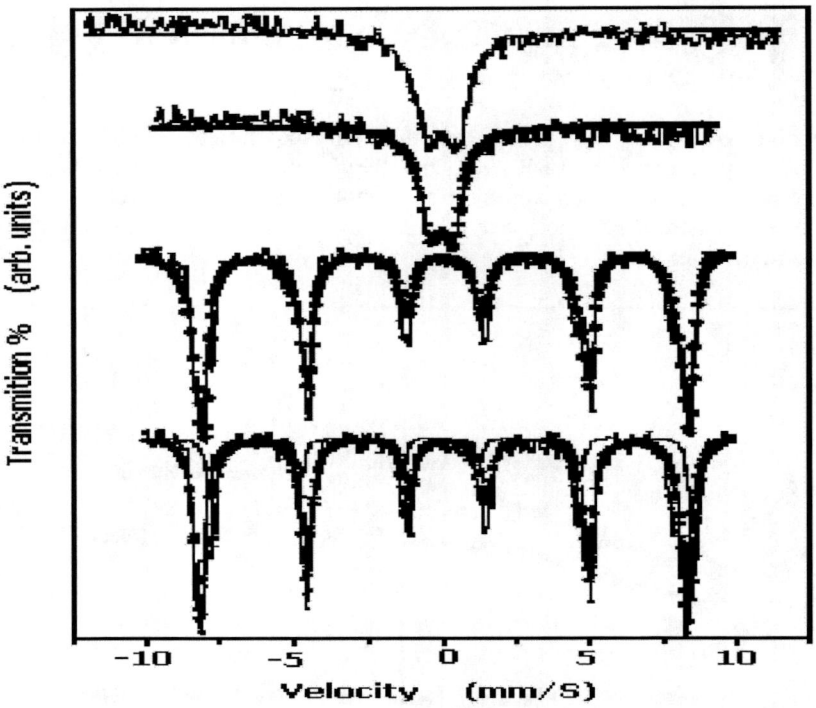

FIGURE 2. Mossbauer spectra for the sample with x=1 which annealed at 350,450, 650,and 900 °C respectively.

TABLE 2. represents the parameters of ME spectra for (a) sample1 with x=1 annealed at 350,450,650 and 1000°C (b) sample 2 with x= 1.25annealed at1000°C (c)sample 3 with x=1.5 annealed at 1000°C.

S. no	Ni : Fe	Ann. temp. C	HF_1	HF_2	IS_1	IS_2	QS_1	QS_2	Lw_1	Lw_2
1	1:2	350	--	--	0.290	--	0.80	--	0.83	--
	1:2	450	--	--	0.270	--	0.740	--	0.76	--
	1:2	900	513	--	0.290	--	-0.104	--	0.55	--
	1:2	1000	506	511	0.180	0.14	-0.210	0.39	0.35	0.42
2	1.25:1.75	1000	483	517	0.160	0.27	0.270	0.37	0.42	0.48
3	1.5:1.5	1000	484	516	0.156	0.26	-0.230	0.17	0.43	0.44

Quadrapole shift reflects the field gradient around Fe nucleus. The deviation from spherically symmetric field is the origin of Q.S. Two interesting features are observed: Firstly, Q.S. values are relatively small and it is almost annealing temperature (particle size) independent. Secondly, Q.S. is independent on Ni content. This is most likely can be accounted for by assuming that is the origin of the deviation of field symmetry is meanly due to 3d.electrons.

HYPERFINE INTERACTION

As has been mentioned all samples annealed at temperatures up to 650oc exhibit strong paramagnetism. However samples annealed at 900 and 1000o c exhibit strong magnetic hyperfine interaction. For annealing at 900o c one sextet is observed having the following parameters: H=51.3T, Q.S.=0.14 , I.S.=0.39, L_w=0.55 . it is worth mentioned that pure iron exhibit $\approx$33T, so the observed high magnetic field in the present study is directly related the high magnetic moments associated with nano ferrites. Samples annealed at 1000oc exhibit two sextets with the parameters:
Q.S1= 0.39 , I.S1= 0.18 , LW_1= 0.35 , HF1= 50.6T
Q.S2 = 0.21 , I.S2 = 0.14 , Lw_2 = 0.42 , HF2 = 54T
This is most likely due to A site and B site respectively.

REFERENCES

1. H. Gleiter, Prog. Mater. Sci. **33**, 223 (1989).
2. M.A. Balbo Bolock, et al., Top. Curr. Chem. **245**, 89 (2005).
3. Y.K. Huang, A.A. menovsky, F.R. de Boer, Nanostruct. Matter.**29**, 505 (1993).
4. R.A. Andrievski and A.M. Glezer, Solid Satate. Comm. **44**, 1621 (2000).
5. I. Safarik and M. Savarikova, Mont. Chem. **133**, 737 (2002).
6. D.O. Yener and H. Giesche,J.Am.Ceram. Soc.**84**, 1987 (2001).
7.Y.Zhang,A.Kolmakov,S.Chretien,H.Metiu and M.Moskovits,Nano Lett.**4** , 403 (2004).
8. S.F. Yin, B.Q. Xu, C.F. Ng. . and C.T. Au, Appl. Catal., B, Environ.**4**, 237 (2004).
9. Xiong Gang et al., Chinese Phys. Lett. **18**, 692 (2001).
10. Z.H. Zhou, J.M. Xuc, H.S.O. Chan and J.Wang , Mater. Chem. Phys. **57**, 181 (2002).
11. L. Wang, Q.G. Zhou and F.S. Li, Phys. Status Solidi, B **241**, 377 (2004).
12. A.H. Morrish and K. Haneda, J. Appl. Phys. **52**, 2497 (1981).
13. A. Verma, T.C. Goel and R.G. Mendiratta, J. Magn. Magn. Mater. **210**, 274 (2000).
14. A.S. Albuquerque, J.D. Ardisson, W.A.A. Macedo, J.L. Lopez, R.Paniago and A.I. C. persiano, J. Magen.Magen.Mater.**226**(2001)1379.

Optical Absorption Spectra of Sodium Borate Cobalt Doped Glasses

M.M. Elokr[1], A.M. Yaseen[2], R. Elokr[3] , M.A. Hassan[1]

1- Physics Dept. Faculty of Science. Al Azhar University. Cairo, Egypt
2-Physics Dept. Faculty of Education . Ain shams University. Cairo, Egypt.
3-National center of Radiation technology. Cairo, Egypt.

ABSTRACT. Glassy system: xNa_2O-$(100$-x-$y)B_2O_3$-yCo_3O_4 has been prepared by conventional melt quenching technique. Optical absorption spectra have been obtained in the range $300 - 2500$ nm at room temperature. An absorption edge was observed in the near UV range, the analysis of which reveals that indirect transition is the dominant absorption mechanism. All prepared samples exhibit blue color, indicating that the Co ions are acted upon by tetrahedral ligand field. Obtained spectra were used to estimate some ligand field parameters.

INTRODUCTION

Oxide glasses are materials of considerable technical importance, their transparency and chemical resistance gave rise to a lot of applications in construction industry, in optics, chemical technology [1]. Bamford [2] studied the application of the ligand field theory to colored glasses in sodium borate glasses containing cobalt. The transition metal ions were in octahedral co-ordination. The increase in the Na_2O content in the glasses increases the proportion of metal ions in the higher valency state. The intensity and wavelength of the absorption bands changed markedly for glasses containing more than 15 % by weight of Na_2O. It was found that the color of the glasses varies from pink through mauve to blue with increasing Na_2O content. The color of the glasses containing cobalt (II) changes from pink to blue when the co-ordination number of cobalt decreases from six to four [3,4]. Also Beaury et al [5], studied the blue allotropic form of Co^{2+} : $Na_2CoP_2O_7$: Optical properties. Two new allotropic forms of this diphosphate have been characterized: the first one is rose and belongs to the triclinic crystallographic system with a site symmetry C1 for Co^{2+} , the second one is blue and orthorhombic with a site symmetry Cs for Co^{2+} .

EXPERIMENTAL TECHNIQUE

Samples of boron sodium and cobalt oxides as: xNa_2O-yCo_3O_4-$(100$-x-$y)$ B_2O_3 where x=20, 25, 30, 35 at y=1.0 mol % (donated by system I), and x=30 at y=0.0, 0.33, 1.0, 1.5 mol % (donated by system II). The samples were prepared by weighting the oxide powders (in the form sodium carbonate-boric acid -cobalt oxide) by using an electrical balance of sensitivity 0.001g. Homogenous mixture was melted in porcelain crucible in an electric muffle furnace at 1000 oC for two hours. The molten stirred twice to get homogeneity of the samples, and then it is quenched in air on stainless steel plate kept at room temperature. The glass surface of all samples was made plane by polishing with fine emery paper to small thickness of order 0.5-0.3 mm. optical measurements were obtained for all studied glassy samples with wavelength between 300-2500 nm, using a spectrophotometer (Carl Zeiss-PMQ), the transmission T %, were measured at room temperature

CP888, *Modern Trends in Physics Research,*
Second International Conference on Modern Trends in Physics Research—MTPR-06,
edited by L. El Nadi
© 2007 American Institute of Physics 978-0-7354-0354-9/07/$23.00

RESULTS AND DISCUSSION

1. Absorption in UV Region

All cobalt doped samples has blue color, while free cobalt sample is colorless [2]. Figure (1) shows the obtained absorption coefficient in cm^{-1} versus photon energy (in eV). The edge extended over wide wave length range, confirming the amorphous nature of the prepared samples. The optical band gap energy $E_{Opt.g}$ and band tail width E_t have been determined in the region 4.01-3.27 eV. The spectral dependence of the absorption coefficient has the form [6]

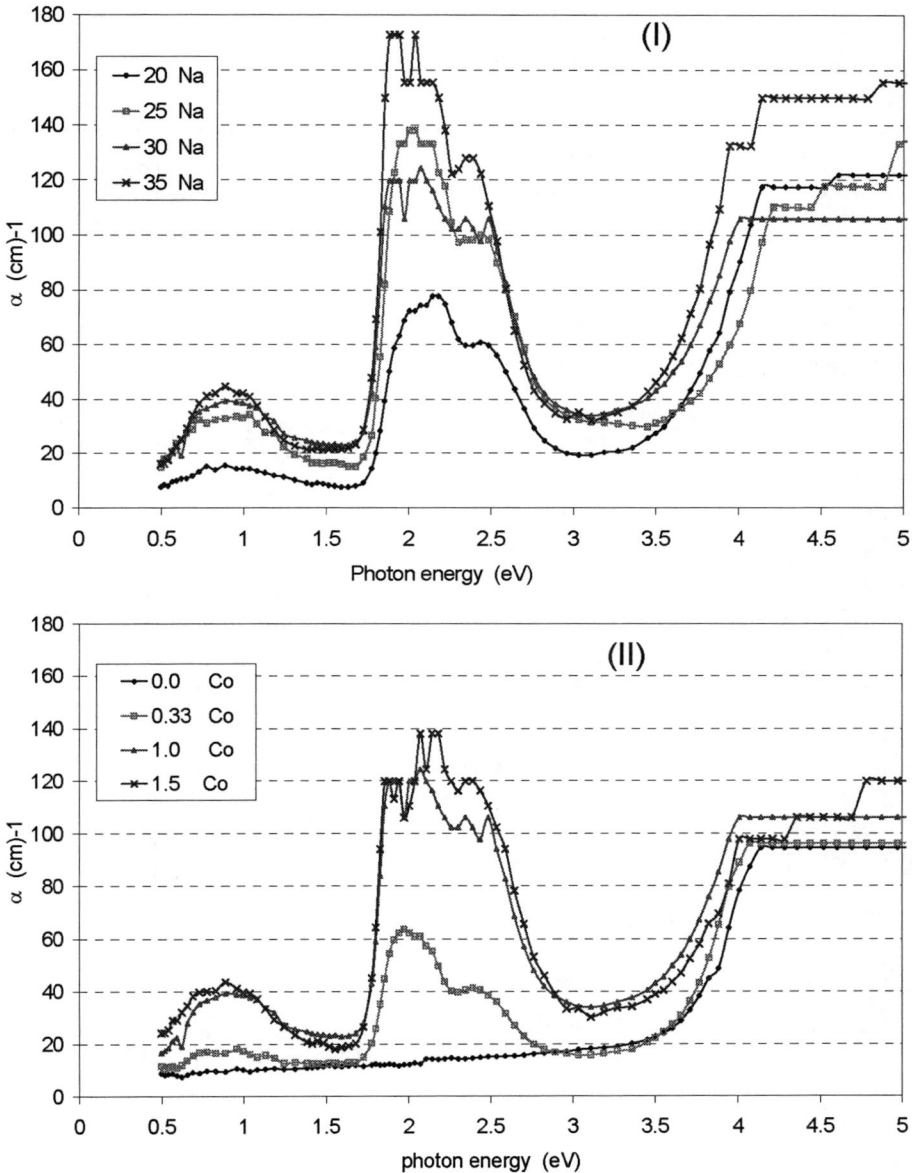

FIGURE 1. the Absorption Coefficient as a Function of Photon Energy.

$$\alpha\hbar\omega = A(\hbar\omega - E_{Opt.g})^n \qquad (1)$$

where A is a constant including the thickness, and $E_{Opt.g}$ is the optical band gap energy and n is a parameter. The best fit in many amorphous or glassy materials was found to be with n=2 this is the characteristic behavior of indirect transition. The later is usually observed in amorphous systems due to the lake of translation

symmetry, where the wave vector k is not good quantum number. The dependence of $(\alpha\hbar\omega)^{1/2}$ on photon energy $\hbar\omega$ .the band gap is estimated by extrapolation to $(\alpha\hbar\omega)^{1/2}=0$ gives the best linear dependence. The estimated values of $E_{Opt.g}$ are listed in table (1), and plotted in figure (2) as a function of composition. It is clear that the band gap decreases with increasing sodium oxide up to $x \leq 30$, then attains constant value up to $x = 35$. This is most probably due to the increase of tetrahedron structure units at the expense of octahedron units by Na_2O addition.

TABLE 1. Band Gap Energy ($E_{Opt.g}$ ev) and Band Tail Width (E_t ev) for All Investigated Samples.

System I			System II		
Na_2O	$E_{Opt.g}$	E_t	Co_3O_4	$E_{Opt.g}$	E_t
20	3.171	0.4115	0.00	3.228	0.3612
25	3.045	0.4585	0.33	3.274	0.2976
30	2.929	0.5034	1.00	2.929	0.5034
35	2.917	0.5058	1.50	2.755	0.5715

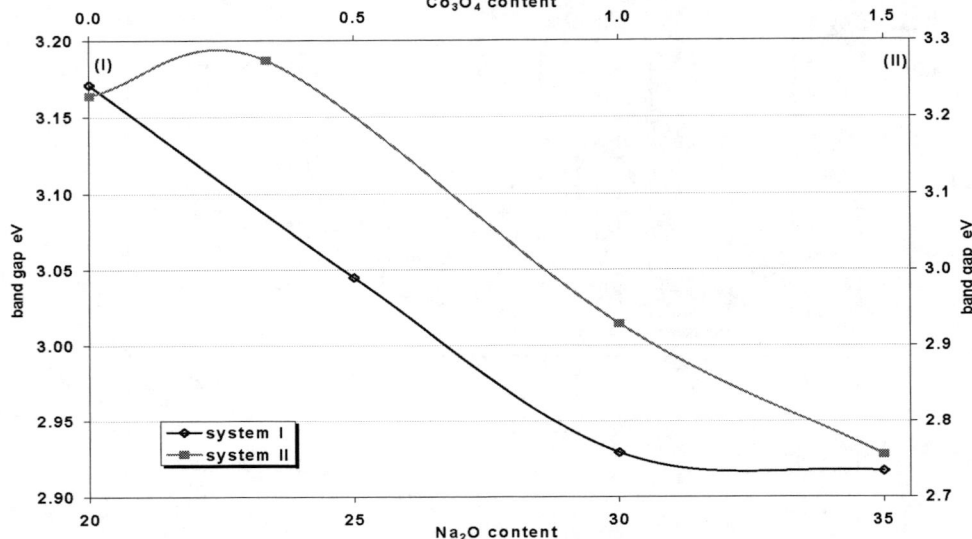

FIGURE 2. the Composition Dependence of Band Gap Energy.

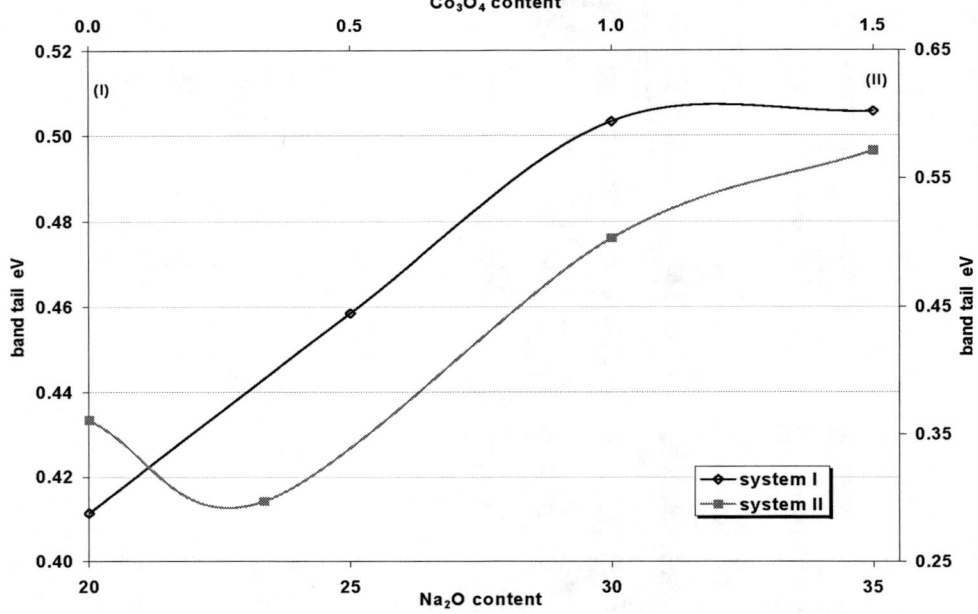

FIGURE 3. the Composition Dependence of Band Tails.

The band gap increases with the addition of cobalt oxide content up to $y \leq 0.33$. This is attributed to the blocking effect, the addition of cobalt affect the oxygen in the glass network, where the cobalt ions occupy the tetrahedral coordination; this is consistent with the observed blue color. And then decreases with increasing cobalt oxide content to follow the same pattern of electrical and IR measurements. Is attributed to the ionic radii of both cobalt cation is larger than that of boron cation) [6]. By more addition of cobalt oxide ($0.33 \leq y$), the observed a linear decrease of optical band gap energy $E_{Opt.g}$, is most likely due to enhancement of electronic hopping conduction. In such case Co ions are distributed among octahedral and tetrahedral sites. Thus, the investigated glasses exhibit mixed conduction phenomenon in which ionic conduction as well as electronic conduction [7].

The width of the band tail was estimated using Urbach's empirical formula [6,8]

$$Ln\alpha = Ln B + (\hbar\omega / E_t) \tag{2}$$

where B is a constant, and E_t is the width of the tail (of localized states) in the optical band gap, the estimated values of $E_t (eV)$ are listed in table (1) and plotted in figure (3) as a function of composition. The band tail shows opposite trends to that of band gap energy.

2. Ligand Field

As shown in figure (1), two absorption bands for all doped samples with cobalt oxide are observed. the first band located between 1.48-0.62 eV (840 – 2000 nm), this band is weak broad band assigned as (V_2). The second band located at higher energy between 3.27-1.8 eV (380 – 690 nm) and assigned as (V_3). The V_3 band is splits into two bands, which may be caused by lowering symmetry around Co ions. The observed blue color may indicate that tetrahedral sites are more equipped with respect to octahedral ones [2, 3, 5, 9-12].

The low energy band (V_2) is known to be due to ${}^4T_{1g} \leftarrow {}^4A_2$ transition and the other (V_3) ${}^4T_1(p) \leftarrow {}^4A_2$ transition. Both transitions are considered as multiple absorption [9]. This explains the observed broadening of such bands. Parameters of the legend field acting on Co ions can be estimated from the values of the crystal field parameter $10Dq$ and the anti-electronic repulsion Racah parameter B which are related only to electronic transition. In the case of d^7 tetrahedral site (octahedral d^3) the following expressions are used to determine the specific values of Dq and B from the allowed experimental transition [5,13]

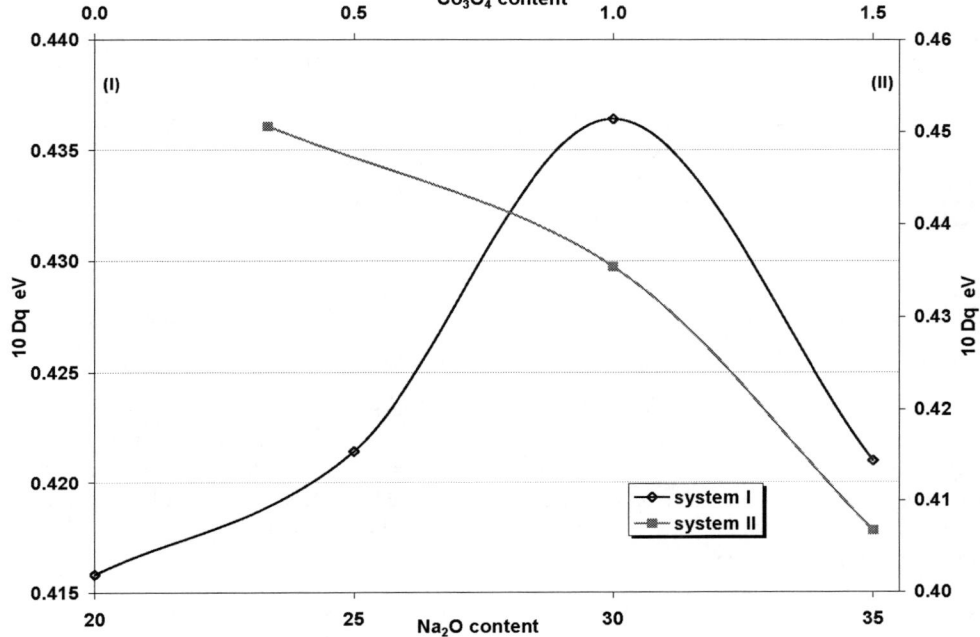

FIGURE 4. The Composition Dependence of Crystal Field Parameter (10 Dq)

$$B = \frac{1}{510}[7(\nu_2 + \nu_3) \pm \{49(\nu_2 + \nu_3)^2 + 680(\nu_2 - \nu_3)^2\}^{1/2}] \qquad (3)$$

$$10\,Dq = \frac{1}{3}(\nu_2 + \nu_3) - 5B \qquad (4)$$

The obtained values are summarized in table (2), and plotted in figure (4) and Figure (5) for both systems.

TABLE 2. Crystal Field Parameter (10 Dq) and Inter-electronic Repulsion (Racah Parameter B) for All Investigated Samples.

	System I			System II	
Na_2O	$Dq\ cm^{-1}$	$B\ cm^{-1}$	Co_3O_4	$Dq\ cm^{-1}$	$B\ cm^{-1}$
20	3345.39	1010.02	0.00	---	---
25	3390.49	1015.24	0.33	3619.89	958.92
30	3510.86	1022.56	1.00	3510.86	1022.56
35	3386.92	1015.42	1.50	2109.10	638.45

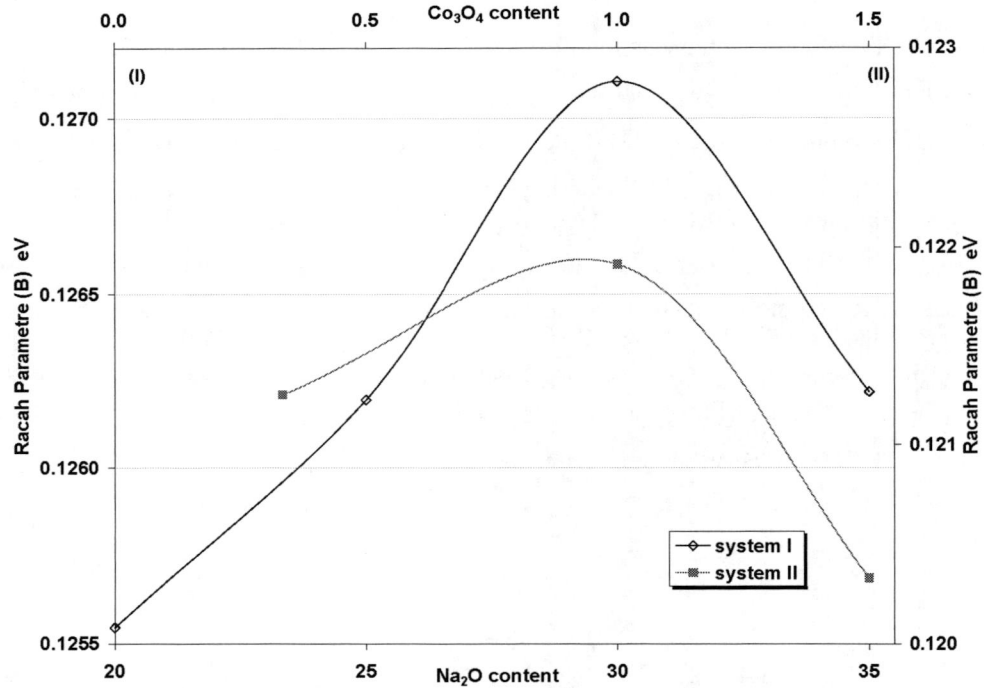

FIGURE 5. The Composition Dependence of Inter-electronic Repulsion (Racah Parameter B).

From the results it clears that, the values of $10Dq$ and B lies in acceptable range. The absorption band were shifted to longer wave number (or higher energy) with increasing sodium oxide content up to $x \le 30$, then the band shifted to shorter wave number at $x = 35$. This is attributed to the increase of compactness. The shift of obtained values to higher energy of B indicates a decreasing covalence [9]. But the absorption bands shifted to shorter wave number (or lower energy) with increasing cobalt oxide content in system (II), which refers to the opening of the structure. While Racah parameter tends to decrease with increasing cobalt oxide content indicating the increase of degree of covalence [9]. It is clear that the change of the legend field around cobalt ions in system I is lower than that in system II.

REFERENCES

1. U. Schoo, H. Mehrer, Solid State Ionics, **130**, 2436 (2000).

2. C. R. Bamford, Physics and Chemistry of Glasses, **3**, No. 6 (1962).

3. J. Lakshmana Rao, G. L. Narendra, S. V. J. Lakshmana, Polyhedron, **9**, 1475 (1999).

4. M. M. Morsi, S. El- Konsol , M. El-Shahawy, J. Non Cryst. Solids **83**, 241 (1986).

5. L. Beaury, J. Derouel, L. Binet, F. Sanz, C. Ruiz-Valero, J. Solid State Chemistry, **177,** 1437 (2004).

6. A.K. Varshneya, "Fundamentals of Inorganic Glasses", Academic press Inc, 1994).

7. E.E. Khawaja, M. Sakhawat Hussein and Khan J. Non Cryst. Solids, **79**, 275 (1986).

8. I.W. Donald, P.W. Millwan, J. Mat. Sci., **13**, 1151 (1978).

9. A.B.P. Lever, Inorganic Electronic Spectroscopy, Elsevier Publishing Company, (1968).

10. L. Galoisy, Laurent, G. Calas, V. Briois. J. Non Cryst. Solids, **293**, 105 (2001).

11. Z. Sun, D. Yuan, X. Duan, X. Wei, H. Sun, C. Luan, Z. Wang, X. Shi, D. Xu, M. Lv, J. Crystal. Growth, **260**, 171 (2004)

12. T. Mimani and Smart Ghosh, Current Science, **78**, No.7, 10 (2000).

13. C. A. Ballhausen, Introduction to Ligand Field Theory ,1962.

Optical Reflectivity and Excitonic Spectra of Nano Ni Ferrite System

M. M. Elokr[1] , Ayser s. Al Aloosy[2] , G. El Zenki[3] M. Ashosh[1],
A. Saadon[1]

1-physics depart. Faculty of science. Al Azhar University. Cairo.
2- college of health studies, Kuwait
3- Faculty of technological studies, Kuwait

ABSTRACT. Nano ferrite system of composition $Ni_xFe_{3-x}O_4$ has been prepared by wet method. Sample has been subjected to annealing at 350, 450 and 650 °C, where the optical reflectivity for annealed samples have been recorded at room temperature. Sharp line spectra have been observed just before the reflectivity edge, these are most likely due to exitonic reflectivity. The obtained results have been used to estimate exitonic parameters such as exitonic radius, confinement energy, number of unit cells contained in exitonic volume and maximum temperature at which the exiton can be observed. It was noticed that the increase of particle size reduces the confinement of exiton.

INTRODUCTION

The study of exitons (exiton and biexiton) exhibit now great importance. They play an important role in the four-wave mixing, where a propagating pulse is strongly altered and exitonic and biexitonic levels show population inversion [1]. In three level systems such as $Cu\,Cl_2$ involved three states are: ground state, exiton and biexiton states. A comparison between experimental results and theoretical predictions has been discussed in terms of exitonic damping [2].

Lately nano-sized crystals find great interest. This is mainly due to the fact that one can control their properties. The energy levels in such systems well separated by $\Delta E > KT$. This in turn increases the degree of localization and hence the conductivity may be reduced. The relatively large energy separation ΔE allows the detection of excitons at high temperatures around the room temperature. In addition reflectivity edge show a blue shift which is usually increase by reducing the crystal size[3].

Magnetic nanoparticles are widely employed in up to date technology. The aim of the present article is to study the optical reflectivity of $Ni_x\,Fe_{3-x}\,O_4$ system, where (x= 1, 1.25, 1.5, 2). The behavior of reflectivity edge and line (quasi-lines) reflectivity spectra have been observed. The obtained results are used to characterize the formed excitons.

EXPERIMENTAL

Mixed ferrite system of composition $Ni_x\,Fe_{3-x}\,O_4$ has been prepared by wet-method [3,4]. Samples were annealed at 350, 450, 650°C. Optical reflectivity has been measured at room temperature. It has been observed that raising the annealing temperature leads to increase in the particle size [4].

RESULTS AND DISSCUSSION

Reflection spectra of all as prepared and annealed samples show a reflectivity edge, just before edges, a well defined line spectra are also observed Fig. Reveals representative spectra. The reflectivity edges exhibit red shifts by increasing annealing temperature (increasing particle size). This could account for by considering the increase of energy separation ΔE by reducing crystal size. The observed red shifts are listed in table [1]. It is worth mention that this behavior is observed for all values of x. Moreover,

CP888, *Modern Trends in Physics Research,*
Second International Conference on Modern Trends in Physics Research—MTPR-06,
edited by L. El Nadi

the edge position and then the optical gap E_{opt} is almost Ni content independent. This may lead to assume that Ni content almost has no effect on optical behavior of Ni ferrites. Similar effects have been observed for ME spectra [4].

EXCITONIC REFLECTIVITY

When an electron in the valence band absorb energy $E=h\upsilon>E_g$ it is exited to the conduction band generating independent free electron and hole. Both move independently and contribute to conduction process. However, if the electron is trapped in one of the localized levels, just below C.B it forms exciton. Both (electron and hole) are bound to each other forming hydrogen like systems. In bulk crystals the separation between trap level and the bottom of the C.B is small (ΔE less than kT). The electron is thermally excited (through phonon) to C.B. though the observation of excitons should be observed, in bulk crystals, at low temperature. On the other hand, in nano-sized crystals the exciton can be detected at relatively high temperature (maybe around and above R.T). The energy levels in hydrogen atom is given by [5,6]

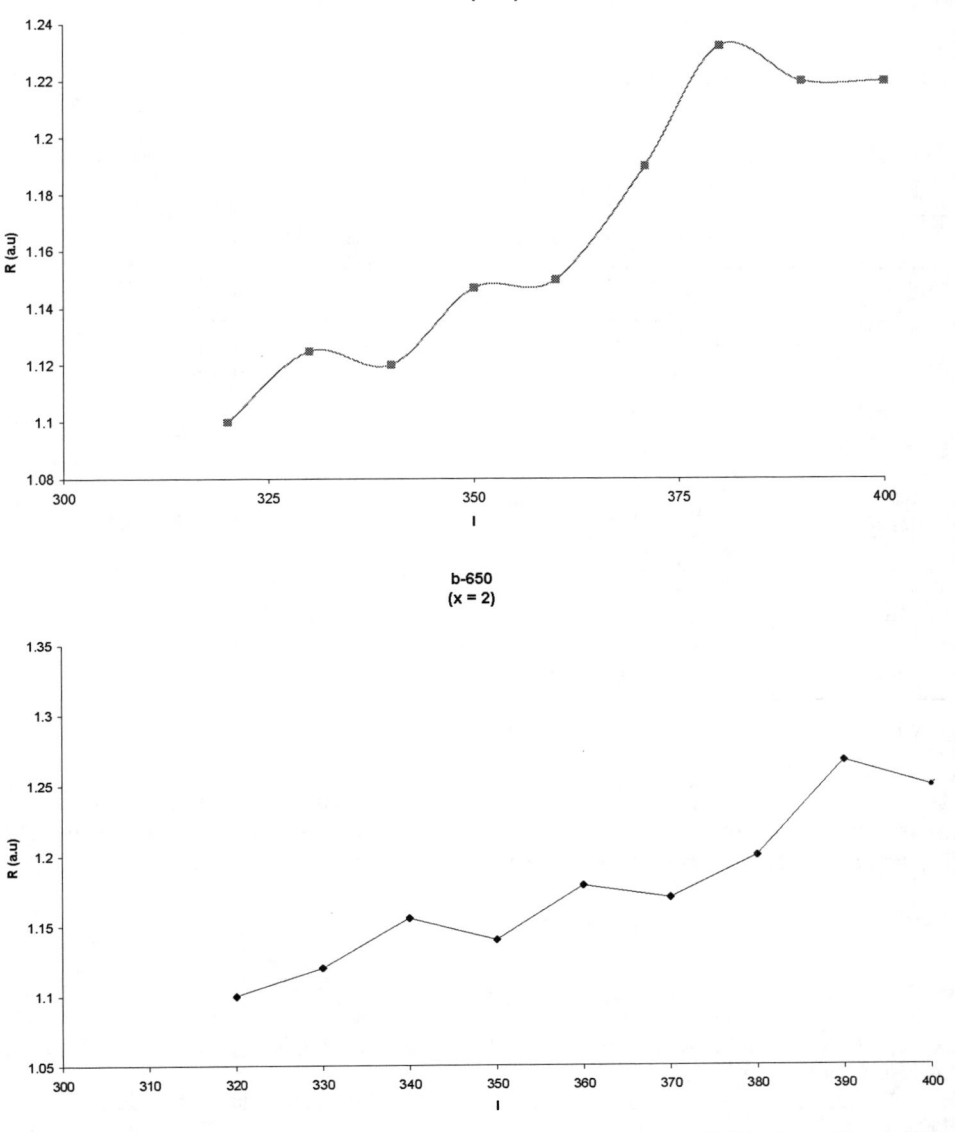

FIGURE 1. Shows the reflectivity spectra of sample x = 1 annealed at 350 $^{\circ}$C (a), x = 2 annealed at 650 $^{\circ}$C (b).

$$E_n = \frac{-R_H}{n^2}$$

Where $R_H = 13.6$ eV is Rydberg's constant and n=1, 2, 3... is the principle quantum number.

For exciton the energy is given by

$$E_n = \frac{-R_X}{n^2} = \frac{-\mu R_H}{\varepsilon_r^2 n}$$

Where R_x is the excitonic Rydberg's constant.

Here ε_r is the dielectric constant of the crystal which can be calculated knowing R_x. drawing a relation between E_n and $1/n^2$, R_x can be estimated and other important parameters.

The radius of exciton $a_x = \frac{\varepsilon_r \times 5.29}{\mu} \times 10^{-11}$

$$V_{ex} = \frac{4\pi a_x^3}{3}$$

The number of unit cells contained within the exitonic volume is N= $V_{ex}/ V_{u.c}$
At the same time the maximum temperature at which exciton can be observed T= R_x/K.

TABLE 1.a: Some excitonic parameters for sample of x=1

Annealing Temperature	R_x, eV	N	T K	E_{Opt}, eV
RT	0.47	2.27715	545.2436	4.1
350	0.428	2.620431	496.5197	4
450	0.28	4.952236	324.826	3.9
650	0.268	5.288545	310.9049	3.6

TABLE 1.b: Some excitonic parameters for sample of x=2.

Annealing Temperature	R_x	N	T	E_{Opt}, eV
RT	0.468	2.291763	542.9234	4.1
350	0.428	2.620431	496.5197	4
450	0.278	5.005774	322.5058	3.8
650	0.27	5.229892	313.2251	3.7

Inspection of table 1 shows that the increase of the annealing temperature T (increase of particle size) reduces N. This is expected according to the degree of confinement, which increases by reducing the particle size. The latter increases by increasing the annealing temperature.

REFERENCE

1. S.S. Montasser, J. Miletic and B. Honerlage. Phys. Rev. B, **40**, 6163 (1989).
2. S.S. Montasser, J. Miletic and B. Honerlage. Phys. Rev. B, **44**, 11490 (1991).
3. A.S Albuquerque, J.D. Ardisson, W.A.A. Macedo, J.L. Lopez, R. Paniago, A.I.C. Persiano. J. Mag. Mag. Materials **226-230**, 1379 (2001).
4. M.M. El-Okr, Ayser S. Al-Aloosy, R.M. El-Okr, M.A. Salem, M. Ashoush. (In the same proceeding)
5. J. Garcia, L.E Bausa and D. Jaque." An introductiom to the optical spectroscopy of inorganic solids" John Wiely & sons, Ltd. 2005, p.139-142.
6. M.M. El-Okr, "A simple approach to explain nano-sized crystals". (In the same proceeding)

II. CHEMICAL PHYSICS, LASER & APPLICATIONS

Plasma-based X-ray Lasers and Applications

A. Klisnick

LIXAM, Bât. 350, Université Paris-Sud, 91405 Orsay Cedex, France

Abstract. We review the status of research on soft X-ray lasers generated from hot and dense plasmas and explain their potential as high-brightness sources for applications in different fields of science.

INTRODUCTION

Two types of amplifying plasmas are used to generate the existing X-ray lasers: (i) those resulting from a fast electrical discharge in a gas-filled capillary, and (ii) those resulting from the interaction of a high-power laser with solid or gaseous targets. Both approches lead to X-ray lasers having distinct and complementary characteristics in terms of output energy, pulse duration, spectral range covered, etc. This paper will mostly focus on the second, laser-pumping approach. Recently the field has been revolutionized by the advent in research laboratories of relatively small-scale ultra-intense (10-100 TW), ultra-short pulse (10 fs - 1 ps) optical lasers, based on the chirped-pulse amplification (CPA) technique. These pump lasers have led to a significant reduction of the energy required to generate an X-ray laser, hence enabling operation at a higher repetition rate.

The reduction in cost and scale and the continuing improvement of the X-ray laser beam characteristics give good prospects for the source to remain competitive in the future and to be complementary with other emerging bright coherent sources, i.e. high-order harmonics radiation, and X-ray free-electron lasers. In particular we will present the main features of the future french LASERIX user-facility, which is currently under construction in Palaiseau, as part of the PÔLA project at Université Paris-Sud.

WHY DO WE NEED X-RAY LASERS ?

Although the lasers which are considered in this paper are widely called X-ray lasers, this is not a fully appropriate denomination: the spectral range in which they emit is situated at the lower energy end of the X-ray region, in a range intermediate between X-ray and UV, usually called the XUV region. This corresponds to typical wavelengths ranging between 2 and 50nm or photon energies between typically 20 and 500eV. Since many primary resonances and absorption edges of low and intermediate Z elements are located in the XUV range, the absorption length is very small (a few μm at solid density). As a consequence XUV radiation, which can propagate over long distances only in vacuum, has been widely used as a tool (e.g. in synchrotron radiation facilities) for elemental matter studies of thin films, surfaces or plasmas.

In the recent years there has been rapid and substantial progress in the development of ultra-short (picosecond to attosecond timescales), ultra-intense sources in the XUV domain. These new sources give access to extreme, unexplored regimes of X-ray interaction with matter, where the study of non-linear processes, warm dense matter or

CP888, *Modern Trends in Physics Research,*
Second International Conference on Modern Trends in Physics Research—MTPR-06,
edited by L. El Nadi
© 2007 American Institute of Physics 978-0-7354-0354-9/07/$23.00

femto- to atto- physics offers new and exciting prospects. Three types of sources are currently under development, with distinct and complementary characteristics: (i) high order harmonics from a femtosecond infrared laser pulse [1], (ii) X-ray free electron lasers [2], and (iii) plasma-based X-ray lasers which are discussed in this paper.

Apart from their short wavelength, coherence and directivity, X-ray lasers have specific properties that are of interest for applications: extremely high brightness and large number of photons per pulse, very high monochromaticity, short pulse duration (a few picoseconds). Further they can be generated at a relatively moderate cost and equipment scale, at least if compared to accelerator-based sources. These characteristics can benefit to two different classes of applications: (i) flash imaging or sampling of matter with high spatial resolution; (ii) excitation or irradiation of matter at high focused X-ray intensity. Many demonstrations of applications of X-ray lasers to various areas of science can be found in the literature, an extensive review can be found in [3]. Among others, X-ray lasers were used to probe dense plasmas [4, 5] which are opaque to usual optical or UV lasers. They were also used to map nanometric deformations of surfaces on which a transient perturbation (electric field, laser irradiation) was applied [6]. When focused to a small spot they were used to investigate ablation of solids and plasma creation [7] by X-rays, which of particular interest in the context of thermonuclear fusion.

HOW DO WE GENERATE X-RAY LASERS ?

The active medium of an X-ray laser is a hot and dense plasma with typical electron temperature $kT_e \sim 50\text{-}500$ eV ($\sim 5.\ 10^5 - 5.\ 10^6$ K) and density $N_e \sim 10^{19} - 10^{21}$ cm^{-3}. The emission of an X-ray laser line results from the existence of a population inversion between some particular excited states of highly-charged ions (Ag^{19+}, Zn^{20+}, ...). X-ray lasers were demonstrated in two types of laboratory plasmas, namely those generated from a fast electrical discharge in a gas-filled capillary (Z-pinch) and those generated from the irradiation of a solid or gas target by a high-power laser pulse. Although this paper will mainly cover the second type of plasmas, the physical bases of X-ray laser generation presented below also largely apply to discharge-based X-ray lasers. More details about discharge X-ray lasers can be found in [3].

Figure 1 shows a typical view of the experimental geometry for the generation of laser-driven X-ray lasers. The key element is the high-power infrared laser beam which is linearly focused at the surface of a solid target, leading to the formation of a small plasma rod. The typical size of the rod is ~ 5 - 20 mm in length and ~ 100 µm in diameter. Under appropriate conditions, population inversions are created in a delimited zone of the plasma rod. In this active zone laser effect takes place, giving rise to two counter-propagating X-ray laser beamlets emitted from each plasma end.

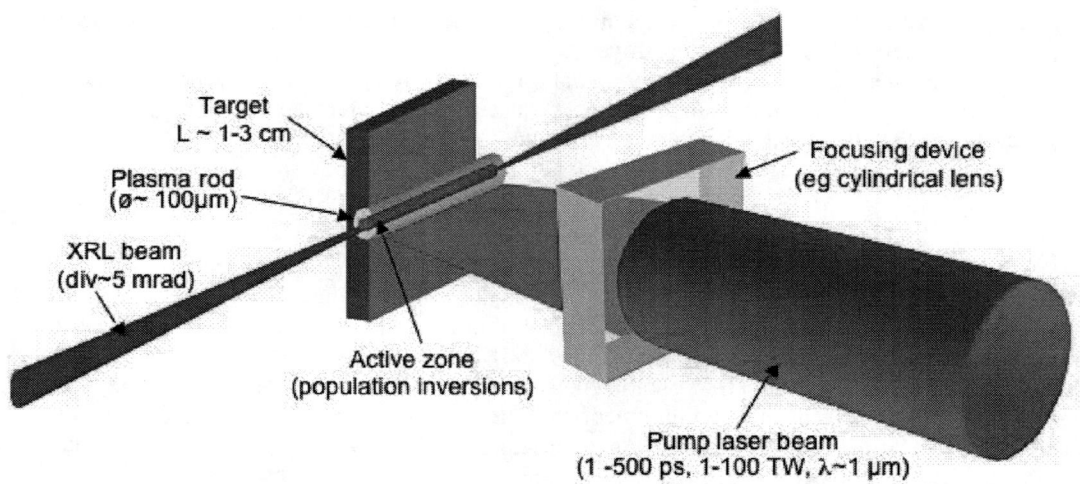

FIGURE 1. Geometry of an X-ray laser experiment showing the pump laser beam which generates the active plasma filament.

The parameters of the driving laser pulse, in particular the peak power ($\sim$1- 100 TW), or the pulse duration ($\sim 1 - 500$ ps), largely control the main characteristics of the X-ray laser through the space and time evolution of the

plasma (density, temperature, ionisation stage, ...). They also control more practical features such as the cost and scale of the X-ray laser system, or the repetition rate. This is why progress in X-ray laser research is also closely linked to the development of high-power laser technology.

Current X-ray lasers operate in the so-called Amplification of Spontaneous Emission (ASE) regime, i.e. they do not rely on a resonant cavity as for usual lasers. Cavity operation is difficult to realize because (i) high reflectivity (close to 100%) mirrors are not available in the XUV range and (ii) the lifetime of population inversions is of the same order of magnitude as the transit time of the amplified photons along the plasma length, hence preventing multipass operation. In the ASE regime, laser effect starts from spontaneous radiation emitted at the laser wavelength at each end of the plasma rod. Due to the existence of a population inversion this radiation is amplified through stimulated emission while propagating along the plasma axis. As the linear gain coefficient produced in X-ray laser plasmas is usually very high ($1 - 100$ cm^{-1}) amplification reaches saturation after only one pass (or sometimes two passes [8]) in the active zone. Operating X-ray lasers in saturated regime is a way to maximise the extraction of energy and to optimize the beam stability. However ASE operation leads to partially coherent, unpolarized X-ray laser beams. This is why, as will be discussed later in this paper, major progress is expected from advanced architectures in which an X-ray laser amplifier is seeded with coherent, polarised radiation.

An outstanding feature of currently operational X-ray lasers is that they are all based on the so-called collisional excitation pumping scheme, which is schematized in Figure 2. The two levels involved in the population inversion, in particular the upper level 2, are strongly populated through excitation from the ground state 0 by collisions between the lasant ions and the plasma free electrons. This process is efficient when the electron temperature of the electrons is of the same order of magnitude as the excitation energy ΔE_{0-2}. On the other hand, the probability of (spontaneous) radiative decay to the ground state is much higher for level 1 (dipole-allowed transition) than for level 2 (dipole-forbidden transition). Hence level 1 can be rapidly depopulated and this allows the existence of a population inversion between levels 2 and 1. In principle this pumping scheme could be applied to any ions with appropriate excited level structure. In practice large gains with the collisional excitation scheme were experimentally demonstrated only in three isoelectronic sequences: neon-like (10 bound electrons), nickel-like (28 electrons) and, to a lesser extent, palladium-like (46 electrons). Due to the closed-shell structure of their ground state these ion species have a high ionization energy. They are thus relatively stable and can be produced with a large abundance in the plasma.

Isoelectronic scaling of the collisional excitation pumping was extensively investigated in Ne-like ions (lasing transition: $1s^2 2s^2 2p^5 3p - 1s^2 2s^2 2p^5 3s$) and Ni-like ions (lasing transition: [Ne] $3s^2 3p^6 3d^9 4d - 3s^2 3p^6 3d^9 4p$) over a broad range of atomic numbers. X-ray laser lines were generated between 60 and 10 nm in Ne-like ions and between 25 and 4 nm in Ni-like ions [3].

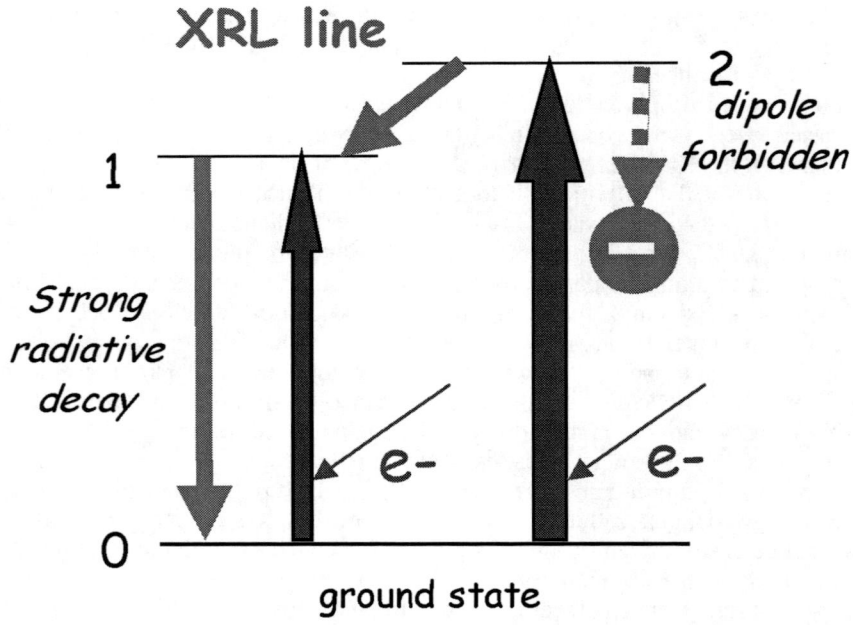

FIGURE 2. Collisional pumping scheme on which all currently operational X-ray lasers are based.

CURRENT STATUS OF X-RAY LASER RESEARCH

Although continuing effort is made by researchers to demonstrate other types of pumping schemes for X-ray lasers, collisional excitation pumping has really allowed considerable progress in the understanding and control of X-ray lasers over the last 15 years. In particular laser-driven X-ray lasers have proven to be flexible systems, able to cover a wide range of beam characteristics, in terms of wavelengths, output energy or pulse duration. It is important to recall that the first convincing demonstration of X-ray laser, reported about 20 years ago [9] at Livermore Laboratory, required more than 3 kilojoules of laser pump energy. Such a high energy was only available in two or three laboratories in the world, and the concept of "table-top" X-ray lasers did appear as an ideal but distant goal. Today X-ray lasers are routinely generated with a few Joules over even less, delivered by pump lasers that become affordable at the University scale. Such major advances in the development of X-ray lasers are the result of both better understanding of the physics underlying plasma pumping, and of the substantial progress in the technology of ultrashort, ultra-intense pumped lasers and of chirped pulse amplification (CPA).

Table 1 summarizes the performances achieved with plasma-based X-ray lasers in 2006. Until recently X-ray lasers were by far the brightest sources in the 20 nm spectral range. They have recently been surpassed by X-ray free electron lasers. However due to their higher monochromaticity and much lower cost and scale, plasma-based X-ray lasers will remain competitive as useful sources for applications.

The characteristics shown in table 1 are covered by three main types of X-ray lasers, pumped by three distinct classes of high-power laser driver (see figure 3). The first and oldest type is referred to as quasi-steady state (QSS) X-ray laser, pumped by relatively long (100-500 ps), high-energy (0.1 – 1 kJ) laser drivers (Fig. 3a). In QSS X-ray lasers linear gains are relatively small (less than 10 cm^{-1}), requiring long plasma length or double-pass operation to reach the saturation regime. Their main advantage is that they provide the largest energy output, up to 10 mJ per pulse [8], although at a low repetition rate of ~ 3 shots per hour.

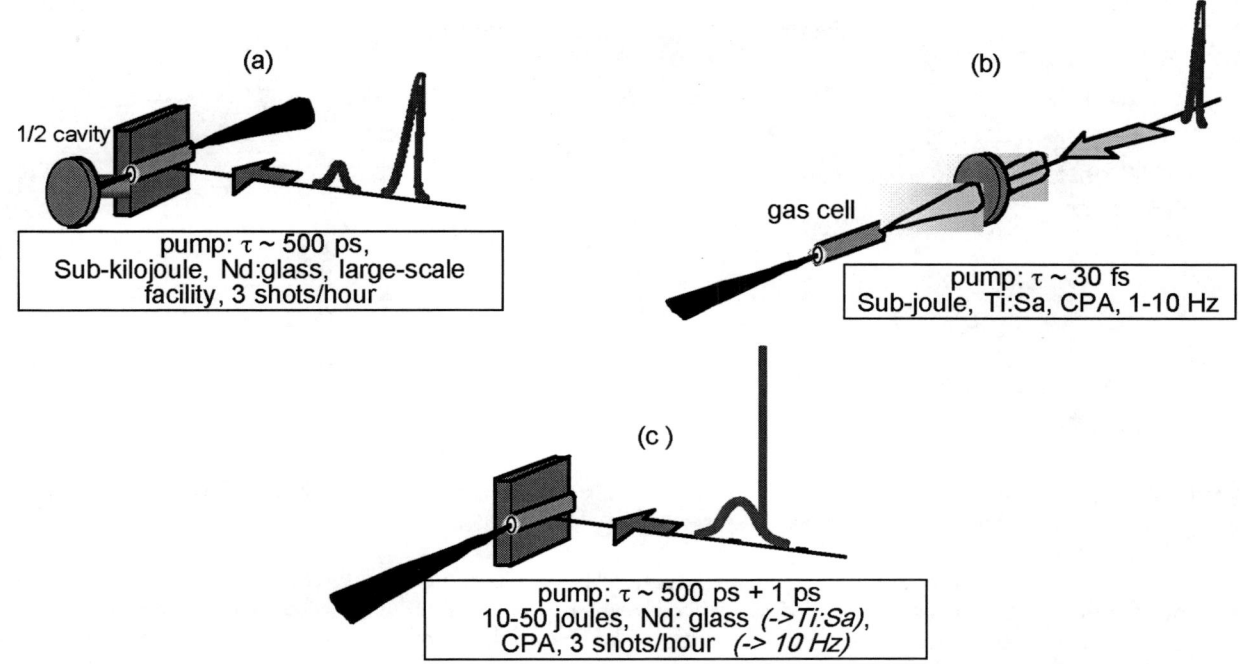

FIGURE 3. Three types of laser-driven X-ray lasers are currently operational, with distinct and complementary characteristics. (a): quasi-steady state (QSS) pumping; (b) optical-field ionization (OFI) pumping; (c) transient collisional excitation (TCE) pumping.

TABLE 1. Demonstrated performances of current X-ray lasers.

Spectral range [nm]	5 – 50 (discrete lines)
Output energy [μJ]	$0.1 - 10^4$
Pulse duration [ps]	$2 - 10^3$
Spectral bandwidth $\Delta\lambda/\lambda$	$10^{-4}-10^{-5}$
Repetition rate	3 shots/hour to 10 Hz
Peak spectral brilliance	$1 - 5 \ 10^{25}$
[ph/s/mm^2/mrad^2/0.1% BW]	

With the rapid development of the CPA technique for Nd:glass and Ti:Sa high-power lasers, substantial progress was obtained through the use of picosecond to femtosecond pump pulses. Using a 30 fs, ~1 J Ti:Sa laser, saturated X-ray lasers were obtained at 42 nm and 32 nm through optical field ionization (OFI) of a gaseous target [10] in a longitudinal pumping geometry (Fig. 3b). OFI X-ray lasers were the first high rep-rate (10 Hz), laser-driven X-ray lasers. Their output energy is relatively low (~10-100 nJ) and their extension to shorter wavelength seems difficult.

Finally transient collisional excitation (TCE) X-ray lasers (Fig. 3c) were first demonstrated in 1997 [11] and actively studied by many laboratories since then [12-15]. In TCE X-ray lasers pumping is realized in two successive steps as schematized in Fig. 4. A first long (~ 500 ps), low-intensity pulse creates a plasma rod containing the lasant ions at a relatively low temperature. A few 100 ps later, a short (~1 ps), high-intensity pulse irradiates the preformed plasma and induces a rapid and transient heating of the free electrons, hence leading to very efficient collisional excitation pumping of the lasant ions. The achieved linear gains reach very high values (~ 100 cm^{-1}) but with a short lifetime of ~10 ps, i.e. shorter than the transit time of the amplified photons along the plasma length (33 ps/cm). This is why the short pulse energy front is tilted by 45° (see Fig. 4) to provide a travelling wave irradiation at the velocity of light c [16].

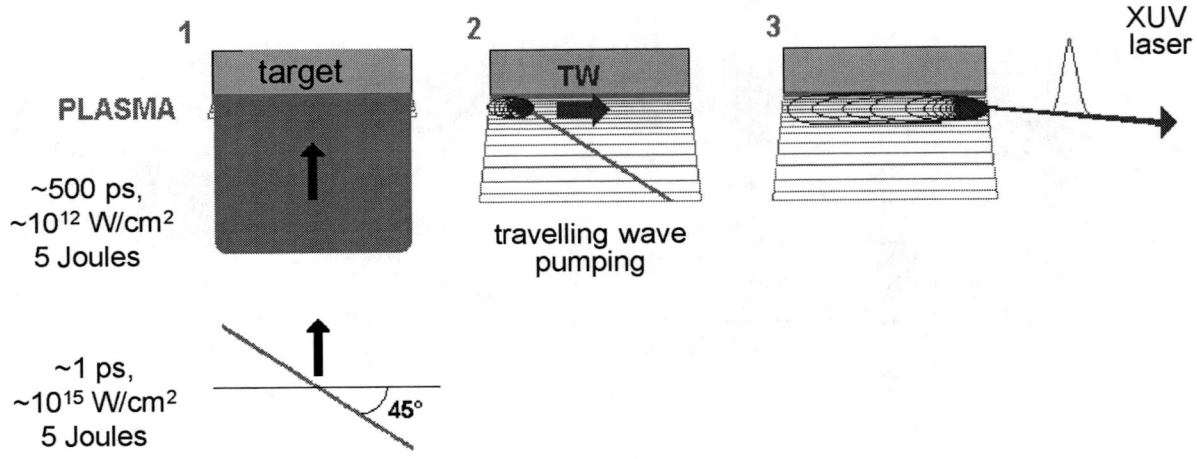

FIGURE 4. Principle of transient pumping of an X-ray laser. Due to the short lifetime of population inversions a travelling wave irradiation is used to heat the preformed plasma and generate X-ray laser.

Compared to QSS X-ray lasers, TCE systems require a lower pump energy (~ 10 – 50 Joules). Further they exhibit specific characteristics that were extensively investigated at LIXAM and in other laboratories over the past years. In particular TCE X-ray lasers were shown to deliver short pulses, down to 2 ps [17]. On the other hand their temporal coherence was measured to be of the order of 3 ps, corresponding to an extremely narrow bandwidth, smaller than 10^{-5} [17, 18]. The source size at the output plane of the plasma rod is very small, typically 25µm x 50 µm, but the spatial coherence is relatively poor (spatial coherence length ~ 5 µm at the output plane). It was shown [19] that the combination of these characteristics explains the existence of highly-contrasted small-scale structures which are observed in the beam cross section of transient XRLs. Figure 5 shows a typical experimental image. The beam is composed of randomly distributed bright spots with elongated shape. We have shown [19] that the size and shape of these spots, similar to speckles, is linked to the size and shape of the source at the output plane, through a Fourier-transform relationship. A better homogeneity of the beam would require either to decrease the temporal coherence, or to increase the spatial coherence, as will be discussed below.

NEW TRENDS AND PROSPECTS

Laser-driven X-ray lasers are now able to offer a wide variety of extremely bright, partially coherent, monochromatic sources for applications. The substantial reduction in their pump energy requirement achieved over the last 10 years has allowed these sources to be generated in an increasing number of laboratories in the world. The directions for further improvements or new developments which are presently considered in the X-ray laser community concern three main aspects: (i) extend the spectral range of XRL to significantly shorter wavelengths ($\lambda \sim 1$ nm or below), (ii) further reduce the pump energy while increasing the repetition rate; (iii) improve the beam quality and allow the control of its coherence and polarisation.

The extension of transient X-ray lasers down to the so-called "water-window" (~2 - 4 nm) should be realized through pumping of heavier elements using high-energy, short pulse CPA lasers which are now becoming available in a number of laboratories. However further reduction to significantly shorter wavelengths (i.e. 1 nm or below) can not be realized with collisional excitation pumping and requires alternative pumping mechanisms. Several schemes have been proposed, such as inner-shell photoionization of atoms by an ultra-short, ultra-intense X-ray pulse [19], or by highly energetic electrons [20].

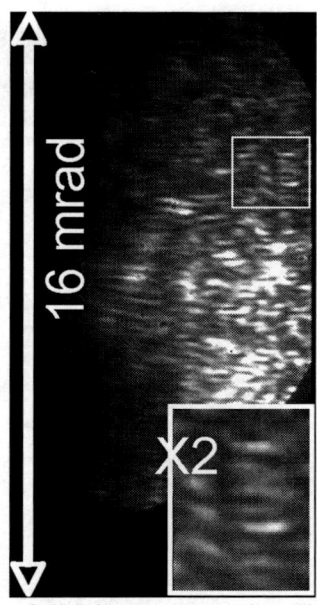

FIGURE 5. Example of an experimental image of XRL beam cross-section. The small-scale structures which are apparent have been interpreted theoretically (see text).

Further progress in the reduction of the pump energy requirement for transient X-ray laser has been realized recently, opening new and promising prospects. This was obtained by irradiating the preformed plasma with the short pulse at grazing incidence angle ($\sim 10 - 20°$), instead of near-normal incidence. This new geometry, called Grazing Incidence Pumping (GRIP) was proposed by Shlyaptsev in 2003 [21] and first demonstrated at Livermore [22]. The short pulse beam is refracted in the density gradient. The energy is mainly absorbed at the turning point of the beam trajectory, at an electron density given by $N_{e, abs} \sim N_c \times \sin^2 \Phi$, where N_c is the critical density and Φ is the grazing angle relative to the target surface. This leads to an increased absorption, hence better pumping efficiency. Further the density at which the short pulse is absorbed (i.e. the plasma is heated) can be matched to the optimum density for gain, by adjusting the grazing angle Φ. Using the GRIP geometry saturated X-ray lasers were demonstrated down to 13 nm [23] with 1 J Ti:Sa laser drivers operating at up to 10 Hz. This increased repetition-rate operation leads to a substantial increase of the average power by more than two orders of magnitude, compared to previous normal-incidence systems. In a recent experiment performed at the Lund Laser Center (Sweden), we were able to deliver up to 30 µW with the GRIP molybdenum X-ray laser emitted at 18.9 nm [24].

Finally a major advance in the field of X-ray lasers has been demonstrated in [25] through the demonstration of strong amplification of high-order harmonic (HOH) radiation at 32 nm in a OFI X-ray laser plasma. The output beam was observed to retain the characteristics of the injected beam, namely a complete spatial coherence and a linear polarization.

In order to extend this seeding technique to both shorter wavelength and higher output energy it is necessary to use solid-target plasma amplifiers. This was first attempted ~10 years ago at the Rutherford Laboratory (UK) using a QSS Ga X-ray laser [26], but with a limited success. More recently preliminary results were obtained at APRC/JAERI (Japan) showing weak amplification of HOH at 26.9 nm in a Ni-like Mn X-ray laser plasma pumped in the transient regime [27].

Amplification of HOH radiation in an X-ray laser amplifier is a complex problem since those radiation sources have distinct spectral/temporal characteristics. HOH radiation is emitted in a very short pulse of $\sim$ 40 fs or less, with a broad spectral bandwidth $\Delta\lambda/\lambda \sim 10^{-2}$. In contrast the shortest duration measured for an X-ray laser is 2 ps [17] but the spectral bandwidth is very small $\Delta\lambda/\lambda \sim 10^{-4} - 10^{-5}$ [17, 18]. Detailed numerical modeling involving Bloch-Maxwell equations is under development at LIXAM to explore this new topic. On the experimental side, seeding a transient X-ray laser amplifier with HOH radiation will be one of the main goal of the new X-ray laser facility LASERIX.

LASERIX will be the first user facility dedicated to the development of X-ray lasers and their applications in Europe [28]. Started in 2002 this ambitious project is now in its final stage of achievement. LASERIX is based on a Ti:Sa laser driver with unique performances: the energy delivered at the final stage of amplification is 40 Joules and the repetition rate is 0.1 Hz, i.e. almost 100 times higher than existing lasers operating at that level of energy.

LASERIX is located at ENSTA (Palaiseau, France) where the project was achieved in collaboration with LOA. The 40 Joules ouput laser beams will be divided in up to 6 beamlines with different energy and pulse duration, as shown in Figure 6. This will allow the generation of various laser-based X-ray sources synchronized with the X-ray laser beam. In 2007 the laser driver will be moved in a new building currently under construction and the experimental area will be implemented. The facility will be then opened to users in 2008.

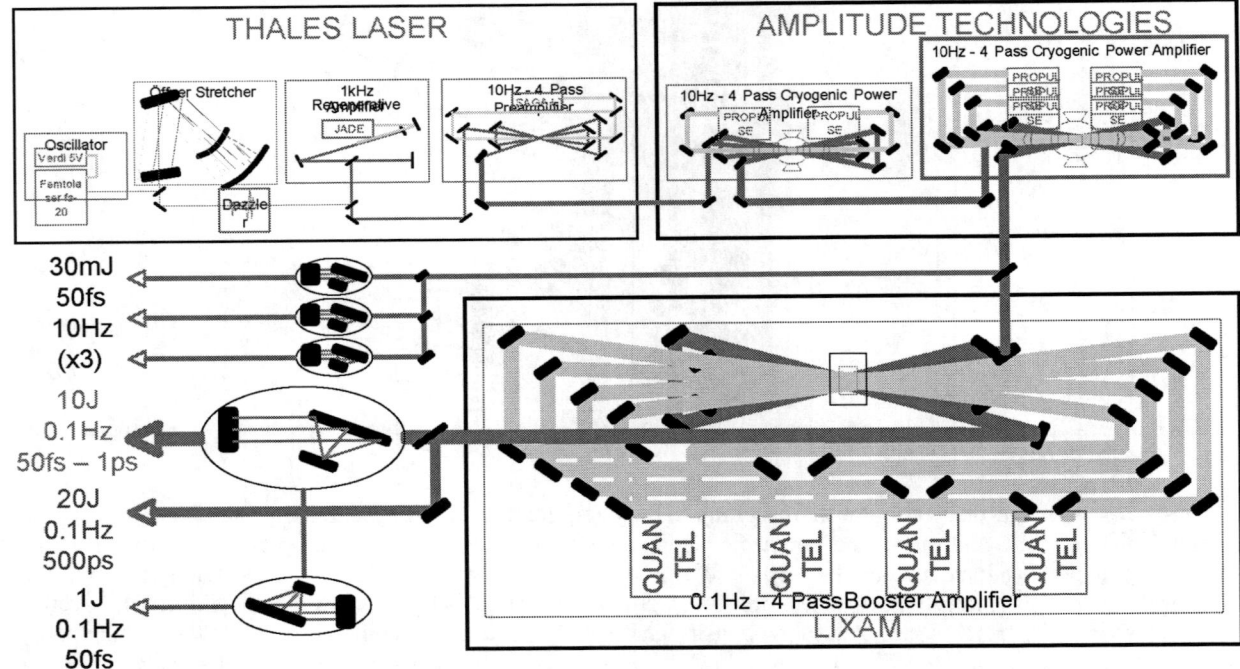

FIGURE 6. Layout of the LASERIX laser driver, which is completely based on Ti:Sa crystal amplifiers. This laser delivers up to 40 Joules at a repetition rate of 0.1 Hz. Up to 6 beamlines will be implemented for X-ray source generation.

CONCLUSION

In conclusion I have shown that considerable progress has been achieved worldwide in the development of high-brightness X-ray lasers, in particular generated from laser-produced plasmas. Among the variety of X-ray lasers demonstrated, those relying on transient pumping with ultrashort, CPA lasers, have led to a significant reduction of the pump energy, allowing for more compact systems. Although the potential of X-ray lasers as a tool for science is high, application experiments have been limited so far to demonstrations of feasibility. This is largely due to the limited access to beamlines and low repetition rates of the source. The recent progress made both in X-ray laser pumping efficiency and in the technology of ultra-intense CPA lasers will allow the construction of dedicated facilities, such as the new LASERIX in France, where reliable and controlled X-ray laser beams will become available to the scientific community, hence stimulating their exploitation for applications.

ACKNOWLEDGMENTS

The X-ray laser research team at LIXAM is gratefully acknowledged, in particular G. Jamelot, D. Ros, S. Kazamias, M. Pittman, J.-C. Lagron and O. Guilbaud. I am also grateful to P. Dhez and P. Jaeglé who pioneered the field of X-ray lasers and initiated collaborations with Prof. Lotfia El Nadi.

REFERENCES

1. S. Kazamias et al., *Phys. Rev. Lett.* **90**, 193901 (2003).
2. V. Ayvazyan et al., *Phys. Rev. Lett.* **88**, 104802 (2002).

3. H. Daido, *Rep. Prog.. Phys.* **65**, 1513 (2002).
4. H. Tang et al. *Appl. Phys. B* **78**, 975 (2004).
5. J.J. Rocca et al., *Phys of Plasmas* **10**, 2031 (2003).
6. G. Jamelot et al, *J. Appl. Phys.* **98**, 044308 (2005).
7. L. Juha et al *Appl. Phys. Letters* **86**, 034109 (2005).
8. B. Rus et al *Phys. Rev. A* **66**, 063806 (2002).
9. D. L. Matthews et al., *Phys. Rev. Lett.* **54**, 110 (1985).
10. S. Sebban et al., *Phys. Rev. Lett.* **89**, 253901 (2002).
11. P. V. Nickles et al., *Phys. Rev. Lett.* **78**, 2748 (1997).
12. J. Dunn et al., *Phys. Rev. Lett.* **80**, 2825 (1998).
13. A. Klisnick et al., *J. Opt. Soc. Am. B,* **17**, 1093 (2000).
14. R.E. King et al., *Phys. Rev.A* **64**, 053810 (2001).
15. T. Kawachi et al., *Phys. Rev.A* **66**, 033815 (2002).
16. J.-C. Chanteloup et al., *J. Opt. Soc. Am. B,* **17**, 151 (2000).
17. A. Klisnick et al., *Phys. Rev.A* **66**, 033810 (2002).
17. R. Smith et al., *Opt.Lett.,* **28**, 1 (2003).
18. A. Klisnick et al., *J. Quant. Spectr. Rad. Transf.* **99**, 370 (2006).
18. O. Guilbaud et al., *Europhys. Lett.,* **74**, 823 (2006).
19. S.J. Moon et al., *Phys. Rev.A* **57**, 1391 (1998).
20. D. Kim et al., *Phys. Rev.A* **59**, R4159 (1999).
21. V.N. Shlyaptsev et al., *Soft x-ray lasers and Applications V*, edited by E.E. Fill and S. Suckewer, SPIE Int. Soc. Opt. Eng. Proceedings vol. **5197**, 2003, pp. 221 - 228.
22. R. Keenan et al., *Phys. Rev. Lett.* **94**, 103901 (2005).
23. Y. Wang et al., *Phys. Rev. A* **72**, 053807 (2005).
24. K. Cassou et al., submitted to *Opt. Lett.* (2006).
25. P. Zeitoun et al., *Nature* **431**, 466 (2004).
26. T. Ditmire et al., *Phys. Rev. A* **51**, 4337 (1995).
27. N. Hasegawa et al., in *X-ray lasers 2004*, edited by J. Zhang, IOP Conference Series 186, Institute of Physics Publishing, Bristol and Philadelphia, 2005, pp. 273-276.
28. G. Jamelot et al., in *X-ray lasers 2004*, edited by J. Zhang, IOP Conference Series 186, Institute of Physics Publishing, Bristol and Philadelphia, 2005, pp. 677-684.

Use of Spectral Lines of a Pure Ti Target for the Spectroscopic Diagnostics of the Laser Induced Plasma in Vacuum

H. Hegazy**, H. Sharkawy*, and Th. M. EL Sherbini*

*Laboratory of Lasers and New Materials, Physics Department, Faculty of Science,
Cairo University– Egypt
**Plasma Physics Department, Nuclear Research Center, Egyptian Atomic Energy Authority

13759 Enchass, Egypt. Corresponding author E. mail: hos_heg@yahoo.com

Abstract. In this paper, we propose a simple method to use the spectral lines which are free from self-absorption to diagnose the plasma generated from a pure Ti target. The method is based on multiplet spectral lines. Multiplets are very useful in the assessment of self-absorption; the intensity ratio of their components is usually well known for atoms and ions in LTE and any deviation will indicate the magnitude of the self-absorption. Ti II spectral lines are developed for measurements of excitation temperature and electron density of plasma generated at different distances from pure titanium target when irradiated by Nd-YAG 750 mJ in 6 ns FWHM at the fundamental wavelength 1.06μm.

INTRODUCTION

The interaction of laser beams with solid targets and the properties of the plasmas produced have been studied for many years [1-6]. Nevertheless, further investigations of the plasmas are still necessary for understanding many of the basic physical mechanisms in different interesting applications such as material processing technology, thin film deposition and synthesis of nano-materials [7-10]. The fabrication of thin films using laser plasma deposition becomes increasingly popular with respect to other techniques such as chemical vapor deposition or ion implantation. One of the most significant advantages of using the laser ablation technique for thin film production is its easy handling, since the laser beam is outside the reaction chamber [11], and its flexibility in the selection of materials; moreover, almost any solid material can be ablated. The emission of the energetic particles during the laser target interaction has an important influence on the layer formation which enables growth of adherent and epitaxial films at lower substrate temperatures than other techniques. Thin films of titanium nitride have many applications in mechanics and microelectronics because of their hardness, good conductivity and their chemical stability [12-13]. This is important for the deposition on doped semiconductor materials for microelectronics, because heating can alter the depth composition and physical properties of the crystal [14].

The laser deposition process can be divided into different stages: laser ablation of the target, plasma generation, plasma expansion, and deposition of the ablated materials on the substrate. However, the properties of the produced plasma and its expansion play a dominant role in the thin film production since it determines the production of the reactive species and high-energy particles. Hence reliable diagnostics of the produced plasma to determine electron temperature, electron density and ion velocity are important for better understanding, improving and developing laser plasma deposition thin film technique, and improving the quality of the produced films. Optical emission spectroscopy is the most widely used technique [15-19] for the diagnostics of the generated plasma. At high plasma densities, the employed methods are usually based on some assumptions, namely the existence of local thermodynamic equilibrium (LTE), and negligible optical thickness of the lines used. Stark broadening of the optically thin spectral lines allows estimating the electron density without any LTE assumption.

The aim of the present study is to propose a simple method to find a group of spectral lines that are free from self-absorption to diagnose the plasma generated from pure targets. Pure target as titanium should be used in laser ablation experiments for thin film production or synthesis of nano materials and diagnostics. The proposed method is based on multiplet spectral lines: their intensity ratio is usually well known for the free atoms and ions and any deviation will indicate the magnitude of the self-absorption. In the present study, the excitation temperature of a laser induced plasma is determined from the Boltzmann plot using Ti II spectral lines at different distances from the Ti target (0.1 - 4 mm) - based on the LTE assumption and after proofing that the used spectral lines are not affected by self-absorption. The electron density of the produced plasma is determined from the Stark width of the Ti II 350.49 nm spectral line at the same distances.

CP888, *Modern Trends in Physics Research,*
Second International Conference on Modern Trends in Physics Research—MTPR-06,
edited by L. El Nadi

EXPERIMETAL SET-UP

The present work is performed using the experimental setup shown in Fig. (1). A Q-switched Nd-YAG Brilliant laser from Quantel delivering 750 mJ in 6 ns FWHM at the fundamental wavelength 1.06 μm at a repetition rate of 10 Hz was used. The plasma is generated in vacuum by focusing the laser beam using a quartz lens having a focal length f_1 = 20 cm, perpendicularly onto the target. The measurements are carried out under a pressure of 4×10^{-6} mbar, produced by Leybold turbo molecular pump.

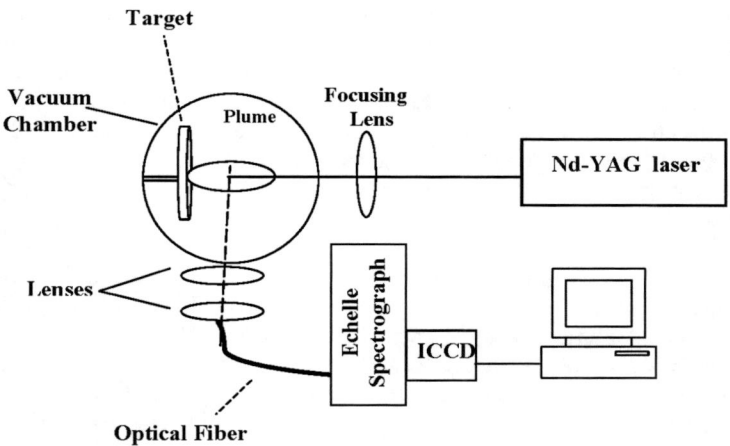

FIGURE 1. Experimental Setup

The emission spectrum is collected perpendicularly with the aid of a two lens system (focal lengths f_2 = 20 cm, and f_3 = 5 cm); both are mounted with the Quartz fiber cable of diameter 25 μm on a x-y translation stage that allows for an easy illumination of the slit of the spectrograph. The emission spectrum was recorded using SE 200 Echelle spectrograph (Catalina Corp.), equipped with ICCD camera (Andor model iStar DH734-18F). The gain of the camera was fixed at a value 250 with binning mode at 1×1. This spectrometer allows for a time resolved spectral acquisition over the whole UV-NIR (200-1000nm) with a constant resolution 4500 over three points. An Oriel low-pressure Hg lamp is used for wavelength calibration of the Echelle spectrograph. The instrumental bandwidth of the system is 0.04 nm measured by low pressure Hg lamp, where the natural width of the mercury spectrum is known.

In the present study, plasma is produced by irradiating pure titanium targets with a switched Nd-YAG laser at its fundamental wavelength λ = 1.06 μm, with a pulse duration of 6 ns and a repetition rate of 10 Hz. The emission intensities are optimized by accumulating the spectra at a repetition rate of 10 Hz of the laser beam and an exposure time of two seconds at a laser energy 750 mJ/ pulse. The emitted spectrum is recorded in a direction parallel to the target surface.

RESULTS AND DISCUSSIONS

1- Determination of Excitation Temperature (T_{exc})

The temperature is considered one of the most important parameters used to characterize the state of the plasma. An accurate knowledge of the temperature leads to understand the plasma processes occurring, namely vaporization, dissociation, excitation and ionization.

The excitation temperature can be determined from the measurement of the intensity of its spectral lines assuming that the population of the energy levels follows the Boltzmann distribution law. This is also can be applied to atomic species in plasmas which are in partial LTE. For full LTE the intensity of the spectral line is given by:

$$I_{ij} = \frac{L}{4\pi} \, h\nu \, g_i \, A_{ij} \, \frac{N_\circ}{U(T)} \, \exp\left\{-\frac{E_i}{KT}\right\} \qquad (1)$$

where L is the thickness of the plasma layer, h is Planck's constant, ν is the frequency=c/λ, A_{ij} is the transition probability, N_o is the total density of atoms or ions, g_j is the statistical weight, U(T) is the partition function, E_i is the excitation energy of the upper level, k is Boltzmann constant, and T is the excitation temperature.

This equation can be applied using the absolute intensity of one spectral line provided accurate values for N_o and $U(T)$ are known. However, in case of a relative ratio of two spectral lines from the same ionization stage, the excitation temperature (T_{exc}) is given by:

$$\frac{I_1}{I_2} = \frac{A_1 \, g_1 \, \lambda_2}{A_2 \, g_2 \, \lambda_1} \, \exp\left\{-\frac{E_1 - E_2}{KT}\right\} \qquad (2)$$

The accuracy of temperature determination is better when the difference between the excitation energies of the two lines is larger than 2 eV[20]. In many cases the energy difference $(E_1 - E_2)$ is below 2 eV. Higher accuracy is achieved by measuring the intensity of a group of spectral lines of the same ionization stage. Plotting $\ln(I\lambda/gA)$ versus E_{exc}, should yield a straight line and the so called Boltzmann plot is obtained with a slope equal $(-1/kT)$. Deviation from a straight line means deviation from the Boltzmann distribution law, which can be interpreted as due to an overpopulation or an under population of the various energy levels. It is well known that the group of spectral lines selected for the construction of a Boltzmann plot should satisfy certain requirements. The most important of these requirements is that the difference in excitation energy should be large enough to ensure an accurate temperature measurement (1-2 eV). Their intensities should be sensitive to temperature changes and their profile should not be influenced by self-absorption. It is convenient to have the lines in a narrow wavelength interval, otherwise a relative sensitivity calibration of the system becomes necessary.

In the literature there are different groups of spectral lines emitted from certain thermometric species which are traditionally used for temperature determination in analytical plasmas. However, their intensities should fulfill the above mentioned requirements. Among these are spectral lines of Fe I, Cu I, Mo I, V I, Mn I, Zn I, and Si I [21-26]. The important point in the selection of the thermometric elements is that its concentration must be low enough to avoid self-absorption problems in their spectral lines. In case of pure targets as used in the present study in the laser ablation process for thin film production, the selected spectral lines should be proven to be free from self-absorption.

The present measurements are carried out at different distances (0.1, 0.6, 2, 3 and 4 mm) from the target. Five Ti II spectral lines are selected to determine the excitation temperature at different distances from the target. The atomic data of the selected spectra lines are taken from *NIST* [27] and are given in Table (1).

TABLE 1. Spectral lines and their atomic data

Transition	λ_{nm} (nm)	E_n (eV)	g_n	A_{mn} ($10^8 s^{-1}$)
2G*-2F	348.36	7.866	8	9.7e-01
2F-4D*	357.37	4.042	4	2.8e-02
2F-4D*	358.71	4.062	8	1.1e-02
2P-2S*	362.48	4.640	2	5.7e-02
2P-2S*	364.13	4.640	2	4.9e-01

As mentioned above the selected spectral lines should be free from self-absorption. A multiplet is very useful in the assessment of self-absorption, since the intensity ratio of its components is well known for LTE and any deviation will indicate the magnitude of the self-absorption. A convenient possibility to analyze quantitatively the total self-absorption of a line is offered by the measurement of a second line from the same upper level. If in addition, the absorption oscillator strength is also much weaker, this transition will hardly be influenced by absorption. A comparison of the measured intensity ratio of both lines with their optically thin limit is required. The optically thin limit in this case is simply given by the branching ratio:

$$\frac{I_1}{I_2} = \frac{A_1 \, \lambda_2}{A_2 \, \lambda_1} \qquad (3)$$

Ti II spectral lines at 362.48 nm and 364.13 nm fulfill the above mentioned conditions. The optically thin limit is 0.59 and the measured ratio confirms this value at different distances from the target as shown in figure (2).

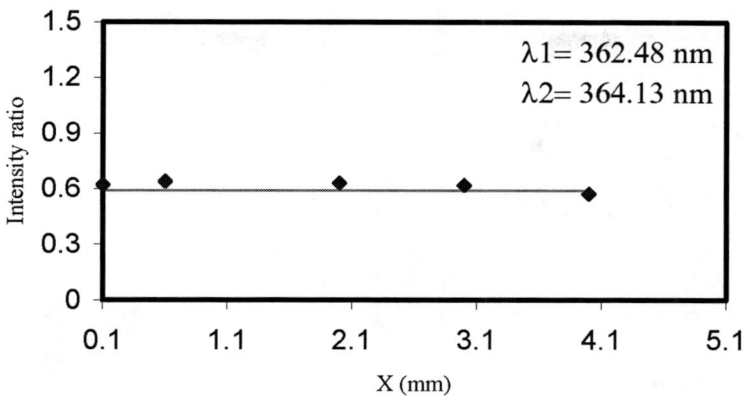

FIGURE 2. Measured intensity ratio of Ti II 362.48 / Ti II 364.13 at different distances from the target, where the solid line shows the optically thin limit calculated according to the branching ratio.

The same possibility is derived for two other spectral lines at 357.37 nm and 358.71 nm, where their upper and their lower levels are very close; they are definitely strongly coupled by collisions. The optically thin limit in this case can be approximated as:

$$\frac{I_1}{I_2} = \frac{g_1 \, A_1 \, \lambda_2}{g_2 \, A_2 \, \lambda_1} \tag{4}$$

The optically thin limit in this case is 1.27 and the measured ratio confirms this value at different distance from the target as shown in figure (3). The fifth spectral line used for temperature measurements in the present study is at wavelength λ = 348.36 nm. This transition arises from an upper energy level which is hardly affected by any self-absorption.

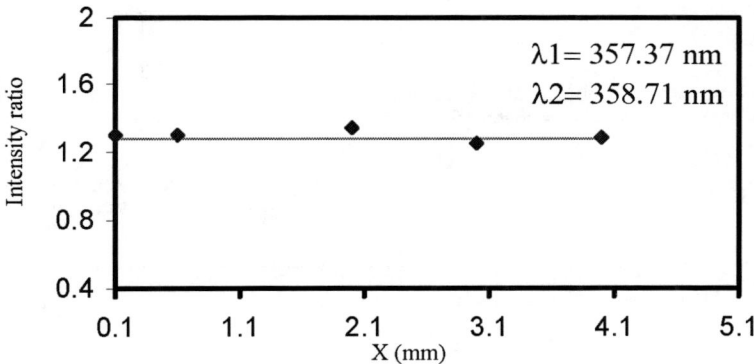

FIGURE 3. The measured intensity ratio of Ti II 357.37 / Ti II 358.71 at different distance from the target, where the solid line show the optically thin limit calculated according to the branching ratio.

Figures (4) shows Boltzmann plots obtained for excitation temperature measurement using the selected Ti II spectral lines measured at different distances from the Ti target which is in vacuum (4×10^{-6} mbar). The obtained Boltzmann plots indicate that the Ti ions follow the Boltzmann distribution law. **Figure (5) shows the variation of the plasma temperature measured at different distances from the Ti target.**

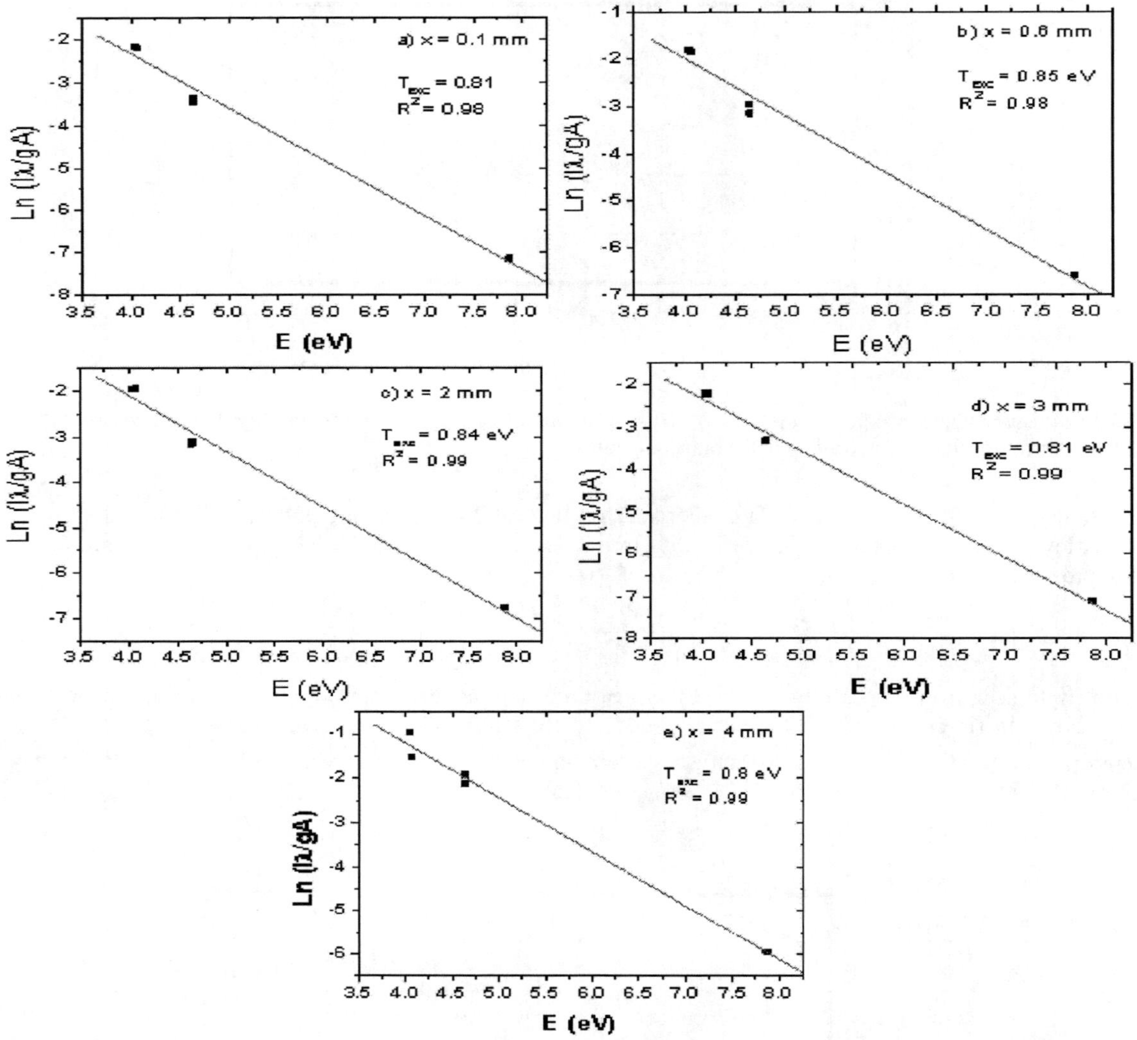

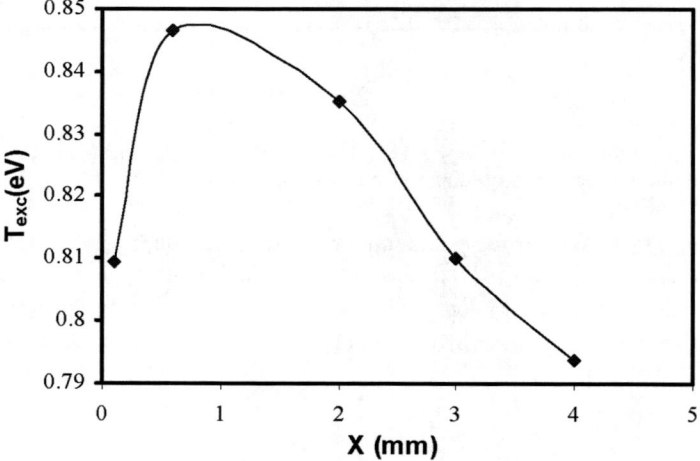

FIGURE 4. Boltzmann plots at distances a) 0.1mm, b) 0.6 mm, C) 2 mm, D) 3 mm, and e) 4mm from the target.

FIGURE 5. Variation of excitation temperature measured at different distances from the target.

2- Measurements of the Electron Density from the Stark Line Width

The measured spectral lines are noticeably broadened above their natural line width. Broadening mechanism in laser plasmas are: Stark broadening, Doppler broadening, Van der Waals broadening and resonance broadening from collisions between like neutral species on strong resonance line[28-29]. In laser plasmas where low temperatures and high densities prevail, the contributions by Doppler, Van der Waals, and resonance broadenings can be neglected [30]. In the hottest region of the laser plasma, the Stark effect leads to a broadening of the atomic and ionic emission lines as well as to a shift in the line-center wavelength. Stark broadening results from collisions by electrons (and ions) and also leads to a shift in the energy levels of the emitting atom. In case of well-isolated lines in neutral and singly ionized atoms, Stark broadening predominantly occurs by electron collision. So the Stark half width of these lines is usually used for the determination of the electron number density. Eq. (5) relates the FWHM of lines ($\Delta\lambda_{1/2}$) with the electron number density n_e [29].

$$\Delta\lambda_{1/2}(\overset{\circ}{A}) = 2w\left(\frac{n_e}{10^{16}}\right) + 3.5\ A\left(\frac{n_e}{10^{16}}\right)^{1/4}\left[1 - 3/4\ N_D^{-1/3}\right]w\left(\frac{n_e}{10^{16}}\right) \tag{5}$$

Where w is the electron impact width parameter or Stark width parameter, A is the ion broadening parameter, and N_D. represents the number of particles in the Debye sphere [31]:

$$N_D = 1.72 \times 10^9\ \frac{[T_e(eV)]^{3/2}}{[n_e(cm^{-3})]^{1/2}} \tag{6}$$

The importance of knowing the magnitude and the temporal evaluation of the electron density arises from the fact that many kinetic reaction rates depend directly or indirectly on this parameter.

In the present study the Ti II spectral line at 350.49 nm is used to derive the electron number density. The Stark width parameter (w) of this line has been experimentally accurately determined by Hermann [13]. Due to the small contribution of the second term of eq.(5) in the present working condition.

$$\Delta\lambda_{1/2} = 2w\left(\frac{n_e}{10^{16}}\right) \tag{7}$$

Analysis of each line profile proceeded in the following way. The instrumental function was convoluted with a Lorentzian function of variable width and fitted to the experimental profile by a least-square fitting procedure.

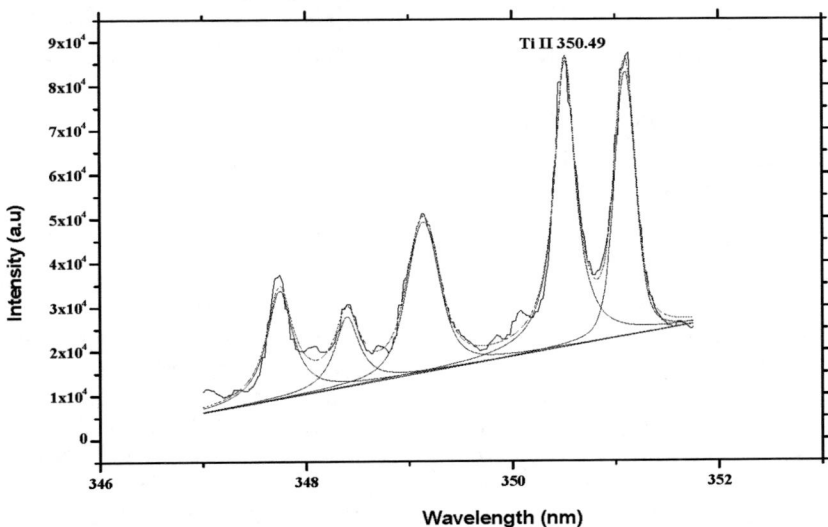

FIGURE 6. Fitting of a Voigt function with the experimental data at a distance of 0.6 mm from the target.

In the present study $\Delta\lambda_{instrument}$ is 0.04 nm (determined by measuring the FWHM of the Hg lines emitted by a standard low pressure Hg lamp as mentioned in experimental setup section). Since the Ti II spectral line at 350.49 nm arises from upper lying energy level with respect to the spectral lines discussed in the previous section, the line is

not subjected to self-absorption. The Stark FWHM of Ti II at 350.49 nm is obtained according to the convolution procedure described below. Figure (6) shows the fitting of the line (350.49 nm) at a distance 0.6 mm from the target. Using equation (7) it is found that the electron density of the Ti plasma produced by Nd:YAG laser (1064 nm) in vacuum has a maximum value (1.8×10^{16} cm^{-3}) at a distance 0.6 mm from the target. The variation of the electron density at the different distance from the target in the range (0.1, 0.6, 2, 3 and 4 mm) is shown in figure (7). The decrease in the density with the increase of the distance above 0.6 mm, can be attributed to the shielding effect of the target by the plasma particles. The shielding of the target by the plasma particles prevents further interaction of laser radiation with the target. Moreover the recombination processes are enhanced at large distance. It is important to note that the Stark broadening is rather weakly independent of the temperature. Therefore, the electron density can be determined with a good accuracy [32]. Moreover the determination of electron number density by this method is independent from any assumption of LTE.

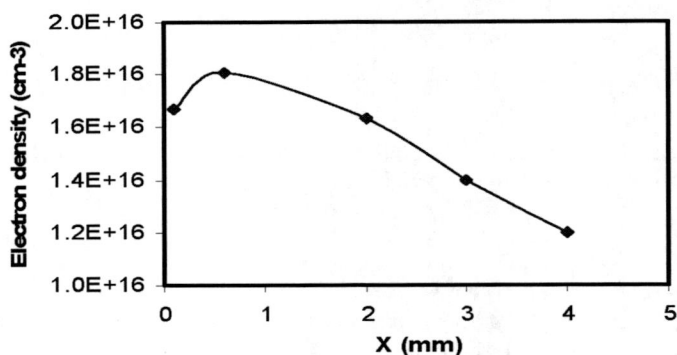

FIGURE 7. Variation of the electron density as a function of distance from the target.

3- Equilibrium Considerations

In order to determine the plasma temperatures, it is necessary to assume that the plasma produced is in LTE soon after its formation. Thus, a Maxwellian distribution of the free electrons is required. Griem considered that when a plasma is relatively dense and of low temperature ($n_e > 10^{16}$ cm^{-3}, kT<5 eV) its electron velocity is nearly always Maxwellian [34]. The measured density and temperature in the present study confirm that the free electron velocity distribution can be Maxwellian. However, LTE conditions require that the collisional excitation and de-excitation processes predominate over the radiative processes. The lower limit of the electron density n_e necessary for LTE between two energy states separated by ΔE, is a function of T_{exc} and is given by equation [32-34]:

$$n_e (\text{cm}^{-3}) \geq 1.6 \times 10^{12} (T_e)^{1/2} (\Delta E)^3 \qquad (8)$$

In the present study, the determined electron density using Ti II spectral line at 350.49 nm is higher than the lower limit of n_e, therefore, the assumption of LTE is valid.

CONCLUSION

In this paper, a simple method is proposed for testing the self-absorption of the spectral lines before using them in determining the temperature and electron density of plasma generated by incidence the laser beam on pure Ti target. The excitation temperature of the core of the plasma plume is measured in vacuum using Boltzmann's plots. It is found that in the core of the plume the excitation temperature T_{exc} is 9900 K in vacuum (at the distance 0.6 mm from the target). The density of electrons is determined from the Stark broadened line width. It is found that the electrons in the core of the plume have a density of 1.8×10^{16} cm^{-3} in vacuum at a distance 0.6 mm from the target. The effect of the dependence of the measured emission lines at different distances from the target on the electron temperature and electron density were investigated. The temperature is increases with distance from the target until a distance of 0.6 mm, after that it decreases. This is due to the enhancement of the cooling rate in the outer part of the plasma. The decrease of the electron density after 0.6 mm may be due to the shielding of the target by the plasma, which prevents further interaction of the laser radiation with the target. Moreover, it might be due to the enhancement of the recombination processes.

ACKNOWLEDGMENT
Authors would like to thank Prof. H.-J. Kunze, and Prof. M. A. Eid for their fruitful discussions, and suggestions.

REFERENCES

1. E. T. Kennedy, Contemp. Phys. **25**, 31 (1984).
2. R. Kelly and R. W. Dreyfus, Nucl. Instrum. Methods Phys. Res. B **32**, 341 (1988).
3. R. Kelly, M. Miotello, B. Braren, A. Gupta, and K. Casey, Nucl. Instrum. Methods Phys. Res. B **65**, 187 (1992).
4. R. Kelly and A. Miotello, Appl. Phys. A **69**, 145 (1993).
5. S. Amoruso, Appl. Phys. A **69**, 323(1999).
6. G. Colonna, A. Casavola, and M. Capitelli, Spectrochimica. Acta Part B **56**, 567 (2001).
7. D. B. Chrisey , and G. K. Hubler (eds.) Pulsed Laser Deposition of Thin Films, Wiley, New York (1995).
8. M Von Allmen and A. Blatter, Laser Beam Interactions with Materials: Physical Principles and Applications Springer-Berlin (1995).
9. P. R. Willmott, and J. R. Huber, Rev. Modern Phys.**72**, 315 (2000).
10. D. P. Yu, C. S. Lee, I. Bello, X. S. Sun, Y. H. Tang, G. W. Zhou, Z. G. Bai, Z. Zhang and S. Q. Feng, Solid State Communications **105**, 403 (1998).
11. J. T. Cheung and H. Sankur, CRC Crit Rev. Solid State Mater. Sci. **15**, 63, (1989).
12. I. N. Mihailescu, N. Chitica, L.C.Nistor, M. Popescu, V. S. Teodorescu, I. Ursu, A. Andreim A. Barborica, A. Luches, M. L. Giorgi, A. Perrone, B. Dubreuil, and J. Hermann, J. Appl. Phys. **74**, 5781 (1993).
13. J. Hermann, A. L. Thomann, C. Boulmer-Leborgne, and B. Dubreuil, J. Appl. Phys. **77**, 2928 (1995).
14. H. Sankur, , W. J. Gunning, J. DeNatale, and J. F. Flintoff, J. Appl. Phys. **65**, 2475 (1989).
15. X. T. Wang, B. Y. Man, G. T. Wang, Z. Zhao, B. Z. Xu, Y. Y. Xia, L. M. Mei, and X. Y. Hu, J. Appl. Phys. **80**, 1783 (1996).
16. B. Y. Man, Appl. Pys. B **67**, 241 (1998).
17. F. Fuso, L. N. Vyacheslavov, G. Masciarelli, and E. Arimondo, J. Appl. Phys. **76**, 8088 (1994).
18. A. de Giacomo, V. A. Shakhatov, and O. de Pascale, Spectrochimica Acta Part B **56**,.753 (2001).
19. A. de Giacomo, V. A. Shakhatov, G. S. Senesi, and S Orlando, Spectrochimica. Acta Part B **56**, 1459 (2001).
20. W. Lochte-Holtgreven, Plasma Diagnostics, AIP Press, New York, (1968)
21. V. Detalle, R. Heon, M. Sabsabi, and L. St-Onge, Spectrochimica Acta Part B **56** 1011 (2001).
22. M. A. Eid, N. Nada, A. A. Mahdy, Z. M. Hassan, and H. A. Hegazy, Optica Pura Y Aplicada **27**, 63 (1994).
23. Jarosz J, .J. M. Mermet and J. Robin, Spectrochimica Acta Part B **33**, 55 (1978).
24. P. W. J. M. Boumans, Anal. Emission Spectroscopy II, Marcel Dekker Inc., N.Y. (1972).
25. M. A. Eid, M. H. Abdallah, A. A. Mahdy, Z. A. El Sayed, and K. A. Eid, Spectrosc. Lett. **27**, 397 (1994).
26. M. Milan, and J. J .Laserna, Spectrochimica Acta Part B **56**, 275 (2001).
27. www.Nist.gov
28. H. R. Griem, Spectral Line Broadening by Plasma, Academic Press, New York, (1974)
29. G. Bekefi, Principles of Laser Plasmas, Ed. G. Bekefi, Wiley Interscience, New York, (1976).
30. I. B. Gornushkin, L. A. King, B. W. Smith, N. Omenetto, J. D. Wienfordner, Spectrochimica Acta B **54**, 1207 (1999).
31. F. F. Chen, Introduction to Plasma Physics, Plenum Press, New York, (1974).
32. M. Sabsabi, and P. Cielo, Appl. Spectroscopy **49**, 499 (1995).
33. H.R. Griem, Phys. Rev. **131**, 1170 (1963).
34. K.J. Grant, and G. L. Paul, Appl. Spectroscopy **44**, 1349 (1990).

Spectroscopic Identification of Lipid, Protein and DNA Changes in Breast Cancer tissues

Y.A. Badr and S.I. Hassab Elnaby

National Institute of Laser Enhanced Sciences (NILES), Egypt, Cairo

Abstract. The FTIR spectroscopy, at the range 4000 – 6000 cm^{-1} showed a clear distinction between normal and cancer tissues. Normal tissues spectra contain a doublet structure at 4258 and 4332 cm^{-1}. This structure is usually on top of a small band that extends from 3950 cm^{-1} to 4400 cm^{-1}. This structure us also observed from pure lipid tissues from control patients. The origin of this structure could be attributed to combinations of lipid lines. This structure is completely absent in cancer tissues, instead a broad intense band appears from 5100 cm^{-1} to 5200 cm^{-1}. The intensity of this band varies from one patient to another. The shape of this broad band indicates that it is the due to random orientation changes in the proteins. This band has a peak at 5164 cm^{-1}, it contains another small kink at 4882 cm^{-1}. This may lead also to the conclusion that this window band is associated with a short half life time energy levels. On The other hand the photoacoustic spectrum of the same tissues , shows that in normal tissues there are three very distinct peaks (namely 1097,1159 and 1232 cm^{-1}) they disappear in malignant tissues and replaced by many weak ripples. Two peaks (1578, 1690 cm^{-1}) changes their position in malignant tissues(1626, 1678 cm^{-1}). A change in DNA markers was also noticed in the range 600-1700 cm^{-1}

INTRODUCTION

In a previous work [1] we showed that FTIR can be used with high degree of certitude to differentiate between normal and tumor tissues spectroscopically in the range 2700-3750 cm^{-1}.

Further investigations of recent results showed a good evidence of changes in the region 4000-5000 cm^{-1} that can be attributed to lipid and protein changes this difference is a very sharp and subtle and anifests itself clearly in advanced phases of breast cancer.

Photoacoustic spectroscopy was used to study the range 400- 2000 cm^{-1}. In direct FTIR spectroscopy this part of spectrum was hidden by the bands originated from the glass slide. This technique has an additional advantage that it doesn't require specific sample preparation [2,3]. Samples can be even opaque, so it can be used in vivo to examine tissues during surgical operations

The region 600-1700 cm^{-1} reveals very important features that can be considered as early indications of the cell transformation from normal to malignancy. It seems that this transformation is a slow one and takes long time before it manifests itself. We believe it can be observed through permanent and regular study of the spectrum in the range 600-1700 cm^{-1}. There may be no direct indicator that can be used , since the spectral analysis of normal tissues shows a good deal of differences between them. However a remarkable difference can be indicated between normal and malignant tissues of the same person. A careful analysis of this band is needed since there are several overlapping bands due to different cells and molecules.

EXPERIMENT AND RESULTS

The system used is based on BRUCKER SS66 FTIR spectrometer equipped with a photoacoustic unit and OPUS software. The analysis was done using in-house developed programs based on Matlab 6.5 and its toolbox.

The photoacoustic spectroscopy of bulk fresh specimens was obtained for 4 normal and 4 malignant persons were used. They were obtained during breast cancer tumor excision operation and examined the same day Focusing the IR modulated beam just on the tumor only is impossible, so usually the spectra should be expected to contain a certain unknown mixture of muscles, connecting tissues and tumor.

Fig(1) shows the obtained spectra between 1000-1700 cm^{-1}. The finger print bands of interest are (a) 1050-1330 cm^{-1} (b) 1330-1400 cm^{-1} (c) 1400-1480 cm^{-1} (d) 1480-1600 cm^{-1} (e) 1600 -1700 cm^{-1}.

In some cases slight change in the position of the specimen can reduce or enhance the spectra, so we decided to keep the mostly reproduced spectra in each case.

CP888, *Modern Trends in Physics Research,*
Second International Conference on Modern Trends in Physics Research—MTPR-06,
edited by L. El Nadi
© 2007 American Institute of Physics 978-0-7354-0354-9/07/$23.00

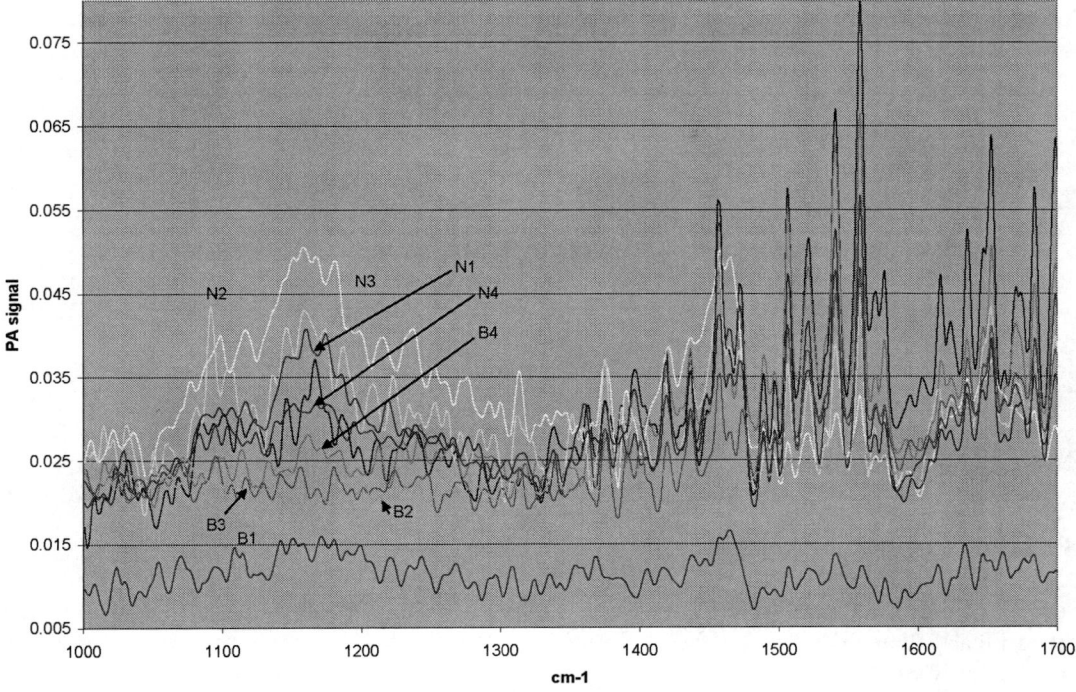

FIGURE 1. Photoacoustic spectra between 1000-1700 cm^{-1} for normal and malignant tissues.

1- Spectral decomposition of the range 1000-1700 cm^{-1}.

The mean principal component fig (3) and fig (4) for normal and tumor specimens, can be fitted using a Gaussian spectral bands using the linear combination

$$\sum_i c_i \exp-\frac{(\nu-\nu_i)^2}{2\ w_i^2}$$

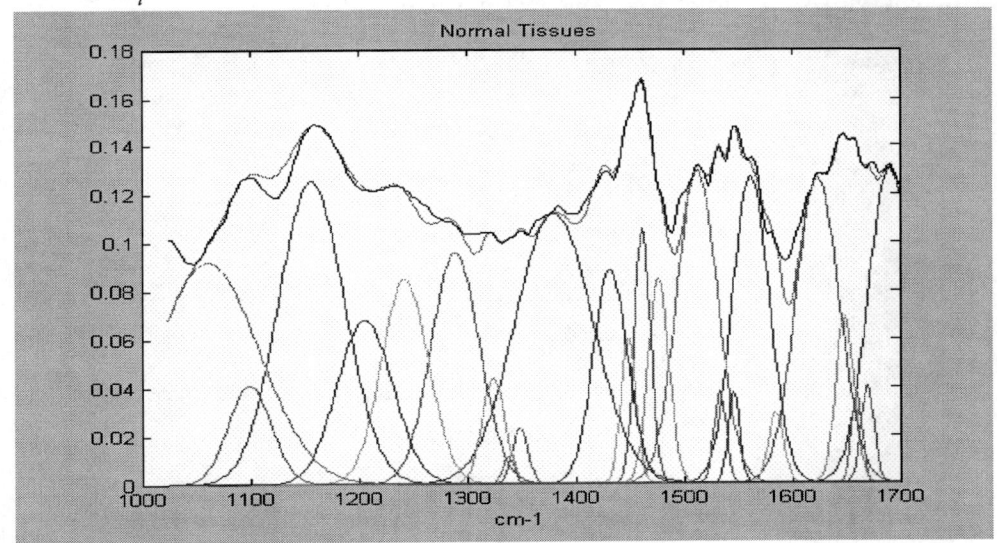

FIGURE 2. Spectral decomposition of normal tissues

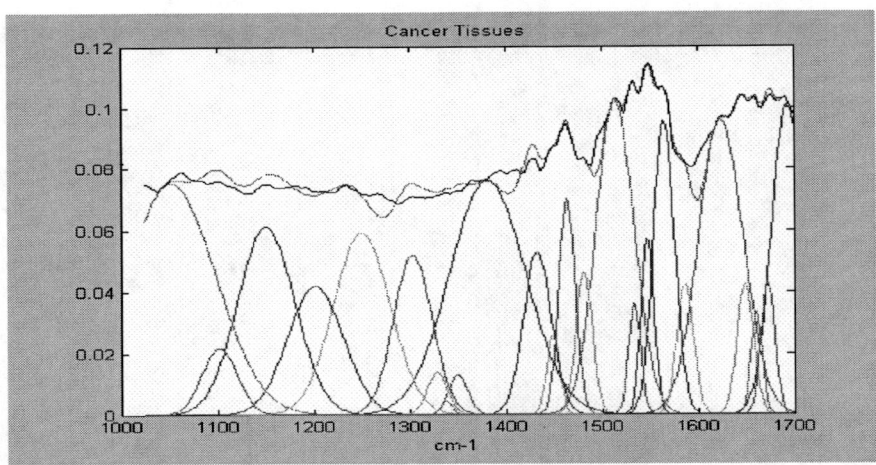

FIGURE 3. Spectral decomposition of cancer tissues

2- Principal Component Analysis.

Principal Component Analysis (PCA) using Singular Value Decomposition (SVD) algorithm showed clearly that the 1st main factor for normal case Fig(2) and malignant case Fig(3). We focus the study of Protein changes in the region 1050-1500 cm^{-1}. The analysis of DNA is based on the range 600- 1800 cm^{-1}. In some cases a shift in the position of peaks will be indicated, while in other cases the peaks are present in the same places without any significant shift. However the intensity is greatly reduced in the tumor cases for the peaks at 1156.4 cm^{-1}, which could be assigned to NH3 or Lysine. The observed shift between normal (1156.4 cm^{-1}) and tumor (1151.2 cm^{-1}). The two satellite band at 1102.1 and 1244.5 cm^{-1} are also reduced in the tumor spectra. The band in 1330-1480 cm^{-1} is also reduced for Tumors. The peak ratios I$_{1463}$/I$_{1244}$ =1.186 for tumors while for normal specimens is 1.5. The peak ratio I$_{1156}$/I$_{1244}$=1.03188 for tumors and the same ratio for normal cases 1.626. This result is in agreement with the conclusion of our previous paper in the study of FTIR MID spectra in the range 2700-3700 cm^{-1} that the lipid concentration is greatly reduced. The reduction is of the order of 55% nearly.

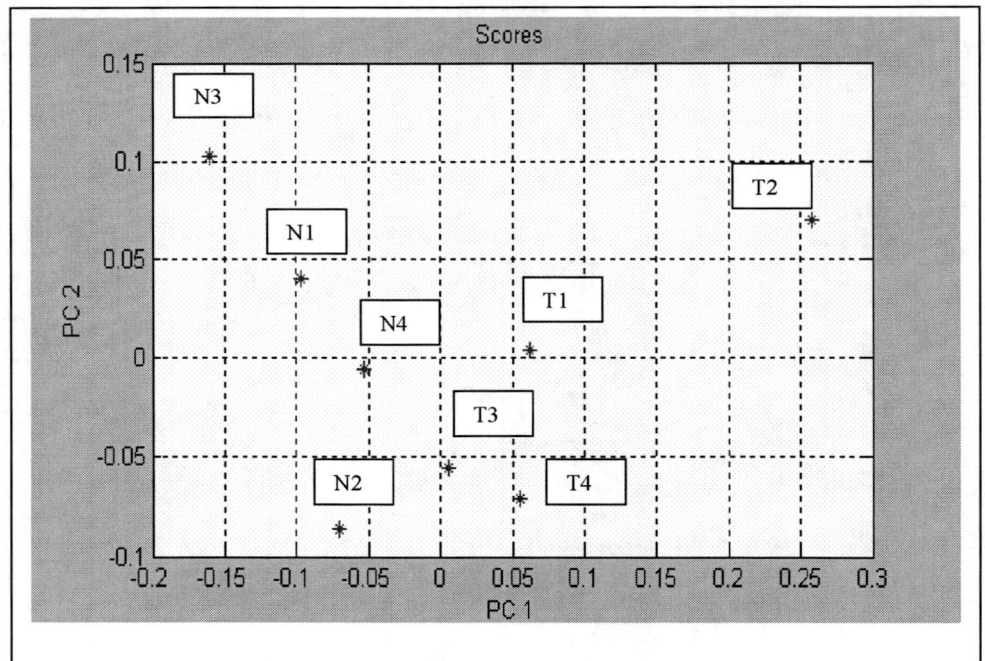

FIGURE 4. Scores of the 1st two factors.

When all of the 8 spectra were input as a mixed spectra to the PCA analysis for the calculation of 4 factor scores and loading , the scores of the 1st two factors are plotted in Fig(2). Table(1) shows the scores of the 4 factors. Normal and Tumor spectra are clearly separated for the 1st factor. All normals has –ve scores while tumors have +ve scores. The second factor scores corresponds to individual differences than categorical differences. Great care should be taken when interpreting the loading factors when the input matrix is mixed, normal or tumor spectra. Also we remark that these spectral matrices are mathematically centered before performing PCA.

TABLE 1. Scores of different spectra

	Factor 1	Factor 2	Factor 3	Factor 4
Normal 1	-0.097345	0.040755	-0.0064157	-0.019558
Normal 2	-0.070484	-0.085987	-0.013175	0.018809
Normal 3	-0.15981	0.10301	0.00041916	0.0085576
Normal 4	-0.052574	-0.0053279	-0.0063476	0.0027138
Cancer 1	0.061551	0.0036395	0.041906	0.0052963
Cancer 2	0.25789	0.070184	-0.016415	0.0042537
Cancer 3	0.0053603	-0.055192	0.003348	-0.012414
Cancer 4	0.055414	-0.071083	-0.0033188	-0.0076579

3- Protein structure changes.

The result of fitting is shown in Fig (3) for normal spectra and Fig(4) for tumor. The main spectral lines and their assignment [4] is shown in table (2). Amide I (1650 cm^{-1}) and Amide II (1550 cm^{-1}) seem to have equal strengths in both normal and tumor spectra. As we mentioned the greatly reduced band in tumor spectra is the P-O stretching band in the 1096 -1180 cm^{-1}. This belongs to phosphate groups in phospholipids. The Amide I line (1600-1700 cm^{-1}) was decomposed to 5 lines assigned to the Lipoprotein structure as follows[5]:

β-strands at 1693 cm^{-1}
Turns at 1672 cm^{-1}
α-helix at 1660 cm^{-1}
Random at 1648 cm^{-1}
β-sheets at 1693 cm^{-1}

A change of the line of β-sheets from 1622.8 cm^{-1} in normal tissues to 1624.1 cm^{-1} in tumor, similar effect has also been observed [5]. The β-strands and β-sheets contributions were increased in tumor (from 28% and 27% resp. in normal to 31% and 30% resp.) the same effect was observed in the case of oxidation by metals [5]. The other two conformations was reduced in tumor tissues.

TABLE 2. Main finger print lines and assumption

Assignment	Normal Tissues			Tumor Tissues			
	Wave number (cm-1)	Linewidth (cm-1)	Amplitude (a.u)	Wavenumber (cm-1)	Linewidth (cm-1)	Amplitude (a.u)	Ratio N/T
Amide I	1693.7	14.544	0.12922	1693.1	16.3	0.10099	1.279533
	1672.9	7.5967	0.068517	1672.4	7.2395	0.042867	1.598362
	1660.6	5.9995	0.059323	1660.6	5.7769	0.033038	1.795599
	1648.4	8.729	0.074255	1649.1	8.588	0.042459	1.748864
	1622.8	21.219	0.12607	1624.1	23.097	0.096009	1.313106
Amide II	1585.1	7.1202	0.053121	1586.6	8.5427	0.042214	1.258374
	1564.6	10.777	0.12588	1564.8	11.074	0.095694	1.315443
Guanosine monophosphate overtone ring	1547.4	6.4593	0.084638	1547.3	6.7255	0.057358	1.475609
Guanosine monophosphate CN,CH stretch	1533.8	6.4793	0.06102	1533.8	6.1379	0.036294	1.68127
	1512.5	18.483	0.13038	1514.2	19.717	0.10247	1.272372
Lipid (CH3)3N3 assymbending	1480.4	8.6843	0.071659	1481.5	8.4162	0.046375	1.545208
Amino acid aniline+++ lipid CH3bending	1463.8	8.2389	0.12003	1464	8.4757	0.070245	1.708734
Amino acid valine	1449.2	7.8516	0.0646	1449.4	6.6943	0.025651	2.51842
	1427.9	18.271	0.12056	1433	14.743	0.052894	2.279276
Amino acid leucine+++ lipid CH3	1382	19.472	0.10755	1381.2	38.058	0.076138	1.412567
CH2	1352.2	9.3286	0.056076	1350.4	9.9781	0.01297	4.323516
	1339.4	6.5754	0.023941	1338.5	5.8404	0.0042776	5.59683
trichlorophenol	1325.7	9.9737	0.044121	1328.6	10.589	0.01382	3.192547
Amide III	1295	28.233	0.096706	1303.4	20.12	0.051916	1.86274
Lipid PO2	1244.5	22.664	0.084353	1249.8	28.979	0.059282	1.422911
Guanosine CN stretch PO3 stretch	1204.8	22.9	0.080179	1202.5	28.471	0.041812	1.917607
NH3Lysine	1156.4	25.658	0.13663	1151.2	29.867	0.061173	2.233502
	1102.1	20.637	0.092454	1102.2	18.616	0.021736	4.253497
	1056.5	26.896	0.08767	1051.6	44.696	0.075347	1.16355

4- Characterization of DNA

We followed the method given in detail [6] which is based on normalizing the spectra for the band 1092-1100 cm-1 of phosphodioxy. Searching for the markers of the four groups

    dA   at 728 cm^{-1} , 1302 cm^{-1}
    dT   at 749 cm^{-1} , 1652 / 1672 cm^{-1}
    dG   at 681 cm^{-1} , 1713 cm^{-1}
    dC   at 783 cm^{-1} , 1531 cm^{-1}

Some deviations has been noticed in some cases the the line 1531 cm^{-1} of the dC group was weak in all spectra. The lines 1652 cm^{-1} of dT and 1713 cm^{-1} of dG manifests themselves very clearly in all cases except one malignant case.

TABLE 3. DNA group markers

	Band	Normal		Malignant	
dA	728	723	1.19- 1.4	730	1.3-1.7
	1302	1299	0.79-0.9	1301	0.84-1
dT	749	748	1.2-1.4	730	1.19- 1.6
	1652	1652	0.94-1	1652	1.6-2
	1672	1673	0.98- 1.3	1672	0.9-1.6
dG	681	680	1.1-1.28	685	1.4-1.7
	1713	1716	0.68-1	1716	1.3-1.5
dC	783	782	0.9-1	786	1.4-1.5
	1531	1533	0.7-1	1533	1.27- 1.4

Tumor tissues show a slight shift to higher wavenumber for the A marker associated with an increase in A marker specially for the line 728 cm^{-1}.
The 1489 cm^{-1} line of Guanine has a line width of 7 cm^{-1} in normal cases it starts broadening and reaches 10-12 cm^{-1} in tumor tissues. Intensity in normal tissues are .06-.09 and in tumors 1-1.2 which indicate that tumors are more rich .
The Adenine residues at 1482 cm^{-1} was absent in all cases. However the line 1578 cm^{-1} was shifted to 1575 cm^{-1} with intensities 1.5-1.6 compared to 0.6-1 for normal tissues. The line at 1511 cm^{-1} is shifted in all cases to 1506 it ranges in normal between .7-.8 and in tumor 1.15-2.
Guanine residues at 1318 cm^{-1} is shifted in normal tissues to 1313-1315 cm^{-1} the intensity doesn't change between normal and tumor tissues.
Strong and sharp line at 1506 cm^{-1} with a shoulder at 1508 cm^{-1} showed an increase in intensity in tumor 1.5-2.2 vs .6-1.2 in normal tissues. Another line at 1456 cm^{-1} with the same behavior but much more symmetric without shoulders.
The Adenine residues near 1482 cm^{-1} does not exit but a line at 1489 cm^{-1} exists it could be Guanine or shifted Adenine line.
Thyamine residues at 1190 cm^{-1} was observed but shifted in normal tissues to 1195 cm^{-1}.
The 1187 cm^{-1} of CP DNA is shifted in one case to 1185 cm^{-1} with increased intensity. The ML DNA observed near 1267 cm^{-1}.
The B-form marker at 835 cm^{-1} is decreased in tumor in two cases and a new line at 825 cm^{-1} was observed.
Cystosine residues at 1180 cm^{-1} exist in all tumor cases

5- FTIR spectroscopy of the range 4000-5000 cm^{-1}.

The MID FTIR spectroscopy using Bruker FTIR spectrophotometer has been done for over than 30 specimens from different patients. A very clear distinction between tumor and malignant tissues was observed. Fig(5) shows this feature in the 4000 – 5500 cm^{-1} spectral band. Tumor tissues show an absorption band wide band that varies in intensity between patients. The peak of the band at 5157 cm^{-1} its width is 429 cm^{-1}, the asymmetry of the band indicates that it is associated with a side band at 4866 cm^{-1}. Usually this region is considered to be combinations and overtones of vibrational bands, however the considerably high intensity of the band may suggest that it is an entirely new bond that appeared only in cancer tissues. Another interesting feature that is present only in normal breast tissues is a double peak at 4319 cm^{-1} and 4245 cm^{-1}. On the contrary these double peak is of low intensity and of width 11 cm^{-1}.

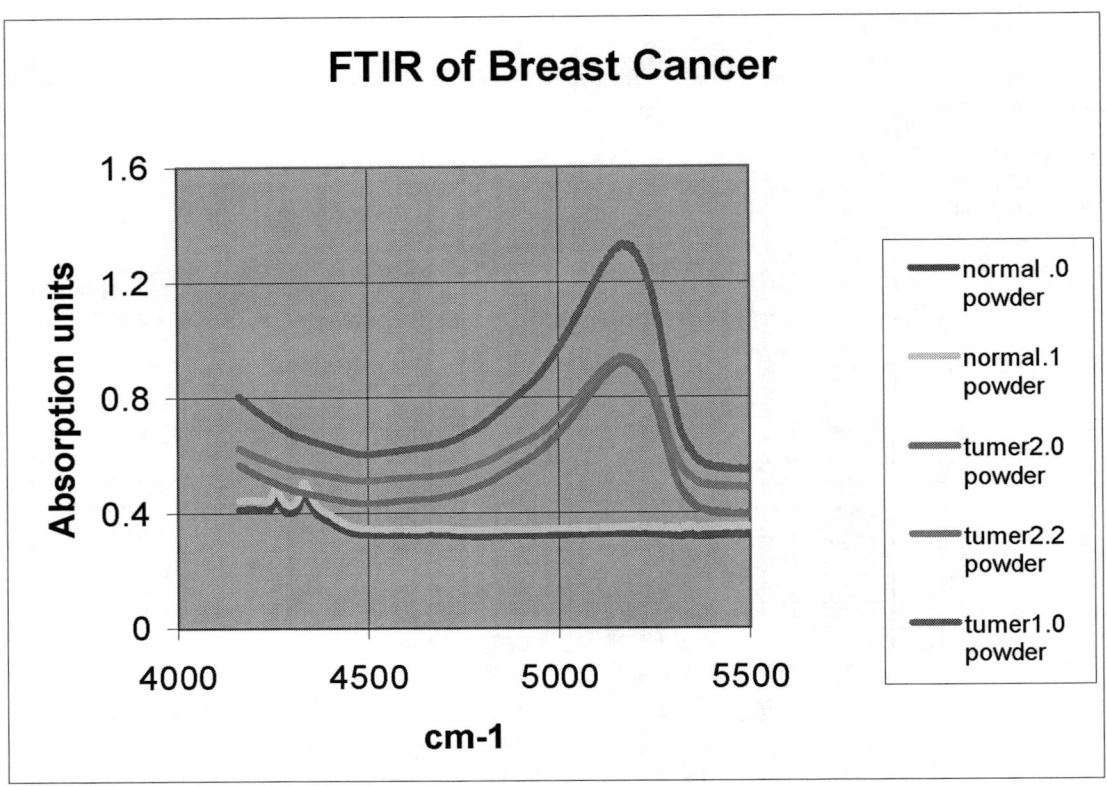

FIGURE 5. IR spectrum in the range 4000 – 5500 cm^{-1}.

6- Differential Photo Acoustic Imaging (DPAI).

There are methods to identify malignant tissues using photo acoustic techniques which depend mainly on the difference in mechanical properties between them. These properties include density, bulk modulus and specific heat. The pervious result suggests the following method. Using laser line at 1939 nm(5157 cm^{-1}) increases the tissue absorption 2-3 times of tumor , then with a system of piezo electric transducers would produce an image. Now using another line at 2.2 μm the obtained signals will be deduced from that obtained with the laser at 1939 nm .

CONCLUSION

A very clear identification between normal and malignant tissues is clear form the doublline 5157 cm^{-1} and 4245 cm^{-1}. The above results suggest the use of a differential photoacoustic imaging (DPAI) as non-invasive method of detecting breast cancer. The straightforward choice will be tunable MID IR pulsed OPO lasers at a reference wavelength λ_1 6.25 μm as having the same behavior for both normal and malignant tissues and λ_2 8.6 μm to differentiate between normal and tumor tissues. The low contrast will indicate tumors while high contrast will indicate normal tissues.

REFERENCES

1. Yehia Badr and S.E.HassabElnaby, ,SPIE **5325** (2004).
2. R.Fainchtein, B.Stoyanov, J.Murphy, D.Wilson,D.Hanely, Proc. SPIE, **3916**, 19(2000).
3. E.Svateeva , A.Karabutov, S.Solomatin, A.Oraevsky, Proc. SPIE, **4618**, 63(2002).
4. D. Naumann, H. Labischinski, P. Giesbrecht, 'The Characterization of Microorganisms by Fourier Transform Infrared Spectroscopy (FT/IR)', in *Modern Techniques for Rapid Microbiological Analysis*, ed. W.H. Nelson,VCH, New York,1991, 43.
5. R. Chehin et al, J. of Lipid Research, **42**,778(2001).
6. Mona Abdelaziz and A Elkhouly Arab Journal of biotechnology, **9** (2006).

Growth of Nanostructure of Metal Oxides by Laser Ablation and by SiO$_2$ Assisted Thermal Evaporation

Lotfia M. EL Nadi, Galila .Mehena, Mgdy M. Omar, Hussein A. Moneim, Fakiha[1] H. A. Taieb, Faried A. Rahiem[2]

Laser Physics Lab., Physics dept., Faculty of science, Cairo university, giza, Egypt.
[1]Chemsitry dept., Faculty of science, Cairo university, Giz, Egypt
[2]Physics Dept., Faculty of Science, Alazhr univ., Egypt.

Abstract We report the results of growing nanostructures of gallium oxide and indium oxide by two methods. In the first one we applied laser ablation in air of pure graphite rod filled with Gallium or Indium metals. The ablated plume then deposited on SS substrates in air. In the second method the oxides were synthesized by thermal heating of the Ga or In metals mixed with powder of graphite and covered with SiO$_2$ plates, supported by ceramic, in high temperature oven. The ablation method produced nanowires of Ga$_2$O$_3$ and nano particles of In$_2$O$_3$ developing in nanowires. . The solid carbon ablated from the graphite rod existing in the ablated plum as fine solid particles mixed with metal Ga or In melt in contact with oxygen gas in air, produced the growth of the metal oxide nano structures by solid –liquid-gas mechanism.

The silica assisted catalytic growth oxides produce only nano particle of each metal. The reaction of the metals with SiO$_2$ melt and graphite produced Si and carbon. The then formed Si carbide can effectively initiate vapor- liquid-solid growth of nano structure metal oxide. It seems that SiO$_2$ in addition to the atmospheric oxygen provide the oxygen source for forming metal oxide nano dots.

INTRODUCTION

Material science and laser technology are introducing revolutionary discipline of nano technology. The advances of producing Nanoparticles or nanostructure for various material imposed many important questions of how to control the growth, how to identify the produced nano structure and how to assemble them into new devices. The extraordinary accomplishments of arranging them into circuits that perform logic operations amplifying signals invert current flows and even perform simple computing tasks, call for new methods for growing and identifying nanostructures [1-5]. Many methods are still developing for preparation of nano meter sized material. One important method vapor-liquid-solid growth either in oven or by laser evaporation or solution–liquid–solid growth at lower temperature, have been used for the synthesis of oxide, carbide and nitride nanostructures [6-9]. Other new growth method known as oxide –assisted nanowire growth, developed Si and Ge nanowires [10-11], during which the decomposition of vapor oxides at high temperature during thermal evaporation or laser ablation helped to grow nano wires. Nanowires of tungsten oxide growth as nano tree like structure has been synthesized by Zhaw et al [12]. By heating tungsten foil covered by SiO$_2$ plate or high temperature Gallium oxide, β-Ga$_2$O$_3$ being a wide band gap [4.9 eV] compound [16]. Has interesting conduction and optical properties therefore its nano structures may also have important applications the fabrication of nano wires of gallium oxide were achieved by physical evaporation [14]. Characterization of photoluminescence properties of Gallium -oxide nanowires grown by carbothermal reduction from gallium oxide powder was reported by X.C. Wu et al [15]. Silica assisted catalytic growth of Ga$_2$O$_3$, GaN and AIN nano wire has been prepared and identified recently [16].

In this study a comparative study for synthesis of nanostructures of Ga$_2$O$_3$ and In$_2$O$_3$ was carried out using two methods in the first one, UV laser ablation of the Ga or In metals embedded in graphite was applied the ablated plumes in air were deposited on SS substrates. While the second method used was the silica assisted catalytic growth of the metal oxide at elevated temperature. Fine powder of graphite mixed with Ga or In metal were heated in atmosphere after covering with SiO$_2$ plate. It has been proved that the laser ablation method enhanced the growth Ga and In oxide nanowires while in the second method, Nanoparticles for both Ga and In oxides were observed

CP888, *Modern Trends in Physics Research,*
Second International Conference on Modern Trends in Physics Research—MTPR-06,
edited by L. El Nadi
© 2007 American Institute of Physics 978-0-7354-0354-9/07/$23.00

EXPERMENT

For the laser ablation method, the confined geometry of the target chamber used for nitrogen laser ablation of graphite described before for formation of the carbon nanowire was applied [17]. A modification has been introduced on the graphite rod in which an axially drilled hole of diameter 2 mm and length 5 mm in the rod was filled with gallium or indium metal as shown schematically in figure.1

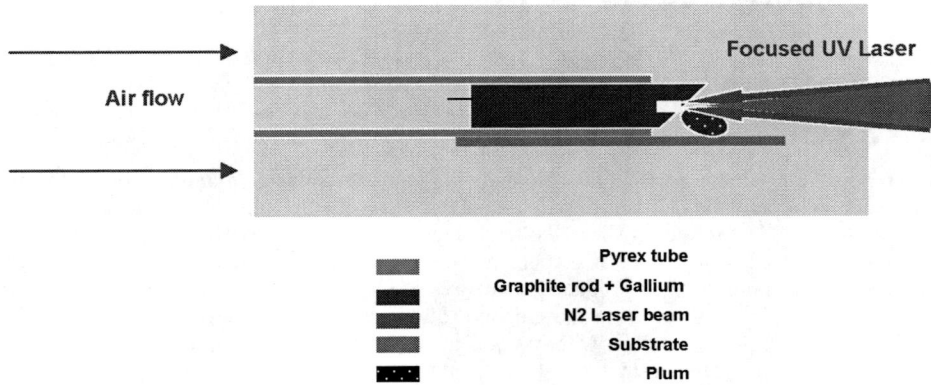

FIGURE 1: Experimental setup with modified graphite drilled rod

The pulsed UV N_2 laser beam was focused on the center of the graphite filled metal rod by quartz crystal. The ablated plum consists of ablated carbon and metal at high temperature. The probability of forming the metal oxide during the ablation process is expected since the laser irradiation process took place in the vicinity of atmospheric air. The irradiance at the target was 25 ± 5 MW/cm^2 per one laser shot summing up to $25\pm$GW/cm^2 for one thousand laser shots. The ablated plumes were allowed to deposit on an area of 1 cm^2 of stainless steel substrates (SS Fe 91%, Ni 4.5%, Co 4.5%) for 1000 laser shots or 2000 laser shots respectively. The substrates were examined by scanning electron microscope (Jeol JXA-840A SEM) provided by electron probe micro analyzer. In the second method of silica assisted catalytic growth of Ga or In oxide nanostructures, 99.99% graphite plays the role of the catalyst. The equal amounts of metal and fine carbon, from graphite, were mixed thoroughly and homogenized in a porcelain mortar for 1 hour. The mixtures were then placed in small porcelain crucibles, each covered with SiO$_2$ Plate. They were then mounted into high temperature small compartment furnace. The temperature was raised to 950 °C during 45 minutes in low atmospheric air flow. When the temperature reached 750 °C, the SiO$_2$ plates melted, at lower temperature than expected (SiO$_2$ Mp=1610°C). The melt mixture of the metals (Ga or In) + SiO$_2$ in presence of solid carbon was formed. The melts were allowed to cool down slowly reaching room temperature during 4 hours. The grayish white crust on the walls of the porcelain crucible, were then collected as fine powder and prepared for imaging by transmission electron microscopy. The samples of the crust powder, dispersed in acetone were pipetted onto Cu grids covered with carbon thin films. The grids were examined by "TEM 10 Zeiss WW EM", and constituents were identified by EDX. The Products left overnight developed in extraordinary hard ingots, sticking to the crucible bottom and it was difficult to turn them in fine powder by conventional methods.

RESULTS AND DISCUSSION

The deposited plumes on each substrate were examined after the laser ablation experiments by SEM. Figure 2a shows the morphology of the plane substrate surface. In figure 2b the image of the deposited plum formed during 1000 laser shots, indicate the presence of Nanowires with few Nanodots of Ga oxide (bright) and few carbon Nanodots (black). The average diameter for gallium oxide nanowires (GaONW) were measured to be 380±20 nm with average length 160 ±9μm and average density $(1.2±0.4)×10^5$ wires/cm². The GaOND are of average diameter 330±30 nm and average density $(3±0.4)×10^4$ dots/cm².

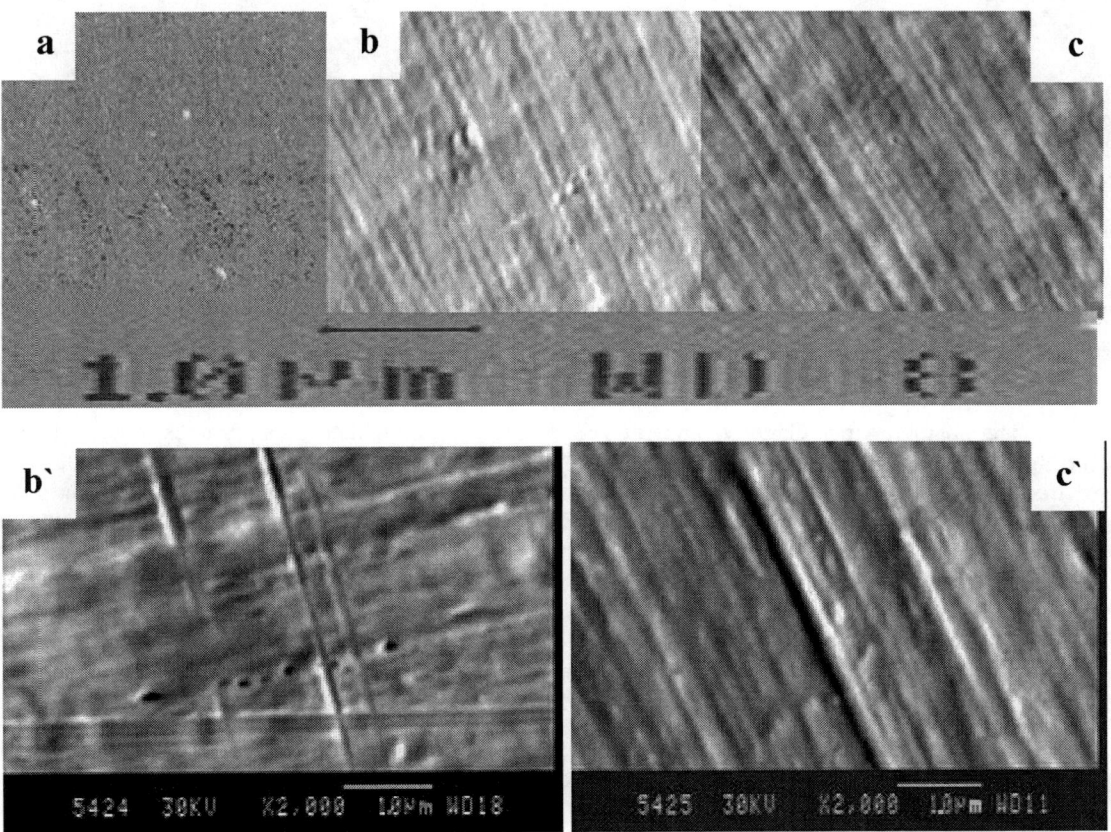

FIGURE 2: Morphology of the substrate surfaces examined by SEM for the deposits of Ga oxide on (a) plane SS substrate surface, (b) formed GaONW after 1000 laser shots, (b`) a spotted magnified area of (b) revealing Nanodots(bright) GaOND, and few carbon Nanodots (Black)(c) formed GaONW after 2000 laser shots and (C`) Spotted magnified area of (C)revealing the parallel nature of GaONW growth.

The deposits due to ablation with 2000 laser shots reveals an appreciable increase in the diameter of the GaONW reaching 420±40 nm while the average length is nearly unchanged as clear from figure 2c where GaOND and the CND nearly vanished. The morphology suggests that the nanowires grow due to accumulation of the GaOND along one axis forming parallel nanowires. Figure 2b` and 2c` are spotted magnified areas of figure 2b and figure 2c confirming the above mentioned suggestions.

Figure 3a represents the morphology of the deposits of indium oxide nanodots (bright) together with carbon Nanodots (black). The InONW seems to be in their early stage of growth for the case of ablation by 1000 laser shots. The SEM Images shown in figure 3b due to ablation with 2000 shots clearly indicate the growth of InONW (bright) in parallel close packing along an incomplete straight fiber like- structure. It is clear that InOND increased in density but the average diameter of 320±30 nm seems to be the same for 1000 or 2000 laser shots within the error limit. One notices that carbon Nanodots(black) increased in density in figure 3b than in figure 3a. An important symptom of

the increase of carbon Nanodots noticed for the In case points to lower utilization of carbon with indium metal in the ablated plum.

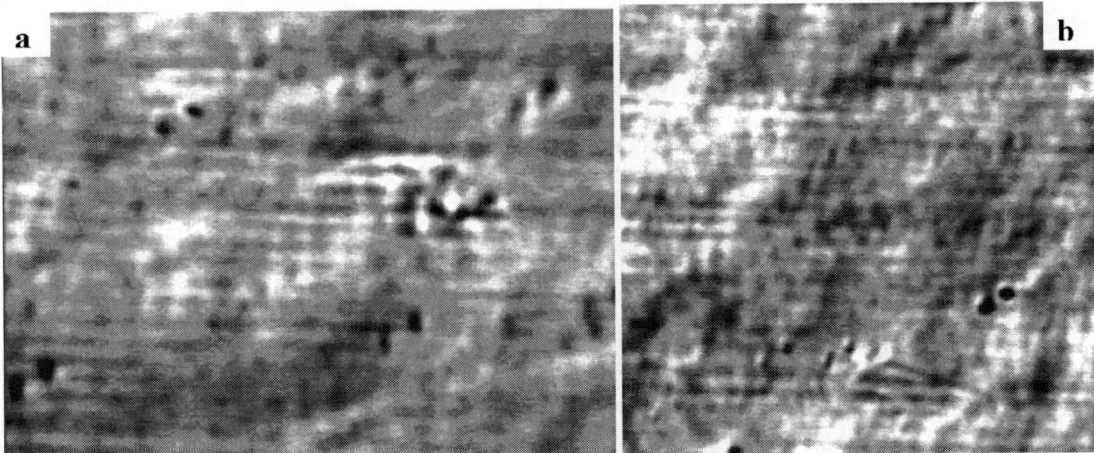

FIGURE 3. (a, b): the SEM images representing the morphology of InOND (bright) and CND(black), (a) the growth for the case of ablation of In by 1000 laser shots, (b) the growth for ablation of In by 2000 laser shots.

The characteristics GaONW and InONW formed from attached nanoparticles indicate that a solid-liquid-gas growth may be the most suitable mechanism for the formation of the metal oxide. The growth process could only be explained if there liquid phase existed that nucleated growth of droplets of the metal in contact with oxygen gas to form oxide wire like growth. Could then build up due to super saturation [18]. This condition could be achieved at higher number of laser shots. The carbon existing in the ablated plumes as solid fine particles as well as the low melting point (Mp) of the metals (Ga Mp=29.78 °C, In Mp=156 °C) support the solid-liquid-gas growth mechanism. The reactants existing in the plume and deposits on the substrate could be represented as

$$\text{Reactant} \quad 4Ga \quad +2O_2 \quad +C \longrightarrow 2Ga_2O + CO_2 \tag{1}$$
$$\text{Liquid} \qquad \text{gas} \qquad \text{solid} \qquad\qquad \text{gas} \qquad \text{gas}$$

The Ga_2O and CO_2 gases can be dissolved continuously into the liquid Ga metal forming drops inside which the following reaction takes place:

$$\text{Products of (1)} \quad Ga_2O + CO_2 \longrightarrow Ga_2O_3 \quad + \quad C \tag{2}$$
$$\qquad\qquad\qquad \text{gas} \qquad \text{gas} \qquad\quad \text{solid} \qquad \text{solid}$$

The solid products deposits on the substrates. When Ga_2O_3 saturates formation of nanowire increase accompanied with nanodots. For the case of indium metal, of higher melting point than Ga, it is suggested that the liquid phase contribution is lower than in the case of Ga. The saturation of the drops with In_2O_3 needs higher dissipation of laser power in order that INONW could grow.

The results of the silica –assisted catalytic growth of Ga and In oxides, are shown in figure 4(a,b) respectively. The TEM images do not indicate growth of nanowires, but only reveal GaOND of average diameter 200±20 nm and InOND of average diameter 320±20nm. The following reaction most probably occur during the heating process of the Ga or In metals with SiO_2 and the carbon from the fine graphite powder in presence of O_2 gas of air in the furnace:

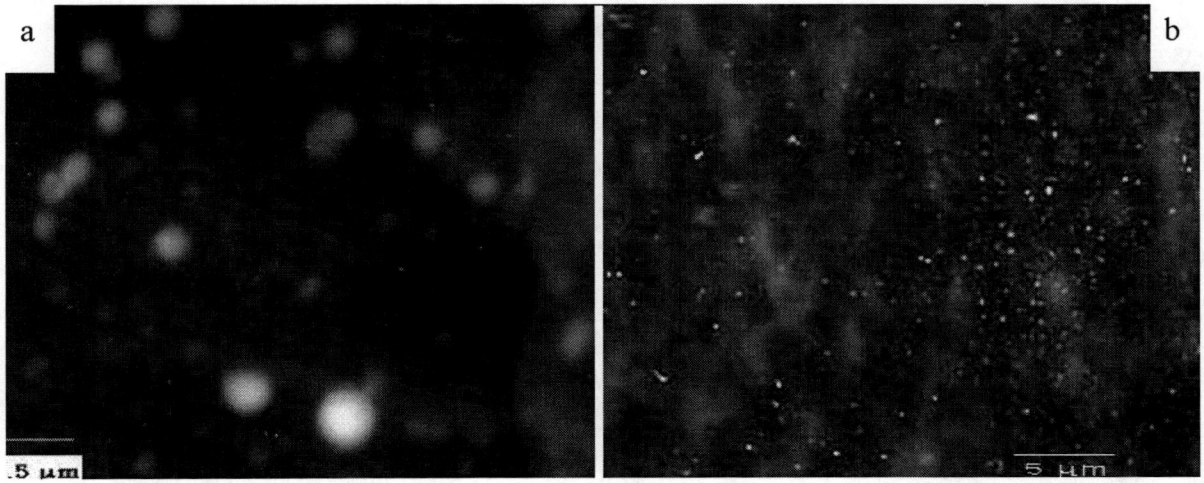

FIGURE 4(a, b): The TEM images indicating growth of nanodots of the metal oxides by the silica assisted growth (a) for GaOND, (b) for InOND.

The reactants

$$4Ga + SiO_2 \longrightarrow 2Ga_2O + Si \qquad \text{products 1}$$
Liquid   Solid   Gas   Solid

The reactants

$$4Ga + C + 2O_2 \longrightarrow 2Ga_2O + CO_2 \qquad \text{product 2}$$
Liquid   Solid   Gas   Gas   Gas

The reactant

$$C + Si \text{ product 1} \xrightarrow{\text{High T}} SiC \qquad \text{product 3}$$
Solid   Solid   Tiny Droplets

The products from 1 and 2

$$3Ga_2O + 2CO_2 \xrightarrow{\text{SiC}} 2Ga_3O_2 + 2C \qquad \text{final product 4}$$
Gas   Gas

The final products Ga_3O_2 +C supersaturates when cooling takes place and solidify in the droplets of SiC(products 3). Nanodots of the metal oxide are then probably formed the same occurs for the indium oxide.

CONCOLUSION

The metal oxides of Ga and In as nanostructure can be synthesized by UV laser ablation of graphite with each metal in air. The deposited plume on stainless steel substrates forms nanowires of gallium oxide and nanodots. The Nanodots in case of In oxide could develop in Nanowire. The solid carbon with the metal liquid melt existing in the ablated plum in contact with oxygen gas of air, suggest that the growth was nucleated through solid- liquid – gas mechanism. In the metal oxides of Ga and In as nanoparticles or dots can also be prepared by silica assisted thermal vaporization, of the mixture of solid carbon. The SiO_2 has an important roles in the catalytic growth of the metal oxide Nanoparticle. It provide Si which forms SiC tiny droplets. The final products in the gas phase. Ga_2O and Co_2 gases supersaturates in the SiC when cooling takes place providing Nanodots of the metal oxide. The metal oxide Nanostructures could effectively grow through vapor (gas)- liquid –solid growth mechanism.

REFRENCES

1- Sab. Suh et al, J.Am.Chem.Soc.**126**, 2198 (2004).

2- S.B. Suh et al, Phys. Rev. **B67**,241402®(2003).

3- H.M.Lee et al, J. Phys. Chem., **B107**, 9904 (2003).

4- Gill Renaud, Aip cop **MTTPR-04**, 748, 63(2005).

5- P.L. Hansen et al, Science **295**, 2053 (2002).

6- X.F. Duan, C.M. Lieber, J.Am. Chem. Soc.**122**, 188(2000).

7- D.P. Yu, Q.L. Hang, Y.Ding, H.Z.Zhang, Z.G.Bai, J.J. Wang, Y.H.Zou, W.Qian, G.C. Xiong, S.Q.Feng, Appl.Phys.Lett.**73**, 3076(1998).

8- N.Wang, Y.H.Tang, Y, F, Zhang, D.P.Yu, C.S.Lee, I.Bello, S.T.Lee Chem. Phys.Lett.**283**, 368(1998).

9- K.W.Wong, X.T.Zhou, F.C.K.Au, H.L.Lai, C.S.Lee, S.T.Lee, App.Phys.Lett.**75**, 1925(1999).

10- N.Wang, Y.F.Zhang, Y.H.Tang, C.S.Lee, S.T. Lee Phys.Rev.**B58**, 1604(1998)

11- N.Wang et al chem. Phys.Lett.**299**, 237(1999).

12- Y.Q.Zhu, W.B.Hu, W.K.Hsu, M.Terrones, N.Grobert, J.P. Hare, H.W. Kroto, D.R.M.Walton, H. Terrones, Chem. Phys. Lett.**309**, 327(1999).

13- H.H. Tippins, Phys.Rev.**140**, A316(1965).

14- H.Z.Zhang, Kong, Y.Z.Wang, X.Du, Z.G. Bai, J.J Wang, D.P.Yu, Y.Ding, Q.L. Hang, S.Q. Feng Solid state comm.**109**, 677(1999).

15- X.C.Wu, W.H.Song, W.D. Huang, M.H.Pu, B. Zhao, Y.P. Sun, J.J. Du, Chem. Phys.Lett.**328**,5(2000).

16- C.C. Tang, S.S. Fan, Marc Lamy de la chapelle, P.Li,Chem.Phys.Lett.**333**, 12(2001).

17- Lotfia ElNadi, Magdy M.Omar, Hussien A.Moniem, **MTPR-06** AIP Conf. Proceeding, *ibid* (2006).

18- A.M.Morals, C.M.Lieber, Science**279**, 208(1998).

Formation of Nano Wires by Laser Ablation of Graphite

Lotfia El Nadi, Magdy M. Omar, Hussein A. Moniem

Laser laboratory, Physics Department, Faculty of Science, Cairo University, Giza, Egypt.

Abstract. We report on preparation of carbon Nano wires by laser ablation of graphite targets using confined geometry in air. Nitrogen laser has been used to provide UV laser beam of wavelength 337 ± 2nm, pulses of duration 15± 1ns and power 1 MW per pulse. Using scanning electron microscopy, reveled that nano dots and nano wires were formed on stainless steel substrate surface at different experimental conditions. The formed nano dots have average density of 600 ± 30 dots cm^{-2} and average diameter 200 ±20 nm. The nano wires were formed with average length 850 ± 9 μm, average diameter of 120 ± 10 nm and average density of 100 ±10 wires cm^{-2}. Random distribution of the formed nano wires was noticed with no specific orientation. Some of the wires seemed to be formed of closely packed nano wires forming straight parallel bundle of nano wires. The effect of substrate material was studied and indicated the importance of using SS substrates (Fe 91%, Ni 4.5%, Co 4.5%). Results will be shown in SEM micrographs and discussed relative to previous investigations.

INTRODUCTION

Nano structures of metals and non-metals gave rise to one-dimensional nano scale materials, which encountered new break through technology of the 21 century. They provide new physical properties that could lead to outstanding nano devices. Recently, developing new efficient growth methods for one-dimensional nano sized materials comprise the main objectives of several scientific groups. Formation of different carbon nano structures produce changes of the electronic properties of carbon [1-5].

Theoretically carbon nano tubes (CNT) can be semi conducting, semi metallic, or metallic depending on their diameters and shapes [6-8]. Different methods have been used to prepare carbon nano structures specially multi-walled nano tubes (MWNT) using catalytic decomposition using a carbon containing gas by nanometers size metal particles supported on a substrates. Carbon single walled nano tubes (SWNT) with small amount of double walled ones have also been reported using the catalytic method [9, 10].

Arc evaporation of graphite rods doped with Y and Ni provided high quality SWNT's [11]. High yield of SWNT's have been grown by laser vaporization of Ni and Co doped graphite rods [12,13]. Chemical vapor deposition techniques were applied to obtain bend junctions of SWNT's [14, 15].

Inspite of the several experimental methods used to prepare nano structures of carbon, the parameters controlling the reproducibility of the growth of certain structures, still need deep further investigation.

In the present study we applied laser ablation of graphite to form carbon nano structures on metallic substrate holders. In order that graphite does not pass into vapor state, pulsed UV nitrogen laser was used to guarantee that the process is mainly ablation process. We also tried to investigate the role of the substrate material on which deposition of the ablated carbon plume took place. The formed structures investigated by scanning electro microscopy SEM revealed the formation of separate nano particles (quantum dots) as well as nano wires depending on the accumulated amount of the ablated carbon plumes. The nano structures were proved to be carbon by x-ray diffraction measurements.

EXPERIMENT

The confined geometry of the target chamber experimentally used for nitrogen laser ablation of graphite targets is shown in figure 1. The chamber simply consisted of open ends outer Pyrex glass tube with an inside diameter of 3

CP888, *Modern Trends in Physics Research,*
Second International Conference on Modern Trends in Physics Research—MTPR-06,
edited by L. El Nadi
© 2007 American Institute of Physics 978-0-7354-0354-9/07/$23.00

cm and length of 40 cm. Another Pyrex tube of inner diameter 1 cm and length of 30 cm was adjusted centrally into the outer tube and supported by two annular Teflon disks. The graphite target (99.999% C) rod of length 10 cm and diameter 8 mm was centrally placed in the inner tube. The rod cross section in front of the incident laser beam was cleaved at an angle of 45° as clear from figure 1. The substrate strip of width 1 cm, length 12 cm and thickness 1 mm was inserted between the inner tube and the annular Teflon support in a geometry that allows its axial motion parallel to the rod axes. The ablated carbon plumes could then impinge on specific parts of the substrate according to its position.

The substrates were either stainless steel (SS Fe 91%, Ni 4.5%, Co 4.5%), or aluminum (Al 99.99 %) or glass slides where each one could be placed separately according to the above geometry.

Pulsed UV laser beams from a nitrogen laser of wavelength 337± 2 nm, pulse duration 15 ± 2 ns, energy 15 mJ/pulse and repetition rate of 15 pulses per minute irradiated the 45° cross sectional area of the graphite rod centrally. Focusing the laser beam by a quartz lens provided spot beam brightness of 25 ± 5 MW/cm^2 on the graphite target. During the experiment the pulse per pulse interval was kept constant and nitrogen gas flow of 0.2 liter /min. passed through the target chamber to decrease the probability of oxidizing the carbon plumes. The experiments were carried out at room temperature with either 300 or 600 laser shots.

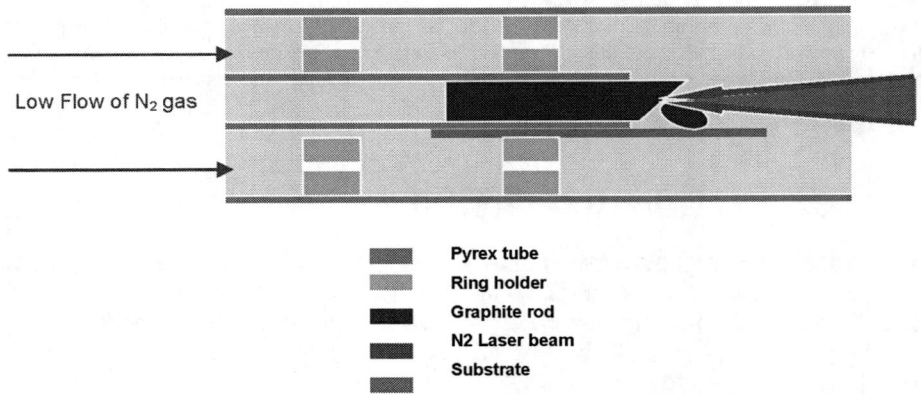

Low Flow of N$_2$ gas

Pyrex tube
Ring holder
Graphite rod
N2 Laser beam
Substrate

Figure 1: The confined geometry of the target chamber experimentally used for nitrogen laser ablation of graphite targets

RESULTS AND DISCUSSION

After each experiment the deposited plumes of carbon on each substrate were examined by the scanning electron microscope (Jewel SEM). Figure 2 (a, b, c, d) show the SEM images of the deposited plumes on SS substrate for different experimental conditions. The formations of carbon nano particles (quantum dots) are the only nano structure obtained for 300 laser beam shots. The average diameter of the dots shown in figure 2a equals 200 ± 20nm with average density of 600 ± 30 dots / cm^2.

Figure 2b is the SEM image for the deposited ablated plumes formed after 600 laser shots. It reveals elongated nano structures with no special orientation. The elongated carbon nano structures do not show nano tubes as clear from figure 2c and 2d but reveals nano wires stacked in parallel forming bundles of nano wires. The average length of the nano wires was found to be 850 ± 9 μm, average diameter of 120 ± 10 nm and average density of 100 ± 10 wires /cm^2.

The comparison between the grown structures when the target was irradiated by 300 shots and that grown by 600 shots suggests that the growth mechanism took place after the deposition of carbon plume on the substrate. For higher total shots the number of carbon dots decreased while the elongated bundle structures of nano wires were formed. This might be explained as due to favorable accumulation of dots merging together to form such elongated structures. The substrates being metallic SS enhances such process.

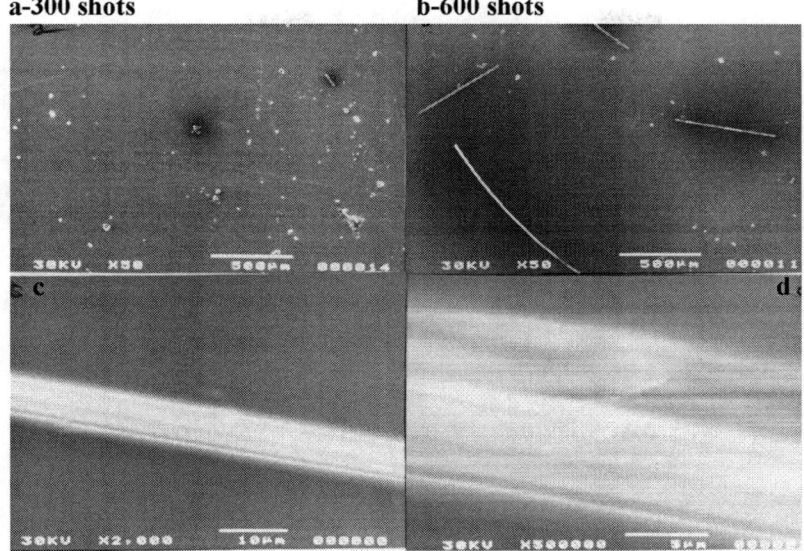

a-300 shots   **b-600 shots**

Figure 2: The SEM images of the deposited plumes on SS substrate. a- Carbon nano particles (quantum dots)for 300 laser beam shots.
b- Elongated nano structures with no special orientation after 600 laser shots. c- Bundles of carbon nano wires. d- Carbon nano wires stacked in parallel forming bundles of average diameter of 120 ± 10 nm

Figure 3a and b shows the SEM images of the deposited carbon plumes on aluminum substrates due to irradiation with 600 laser shots (figure 3a) and with 300 laser shots (figure 3b). The first image reveals that carbon nano dots are of nearly negligible growth while bend irregular structures were produced when 600 laser shots were applied.

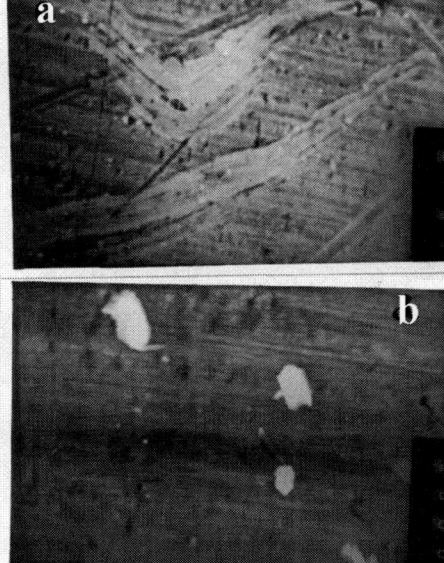

Figure 3.a: The SEM images of the deposited carbon plumes on aluminum substrate due to 600 shots reveals bend carbon irregular nano structures.

Figure 3.b: The SEM images of the deposited carbon plumes on aluminum substrate due to 300 shots reveal few carbon nano dots

The irregular structure could not be easily resolved. This might point to the fact that aluminum substrates are unsuitable to grow carbon nano structures. When using nonmetallic glass slide substrates nano deposits were not recognized.

CONCLUSION

The results described above indicate that the SS (Fe 91%, Ni 4.5%, Co 4.5%) promote the growth of carbon nano wires for higher UV laser ablated graphite. The wires do not stem up from the substrate but are formed parallel to the substrate surface. The observation of nano structure growth on metallic substrate and absence of their growth on nonmetallic substrates might explain the role of the thermal properties of the substrate as well as the thermal processes occurring between the ablated plume and the substrate. This interpretation needs further systematic theoretical and experimental studies. The field of growing nano structures is still open for deeper understanding of the parameters affecting such processes.

REFERENCES

1. Adrian Batchold, Peter Hadley, Takeshi Nakanishi, Cees Dekker, Science **294**, 1317 (2001).
2. H.W.Ch. Postma, T. Teepen, Z. Yao, M. Grifani, C. Dekker, Science **294**, 76 (2001).
3. M.S. Fuhrer *et al*, Science **288**, 494 (2000).
4. Z. Yao, H.W. Ch. Postma, L. Balents, C. Dekker, Nature **402**, 273 (1999).
5. S. J. Tans *et al*, Nature **386**, 474 (1997).
6. C. T. White, T.N. Todorov, Nature **393**, 240 (1998).
7. G. Treboux, P. Lapstun, K. Silverbrook, Chem. Phys. Lett. **302**, 60 (1999).
8. Y.H. Lee, S.G. Kim, D. Tomanek, Phys. Rev. lett. **78**, 2393 (1997).
9. A. Peigrey, Ch. Laurent, F. Dobigeon, A. Rousset, J. Mater. Res. **12**, 613 (1997).
10. H. Jason *et al*, Chem. Phys. Lett. **296**, 195 (1998).
11. C. Journet *et al*, Nature **388**, 756 (1997).
12. R. H. Smally *et al*, Chem. Phys. Lett. **243**, 49 (1995).
13. A. Thess *et al*, Science **273**, 483 (1996).
14. J. Han, M.P. Anantram, R.I. Jaffe, Phys. Rev. **B57**, 14983 (1998).
15. B. Gan *et al*, Chem. Phys. Lett. **333**, 23 (2001).

Biosynthesis of Gold Nanoparticles Using *Pseudomonas Aeruginosa*

M. Abd El-Aziz [‡], Y. Badr [‡], M. A. Mahmoud [†]

(‡) *National Institute of Laser Enhanced Science, Cairo University, Cairo, Egypt.*
[+]*Dept. of Microbiology and Immunology, Faculty of Pharmacy,* [†]*Chem. Dept, Faculty of Science, Zagazig University, Zagazig, Egypt*
Corresponding author: e-mail: mahmoudchem@yahoo.com

Abstract. *Pseudomonas aeruginosa* were used for extracellular biosynthesis of gold nanoparticles (Au NPs). Consequently, Au NPs were formed due to reduction of gold ion by bacterial cell supernatant of *P. aeruginos* ATCC 90271, *P. aeruginos (2)* and *P. aeruginos (1)*. The UV–Vis. and fluorescence spectra of the bacterial as well as chemical prepared Au NPs were recorded. Transmission electron microscopy (TEM) micrograph showed the formation of well-dispersed gold nanoparticles in the range of 15–30 nm. The process of reduction being extracellular and may lead to the development of an easy bioprocess for synthesis of Au NPs.

INTRODUCTION

The nanoparticles have attracted interest because of their unique optical, thermal, electrical, chemical, and physical properties that are due to a combination of the large proportion of high-energy surface atoms compared to the bulk solid[1-5].

Some microorganisms can survive and grow even at high metal ion concentration due to their resistance to the metal. The mechanisms involve: efflux systems, alteration of solubility and toxicity via reduction or oxidation, biosorption, bioaccumulation, extracellular complexation or precipitation of metals and lack of specific metal transport systems[6,7]. A little is known about the resistance against the noble metals[8,9], although it has been stated that gold can serve as a slow acting drug in rheumatology[8]. Silver is highly toxic to most microbial cells and can be used as a biocide or antimicrobial agent[9]. These metal–microbe interactions have important role in several biotechnological applications including the fields of bioremediation, biomineralization, bioleaching and microbial corrosion. Microorganisms considered as potential biofactory for synthesis of metallic nanoparticles such as cadmium sulfide, gold and silver[10].

It is well known that some bio-organisms are able to produce inorganic materials e.g. gypsum and calcium carbonate layers which produced by S-layer bacteria[11,12]. While, magnetite nanoparticles were produced by magnetotactic bacteria[13,14]. Furthermore, siliceous materials were produced by diatoms[15,16].

Fungi was found to be capable of reducing the metals ions into their corresponding nanometals either in intra-cellular or extra-cellular depending of the position of the reduction enzymes. In other words the nanoparticles were formed extra-cellular when the cell walls reduction enzymes were responsible for metal ions reduction as well as when the reduction enzymes secreted extra-cellular[17-20]. Furthermore, the silver nanoparticles were prepared; within the periplasmic space of the bacteria by reduction of Ag^+ using bacterium *Pseudomonas stutzeri* AG259 which isolated from silver mine[21,22]. On the other hand, The Au NPs have been prepared on the surface of bacteria as a result of incubation of the cells with Au^{3+} ions[23,24]. The gold nanoparticles (Au NPs) can be prepared extra-cellularly by parts of whole plants e.g. geranium leaf extract[25].

It is aimed in this present work to investigate extracellular biosynthesis of Au NPs using bacteria, moreover, measuring the UV-Visible spectra of the resulting Au NPs as well the florescence spectra. Moreover, the particles size and morphology will monitored by the transmission electron microscopy.

EXPERIMENT

1. Materials and Methods

Bacterial strain and growth conditions

Two clinical samples of bacterial isolates used in this study are isolated from burns. The isolates were microbiologically and biochemically characterized as *Pseudomonas aeruginosa*[26,27]. Bacteria were routinely cultured in nutrient broth and on nutrient agar plates. In this study, we used *P. aeruginosa (1)* that produce soluble fluorescent pigment pyoverdin and the other *P. aeruginosa (2)* that produce the blue pigment pyocyanin when cultured on cetrimed agar media. *P. aeruginosa* ATCC 90271 was used as standard strain. The isolates used in this study were maintained at Microbiology department, Faculty of Pharmacy, Zagazig University, Zagazig, Egypt.

Biosynthesis of gold nanoparticles

The two isolates and control strains of *P. aeruginosa* were used. The bacteria was grown aerobically in 50 ml nutrient broth media and incubated at 37oC and agitated at 150 rpm for 24 h. After the incubation, the supernatants were obtained by centrifugation of overnight bacterial culture at 5000 rpm for 5 min.

For synthesis of gold nanoparticles (Au NPs); hydrogen tetrachloroaurate was mixed with 50 ml of cell free supernatant to obtain a final concentration of gold ions to be 1 mM, the resulting solution was incubated at 37oC for 24 h. Control (without the gold ions only supernatant) was also run along with the experimental flask. After 24 h of incubation the cell free supernatant containing nanoparticles were obtained.

2. Instruments

The excitation spectra of the samples were measured by UV-Visible spectrophotometer (Bio Cary 50-Varian USA). While, the average diameters and size distributions of Au NPs were calculated by counting over 150 particles from the enlarged transmission electronic microscopy (TEM) (JEOL, 120 kV) photography. The samples were prepared by putting a drop of solution on a cupper carbon grid and setting the drop dry completely in a desicrator. The florescence from the Au NPs in aqueous solution under the He-Cd laser excitation was measured with a photon counting system attached to a 0.5 m double monochromator (Spectrasense Acton. Co. USA).

RESULTS

1. TEM Measurement

The detailed study on extracellular biosynthesis of gold nanoparticles by the *P. aeruginosa* was carried out in this work. *P. aeruginosa* is a gram-negative bacterium that is capable of existing in multiple environmental niches and is an opportunistic pathogen, meaning that it exploits some break in the host defenses to initiate an infection.

The lower part of the Figure 1, represents three test tubes of Au NPs resulting from the incubation of Au+ with cell free supernatant at 37oC for 24 h. On the other hand, the control (without the gold ions only supernatant) was used for comparison in each photo. The color of control is pale yellow which is changed to (pale blue, dark pink, and wine pink) for *P. aeruginosa* ATCC 90271, *P. aeruginosa (2)*, and *P. aeruginosa (1)*, respectively. Moreover, the transmission electron micrographs of the three types of Au NPs are presented in the higher part of Fig. [1], the enlarged micrographs were used to determine the particles size[1]. The particle size and distribution are in the following order 40 ± 10, 25 ± 15, 15 ± 5 for P. aeruginosa ATCC 90271, *P. aeruginosa (2)*, and *P. aeruginosa (1)*, respectively. As the particle size increase the color was found to be shifted from pink to blue due to the surface plasmon of Au NPs[5].

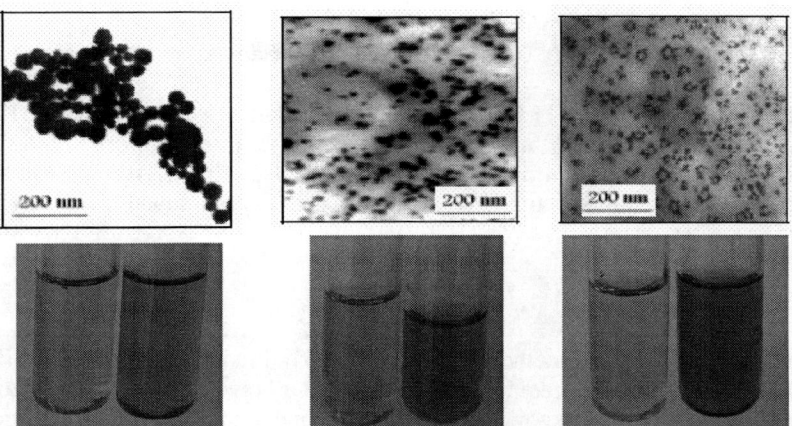

FIGURE 1. The transmission electron microscopy and photo of Au NPs prepared by supernatant of P. aeruginosa ATCC 90271, P. aeruginosa (2) , and P. aeruginosa (1) (from left to right), respectively.

2. Surface Plasmon of Au NPs

The interaction of light having wavelength smaller than the particles size of Au NPs leads to polarization of the free conduction electrons with respect to the much heavier ionic core of Au NPs. Therefore, an electron dipolar oscillation is created and surface plasmon (SP) absorption band is obtained[28]. The UV-Vis. Spectrometer was used to record the SP of Au NPs prepared in case of *P. aeruginosa* ATCC 90271 , *P. aeruginosa (2)*, and *P. aeruginosa (1)*. Strong correlation between the particle size and the maximum absorption peak was observed; the absorption bands at 543, 540, and 531 nm were observed for *P. aeruginosa* ATCC 90271, *P. aeruginosa (2)*, and *P. aeruginosa (1)*, corresponding to particle sizes of 30 ± 10, 25 ± 15, 15 ± 5, respectively. As the particle size increases the maximum absorption was found to be red shifted. Figure 2, shows the absorption spectra of Au NPs prepared with different particle size. No absorption peak corresponding to the control supernatant or gold ion solution in the range of measurement was observed. Moreover, the chemically prepared Au NPs has a characteristic band at 520 nm corresponding 40 nm particle size. Consequently, the Au NPs prepared by bacteria were found to be red shifted, in comparison with that prepared chemically.

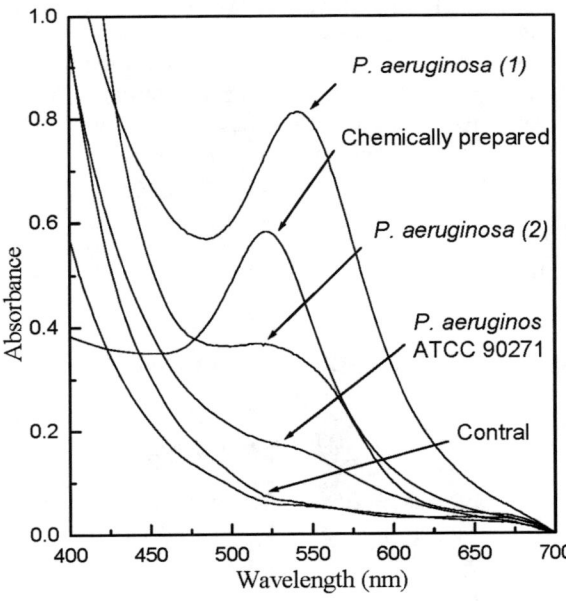

FIGURE 2. The absorption spectra of Au NPs prepared by supernatant of *P. aeruginos* ATCC 90271 , *P. aeruginos (2)*, and *P. aeruginos (1)*.

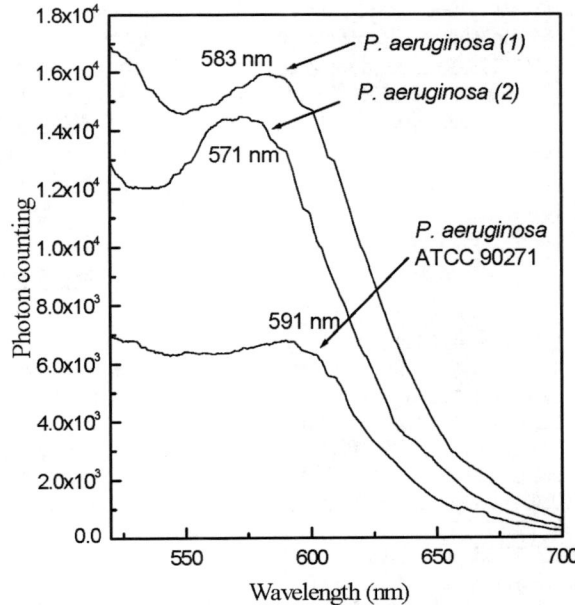

FIGURE 3. The fluorescence spectra of AuNPs prepared by supernatant of P. . *aeruginos* ATCC 90271, *P. aeruginos (2)*, and *P. aeruginos (1)*.

179

3. Florescence spectra of Au NPs

The florescence spectra of Au NPs after 441.5 nm He-Cd laser excitation were recorded as shown in Fig. [3], the florescence spectra of Au NPs prepared by *P. aeruginosa* ATCC 90271, P. *aeruginosa (2)*, and *P. aeruginosa (1)* were found to be centered at 591, 583, and 571 nm, respectively. Moreover, the intensity of florescence was found to increase in the same order of absorption spectra. The position of florescence signal is directly proportional to the particle size of Au NPs[5].

DISCUSSION

There are several physical and chemical methods for the synthesis of metallic nanoparticles that are followed by the material scientists[29]. Nanotechnology requires the collaboration between physicists, chemists, biologists and engineers since both bacteria and fungi have shown ability to reduce metal ions to form nanoparticles.

As shown in this study, the gold nanoparticles were synthesized in the extracellular cell supernatant of the bacteria. This offers a great advantage over an intracellular process of synthesis from the application point of view. Since the nanoparticles formed inside the bacteria would have required additional step of processing for the release of the nanoparticles from the bacteria by ultrasound treatment or by reaction with suitable detergents. Using metal-accumulating microorganisms as a tool for the production of nanoparticles, and their assembly for the construction of new advanced materials, is a completely new technological approach.

Application of the biological systems for synthesis of silver nanoparticles has already been reported earlier[10,22]. However, the exact reaction mechanism leading to the formation of silver nanoparticles by all these organisms is yet to be elucidated. Ahmad et al.[30] have reported that certain NADH dependent reductase was involved in reduction of silver ions in case of F. oxysporum.

Gold nanoparticles are particularly good labels for sensors because a variety of analytical techniques can be used to detect them, including optical absorption, fluorescence, Raman scattering, atomic and magnetic force, and electrical conductivity. These techniques can be used to detect microorganisms and could replace PCR and fluorescent tags used currently.

CONCLUSION

The Au NPs prepared by *P. aeruginosa* ATCC 90271 supernatant have particle size larger than *P. aeruginosa (1 and 2)*, but the particle size distribution in case of *P. aeruginosa (1)* is larger than that in case of *P. aeruginosa* ATCC 90271 and *P. aeruginosa (2)*. The absorption and florescence spectra accorded well with the variation of particle size in a way that, as the particles size increases the absorption and florescence peak position were found to increase. Furthermore, the extracellular synthesis would make the process simpler and easier for downstream processing. In future, it would be important to understand the biochemical and molecular mechanism of nanoparticles synthesis by the cell filtrate in order to achieve better control over size and polydispersity of the nanoparticles.

REFERENCES

1. Y. Badr, M. A. Mahmoud, *J. Molecular Struct.* **749**, 187 (2005).
2. Y. Badr, M. A. Mahmoud, *Spectrochimica Acta Part A: Molecular & Biomolecular Spectrosc.* **63**, 639(2006).
3. Y. Badr, M. A. Mahmoud, *J. Appl. Poly. Sci.*, **99,** 3608 (2006).
4. Y. Badr, M. A. Mahmoud, *Physica B, Physics of Condensed Matter*, **369**, 278 (2005).
5. Y. Badr, M. G. Abd El-Wahed, M. A. Mahmoud, Accepted for publication in *Appl. Surf. Sci.,* May (2006).
6. R. M. Bruins, S. Kapil, S.W. Oehme, Ecotoxicol. *Environ. Saf.* **45**, 198(2000).
7. T. J. Beveridge, M. N. Hughes, H. Lee, K. T. Leung, R. K. Poole, I. Savvaidis, S. Silver, J. T. Trevors, *Adv. Microb. Physiol.* **38**, 177(1997).
8. M. D. Rhodes, P. J. Sadler, M. D. Scawen, and S. Silver, *J. Inorg. Biochem.* **46**, 129 (1992).
9. R. M. Slawson, H. Lee, and J. T. Trevors, *Biol. Met.* **3**, 151 (1990).
10. M. Sastry, A. Ahmad, M.I. Khan, R. Kumar, *Curr. Sci.* **85**, 162 (2003).
11. U. B. Sleytr, P. Messner, D. Pum, M. Sara, *Angew. Chem. Int. Ed.* **38**, 1034 (1999).
12. D. Pum, U.B. Sleytr, *Trends Biotechnol.* **17**, 8-12 (1999).
13. D. R. Loveley, J.F. Stolz, G.L. Nord, E.J.P. Phillips, *Nature* **330**, 252 (1987).

14. H. Spring, K.H. Schleifer, *Syst. Appl. Microbiol.* **18,** 147 (1995).
15. N. Kroger, R. Deutzmann, M. Sumper, *Science* **286**, 1129 (1999).
16. S. Mann, *Nature* **365**, 499 (1993).
17. P. Mukherjee, A. Ahmad, D. Mandal, S. Senapati, S. R. Sainkar, M. I. Khan, R. Parischa, P. V. Ajayakumar, M. Alam, R. Kumar, M. Sastry, *Nano Lett.* **1,** 515 (2001).
18. P. Mukherjee, A. Ahmad, D. Mandal, S. Senapati, S. R. Sainkar, M.I. Khan, R. Ramani, R. Parischa, P.V. Ajayakumar, M. Alam, M. Sastry, R. Kumar, *Angew. Chem.* **113**, 3697 (2001).
19. P. Mukherjee, S. Senapati, D. Mandal, A. Ahmad, M. I. Khan, R. Kumar, M. Sastry, *Chem Bio Chem.* **5**, 461 (2002).
20. P. Mukherjee, A. Ahmad, D. Mandal, S. Senapati, S. R. Sainkar, M.I. Khan, R. Ramani, R. Parischa, P. V. Ajayakumar, M. Alam, M. Sastry, R. Kumar, *Angew. Chem. Int. Ed.* **40**, 3585 (2001).
21. R. Joerger, T. Klaus, C. G. Granqvist, *Adv. Mater.* **12**,407 (2000).
22. T. Klaus-Joerger, R. Joerger, E. Olsson, C. G. Granqvist, *Trends Biotechnol.* **19**, 15(2001).
23. D. Fortin, T. J. Beveridge, Biomineralization, in: E. Baeuerien (Ed.), From Biology to Biotechnology and Medical Applications, Wiley-VCH, Weinheim, 2000 pp 7–22.
24. G. Southam, T. J. Beveridge, *Geochim. Cosmochim. Acta*, **60**, 4369 (1996).
25. S. S. Shankar, A. Ahmad, R. Pasricha, M. Sastry, *J. Mater. Chem.* **13**, 1822 (2003).
26. R. Cruickshan, J.P. Duguid, B.P. Marmion and R.H.A. Swain Medical Microbiology. 12th ed. Vol. II, The practice of medical microbiology. 1975, pp. 170-189.
27. R. E. Buchanan, N. E. Gibbons and Co-Editors, Bergey's Manual of systemic Bacteriology. 9th Edition, Williams and Wilkins Co., Baltimore, USA, 1984, pp. 141-199.
28. S. Link, M. A. El-Sayed, *Int. Rev. in Phys. Chem.* **19**, 409 (2000).
29. A. S. Edelstein, R.C. Cammarata (Eds.), Nanomaterials: Synthesis, Properties and Applications, IOP Publication, Bristol and Philadelphia, 1996.
30. A. Ahmad, S. Senapati, M.I. Khan, R. Kumar, M. Sastry, *Langmuir* **19**, 3550 (2003).

Low Cost Supercomputer for Applications in Physics

Maqsood Ahmed, Rashid Ahmed, M. Alam Saeed, Haris Rashid and Fazal-e-Aleem

Centre for High Energy Physics
University of the Punjab, Lahore-54590, Pakistan

Abstract: Using parallel processing technique and commodity hardware, Beowulf supercomputers can be built at a much lower cost. Research organizations and educational institutions are using this technique to build their own high performance clusters. In this paper we discuss the architecture and design of Beowulf supercomputer and our own experience of building BURRAQ cluster.

INTRODUCTION

Supercomputers, an array of computers, are the most advanced and powerful computers and primarily used for solving complex problems which can not be solved on ordinary computers like weather forecasting, computational fluid dynamics etc [1]. Supercomputer was first introduced in 1964 when Computer Data Corporation developed CDC 6600 supercomputer [2]. In 1976 Seymour Cray, the father of a supercomputer designed the fastest supercomputer Cray-1 with a total cost of 8.8 million US dollars. The peak performance of Cray-1 was 160 Million Floating-point Operation per Second (MFLOPS). Cray Company dominated the supercomputer industry for many years by introducing different models ranging from MFLOPS to GFLOPS [3]. Later on, other companies like NEC, Texas Instruments, Unisys and IBM marketed more powerful supercomputers than Cray-1.

"The concept of parallel processing exists since 1958 and is based on an old saying 'divide and conquer'" [4]. Similarly the idea of parallel processing is to divide a piece of work amongst more than one processor. This idea matured when NASA engineers built the first Beowulf in 1994 for NASA's Earth and Space Sciences project. They used a 10mbit/s Ethernet network to connect 16 Intel 486 personal computers with Linux as the operating system. The first Beowulf style supercomputer built at NASA, introduced a new era of high performance computation at much lower cost [5, 6]. It had a peak performance of 1 billion operations per second. Soon after this innovation, teaching, research and business organization started building their own supercomputers for teaching and research purposes [7, 8]. Using the same technique with low cost commodity computer parts, we built our own supercomputer named "BURRAQ". In this paper, we will give functional details of its working and some of details of the possible applications.

BURRAQ ARCHITECTURE

Based on the design of Beowulf style supercomputer architecture, we made Burraq Cluster for research in high energy and condensed matter physics. Burraq cluster consists of 32 Nodes including master node each having dual Intel Xeon 2.4 GHz processor and 1 GB RAM shared by both processors. Linux, a free version of UNIX operating system is used on each node. The Burraq architecture is entirely based on commodity items and open source software. The master node is configured with two raid controllers, one serving raid-1 consisting of operating system and other Raid-5 to serve as storage system with a capacity of 2.0 Terabytes. The cluster is placed behind a firewall to protect the system from unauthorized access. All the nodes are monitored and operated through a single keyboard, mouse and monitor connected to master node by a KVM switch. The Burraq cluster is shown below.

CP888, *Modern Trends in Physics Research,*
Second International Conference on Modern Trends in Physics Research—MTPR-06,
edited by L. El Nadi
© 2007 American Institute of Physics 978-0-7354-0354-9/07/$23.00

SOME OF THE APPLICATION SOFTWARE FOR BURRAQ

Use of cluster computers is becoming important for scientists, engineers, educationists [9,10,11] for various applications including large scale lattice QCD Monte Carlo calculation [12,13], numerical simulation in material science [14-16], CFD [17, 18] and data mining [19]. The field of high performance computing is developing at an extremely rapid pace. At the same time, molecular modeling methodologies [16, 20] (both quantum and classical) are advancing at a rapid pace. Some of the applications for which the Burraq Cluster can be used and of interest for the conference participants are listed below:

- **WIEN2K:** is used for "electronic structure calculations of solids using density functional theory (DFT)" (http://www.wien2k.at/).

- **GEANT4:** is "a toolkit for the simulation of the passage of particles through matter. Its areas of application include high energy physics, nuclear and accelerator physics, as well as studies in medical and space science" (http://cern.ch/geant4).

- **mpiBLAST:** is "a software tool used in bioinformatics for finding regions of local similarity between DNA sequences. The program compares nucleotide or protein sequences to sequence databases and calculates the statistical significance of matches" (http://mpiblast.lanl.gov/).

- **MILC:** is software for "large scale numerical simulations to study Quantum Chromodynamics (QCD), the theory of the strong interactions of subatomic physics" (http://physics.indiana.edu/~sg/milc.html).

To run these software successfully following additional software will be required:

- **LAPACK:** is a Linear Algebra Package that "provides routines for solving systems of simultaneous linear equations, least-squares solutions of linear systems of equations, eigenvalue problems, and singular value problems" (http://netlib.org/lapack/).

- **ScaLAPACK:** is a library which "includes a subset of LAPACK routines redesigned for distributed memory MIMD parallel computers" (http://netlib2.cs.utk.edu/scalapack/).

- **BLAS:** is a net library of 'Basic Linear Algebra Subprograms' and is hosting 'routines' for performing basic vector and matrix operations. "The Level 1 BLAS perform scalar, vector and vector-vector operations, the Level 2 BLAS perform matrix-vector operations, and the Level 3 BLAS perform matrix-matrix operations" (http://www.netlib.org/blas/faq.html).

- **PBLAS:** is a set of "Parallel Basic Linear Algebra Subprograms" (http://netlib2.cs.utk.edu/scalapack/pblas_qref.html).

The performance of BURRAQ is measured by using LINPACK. A parallel version of LINPACK, downloaded from http://www.netlib.org/linpack, was installed on the cluster. For message passing routine LAM-MPI (http://www.lam-mpi.org/) was installed. Initial tests give a peak performance of 80 GFLOPS. It is expected that the performance will increase by 5-10% after some fine tuning in software and related libraries.

CONCLUSION

Beowulf style supercomputers are growing rapidly due to its low cost, portability and flexibility. With the availability of cheap hardware and free software it is now possible to build high performance, flexible and reliable supercomputer with Terra Bytes of storage media to meet the specific requirements of today's computational need of physicists working in the field of high energy physics, material physics, nuclear and medical physics etc. Beowulf supercomputer is useful for computational problems which need to be solved in a distributed memory environment. A large amount of information is now available on the web and many details have therefore not been included. Authors apologize to all whose scholarly work could not be included.

REFERENCES

1. http://www.cray.com/solutions/earth/index.html, www.cray.com/solutions/scientific/index.htm.
2. http://www.cisl.ucar.edu/computers/gallery/cdc/6600.jsp
3. www.cray.com/about_cray/history.html
4. Hancock J. M. and DasGupta S., (1986) Proceedings of 23 ACM/IEEE Conference on Design Automation, 69-77 and reference therein.

5. Sterling T., Becker D.J., Savarese D., Dorband J. E., Ranawake U. A., and Packer C. V., (1995) Proceedings of the 1995 International Conference on Parallel Processing, 11-14.
6. T. Sterling, D. Saverese, D. J. Becker, B. Fryxell, and K. Olson: *"Communication Overhead for Space Science Applications on the Beowulf Parallel Workstation"*, Proceedings of the Fourth IEEE International Symposium on High Performance Distributed Computing, pp.23-30, August 1995
7. Montante R., (2002) Journal of Computing Sciences in Colleges, 17(6) 10-18.
8. Adams J., Navison C., and Schaller N. C. (2000) Proceedings of 31st SIGCSE Technical Symposium on Computer Education, 65-69.

9. Hyde D. C., (1981) A Parallel Porcessing Cource for Undergraduates. Proceedings of 20th SIGCSE Symposium on Computer Science Education, 170-173.

10. Fisher A. L and Gross T., (1991) Teachng the Programming of Paralle Computers. Porcessing Cource for Undergraduates. Proceedings of 27th SIGCSE Symposium on Computer Science Education, 102-107.
11. Drag D., Juzwick L. and Knox D. (2001) Cluster Computing: Development of Small Scal Cluster and learning Modules for Undergraduates. JCSC 16(4), 329-331
12. Haridass N.D, (2004) Proceedings of International Conference on High Energy Physics (ICHEP 2004), 787-790.
13. Holmgren D. Sing A., Mackenzie P. and Simone J. () Lattice QCD Production on Commodity Cluster at Fermilab.

14. WIEN2K: http://www.wien2k.at/.

15. ABINIT: http://www.abinit.org/about/.
16. Gaussian: http://www.gaussian.com/.
17. Paul R. Woodward, Steven E. Anderson, David H. Porter, Dennis Dinge, Igor Sytine, Thomas Ruwart, Michael Jacobs, R. H. Cohen, B. C. Curtis, W. P. Dannevik, A. M. Dimits, D. E. Eliason, A. A. Mirin, Karl-Heinz Winkler and Stephen Hodson Exploiting the Power of DSM and SMP Clusters for Parallel CFD http://www.llnl.gov/CASC/asciturb/pubs/parcfd99.pdf accessed on 20 Sep. 2006.
18. FLUENT: http://www.fluent.com/.
19. Christen P., Nielsen O. M., Hegland M., and Strazdins P., (2001) Parallel Data Mining on a Beowulf Cluster. Proceedings of the *HPC Asia 2001* Conference, Gold Coast, Queensland, Australia.
20. Gordon Research Group, http://www.msg.ameslab.gov/GAMESS/GAMESS.html.

Parametric Study On The CW Nd: YAG Laser Cutting Quality Of 1.25 mm Ultra Low Carbon Steel Sheets Using O₂ Assist Gas

Hanadi G. Salem[*], Wafaa A. Abbas, Mohy S. Mansour and Yehia A. Badr

National Institute of Laser Enhanced Science, NILES, Cairo- University.

[*]*Department of Mechanical Engineering, American University in Cairo.*

[*]E-mail: hgsalem@aucegypt.edu

Abstract. There are many non-linear interaction factors responsible for the performance of the laser cutting process. Identification of the dominant factors that significantly affect the cut quality is important. In the current research, the gas pressure, laser power and scanning speed were selected as the cutting parameters. Effect of the cutting parameters on the cut quality was investigated, by monitoring the variation in hardness, oxide layer width and microstructural changes within the heat affected zone (HAZ). Results revealed that good quality cuts can be produced in ultra low carbon steel thin sheets, using CW Nd:YAG laser at a window of scanning speed ranging from 1100-1500 mm/min at a minimum heat input of 337watts under an assisting O₂ gas pressure of 5 bar. Higher laser power resulted in either strengthening or softening in the HAZ surrounding the cut kerf. The oxide layer width is not affected by the energy density input but rather affected by the O₂ gas pressure due to exothermal reaction.

INTRODUCTION

The Laser, a clean source of high energy density, has been applied to various types of materials processing. Such processes involve non equilibrium phenomena due to high heating and cooling rates induced by the laser irradiation[1]. The rapid heating and cooling processes due to the laser jet, respectively, affect the material structure, heat affected zone (HAZ) and consequently the surface hardness. Moreover, the number of processing variables and their complex influence on the cutting quality controls the material behavior during laser cutting. This task is further complicated by the fact that laser cutting of the material is an exothermal reaction which is largely dependent on inter-related material properties such as thermal conductivity, melting point, effect of oxidation reactions, reflectivity, viscosity, specific latent heat of transformation, specific heat capacity, density, phase transformation temperature and its dimensions[2]. When using oxygen as assist gas, it will not only drag the melt away but it will also react with the molten material forming the oxide film which then blown into the kerfs covering the molten metal below it [3]. Cutting with this technique will increase the cutting speed, also the use of oxygen as assist gas facilitates the utilization of exothermic energy, which reduces viscosity and surface tension but increases the laser beam absorption. Accordingly, drosses adhering to the bottom of the cutting edge would be some kinds of oxides that could be easily dismast. The laser most often used for this application is the continuous-wave CW CO₂ laser, but Nd:YAG laser is preferred for applications that require narrow kerf width and small HAZ. Because Nd:YAG laser has many selection of waveforms, it can be applied to various processes such as trimming, repairing, scribing, welding, cutting, annealing, and soldering [4]. On the other hand, most carbon steels and stainless-steels absorb CO₂ and YAG beams very well. Since the absorptivity depends on the wavelength [5], Nd:YAG lasers allow higher speeds for cutting of metals compared to CO₂ lasers of the same power [6]. For many years laser cutting technology has been used in industries because of its accuracy and efficiency [7]. Industrial users have developed the experience necessary for the use of the various laser parameters.

The assist gas composition used in laser material processing, as cutting depends on the substrate. From previous research, it has been found that the purity of the gas is affected by the quality of the cut [8-10]. Exothermic reaction is an important heat source in reactive laser cutting, so carbon steel is usually cut with oxygen. When using oxygen as the cutting gas, hydro-dynamical interactions are superimposed by additional effects such as local density and concentration variations of the oxygen at the cutting front, the resulting reaction kinematics and the change in viscosity of the oxide-melt composite [11]. Heat due to chemical exothermal reactions when using O₂ as assist gas in steel alloys processing occur as follows [12]:

CP888, *Modern Trends in Physics Research,*
Second International Conference on Modern Trends in Physics Research—MTPR-06,
edited by L. El Nadi
© 2007 American Institute of Physics 978-0-7354-0354-9/07/$23.00

$$Fe + 1/2\ O_2 \rightarrow Fe\ O \qquad \Delta H= 260\ KJ/mole$$

$$3\ Fe + 2O_2 \rightarrow Fe_3\ O_4 \qquad \Delta H= 1120\ KJ/mole$$

$$2Fe + 3/2\ O_2 \rightarrow Fe_2\ O_3 \qquad \Delta H= 543,000\ Btu/\ Ib\ mole$$

Chen et al.[13], observed that for optimum cut quality a high purity of oxygen is required and a tiny oxygen impurity (1.25%) will reduce the maximum cutting speed by 50%. Gabzdyle et al.[14], showed the significance of gases to the laser industry that gas quality of both purity and composition greatly affect on the quality and productivity of materials processing applications. He also clarified that careful balancing of the laser energy input with the process requirements ensures that the oxidative reaction will be initiated.

The gas pressure is responsible for removing molten material from the cut kerfs, and it protects laser optics from being damaged by the resulting ejected spatters. Also the assist gas works as a cooler to the laser optics. It also helps to blow away the plasma forming during the laser material interactions, which reduce the absorption of the material to the laser energy. Kar et al.[15], concluded that there is a maximum pressure of the cutting gas that can be used for a given set of cutting parameters.

As the incident radiation is absorbed at the surface where sufficiently high heating rates are reached, melting can occur, which could cause structural changes in the material. Surface tension of the material changes with the input of the heat of fusion. Together with the pressure exerted by the evaporating gas, the melting wave propagates away from the laser focus resulting in the material removal from surface. The kinematics of the motion of melting materials depends not only on the incident laser power and scanning speed, but also on the flow and direction of the assist gas used. When using oxygen as assist gas, it will not only drag the melt away but it will also react with the molten material forming the oxide film which then blown into the kerfs covering the molten metal below it. But on the other hand uncontrolled material burning could be included in the case of oxygen as assist gas in the cutting process [16].

The experimental studies carried out by various researchers have been supplemented in recent times by mathematical models to understand the influence of the variable processing parameters on laser cutting quality. Furthermore, establishment of such processing parameters also permits optimization of process parameters like cutting speed, laser power, and assist gas pressure on the basis of actual operating costs. Molian[12] found that, increasing oxygen gas pressure increases the cutting speed or thickness. Hsu and Molian [16] investigated dual-jet laser machining of various kinds of stainless steels to obtain high-speed cutting or thicker section with dross- free edge quality. Also, the dynamics of gas and combustion mechanisms in connection with dual gas-jet laser cutting using a transverse flow type CW 2kW CO_2 laser has been developed by them. Hamoudi [17] stated that, at low speed less than 1000mm/min little dross was adherent at the bottom of the cut edge. But he achieved optimum cut quality at 2000mm/min when using oxygen as assistant gas pressure. HAZ width improved with increasing cutting speed, but the kerf width increased. While, Shariff et al.[18], studied the influence of scanning speed on the cutting process quality of mild steel and titanium. They mentioned that the width of the oxide layer is reduced as the cutting speed increases. However, Quintero et al.[19], mentioned that, increasing the cutting speed leads to reduce the kerf width as the pressure decreased. Rajaram et al.[20], explained that laser power had a major effect on the kerf width and HAZ while cutting speed played a minor role. Li et al. [21], mentioned that a small change in the carbon content in steel alloy could change the microstructure and properties of the alloy.

The aim of the present work is to study the effect of the operating parameters on the cut quality and material properties by monitoring the variation in hardness, oxide layer width and microstructural changes within the HAZ, and hence obtain the optimum ranges of laser power, scanning speed and assist gas pressure.

EXPERIMENTAL TECHNIQUE

In order to achieve the stated objective, laser cutting experiments were carried out using 1.25mm commercial low carbon steel sheets to investigate the effect of laser cutting parameters on cut quality. The size of the specimens cut was 50mm x 50mm with a profile shown in Figure 1. The chemical analysis of the investigated sheets was conducted at the Central Material Research & Development Institute (CMRDI). The average composition values for the alloy are listed in Table 1. A continues wave (CW) Nd:YAG laser at wavelength of 1064 nanometer (Lee LPL series 800 system) was used in the present work. The assist gas used was oxygen with commercial purity, to study the possible influence of different process gas pressures on the cutting process quality. Three main parameters have been selected for the present study. These are laser power, scanning speed and assist gas pressure. The laser power was varied within the range of 337-to-515watt, the scanning speed was also varied within the range of 700-to-1500 mm/min, and the assist oxygen gas pressure range was 3.5-to-5 bar. Testing the effect of one parameter on the cut quality requires

variation of one parameter while keeping the other two parameters at the pre-selected values shown in Table 2. The cut quality and surface structure were tested using several diagnostic techniques, hardness value were measured using a Mitutoyo Vickers micro-hardness tester model HM-112. It was also used for measuring the HAZ width. The cut surface morphology was investigated using a JEOL scanning electron microscopy (SEM) for studying the striation patterns on the cut edge resulting from laser cutting at 30KV. Accumulated energy density values, and microstructural changes within the HAZ region were investigated. An inverter LEICA microscope with a capture system was employed to observe the microstructural changes of the specimens before and after the laser cutting. Several steps were employed in order to facilitate metallographic examination of the samples cross section which including grinding, polishing followed by etching using Nital for 10-12 Sec.

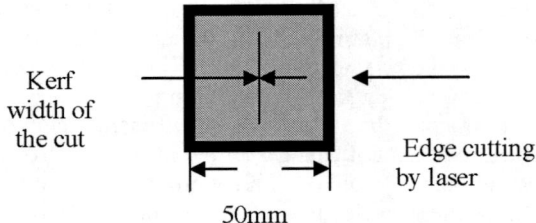

FIGURE 1. Size of the specimen 50x50 mm

TABLE 1. Chemical composition of the material used

C	Si	Mn	P	S	Cr	Mo	Ni	Al	Co
<0.0012	0.002	0.175	0.0053	0.0049	0.0172	0.00166	0.0264	0.0273	0.00447
Cu	**Nb**	**Ti**	**V**	**W**	**Pb**	**As**	**B**	**Fe**	
0.00662	0.00179	0.0424	<0.0005	0.0149	0.00176	0.00218	<0.0001	~99.66	

TABLE 2. Cutting parameters.

Testing Conditions

Power Watt	Scanning speed mm/min	Gas pressure Bar
337	700	3.5
360	800	4
420	900	4.5
465	1000	5
480	1100	5.5
515	1500	6

RESULTS AND DISCUSSION

A. VHN Microhardness

The effect of the laser power on the VHN microhardness is illustrated in Figure 2. The VHN-values were measured along a line parallel to the laser cut through a distance of 70μm from the cut. Average value of VHN-values along this line is shown in Figure 2. The surface hardness increased at high laser power. On the other hand the data show some fluctuations between 337watt and 460watt. Two possible reasons may cause these fluctuations in the surface hardness. These are suggested to be due to the decrease of cooling rate and the reduction of carbon equivalence. The decrease of cooling rate is considered to be caused by the

increase of laser power absorbed on the material surface. But the reduction of carbon equivalence could be due to loss of hardening elements, such as Si, Mn, C and Al, forming oxides. Similar finding were reported by Ono *et al* [22].

Further discussion of this observation will be shown in section 3.B. The effect of laser power on the HAZ width is illustrated in Figure 3. The HAZ width increases by increasing the laser power due to the high heating rate. The fluctuation in the HAZ width refers to the absorption of the laser beam on the surface of the workpiece. Figure 4 shows the effect of scanning speed on VHN-values. It was observed that, the scanning speed affect the VHN-values due to the high and low cooling rates at the high scanning speeds, and low scanning speeds respectively. But the fluctuations observed in the VHN-values could be due to the amount of the absorbed laser intensity at the surface of the workpiece as mentioned earlier. Figure 5 shows that the HAZ width decreases with increasing the scanning speed, due to the decrease in duration time associated with laser cutting. However, at the low scanning speed the adhering dross at the bottom of the cut edge enlarge the HAZ width due to the excess heat absorbed. Scanning speed has the major effect on the HAZ width, but Rajarm *et al.*[20] disagree with that result. This could be explained by the fact that they used a laser cutting on a different alloy of material which could affect the results. The effect of oxygen gas pressure on VHN-values is illustrated in Figure 6. The VHN-values are not affected by increasing the gas pressure at a distance parallel to the cut edge at the HAZ width. Although, the duration time is constant with respect to variable laser power and gas pressure at constant scan speed as shown in Figure 7, the VHN-values increases which could be due to the exothermic reaction.

The effect of oxygen gas pressure on HAZ width is illustrated in Figure 8. It was observed that, at a value lower than 4 bar the HAZ width increased due to the dross adhering at the bottom of the cut edge which makes an additional source of heat. But the HAZ width decrease with increasing the gas pressure, because increasing the gas pressure blows the formed drosses away while in the molten state and hence decreases the possibility of excessive generated heat in the HAZ.

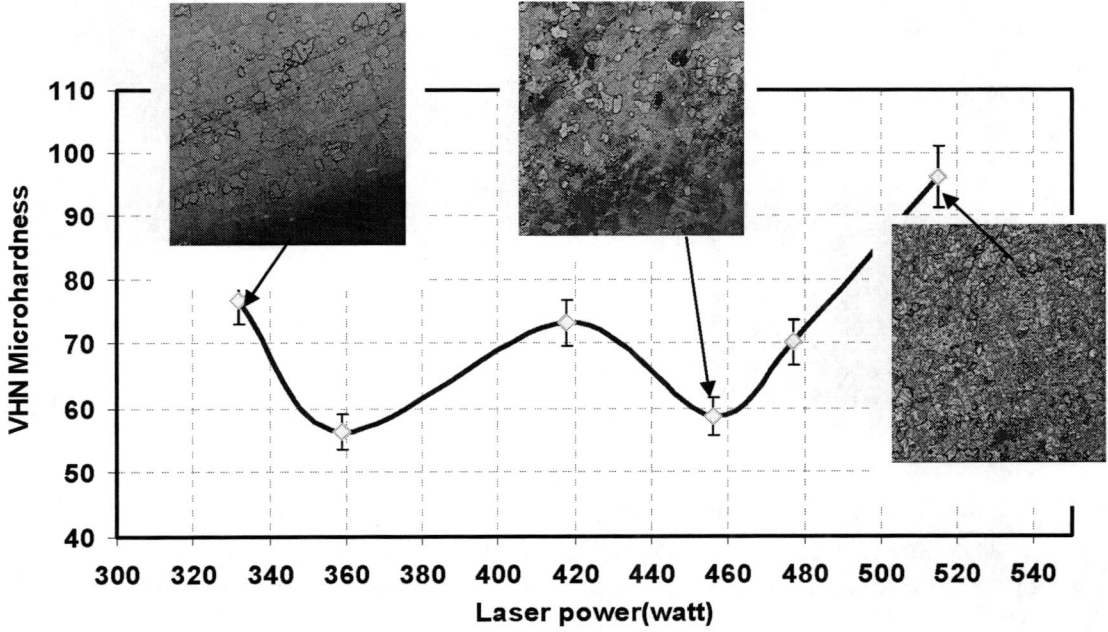

FIGURE 2. Effect of the laser power on the average VHN microharness at a distance parallel to the cut edge,at scan speed of 1500mm/min and gas pressure of 5bar.

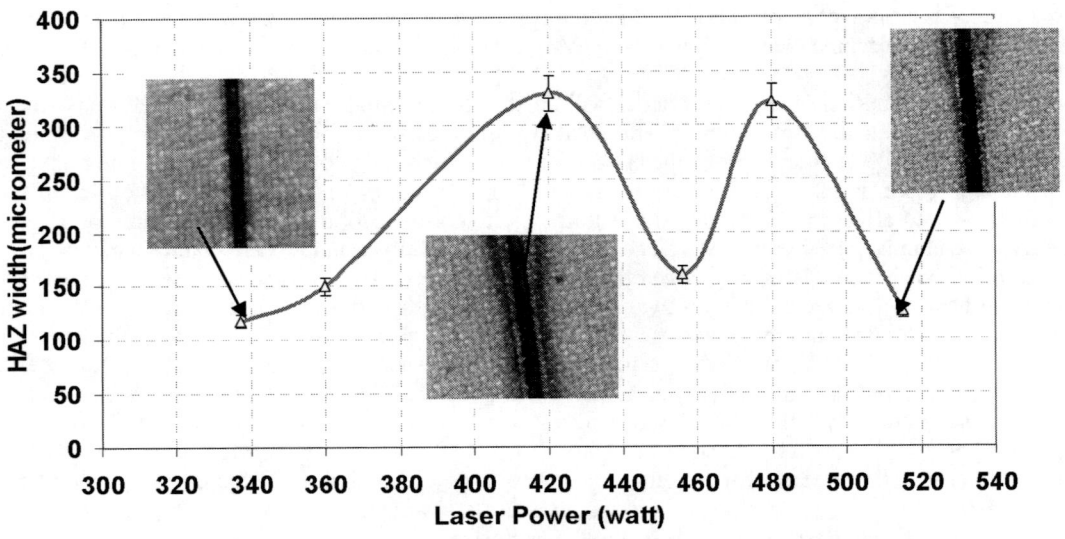

FIGURE 3. Effect of the Laser power on the HAZ width at scan speed of 1500mm/min and gas pressure of 5bar.

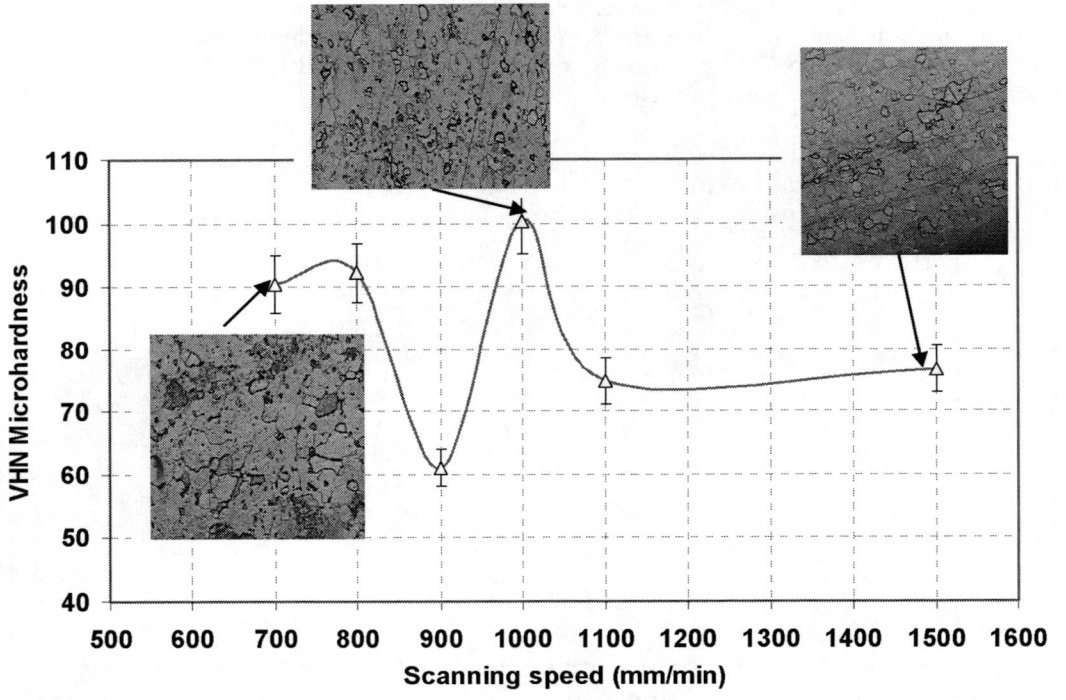

FIGURE 4. Effect of scanning speed on average surface VHN microhardness at a distance parallel to the cut edge, at constant laser power of 337watt and gas pressure of 5bar

B. Microstructural Changes In The HAZ

The body-centered cubic structure of α-ferrite steel is the major constituent in ultra low carbon steel, since carbon has a limited solubility in ferrite [23,24]. In the current research, microstructural evolution as a function of changing scan speed of 700, 1000 and 1500 mm/min, respectively at a constant laser power of 337 watt and gas pressure of 5 bar were compared. From the micrographs shown in Figure 4 it was observed that a scan speed of 1000 mm/min created the finest grain structure compared to 700 and 1500 mm/min. This observation explains the measured highest VHN-values produced at 1000

mm/min. Comparing the microstructure changes as a function of changing laser power of 337, 456 and 515 watt, respectively at a constant scanning speed of 1500 mm/min and gas pressure of 5 bar, it is observed that there was an insignificant change in microstructure of the sheets cut at low and intermediate laser power, while the structure type was superficial affected. From the micrographs shown in Figure 2 it is observed that the VHN-values measured at 337 and 456 watt are almost the same while at laser power of 515 the VHN-values show the highest value. This agrees with the observation made for the microstructure. The suggested significant increase in VHN-values at the highest laser power is due to the phase transformation or solutionzing of Fe_3C phase that could have occurred at high heat input followed by fast rate of cooling. This could have resulted into formation of fine cementite (Fe_3C), which could change the structure. At lower laser power that relatively low heat input is not enough to cause phase transformation but rather could have resulted into coursing of the grains and phase present. This contributed into the decrease in VHN-values measured at low laser power input. This agrees with work done by Kutsuna et al.[25,26] in the case of laser oxygen welding of ultra low carbon steel.

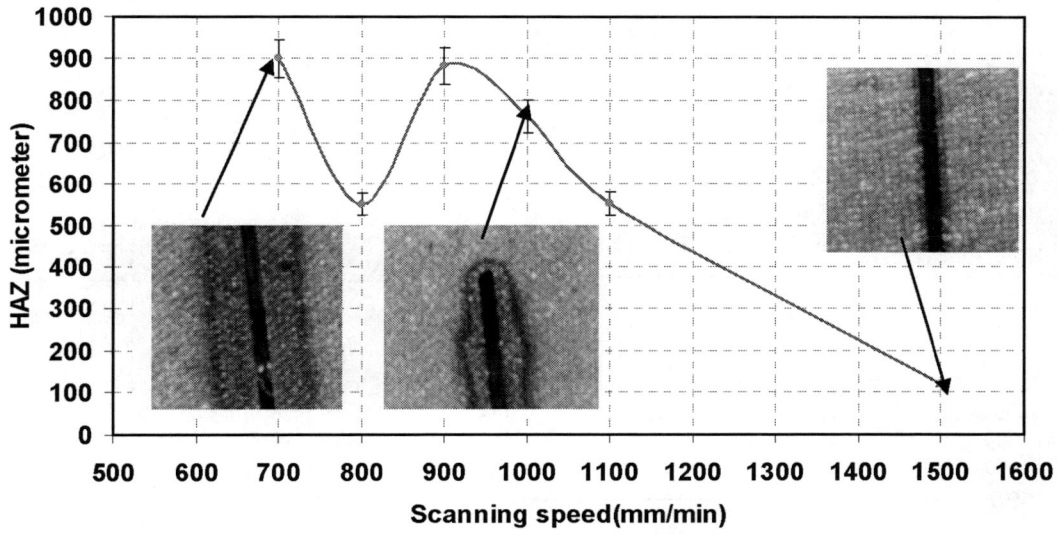

FIGURE 5. Effect of scanning speed on HAZ area at laser power of 337watt and gas pressure of 5bar.

C. Oxide Layers

Oxidation during laser cutting occurs due to the effect of laser energy in the presence of oxygen as an assist gas. Measurement of the surface oxidation mark has been proposed as a common method of evaluating the HAZ width in the case of laser oxygen cutting of carbon steel as it is considered to be a reliable indicator of thermophysical change. The oxide layer width increases by increasing the laser power. While increasing scanning speed leads to a decrease of the oxide layer width due to the decrease in duration time of laser exposure that promotes the oxidation reaction. Oxidation is a chemical reaction and the rate of reaction is proportional to the local oxygen concentration, so increasing the oxygen gas pressure would increase the oxide layers, this agrees with work done by Yilbas[27] and Gabzdyl [14]. Figure 9 depicts the distinct cut surface morphologies using scanning electron microscope (SEM) resulting from the oxygen assist gas cutting of low carbon steel specimens. It shows parallel cut edge at high scan speed of 1500mm/min while the ripples of striations at low scan speed of 700mm/min has been attributed to the cyclic nature of the cutting process involving postulated ignition extinction cycles. Figure 10 shows the diffraction patterns for two samples of ultra low carbon steel cutting by laser power of 337 watt and scanning seed of (a) 700 mm/min and (b) 1100mm/min, respectively. It was found that the X-ray patterns

contain high intensity peaks that belong to α-Fe, besides the iron peaks. It contains some additional peaks that belong to iron oxide phases namely FeO and Fe_3O_4. As can be seen from the figure the oxide peaks have much less intensity than the iron peaks which show that the oxide phases have formed only a thin film on the steel surface. These results agreed with the work of Iordanova [24]. Figure 11 shows the typical diagram of energy dispersive X-Ray system (EDX). It was observed that the oxide layer width was changed along the cut edge.

The accumulated energy density (AED) gives the total amount of energy per unit area that is deposited on the substrate during the process of laser cutting. Higher AED leads to more heat input resulting into more thermal load that is introduced into the surface and the structure of the specimen[28] which could lead to increase the hardness value as shown in Figure 12. So the heat input is an important factor since the exothermic energy contributions vary with energy density that combines both power and speed at each point. Increasing speed decreases the energy input. It was observed from Figure 13 that the oxide layers width is seen to vary with the energy density value, but using oxygen as assist gas in the cutting process, can include uncontrolled material burning, because nearly 60% of the required cutting energy is supplied by the exothermic reaction of the material with oxygen.

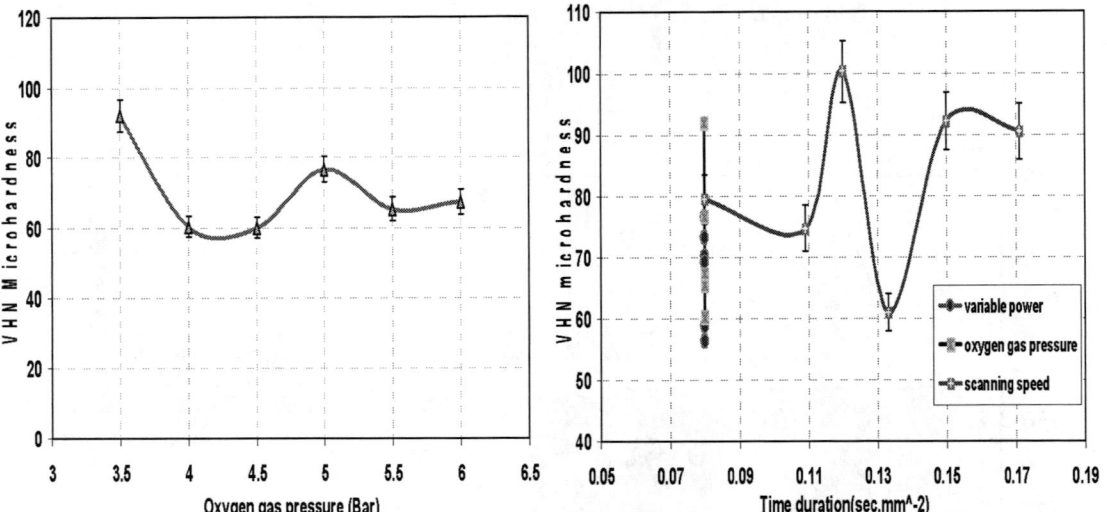

FIGURE 6. Effect of oxygen gas pressure on VHN microhardness at constant laser power of 337watt and scanning speed of 1500mm/min.

FIGURE 7. Plot of duration time of the laser beam affecting the surface of the specimens versus the average value of the VHN

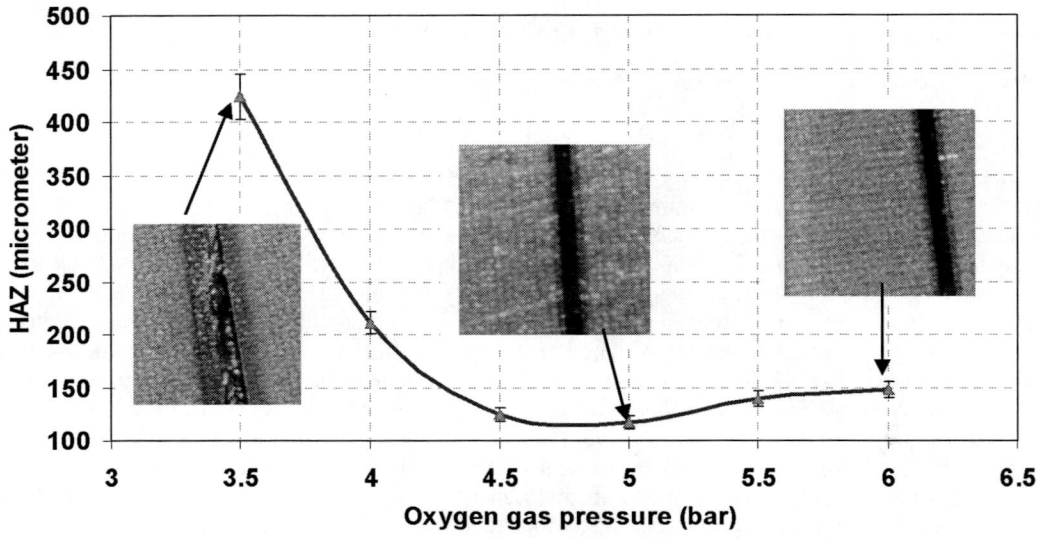

FIGURE 8. Effect of oxygen gas pressure on HAZ area at constant laser power of 337watt and scanning speed of 1500mm/min.

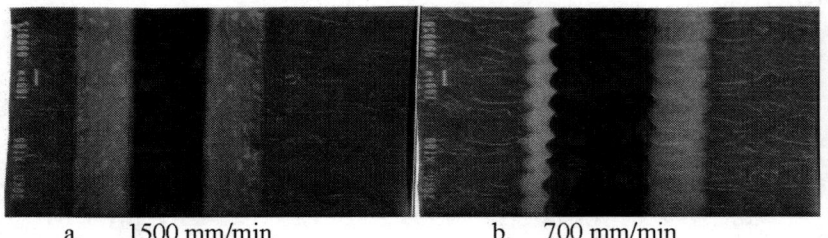

| a. 1500 mm/min. | b. 700 mm/min |

FIGURE 9. Surface morphology of the surface cut edge (used SEM JEOL at 100X, 30kv) at different scanning speed a.1500 mm/min, and b. 700mm/min with constant laser power of 337watt and oxygen gas pressure of 5bar.

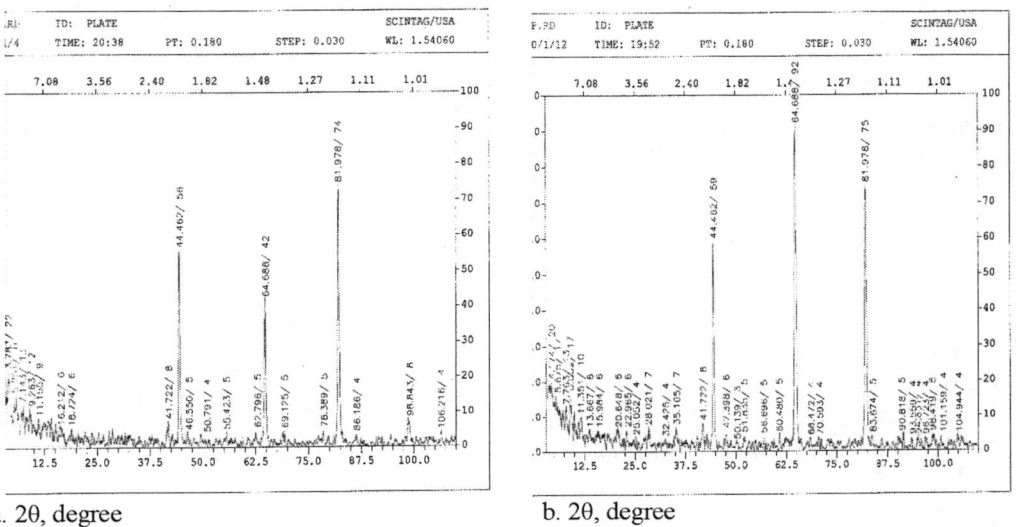

| a. 2θ, degree | b. 2θ, degree |

FIGURE 10. X-R diffraction patterns of ultra low carbon steel specimens cut by CW Nd-YAG with laser power of 337 watt and gas pressure of 5 bar: (a) 700mm/min and (b) 1100mm/min.

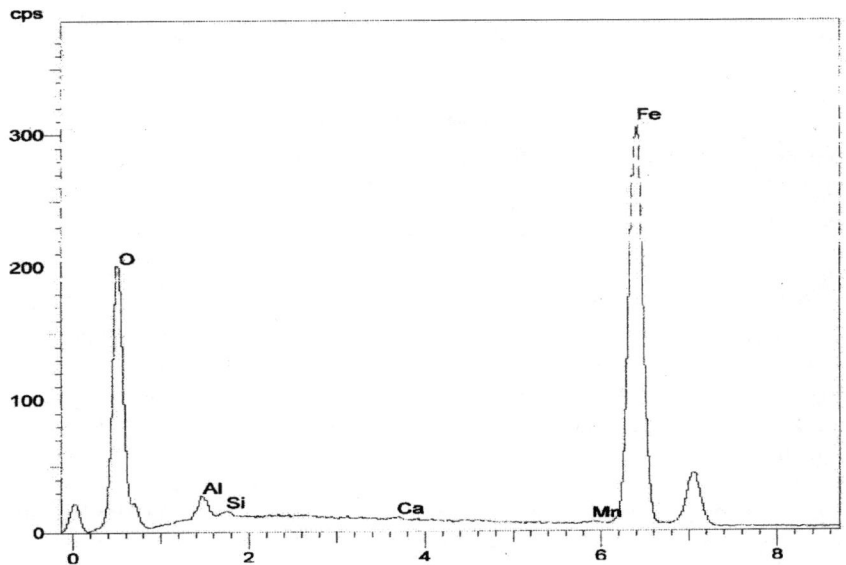

FIGURE 11. The typical diagram of energy dispersive X-Ray system (EDX). JEOL JSM-T200 Scanning Microscope.

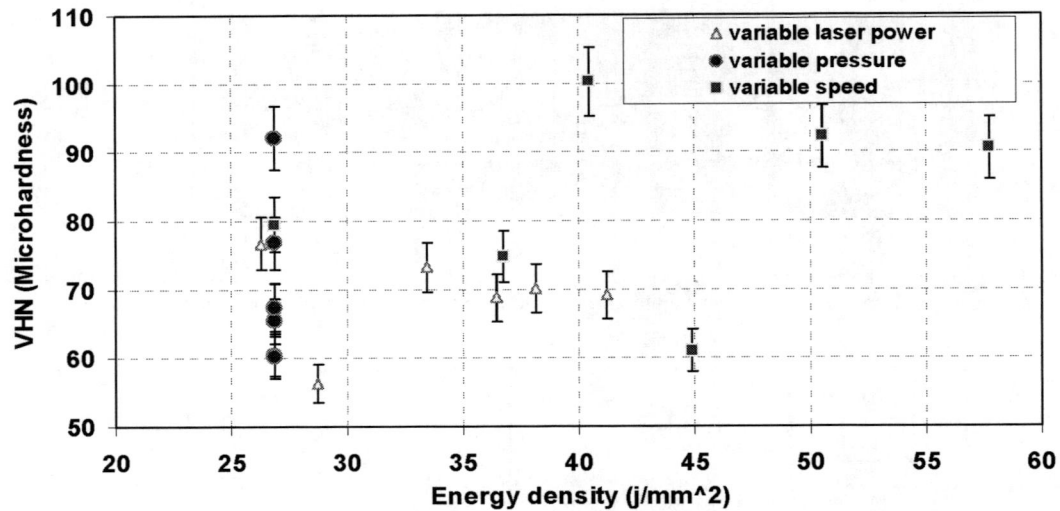

FIGURE 12. Vriation of VHN-values with respect to AED J/mm^2 at several values of laser power, scanning speed and oxygen gas pressure.

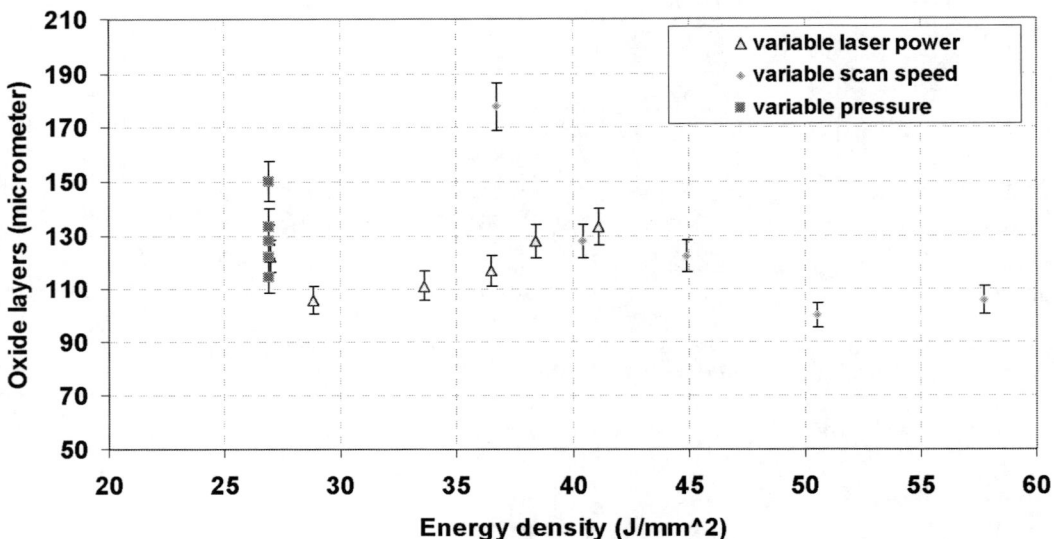

FIGURE 13. Variation of oxide layer width with respect to AED J/mm^2 at several values of laser power, scanning speed, and oxygen gas pressure.

CONCLUSION

The major conclusions may be summarized as follows:

- For a given laser power, an increase in oxygen gas pressure does not enhance oxidation. Thus higher pressure above a certain level is only necessary for the removal of the molten oxide adhering at the bottom of the cut edge.
- HAZ width increases with increasing the laser power and decreases with increasing scanning speed and gas pressure.
- The oxide layer width increases by increasing the laser power due to the excess heat input, and decreases by increasing scan speed due to the less in time that allow the oxidation. It is also increase by increasing the oxygen gas pressure.
- Based on X-ray diffraction peaks a surface oxidation is forms with intensities depending on the cutting parameters.

- The average value of VHN at a distance parallel to the cut edge in the HAZ width increases by increasing laser power and decreasing scan speed while remains around the value of the parent metal with increasing gas pressure within (5.5-6) bar.
- At an intermediate value of the energy density the VHN-values was kept around the base metal value, however the VHN-values increase with increasing the energy density at low scanning speed.
- The HAZ width is affected dramatically with the O_2 gas pressure due to the formation of the adhering dross at the bottom of the cut edge, but not by increasing the energy density.
- The oxide layer width is not affected by the energy density input but it is affected by the oxygen gas pressure due to the exothermal reaction energy.
- Either phase transformation or solutionzing of Fe_3C associated with high heat input followed by fast rate of cooling results into strength of the structure.
- At laser power of 337 watt the heat generated is not high enough to cause phase transformation or solusionzing of the Fe_3C phase which rather results into coursing of Fe_3C phase and hence decrease the hardness value.
- Based on the above conclusions it is recommended to use the laser power of 337 watt and high scan speed within 1000-1500 mm/min when oxygen is used as assist gas at 5 bar, for laser cutting of ultra low carbon steel sheets.

ACKNOWLEDGEMENT

The authors acknowledge gratefully the financial support of The National Institute of Laser Enhanced Science, American University in Cairo, and Faculty of Engineering –Cairo University.

REFERENCES

1. Allmen M. V. and Blatter A., "Laser–Beam Interactions with Materials, Physical Principles and Applications", Second Edition, Springer. (1999).
2. Steen W. M., "Laser material processing", second edition, Springer, (2001).
3. Schuocker D., "High power laser in production engineering", Imperical college press and world scientific publishing Co. (1999).
4. Migliore L., "Laser Materials Processing", Marcel Dekker, (1996).
5. Siegman A. E., "LASERS", copyright © University Science Book (1986).
6. Milonni P. W. and Eberly J. H., "LASERS", A Wiley-Interscience Publication John Wiley & Sons, (1988).
7. Laufer G., "Introduction to Optics and Lasers in Engineering", Cambridge Univ. (1996).
8. Arata Y., Maruo H., Miyamoto I., and Takeuchi S., "New Laser-Gas Cutting Technique for Stainless Steel", Osaka University, Japan, (1998), pp.1-12.
9. Mas C. and Fabbro R., "Steady-state laser cutting modeling", Journal of Laser Applications, **15**,145 (2003).
10. Rathod M. J. and Kutsuna M., "Joining of Aluminum Alloy 5052 and Low-Carbon Steel by Laser Roll Welding", welding research, Department of Material Processing Engineering, School of Engineering, Nagoya University, Japan. (January 2004).
11. Rubahn H. G., "Laser Applications in Surface Science and Technology", John Wiley & Sons, (1999).
12. Molian P. A.," Laser Cutting of Thick Metallic Solids- A Reactive Gas–Flow Approach", Proceeding of LAMP '87, Osaka (May, 1987), pp.245-250.
13. Chen S. L., "The effect of gas composition on the CO_2 laser cutting of mild steel", Journal of Material Processing Technology, 73 (1998), pp.147-159.
14. Gabzdyl J. T., "Effects of gases on laser cutting of stainless steels", Section C., CALEO (1996).,pp.39-44I.
15. Kar A., Rothenflue J. A. and Latham W. P., "Mathematical modeling of thick-section cutting with Chemical Oxygen–Iodine Laser", Section B-, ICALEO (1996), pp.204-213.
16. Hsu M. J. and Molian P. A., "Dual Gas-Jet Laser Cutting of Thick Stainless Steels", Proceeding of LAMP '92, Nagaoka, (June 1992), pp.601-606.
17. Hamoudi W. K., "The effect of speed and processing gas on laser cutting of steel using a 2KW CO_2 Laser", International Journal of Material, (1997), pp.1-9.
18. Shariff S. M., Sundararajan G. and Jeshi S.V., "Parametric influence on cut quality attributes and generation of processing maps for laser cutting", Journal of laser Applications,**11**, 54 (1999).
19. Quintero F., Pou J., Lusquinos F., Larosi M., Soto R. and Amor M.," Cutting of ceramic plates by optical fiber guided Nd:YAG laser", Journal of laser Applications, **13**, 84 (2001).
20. Rajaram N., Sheikh-Ahmed J. and Cheraghi S. H., "CO_2 laser cut quality of 4130 steel", International Journal of Machine Tools and Manufacturing, (2003), pp.351-358.
21. Li S., Hu W. Q., Zeng Y. X. and Ji Q. S., "Effect of carbon content on the microstructure and the cracking susceptibility of Fe-based laser clad layer", J. of Applied Surface Science, **240**, 63 (2005).
22. Ono K. and Adachi K., "Influence of oxide film on weld characteristics of mild steel in CO_2 laser welding" J. of Laser Application, **14**, 73 (2002).

23. Brandon D. and Kaplan D. Wayne, "Microstructure Characterization of Materials", John Wiley & Sons. (1999).

24. Iordanova I., Antonov V. and Gurkovsky S., "Structure modifications during laser treatment of cold rolled low carbon steel", proceeding of SPIE, **4397**, 333 (2001).

25. Kutsuna M. and Uetani K., "Laser Welding of Ultra Carbon Steel", Nagoya University, Dept. of Materials Processing Eng, Japan, (2003), pp.1-9.

26. Rathod M. J. and Kutsuna M., "Joining of Aluminum Alloy 5052 and Low-Carbon Steel by Laser Roll Welding", welding research, Department of Material Processing Engineering, School of Engineering, Nagoya University, Japan. (January 2004).

27. Yilbas B. S., and Sahin A.Z., "Oxygen assisted laser mechanism—a laminar boundary layer approach including the combustion process", Opt. Laser Tech. **27**, 175 (1995).

28. Lobo L., Williams K. and Tyrer JR, "The effect of laser processing parameters on the particulate generated during the cutting of thin mild steel sheet", J. Mechanical Engineering Science, **216** part C-,301 (2002).

Dynamics of Spatially and Temporally Resolved Laser Induced Al-plasma

H. Imam, G. Abdellatif*, V. Palleschi**, M.A. Harith and Yosr E-El. Gamal

National Institute of Laser Enhanced Sciences (NILES) Cairo University, Egypt
*Phys. Dept. Faculty of Science, Cairo University, Egypt
**IPCF/CNR, Applied Laser Spectroscopy Laboratory, Pisa, Italy
Corresponding author: galila_2000@hotmail.com

Abstract. In the present study the temporal and spatial evolution of the plasma produced by interaction of Q–switched Nd:YAG laser pulses at 532 nm with pure aluminum target are investigated via optical emission spectroscopy (OES) in vacuum (10^{-5} torr). Comparison of the spectra taken at different distances from the target surface facilitates discussing fundamental concepts of the Laser Induced Plasma (LIP). Such measurements have been exploited to understand the main processes involved and must be taken into account for the analysis of this kind of plasma. The LIP mean expansion velocity has been determined by measuring the ionic emission temporal profiles usually referred to as the Time of Flight (TOF) profiles. The temporal behavior of the spectral emission has been explained and interpreted in view of the three body recombination processes. Problems concerning the existence of and departure from the local thermodynamic equilibrium (LTE) in the LIP are studied carefully as observed in the performed experiment.

INTRODUCTION

Study of the laser induced plasma (LIP) plays a fundamental role for diagnostic purposes in many applications concerning laser-matter interaction. The phenomenology of energy transport mechanisms encountered in the interaction of laser with solid targets consists of four different stages: the laser ablation of the target; plasma generation; laser interaction with the plasma and plasma expansion. The initiated expanding atomic plasma at high temperature (6000 - 20000 K), is ionized by the inverse bremsstrahlung and the photoionization processes, expanding rapidly (approximately at 10^6 cm/s) perpendicularly to the target surface [1]. During the expansion, the main mechanism of transition of bound electrons from the lower levels to the upper levels and vice versa is driven by inelastic collisions of electrons with heavy particles, while the concentration of charged particles is controlled by the electron impact ionization and three – body recombination of electrons with ions. Radiative processes such as re-absorption, spontaneous and stimulated emission are also important in determining the concentration of emitting levels [2]. The study of laser induced plasma is based on the assumption of local thermodynamic equilibrium (LTE) and optically thin plasma [3,4]. In LTE the Maxwell, Boltzmann and Saha relations are still valid locally, so the electron density, temperature and chemical composition can be easily determined. For the LTE condition to hold, only small variations of the system are admitted so that the times associated to the establishment of kinetic balances are smaller than that of the plasma variations [5]. In fact, this scenario belongs to thermal plasma in stationary conditions. In low of pressure (10^{-5} torr), LIP expanding with supersonic velocity ($10^5 – 10^7$ cm/s), so the plasma parameters can change in such very short time of expansion compared with the time necessary for the establishment balance among elementary processes. This expansion might also results in a departure from LTE. This could be an important point of our study to find out whether the conditions of LTE are satisfied or not in the plasma formation during the expansion phase.

The strong interest in laser induced plasma applications has been leading the development of many experimental techniques to characterize laser ablation plasmas [6]. Among them the most convenient technique, especially for the investigation of the initial stages of LIP, is the optical emission spectroscopy (OES) [7]. Since OES is based on the study of the spectral intensity and broadening of lines, the intrinsic light emission of the LIP and does not need any other excitation sources or intrusive systems, the experimental set-up is very simple and adaptable to automation and remote sensing [8-10].

In the present work, detailed experimental and theoretical study of the dynamics of pure aluminum laser induced plasma in vacuum will be presented. Plasma parameters and composition will be studied

CP888, *Modern Trends in Physics Research,*
Second International Conference on Modern Trends in Physics Research—MTPR-06,
edited by L. El Nadi
© 2007 American Institute of Physics 978-0-7354-0354-9/07/$23.00

spatially and temporally via OES technique. The time of flight (TOF) technique will be exploited to investigate the expansion regimes of LIP in vacuum. The obtained results will be modeled in order to understand different processes involved in the LIP evolution and expansion.

EXPERIMENT

The schematic diagram of the experimental arrangement used for the experimental part of this work is depicted in Fig.1. It consists of a Q–switched Nd: YAG laser (Continuum NY 81-30), a vacuum chamber, a pumping system and an optical emission

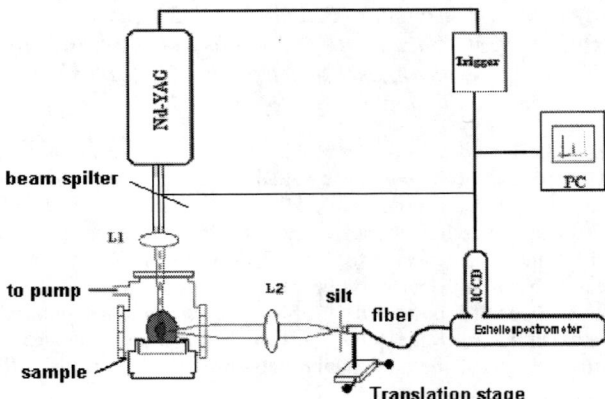

FIGURE 1. Experimental set up.

Laser induced plasma has been produced by focusing the second harmonic (532 nm) of the Nd:YAG laser (100 mJ, 8 ns, and 1 Hz) via a quartz lens of 20-cm focal length onto the target surface. The polished high purity aluminum (99.9999%) target was positioned into the vacuum chamber, which was pumped out up to 10^{-5} mbar by rotary and turbo molecular pumps. The target holder was rotated around an axis parallel to the laser beam between shots, to allow sampling from fresh locations on the target surface and avoiding formation of deep craters, thus improving the reproducibility and accuracy of spectral emission. The plasma emission was collected by a 15cm focal length lens to make a virtual image 1:1 on the optical fiber entrance. The quartz optical fiber used has an aperture of 600μm diameter, and mounted on an x-y micro-translation stage. The aperture of the optical fiber was aligned with the centerline of the plume to ensure that the emission signal was collected perpendicularly with respect to its symmetry axis. The input of the optical fiber was coupled to a slit of 120 μm width for obtaining an effective spatial resolution of 150 μm. The output of the optical fiber was connected to the echell spectrometer (Mechelle 7500) coupled to an intensified charge couple device, ICCD, (DiCAM, Optics, Germany), allowing simultaneous spectral analysis in the range 200 – 900 nm with a constant spectral resolution $\lambda/\Delta\lambda$ =7500.

The gate width and delay time for the spectroscopic data acquisition were controlled by computer. To optimize the signal–to–noise ratio and spectra reproducibility, the acquisition of the spectra was carried out by averaging 50 single accumulated spectra. To investigate the spectral contour and width of the spectral lines the instrumental line shape function of the optical system was measured using the emission lines of a mercury lamp and He–Ne laser. The analysis of the emission spectra was accomplished using the commercial 2D/ and 3D/GRAMS/32 software.

The plasma emission spectrum was recorded at several distances from the target surface while varying the delay time from 0 to 300 ns.

Thus, spatially and temporally resolved optical emission spectroscopy is used to investigate the evolution of the spectral lines intensity and broadening, and to estimate temperature and electron density as a function of time and distance from the target surface (0-2mm).

RESULTS AND DISCUSSION

1- Optical emission spectroscopy of pure Al target

The emission spectra of LIP have been observed at different distances from the target surface along the LIP propagation axis and after different delay times from the laser pulse. By varying the position of the

optical axis of the collecting optics with respect to the target surface, spatially resolved spectra have been obtained that include many neutral and ionized emission lines of the aluminum plasma.

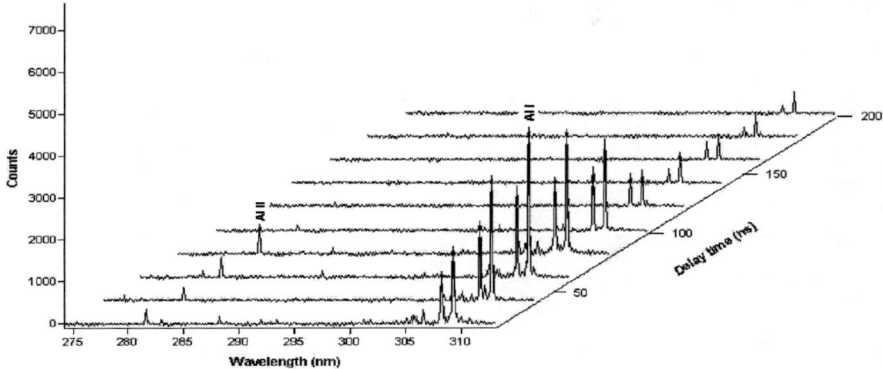

FIGURE 2. Temporal evolution of the emission spectrum at 1 mm distance from target.

From figure (2) it is clear that the maximum intensity of the spectral lines is reached after a characteristic time, depending on the observation distance, and it represents the most populated section of LIP. Such maximum value of the lines intensity maximum is reached when the core of the expanding plasma passes in front of the optical collection system. Depending on the observation distance (with respect to the target surface) the first stage in temporal distribution of the spectra shows a regime where the ionic lines are more intense than the atomic lines. However, on the tail of the temporal distribution - corresponding to the colder part of the plasma - the ionic lines disappear.

Al I (neutral aluminum) at 305.00 nm and Al II (ionized aluminum) at 281.61 nm have been chosen for the spectral analysis as shown in figure 3.

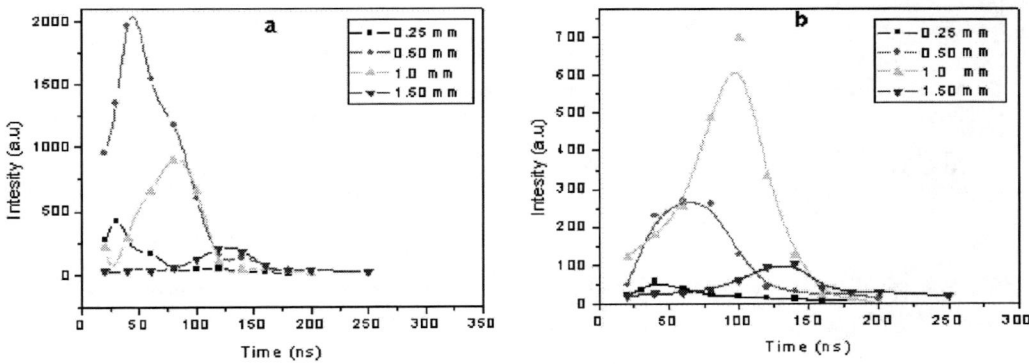

FIGURE 3. Temporal evolution of the emission intensity from Al II transition at $\lambda = 281.61$ nm (**a**) and Al I transition at $\lambda = 305.00$ nm (**b**) at different distances from the target surface.

The transitions of these two emission lines have high lower energy levels and small transition probabilities, which strongly lowers the possibility of their self-absorption in the colder parts of the plasma. From the temporal and spatial distributions of the line intensities at 305.00 nm and 281.61 nm, represented in figure 3, we can observe that Al II emission is confined to a narrower region than Al I emission, corresponding to regions of higher temperature in which the ionization process is enhanced during the expansion process of the plasma. A similar distribution for atomic and ionic lines in plasma was found by Sdorra and Niemax [11].

2- Time of flight (TOF) measurements

The optical emission spectroscopy (OES) is the simplest way to perform TOF measurements in the bright region of LIP (0–2 mm). Study of LIP species evolution has been carried out by temporally and spatially resolved OES at different distances from the surface of the target. As mentioned before, the two

non–resonance lines with high emission coefficients and high lower energy levels, namely Al I at 305.00 nm and Al II at 281.61 nm, have been used for the TOF measurements. For the determination of the expansion mean velocity we considered the ionic emission line because the time of flight of the atomic emission line can present an additional delay due to the three body recombination, as will be explained later.

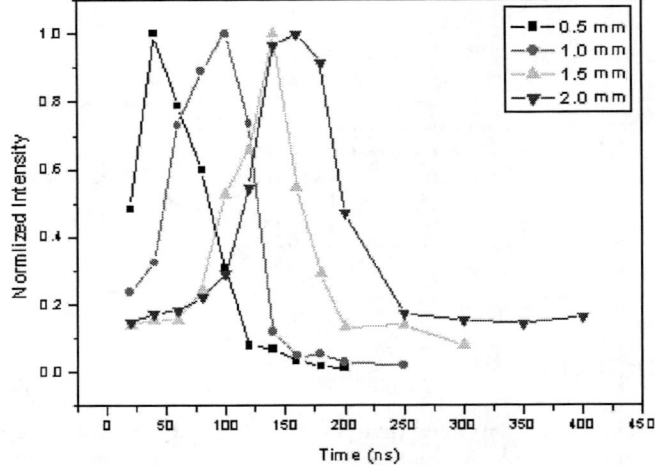

FIGURE 4. TOF profiles of Al II emission intensities at different distances from the target.

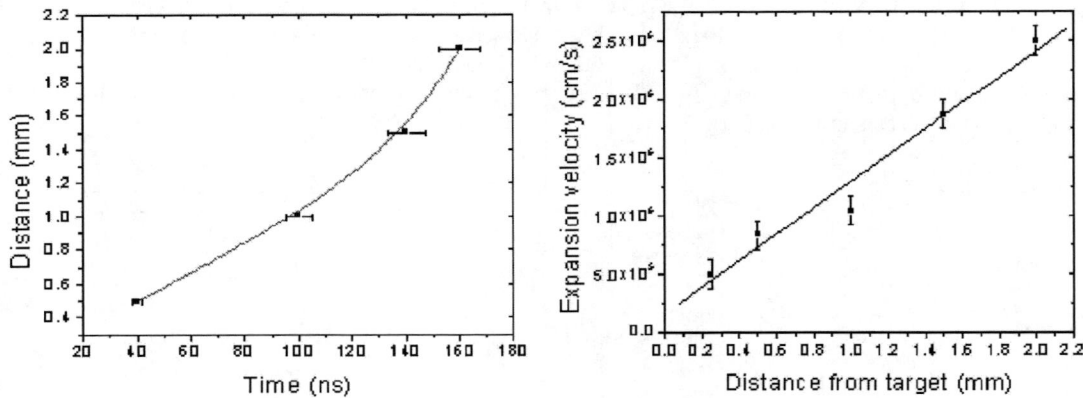

FIGURE 5. Position of the TOF peaks at different delay time.

FIGURE 6. Expansion velocity of the plume as a function of the distance from the target.

Figure 4 shows the normalized TOF of Al II species. By measuring the shift of the peaks of time of flight (TOF) for each distance it is possible to obtain the mean velocities of the LIP along the propagation axis. The corresponding results are shown in figure 5.

In the range of distances investigated experimentally, the velocity of LIP increases with the distance, as shown in figure 6. This trend of LIP velocity is due to the acceleration of the ablated particles from the initial velocity $v_0 = 0$ before the laser pulse energy reaches the ablation threshold, to a maximum velocity v_{max} that is probably reached not far from the maximum distance here considered 2 mm.

3- Determination of the electron density

One of the most powerful spectroscopic techniques to determine the electron density with reasonable accuracy is the measurement of the Stark broadening profile of the atomic emission lines. In the present experiment, the Al II transition at 281.61 nm [$4s^1S(0)$ - $3p^1P^0(1)$] has been chosen for the electron density measurements.

The FWHM of the Stark broadened lines $\Delta\lambda_{1/2}$ is related to the electron density by the expression (for singly ionized ions) [12]:

$$\Delta\lambda_{1/2} = 2w\left(\frac{N_e}{10^{16}}\right)+3.5A\left(\frac{N_e}{10^{16}}\right)^{1/4}\left(1-1.2N_D^{-1/3}\right)w\left(\frac{N_e}{10^{16}}\right)\text{Å} \qquad (1)$$

The first term in Eq. (1) gives the contribution from electron broadening, and the second term is the ion broadening correction. w is the electron impact parameter, which can be interpolated at different temperatures, and A is the ion broadening parameter. Both w and A are weak functions of temperature. N_e is the electron density (cm^{-3}) and N_D the number of particles in the Debye sphere.

The contribution from quasi-static ion broadening (the second term of Eq. (1)) is small in our case. Its value can be evaluated from the extrapolation of the Griem estimation for A and w (for T_e=8000 K and N_e~10^{17} cm^{-3}, its contribution is less than 2%) [12]. Neglecting this contribution, equation (1) can be reduced to:

$$\Delta\lambda_{1/2} = 2w\left(\frac{N_e}{10^{16}}\right)\text{Å} \qquad (2)$$

The temporal evolution of N_e is found to diminish exponentially with time and then level off. The fast decay rate of electron density can be attributed to the plasma expansion, while the slowing and leveling off at longer times are probably due to recombination as shown in Fig 7.

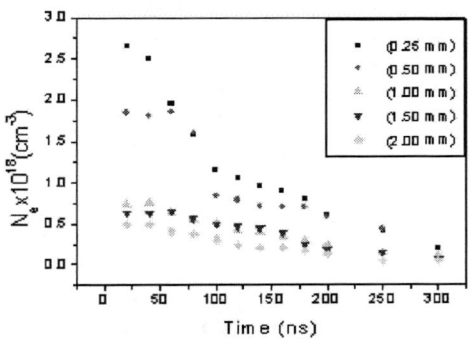

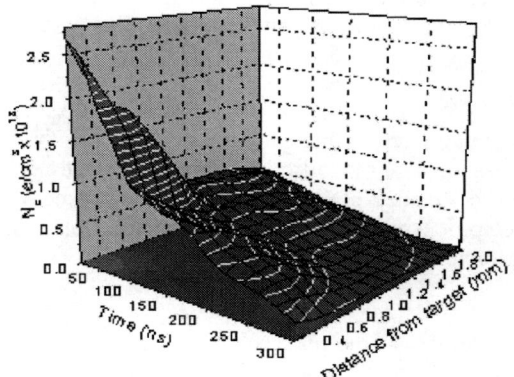

FIGURE 7. Electron density temporal evolution (N_e) at different distances from the target.

FIGURE 8. Spatial and temporal evolution of the electron density N_e.

From figure 8 it is clear that the highest value of the electron density is reached close to the target surface , and then it decreases at longer times and distances from the target. It is also noticeable that the temporal behavior of the electron density is the same at all different distances away from the target surface.

4- The three-body recombination effect on the temporal line profile

Inspecting Fig. (9), it is observable that the TOF peak corresponding to the Al I is time shifted with respect to those of the Al II. This delay could be attributed to the three-body recombination mechanism. Additional amount of neutral atoms produced by the recombination of ions and electrons leads, in fact, to such apparent delay between Al I and Al II TOF maxima.

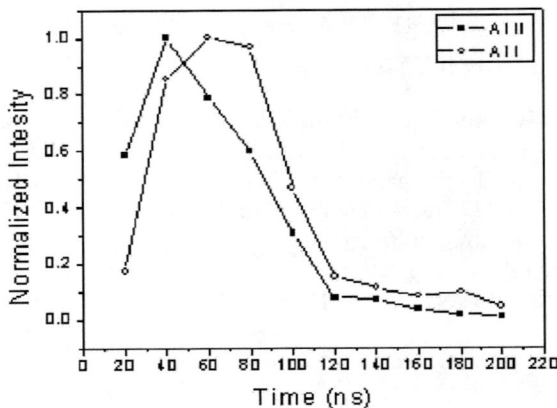

FIGURE 9. Comparison of normalized TOF curves of Al I and Al II at 0.5 mm from the target surface.

During the evolution of the plasma plume, several processes take place that change the relative concentration of the species within the plume. Immediately after the end of the laser pulse, the main mechanism that sustains the ionization process in the plasma is the collisional ionization of neutral atoms by the fast electrons in the plume. This mechanism of ionization can be represented as

$$e^- + A \xrightarrow{k_{ion}} A^+ + e^- + e^-$$

This equation represents the process where a high-energy electron collides with a neutral atom A, producing, with probability k_{ion}, an ionized atom A^+ and another free electron which, in turns, may ionize other neutral atoms through the same process.

The reciprocal process of electron-impact ionization is called three-body recombination, and may be represented as

$$A^+ + e^- + e^- \xrightarrow{k_{rec}} e^- + A$$

In this process, in the presence of a second electron one free electron is captured by an ionized atom, resulting in the production, with probability k_{rec}, of a neutral atom A.

Direct radiative recombination is also possible

$$e^- + A^+ \xrightarrow{k_{rad}} A + h\nu$$

as well as the inverse process

$$A + h\nu \xrightarrow{k_{phot}} e^- + A^+$$

which is just the photo-ionization resulting from the self-absorption process. Atom-atom and ion-atom collisions can be, in general, neglected assuming that electron-atom collision frequency dominates the kinetics of the reactions.

The above mentioned processes may be considered for deriving the temporal evolution of the electron density and the concentration of the other species in the plasma. Let us now apply the theoretical model to our experimental conditions, i.e. the expansion of pure aluminum plasma in vacuum. In these conditions, assuming that the time variations of the plasma electron density due to the LIP expansion are negligible, the following dynamical equation holds for the electron density:

$$\frac{dN_e}{dt} = k_{ion} N_e N_{AII} - k_{rec} N_e^2 N_{AIII} - k_{rad} N_e N_{AIII} + k_{phot} N_e N_{AII} \quad (3)$$

where N indicates the concentration (cm^{-3}) of the particles denoted by the subscripts, k_{ion}(cm^{3}s^{-1}), k_{rec} (cm^{6}s^{-1}), k_{rad} (cm^{3}s^{-1}) and k_{phot} (cm^{3}s^{-1}) are the rate constants of electron-impact ionization, three–body recombination, radiative recombination and photo-ionization, respectively.

Radiative and photo-ionization processes are the dominating processes which occur in the plasma during the laser pulse and immediately after. In fact, such processes are associated with the strong continuum emission which is typical of LIP in its early stage. At longer times, as in our experimental conditions, their influence on the plasma parameters reduces; moreover, these two effects tends to balance out at longer times ($k_{rad} = k_{phot}$), so that they can be safely neglected. Therefore, considering the plasma

neutrality condition $N_e = N_{AlII}$, which holds in the absence of other ionic species that may contribute to the electron density balance, one can rewrite the above equation as:

$$\frac{dN_e}{dt} = k_{ion} \, N_e N_{Al\,I} \quad - \quad k_{rec} \, N_e^3 \tag{4}$$

According to Eq. (4), there are three possible evolution regimes for the electron density

1- $dN_e/dt = 0$ equilibrium condition,
2- $dN_e/dt > 0$ ionization prevails,
3- $dN_e/dt < 0$ recombination prevails.

If $dN_e/dt \neq 0$, then ionization ($dN_e/dt > 0$) or recombination ($dN_e/dt < 0$) prevails and departure from equilibrium occurs [13].

To investigate the time evolution of N_e, it is necessary to take into account that we have to determine the N_e corresponding to the same section of the plasma. However, the plasma is expanding, so that different parts of the LIP will move to different positions with time. In order to sample the same section of the plasma at different times, we can determine N_e from the emission spectrum corresponding to the maximum of TOF signal at different distances. We are thus assuming that the maximum TOF signal is produced by the same portion of the plasma when it passes in front of the detector; because of the expansion velocity, this portion of the plasma will move away from the target and will reach the detector at different times as the distance from the target is increased. By collecting the optical spectrum at the maximum of the TOF emission, both N_e and the corresponding time at which this parameter is measured are known.

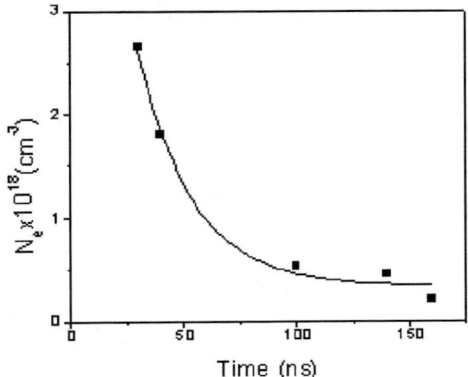

FIGURE 10. Time evolution of electron density corresponding to the maximum Al II TOF emission.

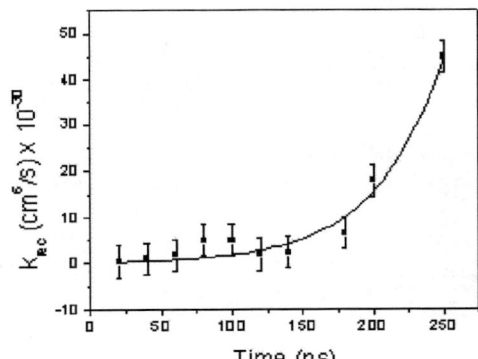

FIGURE 11. Three-body recombination rate constant as a function of time.

Figure 10 depicts N_e as a function of time; it can be observed that $dN_e/dt < 0$. This means that departure from equilibrium indeed takes place, and there is a predominance of the three-body recombination process over that of electron-impact ionization. On the basis of these observations we can conclude that during the expansion, while the excitation and de-excitation by electron impacts are the main mechanisms that govern the distribution of particles among the excited states, three-body recombination is the main mechanism that governs the concentration of charged particles.

From the fitting of the experimental data with a function obtained from Eq. (4), the recombination constant can be estimated as $k_{rec} = 0.9 \times 10^{-30}$ cm^6/s, which is in good agreement with the calculated theoretical values reported in the literature ($k_{rec} \approx 10^{-30}$ cm^{6}s^{-1}) [3].

From our experimental data, the temporal dependence of k_{rec} can also be obtained. Since the TOF profiles are the result of the velocity distribution of the species and of the kinetic processes, considering the time decay of the TOF signal is possible to estimate the recombination constant directly by the temporal evolution of N_e for each distance from the target.

The N_e values have been obtained by Stark broadening for the decay side of the Al II TOF curves between 0.5 and 2 mm. In fact, before the maximum of TOF curves it is not possible to have an exact evaluation of the line contour because of the continuum spectra, as discussed above, while after 2 mm the Stark broadening becomes comparable to the instrumental broadening so that the estimation of N_e becomes too rough [14]. Values of dN_e/dt are obtained by the exponential fit of the temporal evolution of N_e, while the second derivative of dN_e/dt with respect to N_e is calculated using a polynomial fit. In this case we can use all the experimental measurements after the maximum so that an accurate fitting can be achieved (correlation coefficient R > 0.99). The recombination rate constant estimated in this way is

again of the order of 10^{-29} - 10^{-30} cm^{6}s^{-1}, in good agreement with the published data [3]. The corresponding values of the three-body recombination rate constant are reported in Fig. 11.

The recombination time can be estimated by the values of the rate constant of recombination process, using the relation $t_{rec} = (N_e^2 \, k_{rec})^{-1}$ which gives a value of $t_{rec} \sim 10^{-7} - 10^{-6}$ s. Since the velocity of the plasma expansion is in the range $10^5 - 10^6$ cm/s, as determined experimentally, and the typical plasma dimensions considered are of the order of 1 mm, we can estimate an expansion time of $10^{-7} - 10^{-6}$ s. We can then conclude that:

$$t_{ion} > t_{rec} \sim t_{exp}$$

As a consequence of the fast expansion of LIP, we can thus establish that the laser-induced aluminum plasma is in a quasi–equilibrium state. The time of expansion is less than the corresponding time for establishment of Saha balance so the plume expands in quasi-equilibrium state and deviation from equilibrium has a recombination character.

4- Measurements of the Plasma temperature

The plasma temperature can be determined from the relative measured intensities of the emission lines using the well known Boltzmann plot method, provided that the transition probabilities (A_{mn}) from a given excitation state are know. For high electron densities, the main mechanism of transition of bound electrons from the lower level to the upper level and vice-versa is inelastic collisions of electrons with heavy particles (in this case Al I and Al II). Therefore the populations of the excited states follow the Bolzmann distribution and the intensity I_{mn} of the spectral transition line between two bound levels of species z in plasma is given by:

$$I_{mn} = F \frac{g_m A_{mn}}{\lambda_{mn} U_z(T_z)} N_z \exp\left[-\frac{E_m}{kT_z}\right] \qquad (5)$$

By plotting the logarithm of the measured emission intensity of ionic and atomic lines transition of Al as function of the corresponding energy of the upper level, the following equations hold:

$$\ln\left[\frac{I_{mn} \lambda_{mn}}{g_{mn} A_{mn}}\right] = m_z + q_z$$

$$q_z = \ln\left[\frac{N_z}{U_z(T_z)}F\right] \qquad m_z = -\frac{1}{kT_z} \qquad (6)$$

Where I_{mn} is the intensity of the light emitted in the transition, k is the Boltzmann's constant, λ_{mn} is the wavelength of the photons emitted in the transition from the upper excited energy level E_m to the lower energy level E_n, U_z is the partition function of the species z calculated at T_z, N_z is the concentration of the species z in the ground state and F is an experimental parameter which takes into account the optical efficiency of the collection system as well as the plasma density and volume.

For the Evaluation of the plasma temperature via the Boltzmann plot method, it is important to verify that the plasma is not optically thick for the lines used. This was done by checking the ratio of emission intensities of Al II at 281.61 nm, 358.60 nm, 624.33 nm and 466.30 nm and Al I at 256.79 nm, 257.50 nm, 305.00 nm, 305.70 nm, 308.21 nm, 309.27 nm according to the procedure described by Radziemski et al. [15]. The intensities were observed to be in a ratio that is consistent with the ratio of their statistical weights, which indicates that the plasma was optically thin for these wavelengths.

The excitation temperature of the LIP was estimated from the emission intensity of the atomic lines (Al I) and ionic lines (Al II) as mentioned above.

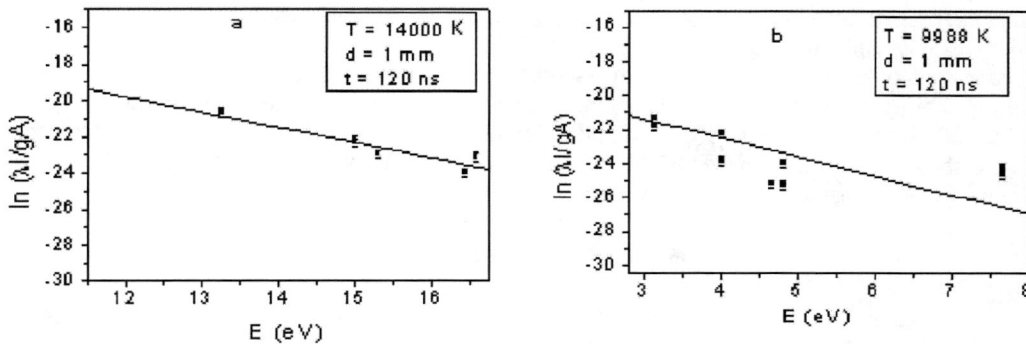

FIGURE 12. Boltzmann plots for Al II (a) and Al I (b) at 1 mm distance from the target and 120 ns delay

Figure 12 (a and b) show Bolzmann plots for Al I and Al II, obtained at 1mm distance from the target surface and the same delay time. The temperature of the species is obtained by the least square linear fit of the experimental data; and its dependence on time, and distance from the target is plotted in figure 13.

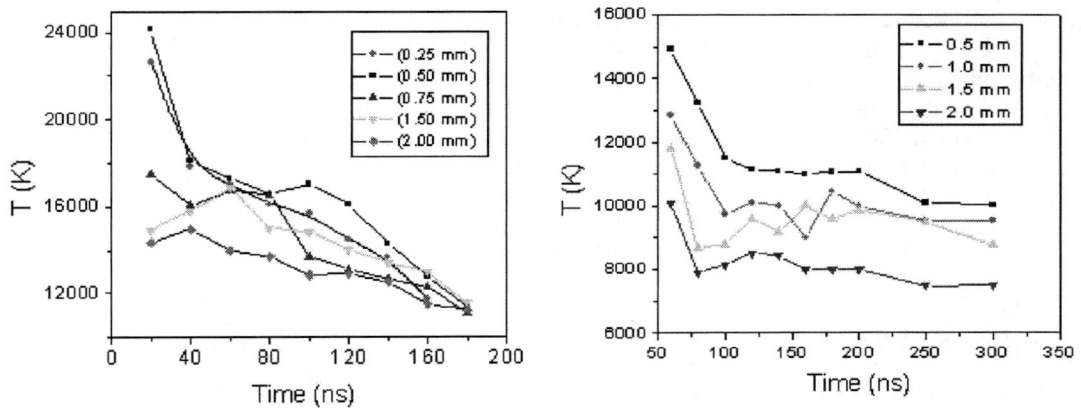

FIGURE 13. Temporal behavior of the plasma temperature obtained via Boltzmann plot method for Al II (left) and Al I (right)

Comparing the obtained temperature values, we can observe that the temperatures of Al II and Al I are appreciably different. The variation of the temperature of the neutral species is almost insignificant in time, as it changes from 12000 to 8000 K, whereas the temperature of the ionized species decreased from 20000 to 11000 K, as shown in Fig. 13. The difference between plasma temperatures calculated for the ionic and atomic species in vacuum increases at shorter delay times and at longer distances from the target, as shown in figure 14.

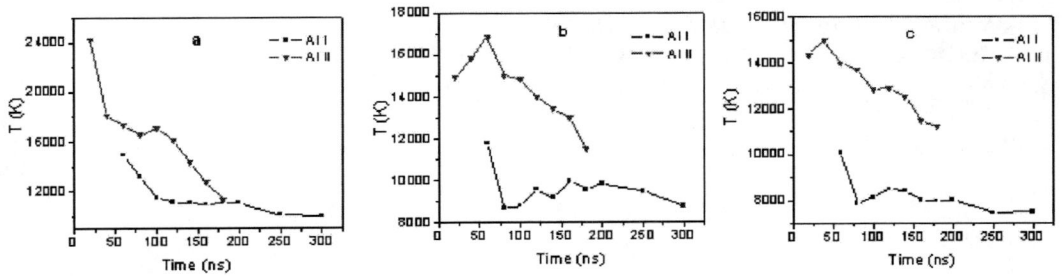

FIGURE 14. Time dependence of the plasma temperature calculated via Boltzmann plot method for Al I and Al II at a) 0.5 mm from the surface; b) 1.5 mm from the surface; c) 2 mm from the surface.

In fact, at shorter times after the onset of the laser-breakdown, the fast dynamics of the plasma does not allow for the system to reach thermodynamic equilibrium, so that the relative concentrations of the different species in the plasma cannot be described with just a single temperature as it should be in the framework of the LTE theory. Moreover, immediately after the onset of the laser breakdown, radiative

processes such as spontaneous emission, reabsorption and stimulated emission are predominant over the collisional effects, thus invalidating the LTE approximation.

In the same way, at longer distances from the target the electron density in the plasma is low, so that the thermal equilibrium between the plasma species cannot be sustained by the electron-ion collisions. At low values of the electron density the three-body recombination effect has also to be taken into account. All these effects lead to the violation of the LTE approximation, thus giving apparently different plasma temperatures when considering the different species in the plasma.

CONCLUSION

In this work, we have reported experimental results on the dynamics of laser-induced plasmas in vacuum. The thermodynamical properties of the plasma, as well as its composition have been studied using the Optical Emission Spectroscopy (OES) technique, involving high-resolution broadband detection of the plasma optical emission with high temporal and spatial resolution. Using the optical Time of Flight (TOF) technique, the expansion velocity of the laser-induced plasma was measured in vacuum. The analysis of the plasma optical emission also facilitated the measurement of time and spatial profiles of plasma electron density and temperature during the plasma expansion stage.

Moreover, modeling and theoretical analysis of the experimental data allowed the study of non-equilibrium processes in the laser-induced plasmas. Investigation of the plasma expansion in vacuum, revealed a departure from equilibrium which has been explained in terms of the three-body recombination effect. The corresponding rate constant of such effect k_{rec} was measured. The obtained results were in good agreement with the corresponding theoretical estimates.

Finally, deviations from the Saha balance were found. An explanation of the phenomenon was given in terms of radiative effects and three-body recombination too.

REFERENCES

1. M.Capitelli, A.Casavola, G.Colonna, and A. De Giacomo, Spectrochim. Acta Part B, **59**, 271 (2004).
2. J. Herman, C. Boulmer-Leborgne, and D. Hong, J. Appl. Phys. **83**, 691 (1998).
3. H. R. Griem, "Principles of plasma spectroscopy" Cambridge University press, 1997.
4. Thuvan N. Piehler, Frank C. DeLucia Jr., Chase A. Munson, Barrie E. Homan, Andrzej W. Miziolek, and Kevin L. McNesby, Applied Optics, **18**, 3654 (2005).
5. J.A.M. Van der Mullen, Excitation equilibria in plasmas; a classification, Physics Reports of Physics Letter, PRPLCM. **191** (2 and 3) 109 (1990).
6. S. Amoruso, R. Bruzzese, N. Spinelli, and R. Velotta, J.Phys. B: At. Mol. Opt. Phys. **32**, R131-R172 (1999).
7. A. De Giacomo, V.A. Shakhatov, and O. De Pascale, Spectrochim. Acta B **56**, 753 (2001).
8. D.A. Cremers, Appl. Spectrosc. **41**, 572 (1987).
9. R. Noll, H. Bette, A. Brysch, M. Kraushaar, 1. Monch, L. Peter, and V. Sturm, Spectrochim. Acta B **56**, 637 (2001).
10. S. Rosenwasser, G. Asimellis, B. Bromley, R. Hazlett, J. Martin T. Pearce, and A. Zigler, Spectrochim. Acta B **56**, 707 (2001).
11. W. Sdorra, and K. Niemax, Microchim. Acta **107**, 319 (1992).
12. H. R. Griem, "Plasma Spectroscopy" McGraw-Hill, New York.1964.
13. M. Capitelli, F. Capitelli, and A. Eletskii, Spectrochim. Acta B, **55**, 559 (2000).
14. NIST Atomic Spectra Database. Available from, www.nist.gov.
15. L. J. Radziemski, D.A. Cremers, and T.R. Loree, Spectrochim. Acta B, **38**, 349 (1983).

The Impact of Receiver Aperture Design and Telescope Properties on LIDAR Signal-to-Noise Ratio Improvements

Yasser Hassebo [1], Khaled El Sayed [2]

[1] *LaGuardia Community College of the City University of New York, Math/Engineering Department, USA*
[2] *Department of Physics, Faculty of Science, Cairo University, Egypt*

ABSTRACT. Range and sensitivities of lidar measurements in daylight are limited by sky background noise power (BGP). This is particularly important for Raman lidar techniques where the Raman backscattered signal is relatively weak. This often restricts Raman lidar measurements to nighttime where BGP is absent. The background noise elimination is particularly important in daytime measurements in case where full overlap between laser beam and receiver telescope field-of-view (FOV) is necessary. Results of numerical simulations for a vertically pointing Lidar show that significant improvements in Lidar signal to noise ratio (SNR) can be obtained, by minimizing the detected sky BGP. This can be, optimally achieved if the receiver telescope aperture is properly designed to track lidar target images, which are range dependant. In this context, the connection between receiver telescope field of view and optimum aperture size are examined. The SNR improvements, which can be obtained in this manner, translate to corresponding improvements in Lidar range for backscatter schemes including Raman and DIAL.

INTRODUCTION

Light Detection and Ranging (lidar) is an active remote sensing instrument that transmits laser and measure the backscatter radiation after interacting with various components of the atmosphere. The impacts of aerosol in the human health with diseases such as lung cancer, bronchitis, and asthma have been essential motivations to record aerosol properties and transportation. Lidars have been applied to study stratospheric aerosols [1], tropospheric aerosols [2] and climate gases such as stratospheric ozone [3] as well as for analyzing the clouds properties [4]. In this work we examine the potential of improving SNR for monostatic lidar systems by analyzing and optimizing detector aperture (field stop) geometry. Monostatic lidar systems can be subdivided into two categories, coaxial and biaxial lidar systems. The main disadvantages in the coaxial lidar systems, where the transmitted laser beam is coaxially with the receiver's FOV, are the detector saturation problem that is occur once the lidar laser beam is shot, the unwanted signal that is detected from reflection of the transmitted light at the transmitter optics, in the top of the receiver telescope, and the part of images, for shorter range, that is blocked by the secondary mirror. Biaxial lidar, where the transmitter and receiver are located adjacent to each other, system is a practical solution to overcome these coaxial lidar systems problems. But, on other hand, the recorded data from the biaxial lidar is negatively affected by the geometrical factor (GF) at shorter range. To realize the effect of GF ' $\xi(R)$ ' in the return lidar signal, the lidar Equation can be written as [5]:

$$P(\lambda_L, R) = P_L \frac{A_o}{R^2} \xi(\lambda_L) \beta(\lambda_L, R) \xi(R) \frac{c \tau_L}{2} \exp \left(-2 \int_0^R k(\lambda_L, R) \, dR \right) \tag{1}$$

Where, $P(\lambda_L, R)$ is the total scatter laser power received from a distance R, P_L represents the average power in the laser pulse, A_o/R^2 describes the solid angle of the receiver optics (A_o is the area of the telescope primary mirror), $\xi(\lambda_L)$ denotes the receiver's spectral transmitter factor, $\beta(\lambda_L, R)$ is the volume backscatter coefficient, $c\tau_L$ represents laser pulse length (c is speed of light, τ_L is Laser pulse rectangular duration), $k(\lambda_L, R)$: Atmospheric extinction coefficient. The smaller the value of $\xi(R)$ is the smaller the return signal and the smaller SNR particularly for short distances. GF can be defined as the ratio of the energy transferred to the photodetector to the energy reaching the telescope primary mirror, E_{det}/E_{scat} [6]. This reduction in the detector response to the return

CP888, *Modern Trends in Physics Research,*
Second International Conference on Modern Trends in Physics Research—MTPR-06,
edited by L. El Nadi
© 2007 American Institute of Physics 978-0-7354-0354-9/07/$23.00

signal is caused by a lack of perfect overlap between the receiver telescope's FOV and the transmitter laser beam. In Section 2, we discuss the overlap effect of the GF including the receiver field stop position and size, and their effect in lidar SNR improvement. Lidar simulation results are introduced in section 3. Also, the telescope best selection to reduce BGP is introduced in Section 4. Conclusions and future works are presented in Section 5.

PROBLEM DESCRIPTION AND ANALYSIS

The standard configuration for most lidars is to place a round aperture in the focal point of the receiver telescope. It is also commonly assumed that once the lidar receiver FOV and transmitted beam are completely overlapping, the efficiency of collection is unity [5]. However, this analysis does not properly take into account the shifted position of the collected backscattering signals on the image plane from the telescope focal point at the receiver. These shifting distance from the telescope focal point and the atmosphere sounding (image) size variation are according to the distance 'b_o' between the laser and telescope optical axis. This image displacement is range dependant as shown in Fig. 1. In fig 1. the telescope is presented by a lens with diameter of t_o and f focal length. The farther the lidar object ($Z=R_{max}$) the smaller the sounding image (Im_1), on the other hand the closer the lidar object ($Z=R_{min}$) the larger the sounding image (Im_2). Numerous papers implemented a wedge like shape aperture design to over come this shifted problems [7, 8]. In this paper we study the effect of changing a round aperture, the realistic shape, size and place in the lidar SNR.

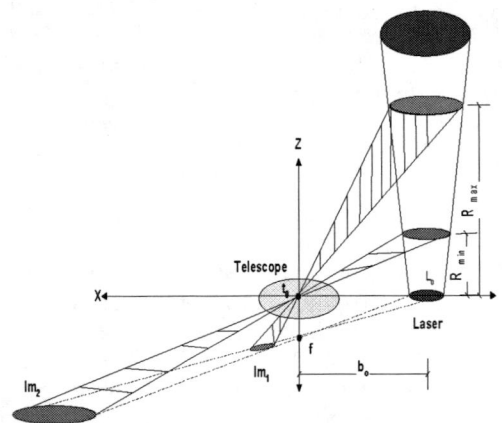

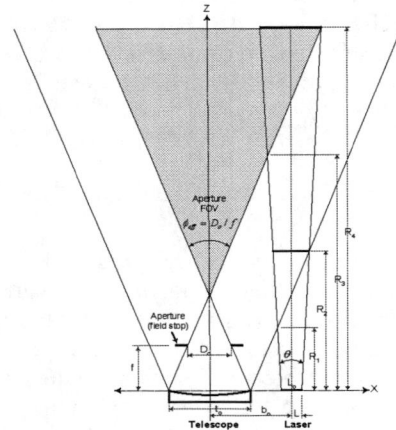

FIGURE 1. Biaxial Lidar, schematic diagram shows transmitter and receiver subsystems, and sounding trace images for lidar objects at heights of R_{min}, and R_{max}

FIGURE 2. Biaxial Lidar: overlapping between effective FOV of receiver telescope (diameter of t_o) and laser beam (initial diameter L_o) and aperture diameter D_o

As shown in Fig. 2, we assume the optical vertical axis z. the ground level (x-axis), where the location of the telescope primary mirror and the transmitted laser beam, is at z=0. The range 'R' increases, there will be a point 'R_1' where the first intersection between the left boundary of the laser beam and the right boundary of the telescope FOV. Then at ($z = R_2$) the complete overlap is formed with the telescope FOV. But this is not the effective overlap function. The effective FOV is based on, D_o, the field stop diameter and f the telescope focal length ($\phi_{eff} = D_o / f$ the shaded area in this case) [9]. The actual overlap started at ($z = R_3$) and finally the effective overlap is completed at ($z = R_4$). At short distances ($z < R_3$) the ratio of the overlapping area (OL_{area}) to the image area (Im_{area}) that formed near to (f) is

$$\xi(R) = \frac{OL_{area}}{Im_{area}} = 0 \qquad (2)$$

where $OL_{area} = 0$. This is making near field observations impossible (effective telescope area: $A_{eff}(R) = A_o \xi(R) = 0$). In the case of a small round aperture ($D_o = 2$ mm diameter) is placed at the telescope focal pint f_o, the overlap function $\xi(R)$ is very small for any object in ranges of ($R_3 < z < R_5$), where at R_5 there is an arbitrary object which has

a large sounding image that is formed very far from f in the imaging plan (see Fig. 1). Design of a unique aperture to cover certain desired ranges becomes feasible.

RESULTS

In this paper we propose a feasible design of a round aperture to house certain desired ranges and minimizing the detected BGP. That can be achieved by moving the commonplace aperture (D_o) center from the origin (f_o) some distance to the left (depend on the object height) and reduces the aperture size from D_o to smaller diameter D_s (i.e., reducing the effective FOV from $\phi_{eff} = D_o / f_o$ to $\phi_{eff} = D_s / f_s$) where f_s is a smaller telescope focal length.

Lidar simulation results for biaxial system are shown in Fig. 3. This simulation is for the following parameters: The distance between the beam and the telescope axis is b_o= 200 mm, laser initial beam diameter L_o= 5 mm, beam divergence $\theta = 0.5 mrad$, telescope primary mirror diameter of t_o=178 mm, and two different telescope focal lengths of f_s=1.7 m, f_L=4 m. We always are taking into account that θ is smaller than the effective telescope's FOV ($\phi_{eff} = D / f$), both D and f (the field stop diameter and the telescope focal length, respectively) have three different values. These values are: (1) $D= D_o$ = 2mm, for commonplace aperture (placed at the telescope focal length $f= f_o$). (2) $D=D_s$ for small telescope focal length ($f=f_s$=1.7 m). (3) $D =D_L$ for telescope with bigger focal length ($f=f_L$=4 m). Obviously, the new aperture position and size reduction are range dependent. For shorter ranges (0.5 – 5 km) the aperture diameter became smaller (i.e., D_s= 1.8 mm for f_s=1.7 m) and the center is shifted by ~0.45 mm, on the other hand, if we use a bigger telescope, a field stop diameter of D_L =4.7 mm (for f_L= m) must be used. This bigger aperture (i.e., bigger BGP) center must be shifted by ~2.4 mm that to housing the entire images. However, for higher lidar range (5-25 km) the field stop size gets much smaller (D_s= 1.4 mm for f_s=1.7 m) that centered approximately at the origin (0, 0), but D_L =3 mm center shifted by ~0.5 mm. We noted that the field stop shifted positions and sizes changing are more significant for the shorter ranges particularly for bigger telescope focal length.

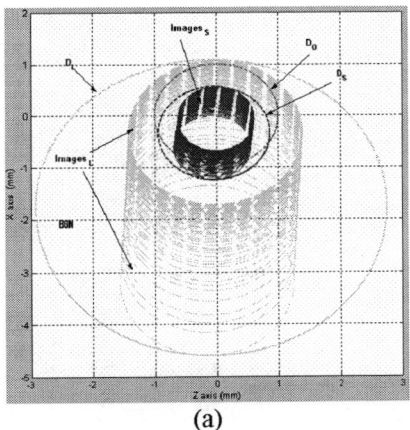
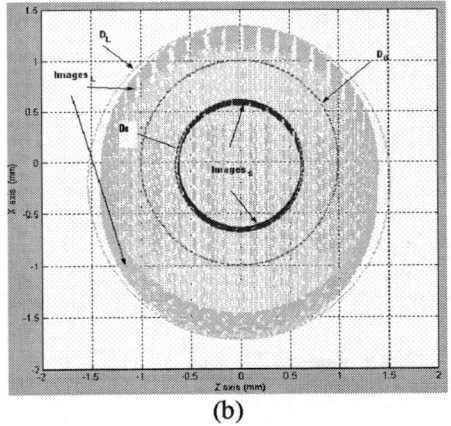

(a)   (b)

FIGURE 3. Lidar images (a) For range 500m-5km. (b) For range 5km-25km. The green images (images $_L$) for 4 m telescope focal length and the blue images (images $_s$) for telescope with 1.7m focal length. The commonplace aperture (D_o=2 mm) placed at the telescope focal point, D_L is a round aperture to accommodate (images $_L$) and D_S is a round aperture to accommodate (images $_s$).

TELESCOPE SELECTION

Typically in lidar measurements a larger VOF$_{eff}$ is desired to decrease the height where the laser beam meets the effective telescope FOV for the first time ($z= R_3$). The VOF$_{eff}$ ($\phi_{eff} = D / f$) enlargement is required increasing the field stop size (D) for the same telescope focal length (f). Yet, the larger the VOF$_{eff}$ (ϕ_{eff}) is the bigger the BGP that reaching the PMT, beside the larger the multiple scattering effect [9, 10, 11]. This trade off can be optimized

using two techniques. (1) If the laser beam is tilted toward the telescope by inclination angel Θ to increase the effective lidar range [12]. (2) Reducing VOF$_{eff}$ by decreasing both field stop diameter (D) and telescope focal length (f), that the BGP increment can be avoided. Where, BGP is proportional to VOF$_{eff}$ in the shot noise regime [13, 14]. As shown in our simulation results (Figs. 3a and 3b), the larger the telescope focal length ($f = f_L$), is the larger the image size (Images$_L$), and the larger the sky BGP mainly in shorter distance. Where (Images$_L$) is the lidar sounding images collected through (f_L = 4m) telescope focal length, and (Images$_S$) represents the lidar sounding images collected through telescope with focal length of (f_S=1.7 m) both from objects at ranges of 500m-5km, and 5km-25km as illustrated in Figs 3a and 3b, respectively. D$_s$, blue circle, and D$_L$, green circle, represent the round shape apertures that housing the entire lidar return signal using a 1.7m, and 4m telescopes respectively. As cab be seen in table 1, numerical results show that as much as a factor of (17.8 %) improvement in lidar signal-to-noise ratio if we used even smaller telescope (f_S=1 m) over conventional large telescope (f_L=4 m). That can be obtained if we assume that the mean value of the photomultiplier tube (PMT) output power (P_d) is proportional to BGP (i.e., $\langle p_d \rangle \alpha BGP$). Meaning the detector operates in the shot noise limit. A system under the shot noise limit if the detected noise amplitude (standard deviation) is proportional to the square root of the mean detected signal at the fare range (i.e. $\Delta P_d = \sqrt{\langle p_d \rangle}$ .The potential of this work becomes visible if we compare the detected SNR with small and bigger telescope focal lengths. In the shot noise regime, the SNR improvement factor (SNR$_{imp}$) can be expressed in terms of BGP corresponding to large and small telescope focal length (BGP^L, BGP^S, where in this case $BGP^L > BGP^S$)[14]:

$$ SNR_{imp} = \frac{SNR_S}{SNR_L} = \sqrt{\frac{P_{sig} / BGP^S}{P_{sig} / BGP^L}} = \sqrt{\frac{BGP^L}{BGP^S}} \tag{3} $$

This formula shows that decreasing the sky background noise from BGP^L to BGP^S using smaller telescope focal length will improve the SNR. This SNR improvement can be translated into improvement in the attainable lidar range.

TABLE 1. Image's size versus telescope focal length for different lidar ranges. Effective FOV (ϕ_{eff}) is shown no big different in the higher range for a variety of f. Normalized ϕ_{eff} with respect to ϕ_{eff} of (f = 4m) is also shown.

Telescope F	Lidar range 'R'		Aperture diameter	$\phi_{eff} = D_L /F$	ϕ_{eff} (Norm)	SNR$_{Imp}$
	From (km)	To (km)	D$_L$ (mm)	(mrad)	%	%
4 m	0.5	5	4.7	1.175	100	0
	5	25	3	0.75	100	0
3 m	0.5	5	3.4	1.13	96	2
	5	25	2.1	0.7	93	3.6
1.7 m	0.5	5	1.8	1.058	90	5.4
	5	25	1.4	0.7	93	3.6
1 m	0.5	5	0.85	0.85	72	17.8
	5	25	0.6	0.6	80	11.8

By comparing the results in table 1, give evident that a lidar system with small focal length (f= 1m) is much better in reducing BGP than is any other system with big (f =4m) focal length telescope particularly for short distances. this deduction of the BGP can be translated to improvement in lidar SNR up to 17.8 %. Also we'll gain a good improvement in the lidar range and the detector's averaging time. Where, we can relate SNR improvement with the detector's averaging time improvement (τ_{imp}^{det}) as: $\tau_{imp}^{det} = (SNR_{imp})^2$. So, an improvement of detector's averaging time (i.e., reducing the required detector's averaging time) of, $\tau_{imp}^{det} = (0.18)^2 \approx 3.2\%$, can be achieved from a 17.8 % lidar SNR improvement.

CONCLUSIONS AND FUTURE WORKS

In comparison with the classical design of lidar receiver subsystem, it does not take into account that in the receiving optics the detected images are placed on a line forming an angle with the imaging plane of receiver telescope, in such a case GF is too small. Based on our proposed design to replace the classical lidar receiver optical design with a new design that attain significant lidar SNR improvements by minimizing the detected sky BGP if we set the receiver round aperture in the proper position with a smaller size. Simulated numerical results for a biaxial lidar have been shown the telescope best selection is the one with smaller f to ensure having the minimum FOV that accepts all return signals for the entire ranges, while at the same time minimizing detected BGP and maximizing lidar SNR and attainable lidar ranges. The improvement in lidar SNR was up to 17.8 %. This in turns lead to a good improvement in the lidar range and the detector's averaging time. A reducing of the required detector's averaging time of, $\tau_{imp}^{det} \approx 3.2\%$, can be achieved.

REFERENCES

1. Zuev V., V. Burlakov, and A. El'nikov, *J. Aerosol Sci.* **29** 1179-1187 (1998).
2. Barnaba F, and G. Gobbi, *J. Geophys. Res.* **106**, 3005-3018 (2001).
3. Douglass L. R., M.R. Schoeberl, S.R. Kawa and E.V. Browell, *J. Geophys Res.* **106**, 9879-9895 (2000).
4. B. Stein, C. Wedekind, H. Wille, F. Immler, M. Müller, L. Wöste, M. del Guasta, M. Morandi, L. Stefanutti, A. Antonelli, P. Agostini, V. Rizi, G. Readelli, V. Mitev, R. Matthey, R. Kivi, and E. Kyrö, *J. Geophys. Res.* **104**, 23983–23993 (1999).
5. Measures R.M ," Laser Remote-Sensing equation," in *Laser Remote Sensing: Fundamentals and Applications*, Publisher, Wiley New York, 1984, pp 237-280.
6. R. Velotta, B. Bartoli, R. Capobianco, L. Fiorani, and N. Spinelli, *Appl. Opt.* **37**, 6999–7007 (1998).
7. Agishev, Ravil R. and Adolfo Comeron, *Appl. Opt.* **41**, 7516-7521 (2002).
8. Y. Hassebo, R. Agishev, F. Moshary, S. Ahmed "Optimization of Biaxial Raman Lidar receivers to the overlap factor effect" *in 3rd. NOAA CREST Symposium*, Hampton, VA USA, April 2004.
9. K. Sassen and R. L. Petrilla, *Appl. Opt.* **25**, 1450– 1459 (1986).
10. P. Bruscaglioni, G. Zaccanti, L. Pantani, and L. Stefanutti, *Int. J. Remote Sensing* **4**, 399–417 (1983).
11. R. J. Allen and C. M. R. Platt, *Appl. Opt.* **16**, 3193–3199 (1977).
12. K. Stelmaszczyk, M. Dell'Aglio, S. Chudzyn´ ski, T. Stacewicz, and L. Wöste, *Appl. Opt.* **44**, 1323-1331 (2005).
13. H. Kuze, H. Kinjo, Y. Sakurada, and N. Takeuchi, *Appl. Opt.* **37**, 3128-3132 (1998).
14. Yasser Y. Hassebo, B. Gross, M. Oo, F. Moshary, and S. Ahmed, *Appl. Opt.* **45**, (in press) (2006).
15. Zuev V., V. Burlakov, and A. El'nikov, *Ten Years (1986-1995) of lidar observations of temporal and vertical structure of stratospheric aerosol over Siberia*. J. Aerosol Sci. **29** 1179-1187 (1998).
16. Barnaba F, and G. Gobbi, "Lidar estimation of tropospheric aerosol extinction, surface area and volume: Maritime and desert-dust cases. J. Geophys. Res., **106** (D3), 3005-3018 (2001).
17. Douglass L. R., M.R. Schoeberl, S.R. Kawa and E.V. Browell, "A composite view of ozone evolution in the 1995-1996 northern winter polar vortex developed from airborne lidar and satellite observations. J. Geophys Res., **106** (D9), 9879-9895 (2000).
18. B. Stein, C. Wedekind, H. Wille, F. Immler, M. Müller, L. Wöste, M. del Guasta, M. Morandi, L. Stefanutti, A. Antonelli, P. Agostini, V. Rizi, G. Readelli, V. Mitev, R. Matthey, R. Kivi, and E. Kyrö, "Optical classification, existence temperatures, and coexistence of different polar stratospheric cloud types," J. Geophys. Res. **104** (D19), 23983–23993 (1999).
19. Measures R.M , Laser Remote Sensing:Fundamentals and Applications, Wiley New York, 1984.
20. R. Velotta, B. Bartoli, R. Capobianco, L. Fiorani, and N. Spinelli, "Analysis of the receiver response in lidar measurements," Appl. Opt. **37**, 6999–7007 (1998).
21. Agishev, Ravil R. and Adolfo Comeron "Spatial filtering efficiency of monostatic biaxial lidar: analysis and applications", App. Opt. **41**, 7516-7521 (2002).
22. Y. Hassebo, R. Agishev, F. Moshary, S. Ahmed "Optimization of Biaxial Raman Lidar receivers to the overlap factor effect" 3rd. NOAA CREST Symposium, Hampton, VA USA, April 2004.
23. K. Sassen and R. L. Petrilla, "Lidar depolarization from multiple scattering in marine stratus clouds," Appl. Opt. **25**, 1450–1459 (1986).
24. P. Bruscaglioni, G. Zaccanti, L. Pantani, and L. Stefanutti, "An approximate procedure to isolate single scattering contribution to lidar returns from fogs," Int. J. Remote Sensing **4**, 399–417 (1983).
25. R. J. Allen and C. M. R. Platt, "Lidar for multiple backscattering and depolarization observations," Appl. Opt. **16**, 3193–3199 (1977).

26. K. Stelmaszczyk, M. Dell'Aglio, S. Chudzyn′ski, T. Stacewicz, and L. Wöste "Analytical function for lidar geometrical compression form-factor calculations", J. App. Opt. **44**, No. 7 (2005).
27. H. Kuze, H. Kinjo, Y. Sakurada, and N. Takeuchi "Field-of-view dependence of lidar signals by use of Newtonian and Cassegrainian telescopes", J. App. Opt.37, No. 15, (1998).
28. Yasser Y. Hassebo, B. Gross, M. Oo, F. Moshary, and S. Ahmed "Polarization discrimination technique to maximize LIDAR signal-to-noise ratio for daylight operations", J. App. Opt. **45**, No. 22 (in press) (2006).

Optical Protection Filters for Harmful Laser Beams and UV Radiation

Osama A. Azim M.

Vacuum Coating Lab. Manager, Arab International Optronics Co., El-Salam City, P.O. Box: 8182 Nassr City, Cairo 11491, Egypt.
E-Mail : osama_azim@intouch.com

Abstract. Due to the rapid growth of radiation protection applications in various devices and instruments, it is essential to use suitable filters for eye protection of the personal working in the radiation field. Different protection filters were produced to protect from four laser beam wavelengths (at 532nm, 632.8nm, 694nm and 1064nm) and block three UV bands (UVA, UVB, and UVC). The design structure of the required dielectric multilayer filters used optical thin film technology. The computer analyses of the multilayer filter formulas were prepared using Macleod Software for the production filter processes. The deposition technique was achieved on optical substrates (Glass BK-7 and Infrasil 301) by dielectric material combinations including Dralo (mixture of oxides TiO_2/Al_2O_3), and Lima (mixture of oxides SiO_2/Al_2O_3); deposition by an electron beam gun. The output transmittance curves for both theoretical and experimental values of all filters are presented. To validate the suitability for use in a 'real world', rather than laboratory test application, full environmental assessment was also carried out. These filters exhibited high endurance after exposing them to the durability tests (adhesion, abrasion resistance and humidity) according to military standards MIL-C-675C and MIL-C-48497A.

INTRODUCTION

Laser Radiation: Exposure to laser beams in the visible (400nm-700nm) and near-infrared (700nm-1400nm) regions of the spectrum may damage the retina. Laser beams in this region are readily transmitted by the eye and focused by the lens to produce an intense concentration of light energy on the retina. The incident exposure on the cornea can be concentrated by a factor of approximately 100,000 times at the retina due to this focusing effect. This energy is converted to heat and may cause a retinal burn resulting in visual loss or even blindness if the optic nerve is damaged. Even low energy laser beams, if concentrated by a factor of 100,000, can cause damage to the eye. For this reason, wavelengths in the 400 nm to 1400 nm range are termed 'the ocular hazard region'. Exposure to the skin from laser beams in the visible and infrared regions may cause photosensitive reactions, skin burns and excessive dry skin. Lasers operating in the visible (VIS) region of the spectrum include Ruby, Neodymium: YAG (doubled), Helium-cadmium, Helium-neon, Argon, and Krypton. Lasers operating in the Near Infra-Red (NIR) region include Neodymium: YAG, Gallium arsenide, and Helium-neon [1-3].

The objective of this paper is to set up and explain the method of producing seven protection filters by using the techniques of optical thin film and the optical properties of the substrates and evaporation materials. These filters can be used in different applications in the civilian and military fields. Five filters are designed for blocking four laser beam wavelengths (Frequency doubled Nd:YAG Laser at 532nm, Helium Neon Laser at 632.8nm, Ruby Laser at 694nm and Nd:YAG Laser at 1064nm). Another Two Filters are designed for blocking three bands of UV radiation [UVA (315-400 nm,), UVB (280-315 nm), and UVC (100-280 nm)].

For Humans, UVB is typically the most destructive form of UV and is not completely absorbed in the atmosphere. It has enough energy to cause photochemical damage to cellular DNA. UVB effects include erythema, cataracts, and exposure can also result in the development of skin cancer. Individuals working outdoors are at greatest risk from the effects of UVB. UVC, far UV and vacuum UV are almost never observed in nature because they are completely absorbed by the atmosphere. Germicidal lamps have been specifically developed to emit UVC because of its ability to kill bacteria. For humans UVC exposure in moderate doses can be absorbed by the outer dead layers of the skin. However, accidental exposure to high doses can cause corneal burns (e.g., welders' flash, snow blindness) or severe sunburn. Although documented evidence suggests that UVC injuries usually clear up in a day or two, they can be extremely painful [3, 4]. Clearly there are risks to humans associated with exposure to various wavelengths of UV radiation.

Optical coatings can be constructed of one through to hundreds of thin layers. Varying the material and thickness of each layer determines the amplitude and phase of the transmitted and reflected light at each interface. By creating regions of constructive and destructive interference, one can adjust the transmittance and reflectance of

CP888, *Modern Trends in Physics Research*,
Second International Conference on Modern Trends in Physics Research—MTPR-06,
edited by L. El Nadi
© 2007 American Institute of Physics 978-0-7354-0354-9/07/$23.00

the optical coating as a function of wavelength [5, 6]. The evaporation processes in vacuum are the prevalent techniques used to get solid materials into a vapor form. Regardless of the deposition technique used, it is necessary in the development of coating equipment and coating processes to have the ability to determine the physical and optical properties of the resultant films. There is a considerable range of sophisticated equipment available to do this in a precisely controlled method in order to achieve the accuracy that is required [7].

In this study, the optical substrates were glass BK-7 [8] and Infrasil 301[9]. The dielectric materials used were Dralo (absorbing part of UV radiation bands) and Lima (mixture of oxides SiO_2/Al_2O_3) (non-absorbing UV radiation bands) [10, 11].

In addition to the required radiation control performance of output filters, adherence to the Standard Specifications MIL-C-675C [12] and MIL-C-48497A [13] was assessed to ensure the products were usable in practical 'real world' applications.

EXPERIMENTAL WORK

The dielectric multilayer deposition of each filter [14] was achieved under High Vacuum Chamber conditions. The equipment used was Type A700QE-Leybold Optics, Germany pressure up $8x10^{-6}$ mbar could be attained. The selected substrate was optical glass type (BK-7) and Infrasil 301 with 1mm thickness. Its surfaces were prepared by very fine polishing and cleaning to obtain high quality filter. The required temperature of the substrate before the deposition of the multilayer stack inside the vacuum chamber was 280°C. The evaporation materials were Dralo as high refractive index material (H) and Lima as low refractive index material (L). The evaporation technique utilized an Electron Beam Gun (EVS-14). The adjusted evaporation rates were 0.5nm/sec for (Dralo), 0.8nm/sec for (Lima). The rate and final layer thickness were controlled by Quartz Crystal deposition Monitor (Type Model IC/5). During the evaporation of the materials, the reactive process with oxygen O_2 was carried out at pressure $2.5x10^{-4}$ mbar. The experimental curves of the transmittance versus the wavelength were obtained by Spectrophotometer 900 (Perkin Elmer, Germany) in spectral range (180nm-3000nm). The correction factor of the machine error (during deposition processes) for the experimental values can be adjusted. This was calculated and applied to achieve matching between the output and the designed values.

RESULTS AND DISCUSSION

The dielectric multilayer system design [15] is based on alternating high and low refractive index layers (H) and (L). The filter of the "stop-band" or area of high reflectivity (HR) is created that is centered on the design wavelength (λ_o). The most basic HR design has been optical thickness of the individual layer equals a quarter of the design wavelength, or a Quarter Wave Optical Thickness (QWOT). The design in its basic form resembles Fig. 1.

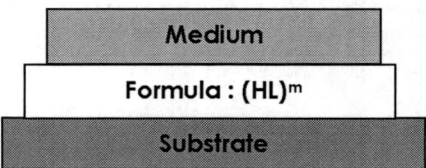

FIGURE 1. Medium/(HL)^m/substrate, where H is high refractive index materials, L is low refractive index materials and m is the number of periods of multilayer stack.

The computer analysis results were presented in seven multilayer filters. The theoretical results of each multilayer filter design being calculated by Macleod Optical Coating Design Software [16]. The design formulas of multilayer stack were prepared by the input data of the substrate type, the evaporation materials (H- Dralo and L- Lima), angle of incident and the center wavelength design (λ_o). The value of the angle of incident was zero degree (normal incident) for the seven design filters. By the series of multiple calculations, multilayer formulas for filters were optimized to achieve the blocking for harmful laser beams wavelengths and UV bands.

Laser Protection Filters:

The deposition of the multilayer processes and four laser beam wavelengths of five laser protection filters on glass BK-7 is shown in the chart Fig. 2.

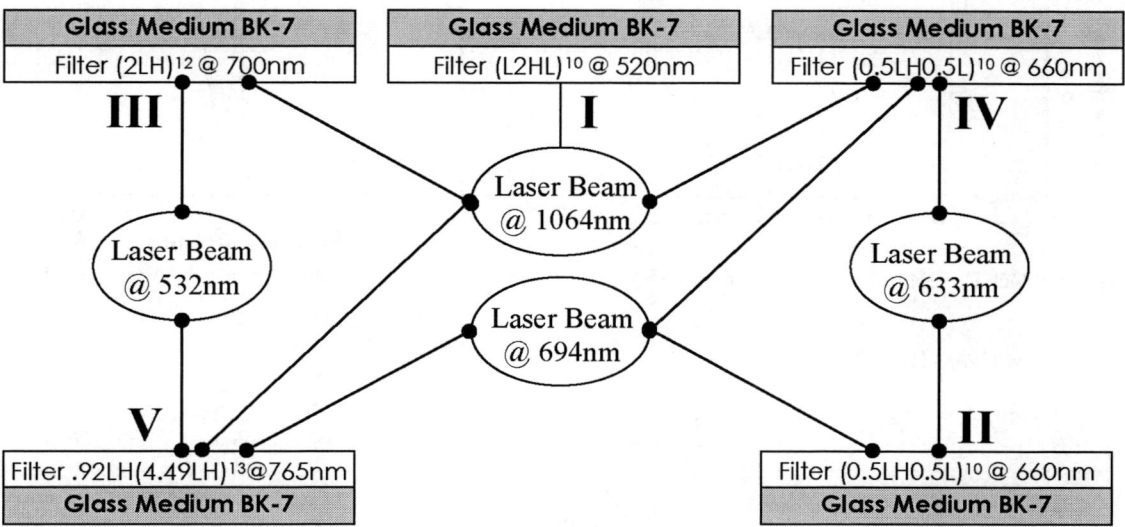

FIGURE 2. The block diagram for four laser beams and five protection filters

The theoretical and experimental transmittance values of the output filters were discussed. For the first five filters, the main analysis was at the laser beam wavelengths and VIS/NIR bands.

The first filter (I):

Fig.(3a) shows the theoretical curve of the transmittance versus the wavelength $T(\lambda)$ of the filter. The multilayer stack of the filter contains 21-layers and its structure $[(L2HL)^{10}]$ at λ_o=520nm. It was deposited on one side of the substrate BK-7. This design achieved $T(\lambda)$ value equal to zero for blocking one laser beam wavelength at 1064nm. Fig.(3b) shows the experimental curve $T(\lambda)$ of the filter (**I**). It shows that the $T(\lambda)$ value was 0.075% at the 1064nm and the average transmittance (T_{avg}%) value was 90.38% in VIS/NIR bands (400nm-825nm).

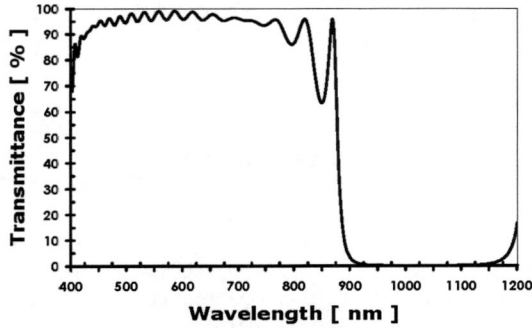

FIGURE (3a) Theoretical curve $T(\lambda)$ for filter (I)

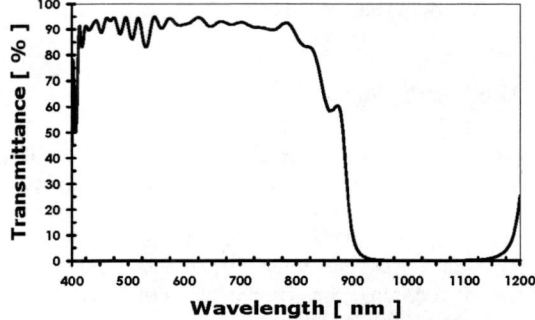

FIGURE (3b) Experimental curve $T(\lambda)$ for filter (I)

The second filter (II):

Fig. (4a) shows the theoretical curve $T(\lambda)$ of the filter. The multilayer stack contains 21-layers and its structure $[(0.5LH0.5L)^{10}]$ at λ_o=660nm. It was deposited on one side of the substrate BK-7. This design protects from two laser beam wavelengths by achieving $T(\lambda)$ value equal to 0.077% at 632.8nm and 694nm. Fig.(4b) shows the experimental curve $T(\lambda)$ of the filter (II). It shows that the $T(\lambda)$ values were 0.133% at 632.8nm and 0.068% at

694nm. The achieved values of T_{avg}% were 89.78% in the VIS band (400nm-560nm) and 82.45% in the NIR band (900nm-1200nm).

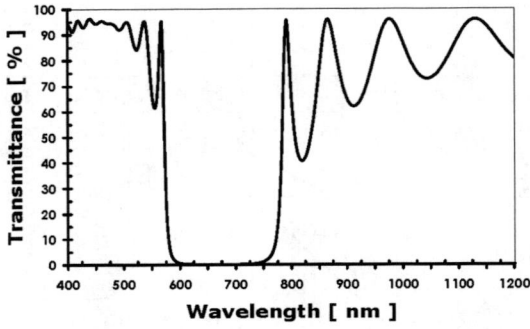

FIGURE (4a) Theoretical curve $T(\lambda)$ for filter (II)   FIGURE (4b) Experimental curve $T(\lambda)$ for filter (II)

The third filter (III):

Fig.(5a) shows the theoretical curve $T(\lambda)$ of the filter. The multilayer stack contains 24-layers and its structure $[(2LH)^{12}]$ at λ_o=700nm. It was deposited on one side of the substrate BK-7. This design protects from two laser beam wavelengths by achieving $T(\lambda)$ values equal to 0.039% at 532nm and 0.063% at 1064nm. Fig.(5b) shows the experimental curve $T(\lambda)$ of the filter (III). It shows that the $T(\lambda)$ values were 0.066% at 532nm and zero at 1064nm. The achieved values of T_{avg}% were 86% in the VIS band (400nm-475nm) and 66.47% in the VIS/NIR band (580nm-925nm).

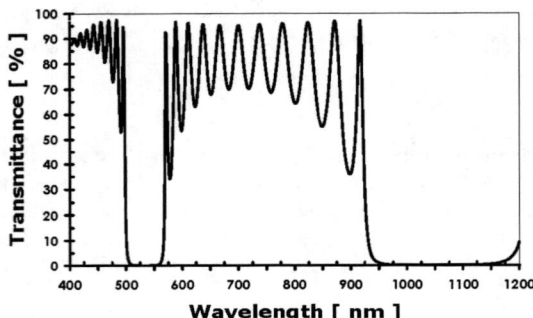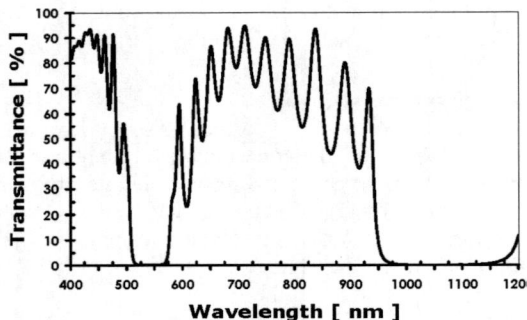

FIGURE (5a) Theoretical curve $T(\lambda)$ for filter(III)   FIGURE (5b) Experimental curve $T(\lambda)$ for filter (III)

The fourth filter (IV):

Fig.(6a) shows the theoretical curve $T(\lambda)$ of the filter. The multilayer stack contains 42-layers with its structure $[(L2HL)^{10}/BK-7/(0.5LH0.5L)^{10}]$. It was produced by 21-layers of filter (I) on one side of the substrate and 21-layers of filter (II) on other side. This design protects from three laser beam wavelengths by achieving $T(\lambda)$ values equal to 0.077% at 632.8nm, 0.078% at 694nm and 0.075% at 1064nm. Fig.(6b) shows the experimental curve of $T(\lambda)$ of the filter (IV). It shows that the $T(\lambda)$ values were 0.126% at 632.8nm, 0.071% at 694nm and zero at 1064nm. The achieved values of T_{avg}% were 86.45% in the VIS band (400nm-575nm) and 44.71% in the NIR band (790nm-980nm).

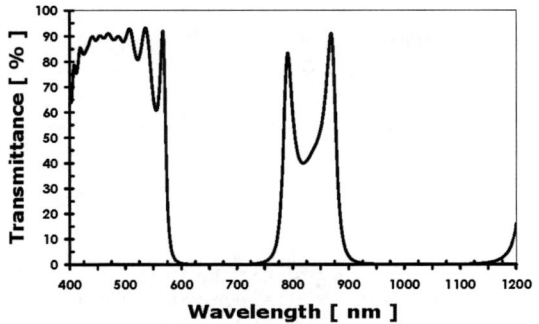

FIGURE(6a) Theoretical curve T(λ) for filter (IV)

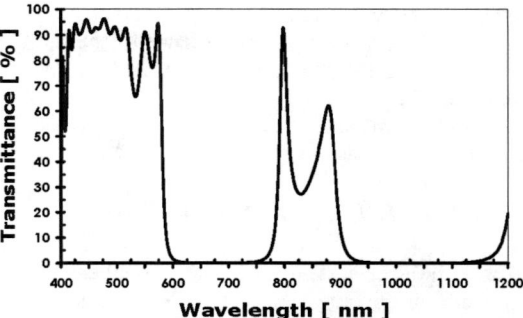

FIGURE(6b) Experimental curve T(λ) for filter(IV)

The fifth filter (V):

Fig.(7a) shows the theoretical curve T(λ) of the filter. The multilayer stack contains 28-layers and its structure [.92LH(4.49LH)^{13}] at λ_o=765nm. It was deposited on one side of the substrate BK-7. This design protects from three laser beam wavelengths by achieving T(λ) values equal to 0.031% at 532nm, 0.013% at 694nm and 0.016% at 1064nm. Fig.(7b) shows the experimental curve T(λ) of the filter (V). It shows that the T(λ) values were 0.083% at 532nm, 0.023% at 694nm and zero at 1064nm. The achieved values of T_{avg}% were 67.93% in VIS band (400nm-500nm), 79.74% in same band (550nm-650nm) and 51.92% in the NIR band (850nm-950nm).

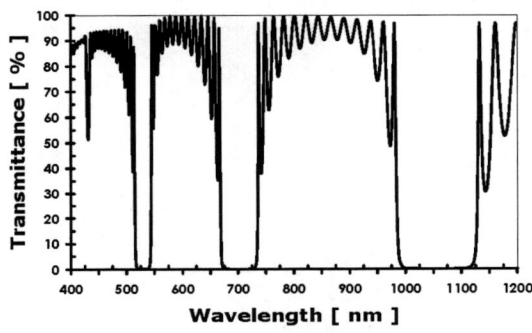

FIGURE (7a) Theoretical curve T(λ) for filter (V)

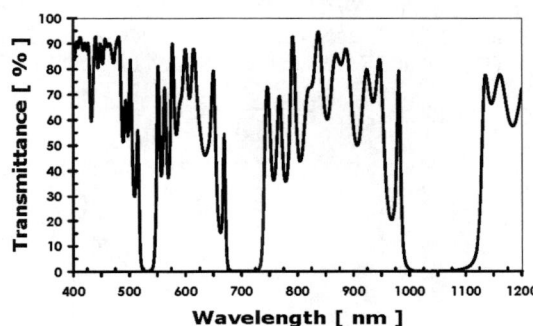

FIGURE (7b) Experimental curve T(λ) for filter (V)

UV Protection Filters:

To produce the UV protection filters, the property of absorbing UV radiation in the substrate BK-7 [17] and the

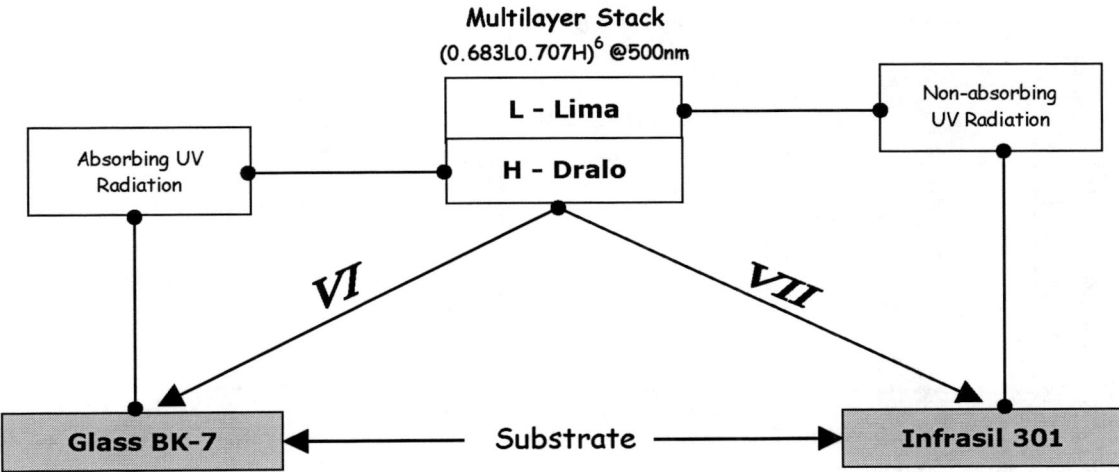

FIGURE 8. The block diagram for two UV protection filters with two different substrate

217

evaporation high index material Dralo were applied. The achieved structure of two UV protection filters depended on the optical interference of the multilayer and the properties of the UV absorbing in evaporation materials. The deposition processes of two UV Protection Filters on glass BK-7 and Infrasil 301 are shown in the chart Fig. 8.

The theoretical and experimental transmittance values of the two filters were discussed. For the sixth and seventh filters, the main analysis was at the three bands of UV.

The sixth (VI) & Seventh (VII) filters:

The multilayer system design of two filters was designed to protect from three UV bands depending on the property of the optical glass BK-7 and Dralo material. It contains 12 layers and its structure $[(0.683L0.707H)^6]$ at λ_o=500nm. It was deposited on two substrates (glass BK-7 and Infrasil 301). The multilayer stack was designed to block the UVA and UVB bands, and using the absorbing property of the optical glass BK-7 and Dralo material to block UVC band.

Fig.9a shows the theoretical curves $T(\lambda)$ for filters (VI) & (VII) (without any effect) at spectral band from 200nm to 800nm. The two curves show that the transmittance values due to the deposition of the system design on glass BK-7 and on Infrasil 301 are close to each other. Fig.9b shows the experimental curve $T(\lambda)$ for UV protection Filter (VI) on Glass BK-7. The absorbing property of (glass BK-7+ Dralo) affected on the transmittance values of UVC band. The achieved values of $T_{avg}\%$ were 0.41% in UVA band, 0.031% in UVB band, zero in UVC band and 71.2% in the VIS band (400nm to 720nm). Fig.9c shows the experimental curve $T(\lambda)$ for UV protection Filter (VII) on Infrasil 301. The absorbing property of the Dralo material affected on the transmittance values of UVC band. The achieved values of $T_{avg}\%$ were 0.28% in UVA band, 0.044% in UVB band and zero in UVC band and 73.4% in the VIS band (400nm to 720nm).

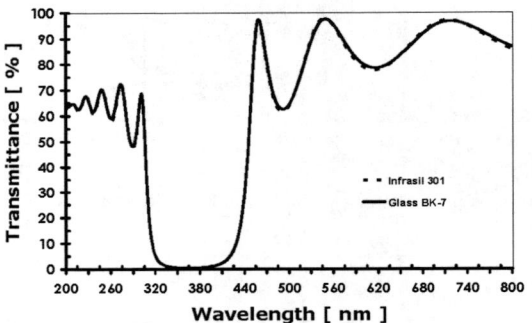

FIGURE (9a). Theoretical curves $T(\lambda)$ of Filter (VI & VII) design on Glass BK-7 and Infrasil 301.

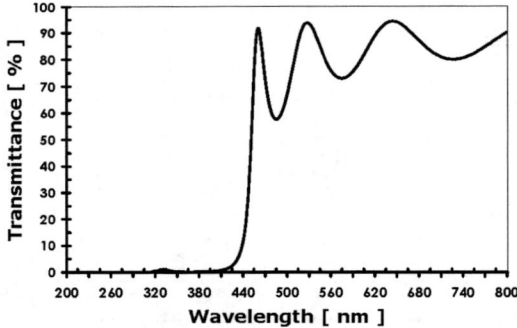

FIGURE(9b). Experimental curve $T(\lambda)$ for UV protection filter (VI) on BK-7 substrate.

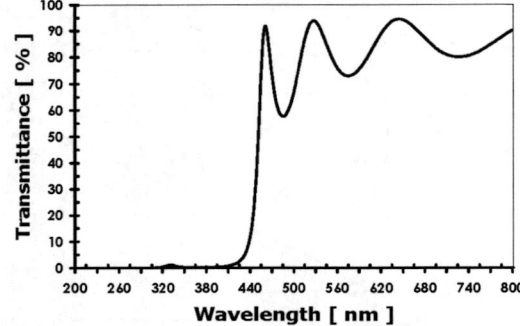

FIGURE(9c). Experimental curve $T(\lambda)$ for UV protection filter (VII) on Infrasil 301 substrate.

For measuring the robustness of the coated surfaces, the standard durability tests (adhesion, abrasion resistance and humidity) were carried out on the output filters. The tests were applied according to military standard MIL-C-675C and MIL-C-48497A. The results showed that the processes of optical multi-layer filters have excellent environmental endurance over the four standard periods (0.1, 24, 200, and 800 hours). This proves the suitability of Lima and Dralo for use in 'real world' applications including the industrial environments which opens up new practical applications in the thin film technology.

CONCULSION

For laser protection filters, theoretical and experimental results achieved the main target for producing laser rejection filters to block one or more designated wavelengths (at 532, 632.8nm, 694nm and 1064nm) while providing the highest possible transmittance between the rejected wavelengths. T_{avg}% value of the produced filters at each laser beam wavelengths was ≤ 0.14% and at the visible band was ≥ 90% for first filter and ≥ 51.5% for other four laser protection filters.

For UV protection filters, the multi-layer processes and the optical properties of glass BK-7 and the optical material (Dralo) were used to block three bands of UV radiation (UVA, UVB and UVC) in spectral range from 200nm to 400nm. They were provided the highest possible transmittance in the visible band. The blocking of the three UV bands was achieved. T_{avg}% value (at three UV bands) was achieved 0.23% for filter (VI), 0.15% for filter (VII) and ≥ 74% in visible band for both filters.

All results confirm the protection from four harmful laser beam wavelengths and UV radiation with suitable visibility for the users during their works in the day light. Under hard durability tests, the result showed that the multi-layer filter applications have excellent environmental endurance according to military specifications MIL-C-675C and MIL-C-48497A.

REFERENCES

1. International Commission on Non-Ionizing Radiation Protection (ICNIRP), Laser, **79**, 431 (2000).
2. MIL-HDBK-828A, Handbook for laser safety on ranges and in other outdoors areas, Dept of Defense, USA, (1996).
3. UC Davis-Office of environmental Health & Safety, SafetyNets #73, "Laser Protection Eyewear", Rev. (02/2003), and SafetyNet #106, "Hazards of Ultraviolet Radiation"), Rev. (05/2005).
4. ICNIRP, Guidelines on Limits of Exposure to Ultraviolet Radiation of Wavelengths Between 180 nm and 400 nm, Health Physics **87**, 171 (2004).
5. H.A. Macleod "Thin-Film Optical Filters", 3rd Edition, Adam Huger Ltd., Bristol, (2001).
6. Sophocles J. Orfanidis "Electromagnetic Waves and Antennas", Multilayer Structures, Rutgers University, 94 Brett Road Piscataway, NJ 08854-8058, (2004).
7. James D. Rancourt, Optical Thin Films: Users Handbook, Society of Photo Optical, (1996).
8. Schott optical Glass Co., "Main Properties of optical glasses, Catalogue, 2000, pp.3111.
9. Heraeus Optics Co., "Main properties of Infrasil", Catalogue (1999-2003).
10. Selhofer, Hubert Müller, René, Thin Solid Films **351**, **180** (1999).
11. Umicore Materials AG., Schlossweg 11, FL-9496 Balzers, Catalogue (2003).
12. MIL-C-675C, (Military Specifications), Coating Of Glass Optical Elements, (1982).
13. MIL-C-48497A, (Military Specifications), Coating, Single, Multilayer, Interface and Durability Requirements, (1982).
14. H.K. Pulker, "Coating On Glass", Elsevier Science; 3rd edition, (1996).
15. R.R. Willey, "Practical Design and Production of Optical Thin Films", Optical Engineering, (Ed. B Thompson), Marcel Dekker Inc., New York, (1996).
16. H.A. Macleod, "Optical Coating Design Program", **86**, 2003, Thin Film Center Inc., 2745 E Via Rotonda, Tucson, AZ, USA.

Biological Studies On The Effect of Laser Radiation on Khapra Beetle Trogoderma granarium (Coleoptera : Dermestidae)

Adel M. El-Nozahy, Salwa M. M. S.Ahmed*, Mahomoud H. Abdel-Kader., Ibtesam A. Khalifa**

Cairo University, National Institute of Laser (NILES)
**Agriculture Center, Plant Protection Research Institute, Cairo*

ABSTRACT. The aim of the present work was to study the effect of Argon-ion laser and carbon dioxide laser radiation on Khapra beetle *Trogoderma granarium* and induced sterility. Radiobiological effects of irradiation were determined on adult stage of resulted 2-3 days-old pupae at LD_{30}. The radiobiological studies induced determination of mortality, of, LD_{30}, LD_{50} emergence , preovipositio period, fecundity, sterility, incubation period , larval duration, pupal duration and emergence of 1st generation. Experiments were carried out to determine the latent effect of irradiation on the wheat grains germination as well as the effects on the chemical constituents. In this respect irradiation of grains had no effect on the above ntioned parameters.

INTRODUCTION

Some investigators have studied the effect of laser radiation against different insect species [1-4]. also, the radiobiological studies investigated against stored grain insects [5] the effects Ruby laser caused death *T. versiolor* [6];irradiated late instar laravae of *T.inclusum and T.variabile* with different energy density and wavelength doses from each of three laser (Ruby laser 1.6-9.5 j /cm^2) and neodymium (4.3-24.9 J/cm^2) lasing killed the darker *T.variabile* larvae at lower doses than those required to kill the polar *T.inclusum* larvae. Carbon dioxide (2.1-12.7 J/cm^2) lasing killed larvae of both species at intermediate and high doses. The response of male female gonads of *Ephestia cautella* to He/Ne laser was examined by Alhakkak et al [7],The effects of laser radiation on wheat grains were observed by Dvernyakove et al., [8] Hassan [9] the objective of the present study is determine the biological and effects of two different lasers (Argon - ion and Carbon dioxide) dosage of $LD_{30 \text{ of the}}$ *T. granarium(* F_1) originally resulted from treated and untreated pupae.

MATERIALS AND METHODS

Laser sources

Argon ion laser and Carbon dioxide laser were used at 488 nm and 10.6μm respectively. Argon ion laser radiation by different energy dose level (200, 400, 800, 1200, 1600 and 2000 mJ) but CO_2 laser radiation energy dose level are (20,80,140,200and 260 mJ).The laser radiation employed in the present study was in continuous flew and energy, and the dosage was determined as a function of time.

Insects and exposure

Test cultures of *T. granarium* were obtained from laboratory strains maintained by the stored product insects research institute, Agriculture Research Center. It was maintained at (32 ± 2) °C and (65 ± 5)% R.H. Rearing medium of grains was adopted to Selium [10]. 2-3 days old pupae were randomly collected into 14 groups, seven groups of each sex (50 pupae in each tube) six groups of each sex were individually irradiated with Argon laser radiation by the above afferemintiad dosage, the remnant groups were kept as control. Four replicates were made for each dose and control. In case of CO_2 laser 12 groups

CP888, *Modern Trends in Physics Research*,
Second International Conference on Modern Trends in Physics Research—MTPR-06,
edited by L. El Nadi

of pupae, six groups of each sex (50 pupae in each tube) five groups of each sex were individually irradiated by above afferemintiad dosage. The remnant groups were kept as control they are all groups were replicated 4 times .Emerged adults were daily recorded until no more adults could emerge and emerged adults were checked the number of normal and abnormal adults eclosed .

In the two laser types, pupal mortality was assessed and obtained data were corrected according to Abbott's formula [11] and the reduction in adults emergence was calculated according Brigg's methods [12].

For LD_{30}, four different pair mating crosses were made between young adults of normal appearance to follow the effect of laser radiation in the next generation and study adult fecundity, fertility and to determining the sterility [13].

$$\% \text{ Calculated Sterility} = 100 - (\frac{axb}{AxB} \text{ x } 100)$$

Where a = No. of eggs/female for treatment, A= No. of eggs/female for control, b= % hatch of eggs/female for treatment B= % hatch for control

Also, the incubation period and the developmental periods of the immature stages and the changes in wet body weight was recorded.

RESULTS AND DISCUSSION

The results Presented in table (1) and (2) show that increasing the dose of the argon and CO_2 lasers radiation produced increasing mortalities of exposed 2 – 3 days old *Trogoderma granarium* pupae of both sexes this effect was accompanied with gradual increases in the percentage of adults eclosed with malformed body and wings. Also table (1) demonstrates LD_{50} and LD_{30} with 578 and 347 mJ for male pupae and 525 and 311 mJ for female pupae, respectively for argon laser.

TABLE 1. Effect of different doses of argon laser radiation (488 nm) on some biological aspects of male and female of *T. granarium* pupae*(2-3 days old)

Energy Dose (mJ)	Males						
	Pupal mortality %	Corrected pupal mortality %	No. of emergenced adults %	Adult emergence %	Reduction of adult emergence %	No. of malformed adults	Malformed adults %
Control	3.00	--	194	97.00	--	0	0
200	20.00	17.53	160	80.00	17.53	1	0.63
400	40.50	38.66	119	59.50	38.66	2	1.68
800	51.00	49.48	98	49.00	49.48	2	2.04
1200	65.50	64.43	69	34.50	64.43	4	5.80
1600	81.50	80.93	37	18.50	80.93	4	10.81
2000	97.00	96.91	6	3.00	96.91	1	16.67
Energy Dose (mj)	Females						
	Pupal mortality %	Corrected pupal mortality %	No. of emergenced adults %	Adult emergence %	Reduction of adult emergence %	No. of malformed adults	Malformed adults %
Control	0	--	200	100	--	0	0
200	21.00	21.00	158	79.00	21.00	4	2
400	43.00	43.00	114	57.00	43.00	6	5.26
800	53.00	53.00	94	47.00	53.00	7	7.45
1200	70.00	70.00	60	30.00	70.00	8	13.34
1600	83.00	83.00	34	17.00	83.00	7	20.58
2000	97.00	97.00	6	3.00	97.00	2	33.34

*200 pupae were tested at each dose

LD_{50} 578 mJ LD_{50} 525 mJ

LD_{30} 347 mJ LD_{30} 311 mJ

In table (2), the calculated LD$_{50}$ and LD$_{30}$ for male and female pupae were 116,62, 100 and 50 mJ, respectively. These results are in harmony with those recorded on the fig moth, *Ephestia cautella* 1 [7] *Culex pipiens* [14] and on the house fly, *Musca domestica*. [4]

TABLE 2. Effect of different doses of CO$_2$ laser radiation (10600 nm) on some biological aspects of male and female of *T. granarium* adults treated as pupae* (2-3-days old).

Energy Dose (mJ)	Males						
	Pupal mortality %	Corrected pupal mortality %	No. of emerged adults	Adult emergence %	Reduction of adult emergence %	No. of malformed adults	Malformed adults %
Control	4.0	--	192	96.0	--	0	0
20	15.0	11.45	172	85.0	11.45	3	1.7
80	30.0	27.08	140	70.0	27.08	4	2.86
140	39.5	36.98	121	60.5	36.98	6	5.00
200	60.5	58.85	79	39.5	58.85	6	7.59
260	92.0	91.67	16	8.0	91.67	2	12.5
Energy Dose (mJ)	Females						
	Pupal mortality %	Corrected pupal mortality %	No. of emerged adults	Adult emergence %	Reduction of adult emergence %	No. of malformed adults	Malformed adults %
Control	1.00	--	198	99.0	--	0	0
20	20.00	19.19	160	80.0	19.19	6	3.75
80	31.50	30.81	137	68.5	30.81	8	5.84
140	38.00	37.37	124	62.0	37.37	10	8.06
200	58.00	57.58	84	42.0	57.58	12	14.29
260	95.00	94.95	10	5.0	94.95	2	20.00

*200 pupae were treated at each dose
LD$_{50}$ 116 mj   LD$_{50}$ 100 mj
LD$_{30}$ 62 mj   LD$_{30}$ 50 mj

Based on the LD$_{50}$ values it is clear CO$_2$ > argon.in this respect it is worth mentioning that CO$_2$ laser is poorly absorbed be tissue pigments and highly absorbed by tissue water. The absorption is so efficient that 98% of the incident energy is absorbed within 0.01 mm of the impact point. These causes rapid localized heat produced immediate boiling and vaporization of intercellular water within the tissue. On the other hand, argan laser possessed on the lowest absorption coefficient for water.

However, its beam is preferentially absorbed by tissue pigments such as chromophoresmeanin and other dark pigments as recorded by [15]. The differential susceptibility between male and female of *Trogdirma granrium* to laser may be due to the different body size. the average body size of the female is approximately three times that of male, hence it is presumed that the total laser energy deposited is correspondingly more in females than makes [15, 16] the abnormalities included failure of complete adult appendage >according to the finings of wiggles worth [17] who reported that the dorsal longitudinal muscles in the anterior portion of the abdomen of *Trogoderma granarium* laravae was damaged by laser irradiation. The biological parameters of the LD$_{30}$ of *T.grangrium* irradiated by argon and CO$_2$ Lasers examined after emergence of different mating crosses of F$_1$, are summarized in table (3).

TABLE 3. Some biological parameters for the different mating crosses with argon and CO_2 laser types at LD_{30} on the *T.granarum*

Meating crosses / Parameters	Argon Laser LD_{30} U♂ X U♀ control)	T♂ X U♀	U♂ X T♀	T♂ X ♀	Carbon Dioxide laser LD_{30} U♂ X U♀ control)	T♂ X U♀	U♂ X T♀	T♂ X ♀
Average preovipostion period (d)	1.25±0.25	1.25±.75	1.50±.29	1.75±0.25*	1.50±.29	1.50±29	1.50±.29	1.75±0.25*
Average no. of eggs Female / day fecundity)	29.3±0.29	26.3±0.3	23.4±0.2	15.4±0.30	25.8±0.5	20.1±2.40	24.6±5.5	12.9±0.87
Average incubation period (d)	5..50±0.17	5.00±0.24	5.33±0.17	4.89±0.20*	5.20±0.13	5.33±0.17	5.22±0.15	4.71±0.42*
Average larva duration (d)	30.00±1.86	32.1±0.99	31.0±2.16	36.6±2.04*	29.6±1.17	29.5±1.65	30.9±0.97*	33.3±048
Average no. of llarval in stars	6.22±0.15	6.00±0.00	5.50±0.34*	5.75±0.48	6.00±0.17	5.83±0.31	6.00±0.27	5.75±0.25*
Average pupal duration (d)	3..89±0.45	3.57±0.20	3.83±0.31	5.00±0.32*	4.33±0.53	3.33±0.21*	31.20±0.37	3.75±0.25*
Av erage total developmental immature stages (d)	39.6±1.86	40.4±0.84	40.0±2.34	46.2±2.15*	39.1±1.27	38.3±1.93	39.2±1.11	42.0±041*
Average udlult emergence rate %0	0.90±0.10	0.78±0.15	0.67±0.17	0.56±0.18*	0.90±0.10	0.67±0.17	0.56±0.18*	0.57±0.20*

The mean difference is significant at the 0.05 level

± SE : standard error, + : increase, – : decrease.

U : adults emerged from untreated pupae T : adults emerged from treated pupae

Data showed that the Pre-ovaposition period was significantly increased with mating cross of T ♂X ♀ with two types of laser. The fecundity of females was greatly reduced in the mating cross if T ♂X T ♀ for both laser types these results are in agreement to those obtained on *Tribolium brevicornis* [18]. This may also agree with the findings which explain this phenomenon as side effect of the visible laser radiation.

TABLE 4. Percentages of egg fertility and sterility resulting from different cross mating combinations between adults *T-granarium* resulted from treated and untreated pupae with LD_{30} of two Laser radiation types.

Mating crosses (LD_{30})	Argon Laser LD_{30} Fertility (% hatchability)	% of Reduction	% Observed sterlity	% calculated sterlity	Carbon dioxide Laser LD_{30} Fertility (% hatchability	% of Reduction	% Observed sterlity	% calculated sterlity
U♂ X U♀ Control	97.19 ± 0.61	——	2.81	——	97.50 ± 0.63	——	2.50	——
T♂ X U♀	92.79 ± 1.82 *	5.26	7.89	14.93	92.76 ± 1.57 *	4.86	7.24	25.88
U♂ X T♀	89.44 ± 2.58 *	7.97	10.56	26.50	91.26 ± 2.17 *	6.40	8.74	10.75
T♂ X T♀	82.39 ± 2.52 *	15.22	17.61	55.44	86.57 ± 2.73 *	11.21	13.43	55.56

The mean difference is significant at the 0.05 level

± : SE : standard error, + increase, – decrease.

T : adults emerged from treated pupae U : adults emerged from untreated upae

Particularly penetrates living tissues initiating photochemical reactions absorbed in approached molecules. the average incubation period was significantly shorter in the T ♂ X ♀ than for the other mating crosses irradiated by two laser types .

223

On the other hand, the larval duration was significantly increased when treatment carried out with two types of lasers treatment and the *Musca domestica* came to the same conclusion [4].

Our results clearly indicate that, the average number of larval instars was significantly lower when both sexes of *Togaderma granarium*. The percentages of water release from pupae due to exposure of males and females were statistically significant, when treated carried out with CO_2 laser than orgon laser, this due to laser thermal effects causing evaporation of pupal body water and these results agreed with the reported that one of the thermal effects of laser on living tissue is evaporation which occurs when the temperature of soft tissue is raised up to 100 C [19,20] and also very good agreement with the results of irradiated *Musca domestica* puae with LD_{30} CO_2 and Nd:YAG [4].

TABLE 5. Effect of laser radiation on body water loss of treated *T-granarium* pupae (25 pupae in 4 replicates were used in each laser type).

Laser type (LD_{30})	Female						Male					
	Weight of pupae before irradiation (mg)		Weight of pupae after irradiation (mg)		Water loss %		Weight of pupae before irradiation (mg)		Weight of pupae after irradiation (mg)		Water loss %	
	Range	Mean ±SE	Range	Mean ±SE	Range	Mean ±SE	Range	Mean ±SE	Range	Mean ±SE	Range	Mean ±SE
Argon Laser	73.0-81.1	78.3±1.80	70.1-80.1	76.9±2.29	0.88-3.97	1.84±0.72	27.8-30.8	29.6±0.68	27.4-30.6	29.2±0.69	0.65-1.64	1.19±0.22
Carbon dioxide Laser	82.7-94.8	89.6±2.88	74.8-88.2	82.6±3.04*	6.90-9.55	7.77±0.62	33.4-38.0	35.9±0.98	32.1-36.7	33.8±1.01*	3.53-8.70	5.93±1.30

REFERENCES

1. W.K. Turner; P.S. Callahanand; F.L Lee, *Annals of the Entomological society of America* **70**, 234 (1977).
2. S Matolin, *Acta-Entomologica Bohem oslovaca* **76**, 93 (1979).
3. Z.Y. LI and Y.K. Lin, *Annu. Rep. Of Guangzhou parasit. Soc.* **7**, 188 (1985).
4. A.A.M .Abd El-Sadek, , " Effect of laser radiation on the biotic and reproductive potential of the House Fly, *Musca domestica* L. " M. Sc. Thesis Faculty of Science, Ain Shams University, 1999.
5. W.H.A. Wilde, *Canadian Entomologist* **97**, 88 (1965).
6. R.W. Kobylnyk, and W.H.A. Wilde, *Canad. Entomol.* **105**, 323 (1973a).
7. Z.S. Alhakkak,.; A.A. Alsufi, and A.M.B. Murad, , J. Biol. Sci. Res., **19**, 95 (1988).
8. V.S Dvernyakov,.; Yu. P.Shalin,; V.L Tikush,.; Timoshenko, V.M.; Pasichnyi, V.V; Teplyakova, O.A. and Rusakov, G.V*Methods of producing breeding material of wheat.* USSR Patent, No. 869703. . (1981).
9. A.M. Hassan, " Biochemical genetic studies on some Environmental Tolerance plants." M. Sc. Thesis, Fac. of Agric. Cairo University 2000.
10. A.A Selim, "A comparative study on the effect of certain insecticides on cereals and certain grain pests." M. Sc. Thesis, Fac. Agric. Ain Shans University. 1963.
11. W.S. Abbott,. *J. Econ. Entomol.,*, **18**, 265 (1925).
12. J.D. Briggs,.. *J. Insect pathol.*, **2**, 418 (1960).
13. Z.A.A. Aly,." Combined effects of Gamma irradiation and insecticides on certain stored-product insects". Ph.D. Thesis, Fac. Sci. Ain Shams University, 1991.
14. T. Ikesoj, *Appl. Ent.zool.* **27**, 227 (1992).
15. F.K. Abdel-Kawy," Studies on the effect of gamma irradiation on the different developmental stages of khapra beetle *Trogoderma granarium* Evert." M.Sc Thesis, Fac of Agric. Ain Shams Univ. 1979.
16. G.C. Carney, *Nature*, **183**: 338 (1959).
17. V.B wigglesworth,. *Q. Ji micros. Sci.* **97**, 465 (1956).
18. M. Hassan *J. Nuclear Agric and Bio.* **27**, 149 (1998).
19. J.A.S Carruth, " The principles of laser surgery". Scott Brown's Otolaryncology. (Alan, G,K, Kerred.), 5th edition 1987 pp. 1 : 513-542.
20. B.M Achauer,.; Vanderkam, V.M. and Berns, M.W" Lasers in plastic surgery and dermatology". Thieme Medical publishers. Inc. New York., 1992, pp 206.

Biochemical Studies Of The Effect Of Two Laser Radiation Wavelengths On The Khapra Beetle Trogoderma Granarium Everts (Coleoptera : Dermestidae)

Mahmoud H. Abdel-Kader, Adel M. El-Nozahy, Salwa M. S.Ahmed, and Ibtesam A. Khalifa*

Cairo University, National Institute of Laser Enhanced Science (NILES)
*Agriculture Center, Plant Protection Research Institute, Cairo

ABSTRACT. The present work was carried out to evaluate the actual effect of subleathal dosage of LD_{30} of two different lasers (Argon-ion and CO_2 lasers) on the main metabolites, phosphatases enzymes, transaminases, acetylcholinestrase and peroxidases in the one day adult stage of *Trogoderma granarium* treated as 2–3 days old pupae. Our results clearly indicated that two different wavelengths of laser radiation increased significantly the total proteins content, whereas no significant changes occurred in the total lipids for the two laser radiation wavelenghts. On the other hand the total carbohydrates were significantly decreased when irradiating using CO_2 laser wavelength which is not the case for the Argon-ion laser radiation. Significant changes of phosphatases occurred for both wavelengths. Inhibition of transaminases GOT (glutamic oxaloacetic transaminases) and insignificant changes of GPT (glutamic pyruvic oxaloacetic transaminases) was observed for both laser wavelengths. Significant inhibition of acetyl cholinestrase was observed using CO_2 laser and insignificant changes were recorded for Argon ion laser radiation where as insignificant decrease of peroxideses was observed for both lasers.

INTRODUCTION

The Khapra beetle *Trogoderma granarium* is one of the most destructive pests of ceral grains allover the world, particularly in hot dry regions. The larvae feeds on many kinds of grains, seeds and other products of vegetable origin. The Khapra beetle *probably* is the most destructive stored product insect in upper Egypt[1]. Physiological and biochemical processes when radiating this insect with gamma, X-ray as well as microwave radiations were demonstrated by many investigators [2].

The present work was carried out to study some biochemical parameters, the effect of two sublethal dosage (LD_{30}) of Argon-ion and CO_2 laser radiations on the adults of *Trogoderma granarium* resulting from irradiating the 2 – 3 days old pupae.

The aim of the present work was to evaluate the effects of Argon-ion laser radiation at a wavelength of 0.488 micrometers and carbon dioxide laser radiation at a wavelength of 10.6 micrometers of *T. granarium* on adult treated as 2 – 3 days old pupae main metabolites (protein, lipids, carbohydrates), phosphatases (Alk. Phosphatases & Acid phosphatases), transaminases (GOT and GPT), acetyl cholinesterases and peroxidases.

MATERIALS AND METHODS

The Khapra beetle *Trogoderma granarium* was reared under laboratory conditions of (32 ± 2) °C and (65 ± 5)% R.H. To study the effect of argon and CO_2 laser radiation at LD_{30} sublethal dosage on main metabolites, phosphatases enzymes, transaminases, acetyl cholinesterase and peroxidases in the one day adult stage of *T. granarium* treated as 2 – 3 days old pupae with the above mentioned laser types at 347 mJ dose for the argon-ion laser and 62 mj dose for CO_2 laser The total proteins were measured photometrically at 595 nm and compared with a standard bovine serum albumin [3]. The total lipids were determined by colourimetric method. A sample of the whole body extract was heated with conc. sulfuric acid and the mixture was then reacted with phosphoric acid-vanillin reagent to give red to purple colour. The intensity of colour was measured by spectrophotometer at 525 nm [4].

The total carbohydrates were determined in acid extracts by the phenol sulphuric acid reaction [5]. The absorbance of characterisitic yellow – orange colour was measured at a wavelength of 490 nm.

Estimation of alkaline and acid phosphatase was assayed according to the methods described by Powell and Smith [6]. The amino-transferase GPT and GOT activities were measured through the use of the technique described by Reitman and Frankel [7].

The acetyl cholinesterase was measured according to the method described by Simpson [8]. The peroxidase activity was measured according to the method described by Vetter and Nelson [9]. The obtained results of biochemical

CP888, *Modern Trends in Physics Research*,
Second International Conference on Modern Trends in Physics Research—MTPR-06,
edited by L. El Nadi
© 2007 American Institute of Physics 978-0-7354-0354-9/07/$23.00

parameters of the insect were statistically analyszed adopting the ANOVA variance analysis using costat computer program.

RESULTS

1- Effect of two laser radiation types on the main metabolites (protein, lipid and carbohydrates)

Results presented in Table (1) show that the total protein content in one day old $T.$ $granarium$ was increased significantly at a level of 1% in the case of adults treated as pupae with argon and CO_2 lasers at LD_{30}. The total protein content was 17.82 and 16.62 mg/g.b.wt respectively as compared with 14.93 mg/g.b.wt. in the control. The total lipid content was insignificantly changed by 15.26% increasing with argon and 4.75% decreasing. Likewise insignificant effect on the total carbohydrates when treatments were carried with argon in laser on the other hand in the case of CO_2 laser treatments, it was decreased significantly. This value was 24.15 mg/g.b.wt compared with 34.12 mg/g.b.wt in the control.

TABLE 1. Effect of two laser radiation types on the main metabolites of one day old T. granarium adults   originally resulted from treated and untreated pupae at LD_{30}

| Laser types (LD_{30}) | Main metabolites (mg/g.b.wt.) | | | | | |
	Total proteins Mean ± SE	%	Total lipids Mean ± SE	%	Total carbohydrates Mean ± SE	%
Control	14.93 ± 0.24	-	36.82 ± 1.20	-	34.12 ± 1.14	-
Argon (347 Mj)	17.82 ± 0.36**	+19.36	42.44 ± 2.39^{ns}	+15.26	33.45 ± 0.68^{ns}	-1.69
CO_2 (62 mj)	16.62 ± 0.23**	+11.32	36.07 ± 0.83^{ns}	-4.75	24.15 ± 0.96**	-29.22

2-Effect of two laser radiation types on the phosphatases (Alkaline and acid phosphatases)

The statistical analysis of the data presented in Table (2) indicated that Alk-pase activity was significantly activated at the level of 5% when treatment was carried out with argon ion laser radiation, the percentage of increase was 8.61% over the control group, however, inhibition in the activity of this enzyme was observed when treatment was carried out with CO_2 laser, it was 43.05% and this inhibition in the activity was also statistically significant at 0.1% level.

Regarding to Ac-pase activity, the data showed that the activity of this enzyme was significantly inhibited at a level of 5%. The values of this inhibition were 28.82% and 27.3% than the control for argon group and CO_2 lasers, respectively

TABLE 2. Effect of two laser radiation types on the phosphatases enzymes (Ac-pase & Alk-pase) expressed as µg phenol released g.b.wt/minute of One day old T.granarium adults originally resulted from treated and untreated pupae at LD_{30}.

| Laser types (LD_{30}) | Phosphatases (µg phenol/min/g.b.wt) | | | |
	Alkaline phosphatase Mean ± SE	%	Acid phosphatase Mean±SE	%
Control	604 ± 13.96	-	145.66 ± 8.41	-
Argon (347 Mj)	656 ± 9.26*	+ 8.61	103.68 ± 4.97*	-28.82
CO_2 (62 Mj)	344 ± 4.17***	-43.05	105.88 ± 6.48*	-27.31

Deviations in the table indicate the standard errors of means
ns: non significant   ***:** significant at level 5%.   ****:** Significant at level 1%   *****:** significant at level 0.1%.
% Percentage activation   (+) or inhibition (-) than the control.

3-Effect of two laser radiation types on the transaminases (GOT, GPT)

Results are depicted in table (3). It indicates that there was significant inhibition of GOT, it was 17.06 and 26.81% for the argon laser treatment and the CO_2 laser treatment respectively. With repsect to GPT, the activity in the two types of laser radiation at LD_{30} showed insignificant activation when treatments were carried with argon ion laser, however, the activity was inhibited in the case of CO_2 laser treatment.

The GPT activity was activated (+ 2.30%) in adults treated as pupae with argon ion laser and in the case of treatment with CO_2 laser the GPT activity was inhibited by (12.64%).

TABLE 3. Effect of two laser radiation types on the transaminases enzymes (GOT & GPT) of one day old

T.granarium adults originally resulted from treated and untreated pupae at LD_{30}

| Laser types (LD_{30}) | Transaminases (µg pyruvate/min/g.b.wt) | | | |
	GOT Mean ± SE	%	GPT Mean ± SE	%
Control	126.6 ± 1.76	-	87 ± 3.52	-
Argon (347 Mj)	105 ± 1.85**	-17.06	89 ± 2.75^{ns}	+2.30
CO_2 (62 mj)	92.66 ± 3.33***	-26.81	76 ± 4.33^{ns}	-12.64

4-Effect of two laser radiation types on the acetyl cholinesterase

The statistical analysis of the data presented in table (4) indicated that acetyl cholinesterase was significantly inhibited by 11.23% than the control at the level 5% with CO_2 laser on the other hand there was insignificant inhibition by treatment of Argon laser by 4.45% than the control.

TABLE 4. Effect of two laser radiation types on the acetyl cholinesterase expressed as µg Ach Br hydrolysed / g.b.wt/minute of one day old *T.granarium* adults originally resulted from treated and untreated pupae at LD_{30}

Laser types (LD_{30})	µg Ach Br/g.b.wt. minute mean ± SE	%
Control	231.3 ± 3.18	-
Argon (347 Mj)	221 ± 2.51 ^{ns}	-4.45
CO_2 (62 mj)	205.33 ± 7.79 *	-11.23

Deviations in the table indicate the standard errors of means

ns : non significant ***:** significant at level 5%. ****:** Significant at level 1%. ***** :** significant at level 0.1 % ..

% Percentage activation (+) or inhibition (-) than the control

5-Effect of two laser radiation types on the peroxidases

Peroxidases activity which was measured by the optical density is illustrated in Table (5). The data showed that peroxidases optical density were insignificantly decreased at the LD_{30} of argon and CO_2 laser radiations, and the calculated decrease for these treatments was 8.62 and 4.02% for argon and CO_2 respectively as compared with control group.

TABLE 5. Effect of two laser radiation types on the peroxidases of one day old T.granarium adults originally resulted from treated and untreated and untreated pupae at LD_{30}.

Laser types (LD_{30})	Peroxidases (O.D. units x 10^3 /min/g.b.wt.) mean $\pm$ SE	%
Control	522 ± 8.19	-
Argon (347 mJ)	477 ± 19.63^{ns}	-8.62
CO_2 (62 mJ)	501 ± 8.14^{ns}	-4.02

Deviations in the table indicate the standard errors of means:

ns: non significant at level 5%.   **% :** Percentage inhibition (-) than the control.

DISCUSSION

Quantitative essays of proteins in the haemolymph are of considerable importance for understanding the different physiological processes associated with reproduction.

In insects, changes in proteins are prominent during stages under going marked development and tissue differentiation such as the case during metamorphosis [10] and sexual maturity of the reproductive organs. The total protein content is the net result of protein biosynthesis [11].

The present research revealed that there is an increase in the total protein content of *T. granarium* development irradiated pupae at LD_{30} of argon ion and CO_2 laser. The significant increase in this criterium may be attributed to the protein which is not utilized distinctly in morphogenetic process to exceed that of the control.

The same results were obtained on *Ephestia kuehniella* and adult of *Musca domestica* which was irradiated with γ–ray [12] were originally resulted from irradiated at sublethal dosages of X-radiation [19].

In present study, there is a significant increase in total proteins of *Trogoderma granarium* adults resulted from irradiated 2 – 3 days old pupae with the two forementioned laser types.

There are numerous publications that deal with the effect of irradiation from different sources on the total protein content in different insect species, for example when the first instar larvae of *Parasarcophaga argyrostoma* irradiated with γ-rays the total crude protein content in the haemolymph of last larval instar was clearly affected by irradiation. Subleathal doses induced some changes in the total protein content in the haemolymph of the last instar larvae [13].

Our observations clearly indicated that there is no significant increase in the total lipids of *T. granarium* when treatments were carried out with the two different types of lasers (argon-ion and CO_2) at dosage of LD_{30}. The variation in the cuticular lipids was observed in the different insect species or even in different stages of same insects and these variations in the composition could account for different permeability and transition points [14].

The obtained results dealing with the above mentioned parameters on *T. granarium* was in full agreement with those obtained on the same insect species when experiments were carried out on the last instar larvae and adults using dosage of 600 kv/m at different exposure periods at 3, 6, 12 or 24 hrs of electric field (EF) [14].

Our results clearly indicated that the total carbohydrates in adults of *T. granarium* was decreased as a result of irradiation of 2 – 3 days old pupae at LD_{30} of both CO_2 and argon-ion lasers.

Generally, ionizing radiation likewise causes denaturizing or inactivation of enzymes. It can also induce other alterations in enzyme solutions such as changes in absorbance sedimentation, velocity and solubility. The mechanism here can be either direct (effect on the enzyme molecules itself) or indirect (effect on the solvent with secondary effect on the enzyme). The actual change brought about in the protein is not clearly established, but ionizing radiation, is capable to cause bond disruption, cross linking between molecules and energy transfer with the molecules [16].

The laser irradiation effects are suggested to be realized by means of conformational changes in the protein molecules and inter molecular association dissociation processes. Thus affecting the structure of the active site [17]. Laser irradiation of enzymes causes significant changes in their activity. It has been shown that achromophore associated laser in activities 93% of β-galactosidase activity and 80% of alkaline phosphatase as well as 87% acetyl cholinesterase activity. Although thermal denaturizing and photochemical mechanisms for proteins in activation were postulated [18], the precise nature of laser mediated damage to the protein function has not been established [17].

The present research which dealer with the phosphatase activity of *T. granarium* Alk-pase and Acid-pase activity, showed significant inhibition in all treatment except that of the Alkaline phosphatase activity of *T. granarium* developed from pupae treated at LD_{30} of argon ion laser. The laser energy caused different changes in metabolic activity, eggs of *Schistocerca gregria*, for which acid and alkaline phosphatase activities showed an increase, after an exposure to 60 min of He-Ne laser followed by a drop after exposure to 120 min. Thus due to affected

photochemical interaction of He-Ne laser beam with the tissue which may cause a change in the Mg^{++} and Zn^{++} that activates the alkaline phosphatase. The reduction occurring in both enzymes after 120 min may be due to the activation of the tissues to control the activities of these enzymes [19].

The same results were obtained in the case of treatment larvae of *Spodoptera littoralis* with cypermthrin. In most insects, GOT appears to be more active than GPT as in *Tenebrio, Bombyx, Aedes, Calliphora, Periplaneta* and *Schistocerca* [20]. Our results clearly indicated that the laser treatments on *T. granarium* induced significant decrease in GOT activity and remarkable increase in the GPT activity of adults resulted from treated pupae and one can conclude that argon-ion and CO_2 laser radiations at 0.488 and 10.6 micron respectively, affected the amino-transferase pathway in the developing embryo and had a direct effect on the energy metabolism. The same conclusion came when the eggs of *Schistocerca gregria* treated with He-Ne laser [19].

Acetylcholinesterase (AchE) has a vital role in the maintenance of the nerve activity by removing ACHE released in the passage of an impulse in the synapses and also possibly along the axon.

Concerning the effect of argon-ion and CO_2, lasers at LD_{30}, it was found that there was an inhibition in the AchE of *T. gramarium* adults treated as 2 – 3 days old pupae. The inhibition of the activity of AchE enzyme of *T. granarium* adults treated as 2 – 3 days old pupae with argon-ion and CO_2 lasers at LD_{30} was similar to the findings reported that γ radiation at a dose level of 50 and 100 Gy applied to full grown pupae decreased the activity of choline esterase in the whole body extract of the females of *Spodoptera littoralis* [21]. They added that, this decrease might be attributed to the effect of irradiation on the nuceloprotein metabolism in the cell and disturbance in some physiological activities of the enzyme after irradiation may be attributed to "metabolic rate"[22].

The activity of peroxidase showed insignificant decrease for developed irradiated pupae, at LD_{30} of argon ion and CO_2 lasers.

Many enzymes including peroxidase were exposed to laser radiation ruby laser of 45 – 85 J exit energy, a neodymium laser, and ruby laser with Q-switching were used. Only peroxidase showed evidence of inactivation [23].

REFERENCES

1. M.Y.Y. Ahmed ,Z.A. Abou-Donia ,S.A. and A.A.. Salem, *Soc. Entomol. Egypt* **45** ,pp. 1-5 (1980) .

2. A.M.R. Afify, O.H. Gharib, and M.A. Shallan, *Ann. Agri. Sci. Moshtohor* **33** (4) PP1517- 1527 (1995).
  3. J.A. Knight, S. Anderson and J.M. Rawle, *Clin. Chem.*, **18**, PP 199-202. (1972)

4. M.M. Bradford,. *Ann. Biochem*, **72**, PP. 248-254. (1976) .

5. M . Dobois, K.A. Gilles, J.K. Hamilton, P.A. Rebers, and F.Smith, *Ann. Chem.* **28**, PP 350-356. (1956).

6. M.E.A. Powell and M.J.H. Smith, *J-Clin. Pathol.*, **7**, PP245-248 (1954).

7. S.Reitman and S. Frankel, *Amer. J. Clin. Pathol.*, **28**, 56. (1957).

8. D.R. Simpson, D.L.Bull, and D.A. Linquist, *Ann. Ent. Soc. Amer.*, **57**, pp.367-371 .(1964).

9. J.L.Vetter, M.R Steinborg and A.L.Nelson, *Agric. and food chemistry*,**6** (1) pp. 39-40 (1958).

10.H.F. Alrubaei, and T.A.Gorell. *Insect Biochem.*, **12** (2), PP171-175. (1982).

11. M.I0 Mohammed, *J. Egypt. Ger. So. Zool.*, **18** (E) PP291-312. (1995).

12. K.A.Abdel-Salam, "Effect of gamma-radiation on the Mediterranean flour moth, *Ephestia kuehniella*" Ph.D. Thesis , Fac. Agric, Cairo Univ., Egypt ,1983.

13. N. Zohdyand ,A.M., El-Gindi, *J. Egypt. Ger.Soc. Zool.* **16** (E) 1-12. (1995).

14 W. Mordue, G.J. GoldsWorthy, J. Brady, and W.M. Blaney, *"Physico-chemical basis for the transition phenomenon" In :* *Insect physiology"* Blackwell Scientific publications, Oxford, London, Edinburg, Boston, Melbourne. (1980).

15. M.M. Abd-Elrahman, "Biophysical study on all stages of *Trogoderma granarium* Everts." ,Ph.D. Thesis , Fac. Sci. Bioph. Cairo Univ., Egypt, 1998.

 16. L.G Augenstine,*Ad van. Enzymol.* **24**: 359. (1962).

17. S.A. Ostrovtsova, A.P. Volodenkov, A.A. Maskovich, I.M. Artsukevich, S.S. Anufrik, A.F. Makarchikov, I.P. Chernikevich and V.I. Stepuro*Effect of the laser irradiation on the functional activity of enzymes with different structural complexity.* Proceedings of applications of ultra short-pulse lasers in medicine and biology Joseph Noev chais/Editor , 1998

18. J.Liao, J. Roider and D.Jay.: *"Chromophore-assisted laser inactivation of proteins is mediated by the photogeneration of free radicals"* Proc. Natl. Acad. Sci. USA 91, 1994pp. 2659-2663.

19. A. M. El-Gindi W.G. Osiris, and N. El-Kes, *J. Egypt. Ger. Soc. Zool.* **30** (E) pp.271-279 (1999).

20. R.M. Desai, and B.A. Kilby, *Archs. Int. Physiol. Biochem.* No. 66, PP.248. (1958).

21. N.M. Shanbaky, B.M. El-Sawaf, Z.H. Zidan, S.R. Souka and N.A. El-HalaFawy, *"Effect of irradiation on cholinesterase activity in the cotton leafworm spodoptera littoralis* (Boied)." (2nd Nat-Conf. Pest. Dis. Veg. Fruits, Ismailia, Egypt, pp 1: 2242-2256. (1987).

22. S.A. Aly, "Observation on the influence of stressors on aspects of the physiology of *Chironomus riparius* (Miegen)", Ph.D Thesis , Ain Shams Univ., Cairo, Egypt,PP 175 1993.

23. J.M. Igelman, T. Rotte , E .Schecter,. B .Blaney, *Ann. New York Acad. Sci.* (1965) .

Photostability of Uranine Via Crossed-Beam Thermal Lens Technique

M. Zein El-Din, K. Elsayed*, S. Al-Sherbini M., and M. A.Harith

National Institute of Laser Enhanced Sciences (NILES), Cairo University, Egypt
*Physics dept. faculty of science, Cairo university
Corresponding author: mharith@niles.edu.eg

Abstract. Uranine is a diagnostic aid in ophthalmology and used as immuno-histological stain. Photostability study on such important compound using crossed-beam thermal lens (TL) technique was carried out. The study is based on the photodegradation (PD) behavior and rate regarding some parameters such as the incident laser power, wavelength, modulation frequency and sample concentration. The effects of such parameters on the TL signal and PD rate are discussed in details. The rate of PD is found to be proportional to the power of the pumping laser and concentration of the sample within the investigated range. The modulation frequency is found not to influence the PD rate. The photochemical quantum yield has been measured using potassium ferrioxalate actinometry and it was found to be very low.

INTRODUCTION

Importance of knowledge about drug photostability is a loss of potency of the drug product, which leads to a therapeutically inactive drug, not yet, but also, can lead to adverse effects due to the formation of minor degradation products during storage and administration [1]. In addition, the drug can cause light-induced side effects after administration to the patient by interaction with endogenous substances.

Fluorescein sodium, resorcinol phthalein sodium or uranine, is a highly fluorescent chemical compound. It absorbs visible light in the blue range with peak absorption and excitation band occurring at wavelengths between 465-490nm. The fluorescence emission occurs in the yellow-green range from 520 – 530 nm. These fluorescent properties have made fluorescein useful in a variety of medical applications such as ophthalmologic diagnostic (corneal trauma indicator, ophthalmic angiography, contact lens fitting) in the form of strips and intravenous injection to determine circulation time [2,3].

The Photothermal lens (PTL) effect was discovered by Gordon et al., [4] when they observed transient power and beam divergence changes in the output of a He-Ne laser. Because of its high sensitivity and versatility this technique has been widely used in spectroscopy, micro volume, trace analysis and photodegradation, as well as, thermo-optical characterization of solid, liquid and gas samples [5.7].

The PTL results from optical absorption and heating of the sample in region localized to the extent of the excitation source. This lens-like element arises from the temperature dependence of the sample refractive index. Moreover, because of the fact that most materials expand upon heating and the refractive index is proportional to the density, this lens usually has a negative focal length. This negative lens causes beam divergence and the signal is detected as a time-dependent decrease in power at the center of the beam [8].

The present study aimed to investigate of the photostability based on photodegradation (photobleaching) behavior of sodium Fluorescein in its aqueous solution using a crossed-beam TL set up.

CP888, *Modern Trends in Physics Research*,
Second International Conference on Modern Trends in Physics Research—MTPR-06,
edited by L. El Nadi

MATERIALS AND METHOD

10^{-3} mol/lt stock solutions of sodium Fluorescein (Uranine) puriss (Fluka) were prepared in double distilled water. Different concentrations were prepared to study the photodegradation. The photochemical quantum yield was determined by using potassium ferrioxalate, 1, 10-phenanthroline and sodium acetate buffer solution, puriss (Fluka). Photochemical quantum yield was determined by the conventional ferrioxalate actinometry as described by Hatchard, Parker, and Murov [9,10]. UV-Visible absorption measurements were carried out using spectrophotometer (Perkin Elmer Lambda 40). Laser beam of 488 nm wavelength from an argon-ion laser (Coherent Innova 400) was used to irradiate the sample within a (3 x 1 x 1 cm) quartz cuvette.

The experimental set up of the crossed-beam thermal lens employed in the photostability study is shown in figure (1). The 488 and 514 nm excitation wavelengths from an argon-ion laser (Coherent Innova 400) were used as pump sources to generate the TL in the medium. Radiation of wavelength 632.8 nm from a He-Ne laser source (Melles Griot, 10.0 mW) was used as the probe beam. The pump beam is intensity modulated at different frequencies using a mechanical chopper (SR-540). The two laser beams are focused into the sample cell and strike it perpendicularly, an interference filter was placed in the path of probe beam which allows only the 632.8 nm wavelength to reach the photo-diode detector and the signal was then processed using a dual-phase Lock-in amplifier (Stanford Research Systems SR-530).

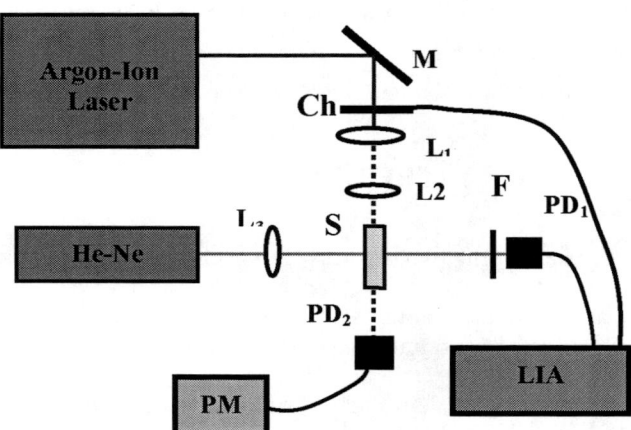

FIGURE 1: Schematic diagram of Crossed-beam TL spectrometer. Where, L2, L3 are convex lenses & L1 is a cylindrical lens, SC is the sample cell, F is the filter, PD1, 2 is the photodiode, PM is the power meter, Ch is a chopper, M is a mirror, and LIA is a lock-in amplifier.

RESULTS AND DISCUSSION

It is interesting to discuss the reasons for choosing an aqueous solution of Uranine substance although its small change in the temperature dependent refractive index (dn/dT) and high thermal conductivity (K). A consequence is a relatively low TL signal with respect to other solvents. The major reason is that to simulate the same conditions of medical use of such substance where water is the most encountered solvent in medical diagnostics. Another reason, water belongs to the non-absorbing liquids. These liquids are well transparent in visible region and their thermal properties are well known. Therefore, the absorbed light energy is then due to the dissolved substance only.

Figure (2) shows the absorption spectra of 133×10^{-6} mol/lt of sodium fluorescein irradiated by 488 nm of argon laser. The figure demonstrates significant decreases in the optical density with increasing the exposure time without any changes in the maximum absorption band (λmax.= 481 nm). The decrease in the optical density at the maximum wavelength may be due to the photodegradation of the uranine dye.

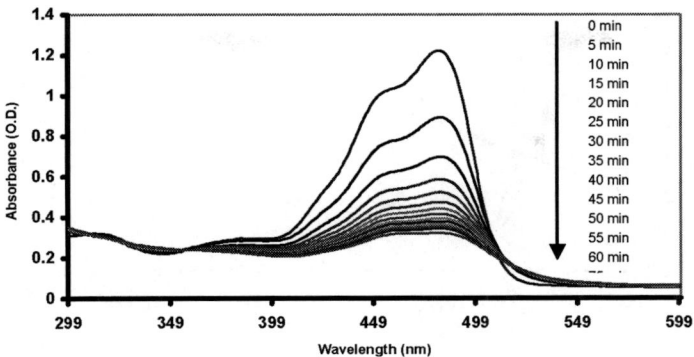

FIGURE 2: The absorption spectra of irradiated sodium fluorescein by argon laser of 488 nm at different exposure time intervals and sample concentration of 133×10^{-6} mol/lt.

The photochemical quantum yield (φ_c) was measured using potassium ferrioxalate actinometry method [9, 10] in case of the sample irradiated by argon laser of 488 nm. Some constants used as: the photochemical quantum yield (φ) of actinometer is 1.08 (at 488 nm, .0816 mol/lt, 23°C) and the excitation coefficient of ferrous 1,10-phenanthroline complex at 510 nm is ($\sim 1.11 \times 10^4$ l/mol.cm). The result photochemical quantum yield was 2.2×10^{-5}. The photostability based on the photodegradation (PD) behavior and the rate regarding on some parameters will be discussed according to the following:

Effect of the concentration

Figure (3) shows the dependence of photodegradation (PD) rate with the different concentrations. On contrary, the earlier reported Photothermal studies concerning with PD measurement on laser dye doped polymer such as poly (methyl methacrylate PMMA)[7] show inverse proportionality of PD rate with concentration. This discrepancy can be ascribed mainly to the host of the dye whether liquid or solid. This may be due to that in case of the dye doped polymer i.e. solid matrix, the number density of absorbing species (undegraded) of low concentration levels is less than that of higher levels. Moreover, the absorbing species in laser-exposed area will not be replaced after degradation due to the static nature of such species in a solid matrix. Consequently, the low concentration levels are logically faster in PD than the high levels. On the other hand, in case of solutions, the species in the exposed area are dynamic in laminar flow due to laser-induced heating i.e. undegraded molecules replace those degraded. So that at higher concentrations, the released heat will be greater than that in case of lower concentrations which can be observed evidently in the irregular decay of the lower concentration (16×10^{-6} mol/lt), figure 4. Therefore, the convection current will be increased leading to increasing the vertical laminar flow of solution within the sample cell, this in turn increases the PD rate of molecules. In this study the levels of concentrations are in the micromoles scale.

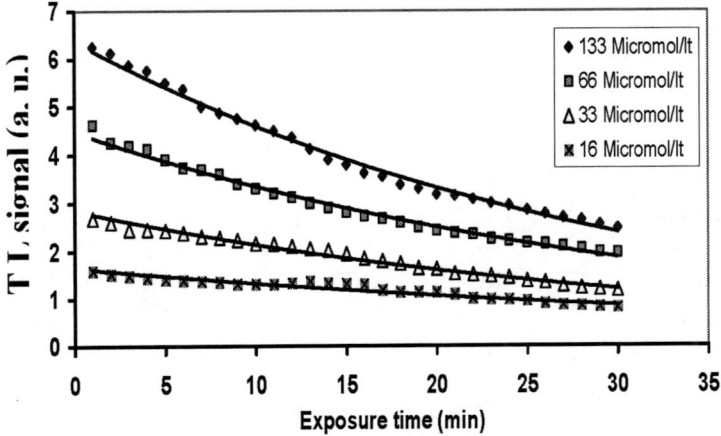

FIGURE 3: PD behavior of different concentrations represented in TL signal.

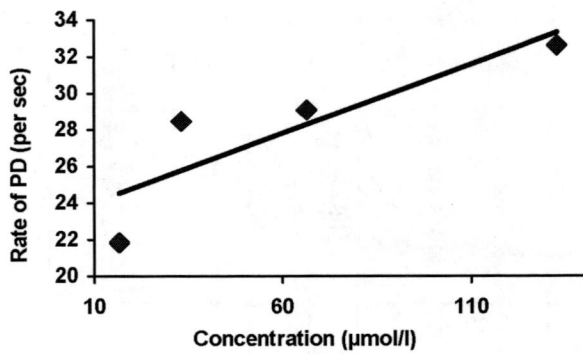

FIGURE 4. Variation of the rate of PD of Na-Fluorescein with concentration.

So, there are slight differences between the rates of PD over the selected levels of concentration. However, they can be clearly observed as shown in figure (4), where, the PD rate at a concentration 133×10^{-6} mol/lt is found to be 1.12 times greater than that at a concentration of 66×10^{-6} mol/lt, 1.15 times greater than that at a concentration of 33×10^{-6} mol/lt and 1.5 times greater than that at a concentration of 16×10^{-6} mol/lt.

Effect of the pumping laser power

The variation of PD rate with the power of the pumping laser is investigated at four levels namely 220, 320, 420 and 520 mW, constant conditions of power are carried out at conditions of 133 $\times 10^{-6}$ mol/lt of sample concentration, 1.0 kHz of modulation frequency and 488 nm of laser wavelength. The observed variations with time for such power levels are shown in figure (5-a , b).

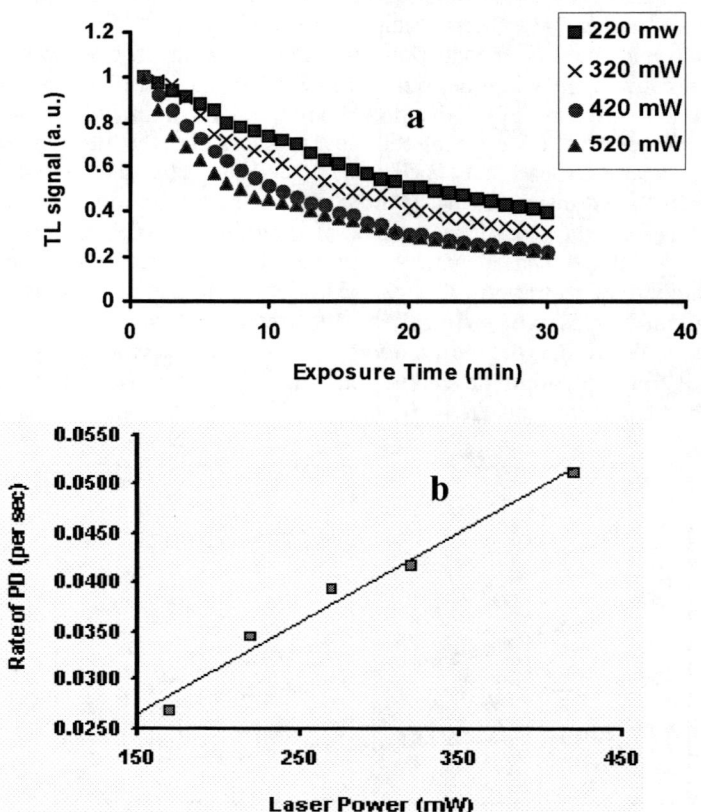

FIGURE 5 : Variation of the rate of photodegradation with incident laser power over 30 minutes exposure time (a) the PD behavior and (b) is the absolute values of PD rate.

234

This revealed a clear proportionality between the incident power and PD rate. Furthermore, the increase in laser power is accompanied with an enhancement in the TL signal. These results are in a good agreement with that of Achamma et al. [7].

Effect of the pumping laser wavelengths

The effect of the pumping wavelength on PD rate and TL signal is studied adopting two different lines of argon ion laser at 488 and 514 nm in order to investigate whether these distinct lines have a major effect on the PD rate and TL signal of Na-Fluorescein. The measurements are carried out while the other parameters are kept constant at 220 mW of pump laser, 133×10^{-6} mol/lt of sample concentration and 1.0 kHz of chopper frequency.

From the absorption spectra of Na-Fluorescein figure (2), one can observe that its absorption band falls in the blue-green region peaking at 489 nm, which is much closed to the blue line (488 nm) of argon laser and away from the 514 nm line.

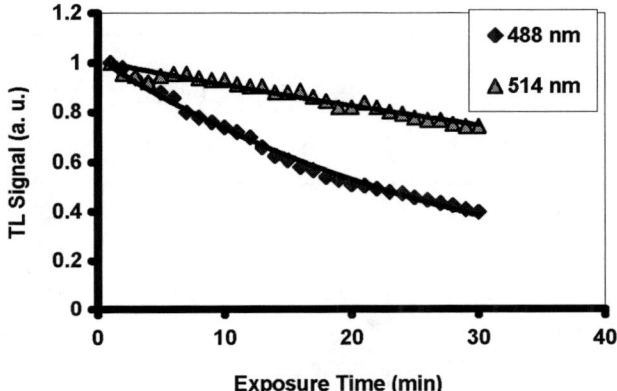

FIGURE 6. Variation of PD behavior of Na- Fluorescein with wavelength.

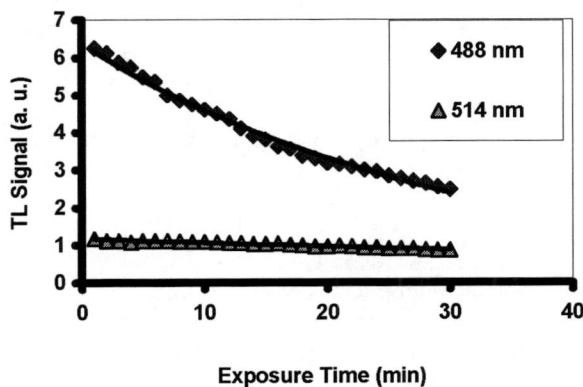

FIGURE 7. Variation of TL absolute signal of Na- Fluorescein with wavelength of pump laser.

So, the PD rate at 488 nm of pump laser is higher than that at 514 nm as shown in figure (6). Where, the PD rate at 488 nm is found to be nearly 3.3 times greater than that at 514 nm. In addition, the extreme higher signal of 488 nm than of 514 nm is due to relative higher absorbance of Na-Fluorescein at 488 nm than at 514 nm as shown in figure (7). Consequently, the line of 488 nm of argon laser has a major effect on PD rate and TL signal of Na-Fluorescein substance.

Effect of the chopper frequency

The effect of modulation frequency of the pumping laser beam on the PD rate and TL signal of Na-Fluorescein is investigated at five distinct chopping frequencies, 500 Hz, 750 Hz, 1.0 kHz, 1.25 kHz

and 1.5 kHz. Such measurements are carried out at the same conditions of 488 nm, 220 mW of the pumping laser and 133×10^{-6} mol/lt. Figure (8) illustrates that within the range selected, the modulation frequency of pump laser has no obvious effect on the PD rate of sample molecules. This denotes that the PD process of Na-Fluorescein depends only on the total incident energy per unit time on the sample, which is the same for all the chopping frequencies [11]. On the other hand, as predicted, TL signal has inverse proportionality with the modulation frequency as shown in figure (9).

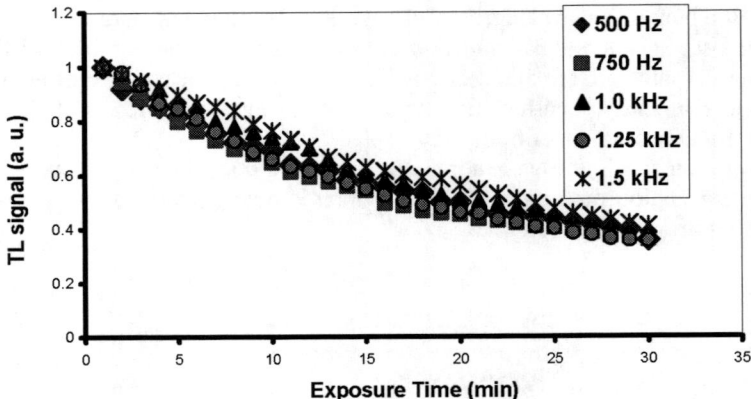

FIGURE 8. Effect of modulation frequency on PD rate of Na-Fluorescein.

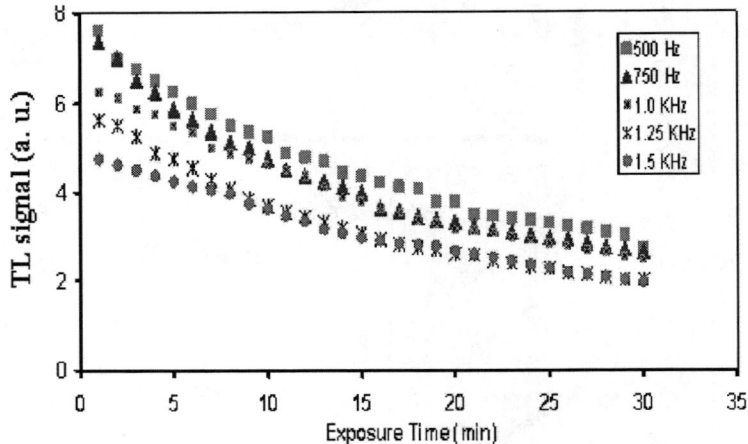

FIGURE 9. Variation of TL absolute signal of Na- Fluorescein with modulation frequency.

CONCLUSION

The dual-beam thermal lens technique has been found to be useful to study the photostability of sodium fluorescein in aqueous solution using crossed-beam set up. The results clearly show that the rate of PD and TL signal amplitude of Na-Fluorescein are found to increase with increasing of pump laser power and the concentrations of sample. In the case of modulation frequency, PD rate is found to be independent on it. While TL signals amplitude is found to be inversely proportional to the modulation frequency. With respect to wavelength effect, the PD rate and TL signal amplitude are found to be higher using 488 nm than 514 nm.

The photochemical quantum yield (φ_c) was measured using potassium ferrioxalate actinometry method for sample irradiated by argon laser of 488 nm wavelength. The resultant photochemical quantum yield was 2.2×10^{-5}.

REFERENCES

1. H. De Vries, Beijrsbergen Van Hangouwen G. M. J. and Huf F. A., Int. J. Pharm. **20**, 265-271 (1984).
2. O. W. Charles, Ole G. and Robert F. D., Textbook of organic medicinal and pharmaceutical chemistry, J. B. Lippincott Co., 7th Ed., PP. 950-951 (1977).
3. Merck index & Co., Inc., SODIUM FLUORESCEIN; C.A.S.#: 518-47-8, Whitehouse station, NJ, USA (1996).
4. J. P. Gordon, R. C. Leite, R. S. Moore, S. P. S. Porto, and J. R. Whinnery, J. Appl. Phys. **36**, 3 (1965).
5. R. D. Snook and Low R. D. , Analyst. **120**, 2051 (1995).
6. M. Franko and Tran C , Rev. Sci. Instrum. **67**, 1 (1996).
7. K. Achamma, Nibu A. G., Binoy P., Nampoori V. P. N. and Vallabhan , Laser Chem., **20**, 2-4 (2002).
8. S. E. Bialkoski , Photothermal spectroscopy methods for chemical analysis, Wiley, New York, P. 12 & 213 (1996).
9. G. G. Hachard and parker C. A., Handbook of photochemistry, Proc. Ray. Soc. A, 235, 518, Marcel Dekker, Inc., New York (1965).
10. S. L. Murov , Handbook of photochemistry, New York, Ch. 13 (1973).
11. A. Philip, Radhakrishnan P., Nampoori V. P. N. and Vallabhan C. P. G. Int. J. Optoelect. **8,** 501 (1993).

III. NUCLEAR, PARTICLE PHYSICS & ASTROPHYSICS

Elastic Scattering – Past, Present and Future

Fazal-e-Aleem, Haris Rashid and Sohail Afzal Tahir

Centre for High Energy Physics
University of the Punjab, Lahore-54590, Pakistan

Abstract.   Various aspects of elastic and diffractive scattering have been studied at Fermilab and CERN. Search for more results is ongoing at RHIC and planned at LHC. In this talk, we review the progress made so far and elaborate future prospects. Theoretical study focuses on the analysis of the available data in the light of predictions of various models with special emphasis on Eikonal picture and QCD inspired models. In the light of this analysis, various possibilities have been explored with reference to RHIC and LHC measurements.

Keywords: Elastic Scattering

PACS: 13.85 Dz

INTRODUCTION

Nucleon- nucleon elastic scattering is one of the most studied reactions in high energy physics. Its importance is evident from the fact that almost one fifth of the contribution to the total cross section comes from elastic scattering. Along with the understanding of diffractive processes, it helps us understand the process as a shadow of many inelastic channels present at high energy [1].

In the past, there had been many discoveries at ISR, CERN-SPS, and FNAL and we hope and expect that more are in store for us when we have complete results from Relativistic Heavy Ion Collider (RHIC) and Large Hadron Collider (LHC) [1]. *In this talk, we give an overview of various aspects of total and elastic cross sections in the light of various theoretical studies.*

EXPERIMENTAL MEASUREMENTS

Current status of the measurements for various parameters along with ongoing measurements at RHIC and proposed experiments at LHC are considered in these sections.

Past

The total and differential cross section, σ_T and $d\sigma/dt$, elastic cross section, σ_{el}, the local slope parameter, B and ratio of the real and imaginary parts of the scattering amplitude, ρ have been measured at CERN-ISR, CERN-SPS, and FERMILAB [1-2]. Measurements from the Cosmic ray data corresponding to LHC energy have also been reported for *pp* elastic scattering [2]. Figs. 1-4 depict the representative results.

There has been a general consistency of the experimental data measured at different colliders except the **CDF** results at FERMILAB which are significantly higher than **E710** results. Recent results from **E-811** further confirm discrepancy with CDF [1-2].

Present

Measurements are ongoing at **PP2PP** [3] experiment at RHIC. This experiment [3] will study *pp* total and elastic scattering in c.m energy range from 60 GeV to 500 GeV at **RHIC,** BNL in the two kinematical regions. In CNI (Coulomb Nuclear Interference) region $0.0005 < -t < 0.12$ $(GeV/c)^2$, σ_T, σ_{el}, ρ and B will be measured. In the medium -t region, $-t < 1.5$ $(GeV/c)^2$, a study of the evolution of the dip structure with $\sqrt{s}$ is planned. These

CP888, *Modern Trends in Physics Research,*
Second International Conference on Modern Trends in Physics Research—MTPR-06,
edited by L. El Nadi
© 2007 American Institute of Physics 978-0-7354-0354-9/07/$23.00

measurements will provide a unique opportunity to compare the results for proton-proton and proton-antiproton at 63 and 540 GeV. Initial results have been reported and are now being analyzed [1,3].

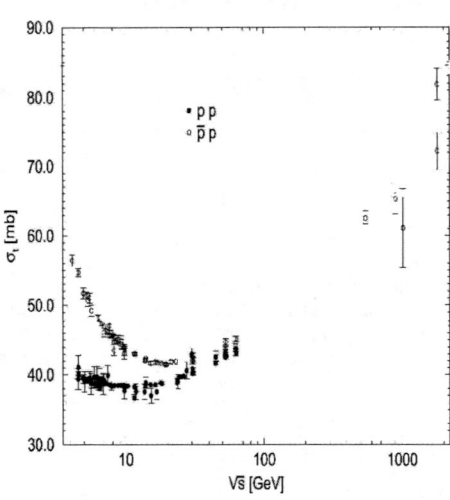

Figure 1:   Experimental data for total cross sections

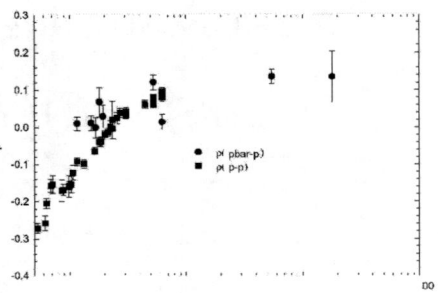

Figure   2:   Experimental measurements for the ratio of the real and imaginary part of the forward scattering amplitude

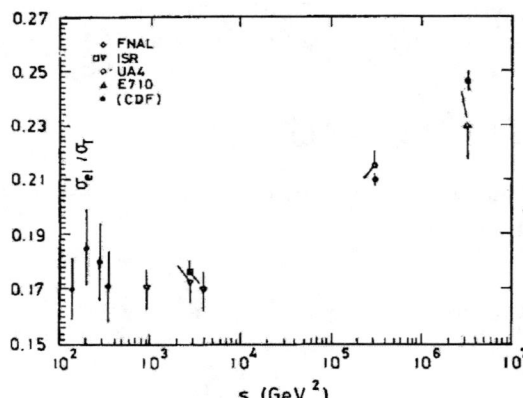

Figure 3   Ratio of the elastic and total cross section as measured by various authors

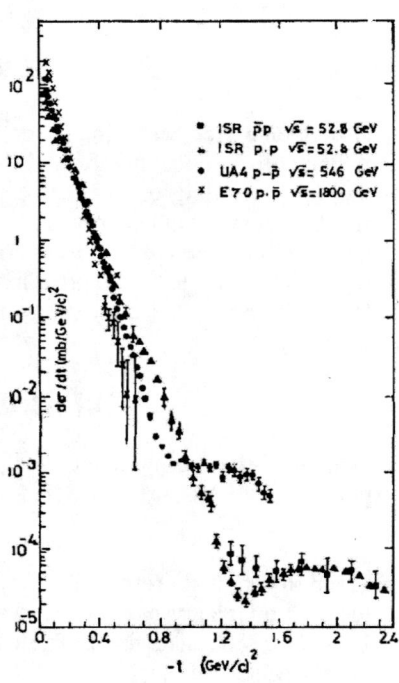

Figure 4   Differential cross sections at ISR, SPS and Tevatron

Future

Measurements in the future are planned at CMS [4], FELIX [5] and TOTEM [6] experiments at LHC.

CMS (Complex Muon Solenoid) [4] will study pp collisions at **LHC**. The Collider running at reduced c.m. energy of 1.8 TeV will provide an opportunity to compare the results with $\bar{p}p$. The experiment will also check $\sqrt{s}$ dependence of the total and elastic scattering in going from 1.8 TeV to 14 TeV.

At **FELIX** [5], experiment is planned at $\sqrt{s}$ = 14 TeV and luminosity of 10^{34} cm^{-2} sec^{-1}. Measurements will be made in the extreme forward directions. Motivation for the "Forward Physics" is for the reason that it has never been taken up in the past at the extreme forward direction. Physics agenda also includes elastic scattering and measurements of the total cross section. At the same time, **TOTEM** [6] collaboration proposes to measure the total and elastic scattering over a large range of -t along with single diffractive scattering and double Pomeron exchange cross section in *pp* collisions at 10-14 TeV.

THEORETICAL MODELS

Theoretical description of the data has been multidimensional [1, 7–51]. Various aspects of these theoretical studies are discussed below:

a) Considerable effort in the past had been devoted to the analysis of forward elastic scattering measurements in a model independent manner assuming only *dispersion relations*. Such models [7-16] are essentially based on the respect for Froissort-Martin bound and other asymptotic theorems. Dispersion relations [17] were derived when a certain asymptotic behaviour of the scattering amplitude is assumed. Various forms of the asymptotic parameterisations of the cross section take account of experimental data.

b) A lot of work [2, 18-25] has been carried out within the framework of Regge pole theory involving the dominance of Pomeron, commonly known as soft Pomeron, at high energies. If Regge pole exchange is the dominant mechanism at high energy, then, the amplitude at large *s* is dominated by the trajectory α (t) with the largest intercept at t = 0. For $\sqrt{s}$ ≥ 10 GeV, the total and differential cross sections were found to vary slowly with energy. This variation was associated with the exchange of *Pomeron* [18-28]. Scores of papers have been written on the meaning, origin and structure of the Pomeron, which is assumed to mediate the high energy soft physics. The object Pomeron gained significant attention after the HERA experimental data providing strong link of soft physics with low x physics in deep inelastic scattering [26]. At present picture of the Pomeron is rather blurred with the presence of two Pomerons. Soft Pomeron, is used to explain diffraction phenomena while hard Pomerons accounts for small x phenomena in Deep Inelastic Scattering (DIS) and is calculated from perturbative quantum chromodynamics, 'pQCD'. Pomeron may be visualized as a complicated entity which in different dynamical situations may have different manifestations but whose origin is always the same, diffraction [20,26]. Evidence for the Pomeron [23] is being explored in the experiments measuring diffractive cross sections and results have been reported in finding its origin in QCD [1].

Simpler picture of Regge theory is a good approximation to the experimental data in the small -t region. However, in the dip region and beyond, one needs further modifications [27]. Since the exchange of Pomeron leads to identical behaviour for proton-proton and proton-antiproton scattering, the difference between *pp* and *p̄p* is accounted for by the exchange of mesonic trajectories.

In order to explain the experimental data beyond the dip for the differential cross section, *Regge cuts* arising from a two Pomeron exchange are introduced [27,28]. In addition, an odd charge conjugation 3 gluon exchange or *Odderon* [18,29-30] was employed. Odderon is the C = -1 partner of the Pomeron. In lowest order it can be understood as the exchange of three gluons in a symmetric colour singlet state. The Odderon intercept is expected to be close to one in contrast to the intercept of C = -1 reggeon exchanges, which is around 0.5. The Regge cuts give rise to a flatter dependence while the Odderon is used to account for the difference between *p̄p* and *pp* at high energies. However, at present, the experimental evidence for the existence of Odderon is not yet conclusive [18]. The ambiguity is for the reason that its contribution is very small compared with the dominant C = +1 exchange contribution. The idea of multi-Pomeron structures has also been extended to three Pomeron [31] to account for soft physics. The explanation does not give an optimum in the number of "relevant" Pomerons as well as underlying physics of such a construction.

Initially when the idea of Pomeron was floated, emphasis was on explaining and correlating the experimental data on the basis of general principles– analyticity, unitarity, crossing symmetry etc. However, in the recent past, physicists have begun to incorporate the ideas of perturbative QCD. Confinement is largely ignored. Most important thing is to learn to reconcile the two approaches. We are now beginning to learn as to how non-perturbative effects can be included in to the perturbative formalism. Attempts have also been made to try to find an origin of the Pomeron in QCD. In order to account for increase in total cross section, the exchanged gluons must interact with each other [1,32].

c) Third approach is based on Eikonal or Geometrical picture [1, 33-46]. The model first proposed by Chou and Yang received boost when its prediction regarding a structure in differential cross section was confirmed by the experiment. Since then it .has undergone several changes and is now significantly developed. It was generalized by using multiple diffraction theory [34]. The model assumes that for large -t, the central partons dominate the process. Multiple scattering therefore occurs by the successive collisions of two or more partons of the colliding particles before leaving the scattering region. Since then, it has been applied to a variety of collisions involving strong interactions. The Geometrical picture was further extended [36] to explain *p(p⁻)p* elastic scattering. The model relied on two experimental observations: the relation between the shape of dσ/dt and the hadronic form factors and a relation between σ_T and B. The hadronic radius is viewed as an interaction radius, which increases with energy because more inelastic channels open up and new degrees of freedom of the colliding hadrons contribute.

d) Fourth, are the recent developments based on QCD-inspired models [1,40-43]. These models incorporate semihard scattering of quarks and gluons or partons in the nucleus. These models mostly differ in their treatment of QCD and also whether or not they respect the constraints imposed by unitarity. In one such work [44] a dynamical description of small angle elastic scattering of light hadrons is given. They calculate elastic hadronic amplitude using the non-perturbative light-cone dipole representation for gluon bremsstrahlung. Their calculations reproduce well the total cross sections and elastic slopes. Their main observation is the existence of small gluonic clouds surrounding the valance quarks in light hadrons.

e) We now know that soft and hard Pomerons dominate the diffraction and small x phenomena in DIS respectively. This gives rise to an apparent dichotomy between the soft and hard aspects of Pomeron and awaits a possible explanation. Theorists are now beginning to develop formalisms which encompass the transition between hadron and quark and gluon degrees of freedom. QCD also predicts the existence of an Odderon based on three-gluon exchange, which through interference with the Pomeron exchange leads to remarkable charge asymmetries in diffractive reactions [1].

What will we look for at RHIC and LHC?

Let us now see as to what do we know about the total and elastic scattering in the light of models discussed above and what is expected at RHIC and LHC energies?

Forward scattering amplitude parameters

Total Cross Section
In almost all the models discussed above forward scattering amplitude parameters, σ_T and ρ are fitted well at the ISR energies. As we move to higher energies predictions begin to differ. This difference becomes visible at LHC/Cosmic Ray energies. We will discuss the same in the following sections.

A typical *dispersion relation* result given by Augier et al [16], favours log^2s dependence of σ_T as compared to *log*s. This kind of behaviour corresponds to the maximum rate of rise of energy allowed by the analyticity and unitarity and is close to the Froissart bound. The extrapolated values for 10 TeV and 14 TeV are 103 ± 7 mb and 112 ± 10 mb respectively.

In *Regge models*, increase in the total cross section is approximated by the intercept of the Pomeron trajectory. High energy data is fitted well by this approximation although at ISR contributions from mesonic trajectories are needed [27]. The predicted cross section at 1.8 and 14 TeV is 75 and 95 mb respectively and is consistent with *log*s behaviour. The σ_T value is predicted to be significantly higher when Odderon is taken in to account [1]. Predictions differ in the RIHC and LHC region (Fig. 5).

In a typical *geometrical picture,* [36] total cross section is described by the shape of the colliding hadrons, which varies with energy. The geometrical picture also gives a good fit to the experimental data for $\sqrt{s}$ > 20 GeV. Real part of the radius (which has been taken as energy dependent) increases linearly with *log*s, which makes predictions to higher energy straightforward. The model predicts σ_T = 73 and 95 mb respectively for 1.8 and 14 TeV respectively. Other geometrical models make similar predictions. Measurements of RHIC will therefore give us a good indication of the trend for the total cross section. However, measurements in the near forward direction would

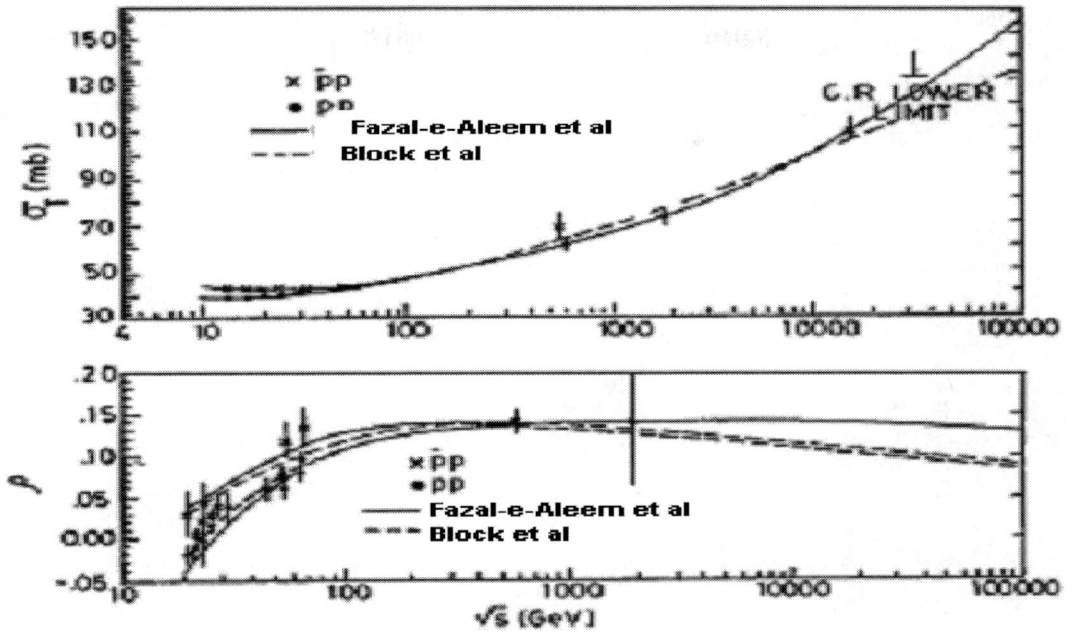

Figure 5: Fit to total cross section as measured at various energies against theoretical predictions of Refs.18 and 41.

be of significant importance at LHC as it would unambiguously establish or definitely contradict $(log\, s)^2$ behaviour, which emerges as a consequence of Odderon. QCD based models generally predict total cross section between 100 and 110 mb at LHC. This is significantly different from the predictions of Odderon-based models.

Fits of a large class of analytic amplitude models for forward scattering against the comprehensive data [11] for all available reactions has also been reported. The analysis favours models with a universal $log^2 s$ Pomeron term.

Results of several parameterizations [39] to two different ensembles of data on $p\,(\bar{p})\,p$ total cross sections at the highest centre of mass energies including cosmic-ray information have also been reported. From one ensemble the prediction for σ_T for $\sqrt{s}$ =14 TeV is 113 ± 5 mb and from the other 140 ± 7 mb. In both cases good descriptions of the experimental data is obtained which is due to large error bars of the cosmic ray measurements. This once again reiterates the need for measurements at RHIC and LHC. Similar results have been reported by other authors [47-49].

From the above discussion we find that at LHC predictions of different approaches are significantly different. A comparison of these models thus reveals that total cross sections for pp and $\bar{p}p$ will begin to differ from the RHIC energies. This difference will become very prominent at the LHC energies in case of the Odderon contribution. We also observe that the value of total cross section for different models varies from about 95 to about 145 mb. Although cosmic ray data, due to large error bars, accommodate these values, accurate measurements at LHC will be very important.

Ratio of the real and imaginary parts of the scattering amplitude

The ratio ρ is also of major interest. This quantity will be measured at RHIC and LHC energies. The kinematical range to be covered corresponds to the Coulomb-nuclear interference region. Measurement at smallest possible -t value will therefore minimize the extrapolation error. Only the models incorporating Odderon predict high ρ value (~0.2) at RHIC and LHC [16]. In Regge picture [28], a constant value of ρ = 0.12, consistent with data, is predicted. In the geometrical (including QCD inspired) models, this value is predicted to be ≈ 0.14 at SPS, FERMILAB and LHC. The results of UA4/2 [1] are consistent with geometrical models.

In the Eikonal models, the dip of the differential cross section is very sensitive to the ρ value. This is shown in Fig.6 where differential cross section for proton antiproton at 546 GeV is plotted for ρ = 0, 0.14 and 0.24. This clearly suggests that in case of higher measured value of ρ at RHIC and LHC, the structure in dσ/dt would disappear and turn into a shoulder. It can be seen that current data for differential cross section does not support a higher value of ρ at RHIC and LHC within the framework of geometrical picture.

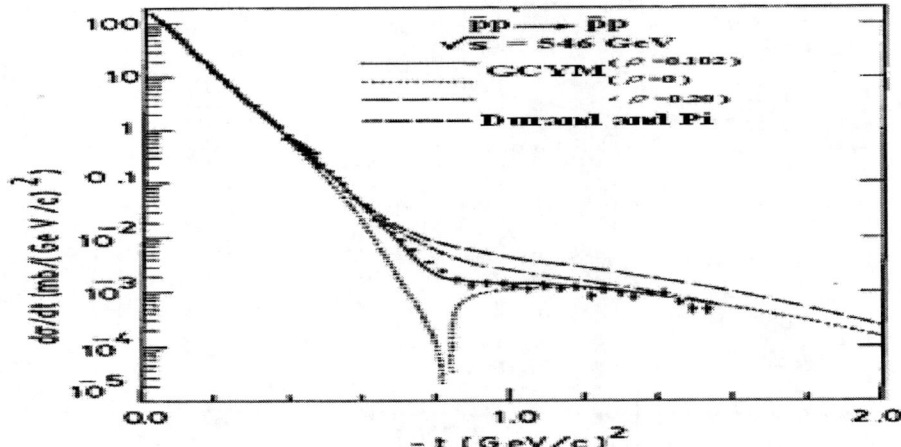

Figure 6: Differential cross section for $p\bar{}p$ elastic scattering at 546 GeV against the predictions of [34] and [40].

DIFFERENTIAL CROSS SECTION

a)   *Shrinkage of the diffraction peak*

The shape of the forward cross section appears to have an exponential slope. The slope of the exponent (called *B*) increases with an increase in energy thus resulting in what is called *shrinking phenomenon*. Shape of the slope is in fact not exactly exponential at ISR and SPS energies in the extreme forward region. It has slightly concave curvature. At Tevatron energy the curvature seems to have disappeared. Measurements at RHIC will therefore be very interesting from this point of view. In the simple Regge pole picture [28], t dependence of differential cross section is of a constant slope with no curvature. The model can be modified so as to exhibit a positive curvature at ISR and SPS Collider. But then it cannot at the same time explain the vanishing of this curvature at Tevatron energy. In the Eikonal picture this emerges as a natural consequence [36]. Models based on impact picture predict a convex curvature at LHC and higher energies. This further enhances the need for measuring the forward scattering at extremely small angle.

b)   *Dip Structure*

Measurements in the dip region are another testing ground for the theoretical models. Interesting observations are being made in regard to current and future measurements:

- The dip structure observed at ISR for $p\bar{}p$ scattering is moving toward -t = 0. This dip structure seems to disappear with an increase in energy. At Tevatron, it appears to have turned into a near shoulder. Simple Regge pole picture of Pomeron with a trajectory of 1.08 + 0.2t proves to be good approximation *only* in the near forward elastic scattering from ISR to Tevatron. In order to account for dip structure, contribution of pole plus cut is needed [50]. Unlike Regge models, the dip mechanism emerges naturally in the Eikonal picture. At RHIC, Tevatron, and LHC the dip is predicted near - t = 0.8, 0.65 and 0.4 (GeV/c)^2 respectively. As pointed out earlier, at RHIC and LHC, with ρ = 0.2, this dip will be filled and turn into a shoulder. Thus with an increasing contribution of the real part of the scattering amplitude, Eikonal models predict flattening/ disappearance of this dip.
- Eikonal models at LHC energies also predict the appearance of another structure at large -t [41,46]. Similar structure is observed in the Regge like models [51] but the position of the dip structure is different from those predicted by the Eikonal models. Comparison of the same has been made in Figs. 7 and 8. Position of

the minima differ in different models for both RHIC and LHC energies. It will therefore be interesting to focus on the position of minimum (or minima in case of a multiple dip structure).

- Another important observation in the region of dip is the difference observed between pp and $\bar{p}p$ cross sections at ISR. This difference is naturally explained in the Eikonal picture by the difference of the ratio ρ for two reactions. Another approach takes account of $P + P \otimes P$ and 3 gluon exchange contribution [28]. Will this difference persist or disappear at RHIC or LHC? This is another important observation to look for.

- Beyond the second maximum of the differential cross section, the interaction dynamics enter into a limit where pQCD can be applied [32]. In this region, the three gluon exchange multiple scattering interaction is shown to account for t^{-8} dependence of the angular distribution at ISR energies. At LHC it predicts a continuous decrease of the differential cross section over the large t region. This behaviour is different from the one predicted by the Eikonal picture and Regge type analysis.

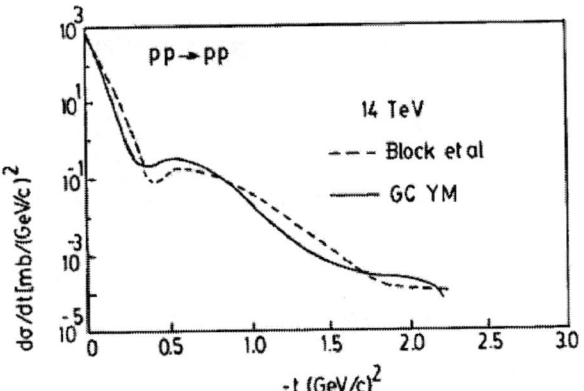

Figure 7: Predictions of Refs. 34 and 41 at LHC energy

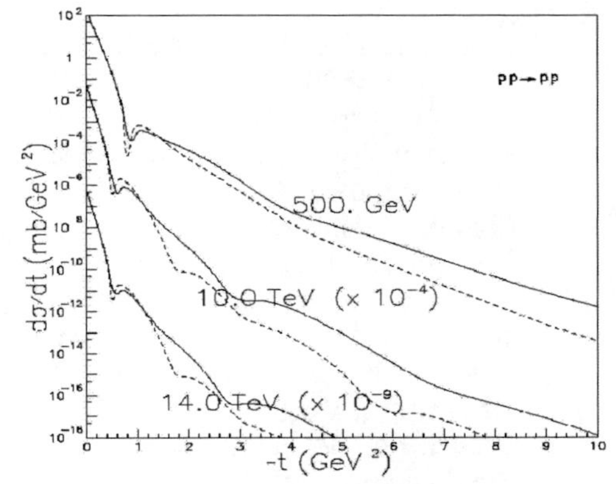

Figure 8: Predictions of Ref. 51 at RHIC and LHC.

Thus measurements at RHIC and LHC in the region of dip and beyond up to large momentum transfer would be very sensitive to the exchange mechanism. This will throw more light on the long-standing question about the nature of the minimum. We will know more as to whether it is purely diffractive minimum or a consequence of the Regge type Pomeron and two Pomeron exchange amplitude. In view of the scope of the conference, many details have not been given. More details are available in the most recent conference proceedings [1] on this topic as well as in excellent review articles of various authors which are available on hep archives [52]. Authors apologize to all those whose scholarly work have either been cited partially or could not be included due to representative selection of the literature. A detailed article, to be published separately on the same subject, will include all such work.

REFERENCES

1. D.E. Groom et al, Europhys. J., **C15,** 1 (2000). Proceedings of the "11th International Conference on Elastic and Diffractive Scattering: Towards High Energy Frontiers": The 20th Anniversary of the Blois Workshops, Chateau de Blois, Blois, France, 15-20 May 2005 (http://lpnhe-theorie.in2p3.fr/EDS05Accueil.html) ; 12th International Workshop on "Deep Inelastic Scattering" (DIS 2004), Strbske Pleso, Slovakia, 14-18 April 2004; Proceedings of "10th Blois Workshop On Elastic And Diffractive Scattering", 23-28 June 2003, Hanasaari, Helsinki, Finland; http://cerncourier.com/main/article/45/8/22 ; Fazal-e-Aleem et al, Int.J.Mod.Phys.**A19**, 4455 (2004).
2. S. Torii, 6th Blois Workshop France, 9 (1995) (Editions Frontieres); A.A. Arkhipov, hep-ph/0108118; http://www.cosmic-ray.org/date/papers05.html.
3. PP2PP collaboration, http://www.rhic.bnl.gov/pp2pp/.
4. CMS Collaboration, http://cms.cern.ch/.
5. FELIX Collaboration, http://felix.web.cern.ch/FELIX/.
6. TOTEM Experiment: http://totem.web.cern.ch/Totem/.
7. Kolar P and Fischer.J, hep-th/ 0110233.
8. A. Martin, hep-ph/0103296 and references given therein.

9. N. N. Khuri, 6th Blois Workshop, France, 463, (1995).
10. V. Kundrat, hep-ph/0001047.
11. K. Kang et al, hep-ph/0111360; hep-ph/0111025.
12. M. Block, Neutrino telescopes, vol. **2**, 513, (2001).
13. P. Gauron et al., 6th Blois Workshop, France, 55 (1995).
14. S.V. Goloskokov et al., hep-ph/9707219.
15. O.V. Selyugin, Nucl. Phys. Proc. Suppl. **A99**, 60 (2001); Phys. Lett. **B333**, 245 (1995);
16. C. Augier, Phy. Letts. **315 B**, 503 (1993).
17. P. Söding, Phys. Lett. **8**, 285 (1964).
18. Fazal-e-Aleem et al, "ICRC99" Vol.1, p186, Uttah, , USA (1999).
19. H.G. Dosch, " *Theory – Summary Talk* " 9th Blois Workshop, Pruhonice, Prague, Czech R., 9-15 (2001).
20. J. R. Forshaw and D. A. Ross, Quantum Chromodynamics and the Pomeron, Cambridge University Press (1997).
21. P. V. Landshoff, Nucl. Phys. Proc. Suppl., **99A**, 311 (2001); Nucl. Phys. (Proc. suppl), **B12**, 397 (1990).
22. J R Ellis et al, Eur. Phys. J., **10C**, 443 (1999).
23. A. Rostovtsev, [hep-ph/0108019].
24. M. Bertini et al., Revista del Nuovo Cimento, **19**, 1 (1996).
25. H. Abramowicz, Invited talk; Proceedings of "ICHEP 96", Warsaw Poland, 25-31 July (1996).
26. L.N. Liptov, "Strong Interactions at Long Distances"; L.L. Jenkovszky, Hadronic Press, Palm Harbor, FL U.S.A. pages 375-386, (1995).
27. M. Saleem and Fazal-e-Aleem, Hadronic J. **6** 699 (1983).
28. A. Donnachie and P.V. Landshoff; hep-ph/0111427 DAMTP, Cambridge U. Preprint 96/66 (December 1996); Physics Lett. **B296**, 227 (1992); Nucl. Phys., **B348**, 297 (1991); Particle world, **2**, (1991); Nucl. Phys. **B267**, 657 (1986); **B231**, 189 (1984).
29. M.A. Braun, hep-ph/9805394; DESY-98-055 (1998).
30. H.G. Dosch, hep-ph/0201294.
31. V.A. Petrov and A.V. Prokudin, hep-ph/0203162.
32. A. Donnachie Cern Courier, 39, 29 (1999).
33. T.T. Chou and C.N. Yang, Phys. Rev. **170**, 1591 (1968); Phys. Rev. Lett. **20**, 1213 (1968); Phys. Lett **B244,** 113 (1990).
34. Fazal-e-Aleem et al., J. Phys. **G16**, 269L, (1991); Phys. Rev. **D44**, 81 (1991); Fazal-e-Aleem and M Saleem, Monograph on "*Chou-Yang model and Elastic Reactions at high energies*" Hadronic Press, FL, USA (1992).
35. C. Bourrely, J. Soffer et al, hep-ph/9903438; Phys.Lett. **B442** 479 (1998); Mod. Phys. Lett. **A6** 2973(1991).
36. J. Hufner and B. Povh, preprint MPIK-V29 (1991); Phys. Lett. **B215**, 772 (1988) Phy. Rev. Lett **58**, 1612 (1987).
37. B. Povh, hep-ph/9908233; hep-ph/9806379.
38. E. G. Luna and M. J. Menon, hep-ph/01055076 (2001).
39. R. F. Avila, E. G. S. Luna and M. J. Menon, Braz.J.Phys. 31 567 (2001); M. J. Menon, Phys. Rev. **D61** 034015 (2000); **51**, 1427E (1995); Phys. Rev. **D48** 2007 (1993).
40. L. Durand and H. Pi, Nucl. Phys. B Proc. Suppl. 12, 379 (1991); Phys. Rev. **D40** 1436 (1989); Phys. Rev. Lett. **58** 303 (1987).
41. M. Block et al, Eur.Phys.J. **C23** 329 (2002); Phys. Rev. **D62** 077501 (2000); 9th Blois workshop, Pruhonice near Prague, June 2001 hep-ph/0003226.
42. C. S. Lam; hep-th 9804.463.
43. E. Levin, TAUP 2650-2000 October 2000.
44. B.Z. Kopeliovich, I.K. Potashnikova, B. Povh, E. Predazzi, Phys.Rev., **D63** 054001 (2001).
45. *DIFFRACTION 2000* Grand Hotel San Michele Cetraro [axpcs5.cs.infn.it/ ~diff2000/].
46. Fazal-e-Aleem and Haris Rashid "*Strong Interactions at Long distances*", Hadronic Press, p21, Palm Harbor, Fl., U.S.A (1995); Proc. Of 2nd Rencontres Du Vietnam "Physics at the Frontiers of the Standard model*", 21-28 October 1995.
47. Giorgio Giacomelli, hep-ex/0006038.
48. R.M. Godbole et al, hep-ph/0104015.
49. H. G. Dosch et al, hep-ph/0201294.
50. Fazal-e-Aleem and M Saleem, Pramana, **31,** 99 (1988).
51. P. Desgrolard et al., hep-ph/9811384.
52. http://www-spires.slac.stanford.edu/spires/ ; http://weblib.cern.ch/share/hepdoc/

Origin Of The Light Neutral Boson Observed In Heavy Ion Collisions

M. S. El–Nagdy[1], A. Abdelsalam[2] and B. M. Badawy[3]

(1) *Phys. Dpt. Faculty of Science, Helwan University, Helwan, Egypt.*
(2) *Phys. Dpt. Faculty of Science, Cairo University, Giza, Egypt.*
(3) *Reactor Phys. Dpt., Nucl. Research Center, Atomic Energy Authority, Egypt*
E. Mail: physicshelwan@yahoo.com

ABSTRACT. We report the results of ($e^+ e^-$ pairs) produced during the interactions of 200A GeV ^{32}S with emulsion nuclei. The results for the electron pairs suggest that they originate from light neutral bosons emitted during the collision. The origin of such neutral bosons could be due to de–excitation of the produced fragments ^{4}He, ^{8}Be and ^{12}C resulting in ^{32}S fragmentation. The masses of the neutral bosons were estimated from electron kinematics and found to be equal 1.51±0.14 and 9.88±2.85 MeV/c^2 and life time of orders 10^{-16} – 10^{-15} s. The data and results obtained could explain and put conclusion to the puzzles which were going on during the last 50 years around the anomalous mean free path of α–particles produced during high energy particle collisions. The depression of average shower particle multiplicities produced in the collisions of secondary helium fragments as compared to those of primary helium at similar energy signs the possibility of formation of the neutral boson.

INTRODUCTION

An existence of a new pseudo scalar neutral boson was first suggested by Weinberg[1] and was given the name axion by Wilczek[2]. He suggested that the axion results from the breaking of the U(1) symmetry. The U(1) problem of the standard model was solved by t' Hooft[3] by demonstrating that the instanton should be taken seriously. Interest in the possible existence of this neutral boson has been revived with surprising observations at Gesellshaft für Schwerionenforschung (GSI), Darmstadt of correlated ($e^+ e^-$ pairs) emitted in super heavy collision system[4].

On the other hand, El–Nadi and Badawy[5] found several experimental results where ($e^+ e^-$ pairs) were produced in the interactions with emulsion fragments resulting from collision of ^{12}C and ^{22}Ne projectiles at ~ 4.5 GeV/c/nucleon. They used stack of nuclear emulsion rather than using modern electronic detectors joined to a huge heavy ion accelerator. These results suggested the existence of neutral boson with mass 1.51±0.14 MeV/c^2 or (2.95±0.27)m$_e$ and life time τ = 1.5 x 10^{-15}s. It is also noticed by de Boer and van Dantzig[6] that three bosons are probably present in data obtained in Cairo.

In short, the classical form of standard model necessitates the Higgs. However, the Higgs has a major drawback. Despite countless dedicated and extensive experimental efforts to find the Higgs, no one observed its existence[7]. Although recently, some arguments given by El–Naschie[8] claims the existence of three and five rather than one Higgs particle.

In addition, El–Nadi[9] suggested the subsequent decay of four short–lived neutral bosons of average masses 1.8, 2.3, 2.6 and 12.2 MeV/c^2 with lifetime of order 10^{-15} – 10^{-16} s in the interactions of 200A GeV ^{32}S ions with emulsion nuclei.

Here, we report the results of neutral boson through the kinematical analysis of 121 measured ($e^+ e^-$ pairs) found in 200A GeV ^{32}S interactions with emulsion, as well as pairs from collisions of ^{12}C and ^{22}Ne at 4.5A GeV/c[5]. A clear example of 8 events is selected to explain explicitly the production of these pairs in view of the associated projectile fragments having notable charge. We also considered the behavior of high cross–section α–fragments produced during the heavy ion reactions. We noticed low average shower particle multiplicities and consequently low kinetic energies of the α–fragments as compared to those of primary α–particle beams at similar energies. This can be taken as additional evidence for the presence of new neutral boson which would be the accordance with anomalous phenomena.

CP888, *Modern Trends in Physics Research*,
Second International Conference on Modern Trends in Physics Research—MTPR-06,
edited by L. El Nadi
© 2007 American Institute of Physics 978-0-7354-0354-9/07/$23.00

EXPERIMENTAL SETUP

This experiment (EMU 03) was performed in a stack consisting of 24 Fuji films exposed horizontally to 200 GeV/nucleon ^{32}S ions at the CERN SPS. Details of the experimental setup, scanning of the pellicles, and event classification can be found in Refs [9, 10]. The microscope used is of type KSM–1. The measurements were carried out for charge Z of projectile fragments using the lacunarity measurement method for Z = 2 fragments and δ–rays method in case Z ≥ 3 fragments. The observed interactions were carefully looked for the direct pairs in the forward cone of the primary beam under 100 x magnification. The following criteria must be verified during the measurements,

(a) The track length of neutral particle undergoing decay should be longer than 20μm to overcome the uncertainty in the location of the pair origin. The (e^+ e^- pairs) tracks should come from the center of the interaction or located away from this center by no more than a distance L = 3μm. This restrictions[6] is consistent with the Daliz process taking into account the π° lifetime distribution.

(b) The ionization density for electron tracks should be less than or equal to the plateau value for relativistic singly charged–particle tracks (~ 30 grains per 100μm).

(c) Coulomb–Scattering measurements on the electron tracks must be carried out with two different cell lengths to eliminate noise and spurious scattering. Other sources of errors are avoided following ref.[11].

Further checking on few selected pairs for energy – momentum balance, were used only in the calculations of the production cross section and excluded from the analysis of the data. This is due to the unstable physical conditions of the emulsion in the vicinity of the electron pairs and hence the energy determination might be affected for either one or both of the tracks of such pairs. The angles for all the analyzed electron pair tracks were measured with reference to the primary beam direction. The accuracy in the angle measurement was found to be about 0.05° when calculated according to ref.[12].

RESULTS

The obtained results were analyzed putting in mind three criterion; Dileptons (e^+ e^- pairs) production, pion production, and anomalous behavior of Z = 2 fragments.

Dileptons (e^+ e^- Pairs)

Cairo high-energy group reported a previous work [9, 10] in which 1351 inelastic interactions of 200A GeV ^{32}S with emulsion nuclei were considered. In this work we studied 121 fully measured (e^+ e^- pairs) of the previous work as well as pairs from collisions of ^{12}C and ^{22}Ne at 4.5A GeV/c[5]. We selected a clear examples of 8 direct (e^+ e^- pairs) for which the charges of projectile fragments and its energy level have been carried out. Detailed information using kinematical analysis of the 8 pairs is obtained in table(1). The mass – lifetime distribution plot is given in Fig. (1). It suggests the emission of intermediate neutral bosons of possible average masses 1.51±0.14 and 9.88±2.85 MeV/c^2 and lifetime τ = 10^{-16} –10^{-15} s after which they decay into the detected (e^+ e^- Pairs). The data indicates the production of clusters "A" and "C" obtained before by De Boer and van Dantzig[9], discussion of the Cairo[5, 9] and Bristol data[13].

Figure (2) represents the energy partition asymmetry parameter distribution Y = ΔE / E, ΔE = |E_+ – E_-|, E = E_+ + E_- which is similar to the our previous data and to Bristol results showing forward peaking and deviating largely from both the calculated distribution for Dalitz pairs as well as the ~ 300 MeV γ – ray [14] and phase space two body decay.

The correlation between the emitted projectile fragments Pfs and the suggested neutral bosons indicates that these neutral bosons could be emitted in the de–excitation of excited fragments ^{4}He, ^{8}Be, ^{12}C produced in the ^{32}S fragmentation. The detection of visible tracks for the excited nuclei depends on their lifetime i. e., on their level widths Γ.

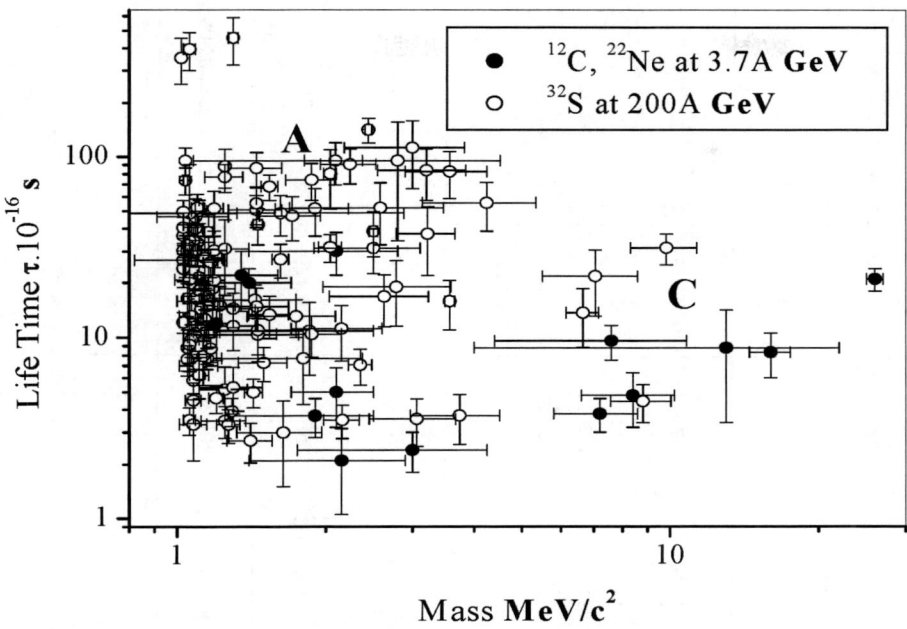

FIGURE 1. Mass – Life Time Plot Clusters A and C of Detected in Previous Cairo Data [5, 6], Clearly Coincide with Present Data Cairo EMUO3 – Exp. for A and C.

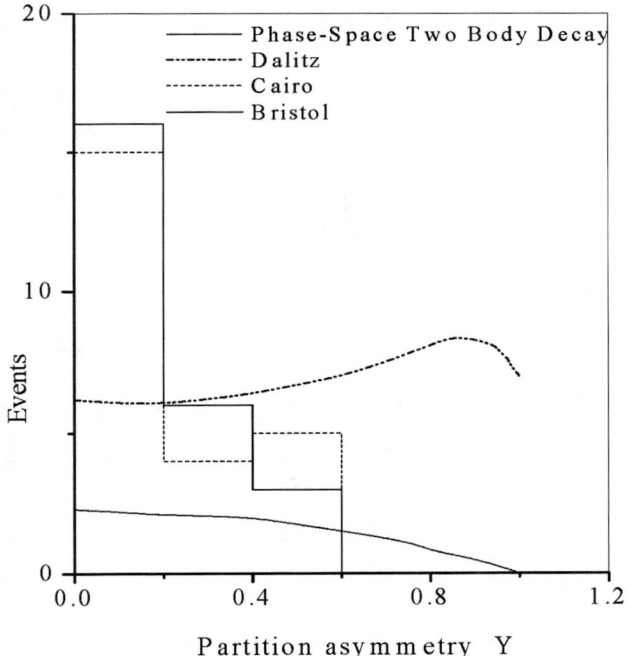

FIGURE 2. Energy Partition Asymmetry $Y = \Delta E/E$, $\Delta E = |E_+ - E_-|$ for Bristol[13], Cairo[15] and the Present Data 200A GeV ^{32}S EMUO3 – Exp. Dalitz Pair and γ–Conversion, as Well as Two Body Decay Corresponding Curves are shown.

TABLE 1. Detailed Information of the Measured e^+e^- Pairs

Star No.	P MeV/c	$\theta_p°$	$\theta_o°\pm0.1°$	M_c MeV/c²	$\tau\cdot10^{-16}$ s	$S(\mu)\pm2\mu$	Y	$E_{exc.}$ MeV
1	109.900±21.300 175.800±34.100	0.418±0.070	14.730	1.440±0.300	2.160±5.000	69	0.230	$^6Be^*$ 8.140±2.000
2	1st Pair 103.200±12.600 204.700±23.200	0.541±0.180	14.460	1.740±0.350	13.100±4.500	69	0.310	$^{12}C^*$ 10.300±0.500
	2nd Pair 163.550±16.230 85.270±10.400	0.147±0.070	23.750	1.150±0.080	7.700±1.600	50	0.330	$^{12}C^*$ 7.740±0.750
3	157.100±39.700 360.900±38.400	0.005±0.001	11.700	1.170±0.100	7.000±1.000	89	0.150	$^{12}C^*$ 10.300±0.500
4	149.140±42.160 121.340±13.460	0.181±0.070	8.260	1.120±0.120	9.500±3.540	115	0.100	
5	144.750±33.430 100.150±17.490	0.481±0.060	23.580	1.450±0.230	11.500±5.800	75	0.180	$^4He^*$ 17.120±2.080
6	218.340±20.310 147.090±12.760	0.330±0.070	8.870	1.460±0.300	10.400±3.400	78	0.190	$^8Be^*$ 5.600±1.000
7	1st Pair 288.500±32.320 225.250±24.750	2.200±0.200	17.310	9.820±1.520	31.200±9.600	49	0.120	$^8Be^*$ 18.100±1.000
	2nd Pair 111.220±28.140 504.770±127.700	0.310±0.050	12.350	1.490±0.650	10.900±6.200	108	0.640	$^{12}C^*$ 10.290±0.800
8	162.140±31.460 239.290±21.270	0.646±0.180	4.420	2.500±0.610	30.600±12.110	150	0.190	

P = Momentum of e^+ and e^-, θ_p = Opening angle of the pair on degrees, θ_o, M_c, τ, S: the emission angle, Mass, lifetime and decay time distance of the neutral particle, Y = Energy partition asymmetry of pair. E_*, is the energy of the excited level of the produced excited fragment.

Fig. (3) shows a clear example of the production of two ^{12}C track fragments, the first is in the excited 10.30 MeV (0^+, 0) level which then decays to its g. s. (0^+, 0) emitting a neutral boson which decays after 69μm into an (e^+ e^- Pairs). The path length of this excited ^{12}C–fragment is undetected due to the short lifetime of the 10.30 MeV level (Γ = 3000 keV). The second ^{12}C fragment is produced at "A" probably in the 7.66 MeV (0^+, 0) level which decays after a detectable distance δ ≈ 5 μm at the point "B" into its g. s. emitting a neutral boson decaying into an (e^+ e^- pair) after 50μm. The δ distance corresponds to the 0.05 fem s lifetime of the 7.66 MeV level[15]. The probability for such pairs to be due to Dalitz decay $\pi^\circ \rightarrow \gamma + e^+ + e^-$ is kinematically improbable for ranges higher than 49μm in table(1).

One can conclude that new light neutral bosons of average masses 1.51±0.14 and 9.88±2.85 MeV/c^2 and life times of orders τ = 10^{-15} – 10^{-16}s are produced in 200A GeV ^{32}S collisions in emulsion. These particles may be strongly correlated to the particles producing the GSI pairs[4].

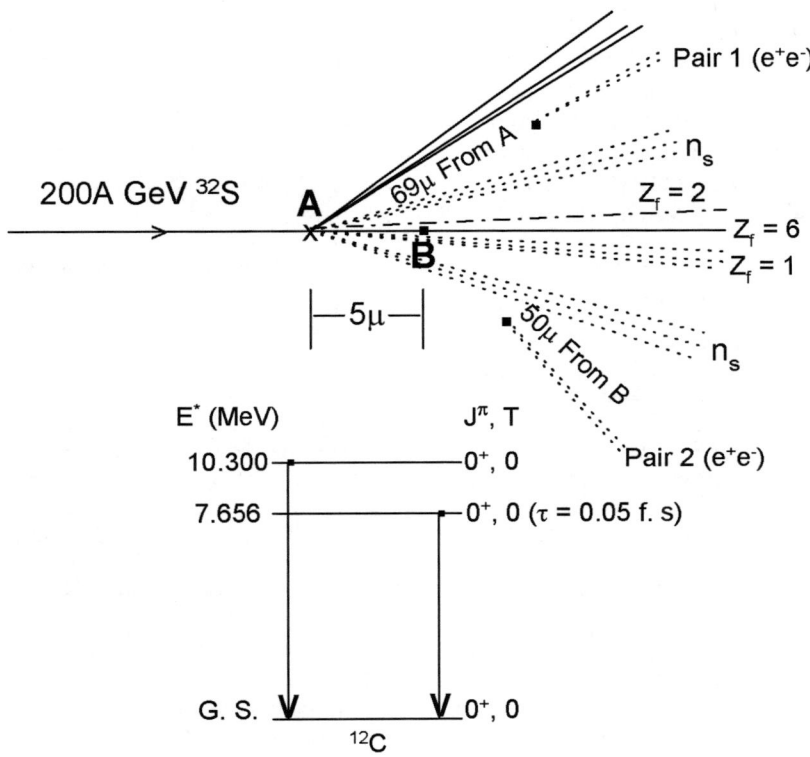

FIGURE 3. Schematic representation of 200A GeV ^{32}S Nucleus Fragmenting at "A". A Relativistic ^{12}C Pf Excited to the 7.654 MeV (0^+, 0) Level (τ = 0.05 f.s.)[14, 15] Decays Into Its g. s. (0^+, 0) at a Point "B" After a Corresponding Distance of ≈ 5 μm, Via the Emission of a Neutral Boson of Mass 1.51±0.14 MeV/c^2 Which Subsequently Decays Into an (e^+ e^- Pair2) After a Distance of 50 μm (τ =(7.7±1.6) x 10^{-16} s). The Second (Pair1) Which is 69 μm Distance From Point "A" is Due to a Neutral Boson of Mass 1.74±0.35 MeV/c^2 Emitted from a ^{12}C Relativistic Pf Excited to the 10.3 MeV Level (0^+, 0), Γ = 3000 KeV[14] Followed by Interaction at Point "A".

Pion Production

Pions are interesting tool for studying light and heavy ion interactions since they are mainly produced in the decay of Δ. Most of the energy spent on the particle creation during high energy nuclear collisions is used for pion production. Their numbers are fixed since, after the decay of the baryons, pions are produced preliminarily. This means, as will be clear latter, that pions get a momentum according to a given P⊥ which determines also its total

energy according to E_{meson}[16]. For $E_{av} - E_{meson} > 0$ (where E_{av} is equal to the total energy in CMS " $\sqrt{S}$ " minus the rest mass of participant nucleons "m_o". Pions are taken as real particles and E_{av} is reduced by E_{meson}. Therefore, one

can say that the multiplicity of pions "n_s" plus other neutral mesons are directly due to energy conservation. Neutral pions $\pi°$ are identified by their decay $\pi° \rightarrow 2\gamma$ with 98.2 % and $\pi° \rightarrow e^+ + e^- + \gamma$ (Dalitz pair) with rate of 1.8 %.

Here, the data reported from different laboratories for nucleus–nucleus interactions at different energies (1.4 – 200.0A GeV) are used concerning the average multiplicities of shower particles $< n_s >$. These are assumed to be mostly produced pions (π^+ and π^-) having relativistic velocity $v > 0.7c$ and interacting proton isotopes produced from the inelastic nuclear reactions. Also, included in these data $< n_s >$ produced from the interactions of secondary projectile fragments with charge $Z = 2$ emitted from the interactions of various beams at energies (1.4 – 200.0A GeV) with emulsion. Identification of proton isotope tracks and produced π–mesons (n_s) can be made on the basis of measuring their grain densities and taking the concept of fragmentation cone as discussed in ref. (17). The helium fragments produced from parent stars of heavy ion beam were easily identified in all labs by Lacunarity method.

In order to see the developments of shower particles n_s with the beam mass numbers A and total kinetic energy E_T. In our opinion, $< n_s >$ is introduced where the onset of meson production is important related to or strongly dependent on both the energy and atomic mass number of the projectile beam. We show in Table (2) the average multiplicities from ^{4}He beams at (1.4, 366.0 and 12.0A GeV)[18–20] as well as those produced from secondary helium fragments emitted from 1.8A GeV ^{56}Fe[21], 3.66A GeV ^{12}C[22] and ^{6}Li[19], and 14.5A GeV ^{28}Si[23]. Also included is the ratio $< n_s >_{sec.} / < n_s >_{pri}$. Nearly at each available energy, the values of $< n_s >$ are given for primary and secondary helium nuclei. It should be noted, however, that in all cases $< n_s >$ of primary ^{4}He beam is higher than the corresponding values of secondary He–fragment. The ratios $< n_s >_{sec.} / < n_s >_{pri.}$ are nearly constant for all given energies. Judek[24] also observed a lower value of $< n_s >$ for fragments originating from cosmic rays interacting with emulsion than for primary interactions of singly and doubly charged particles.

TABLE 2. Average Values of Shower Track Multiplicities Produced from the Interactions of Primary and Secondary Projectile Helium Nuclei in Emulsion from Different Experiments.

Energy of Primary ^{4}He A GeV	$< n_s >_{primary}$	Parent Nuclei Producing Secondary ^{4}He	$< n_s >_{secondary}$	$\dfrac{< n_s >_{secondary}}{< n_s >_{primary}}$	Ref.
1.4	3.36±0.05	^{56}Fe (1.8)	2.66±0.07	0.79±0.03	18, 21
3.7	4.54±0.11	^{12}C (3.7)	3.61±0.12	0.80±0.03	19, 22
		^{6}Li (3.7)	3.65±0.22	0.80±0.06	19
12.0	8.33±0.27	^{28}Si (14.5)	6.80±0.30	0.82±0.05	20, 23

The observed decrease in $< n_s >$ for secondary helium intersections is about 20 % as compared to the primary α interactions at similar energy. Production of neutral boson would imply a source to that decrease in addition to the expected decrease due to the existence of ^{3}He species among the $Z = 2$ fragments. Thus it is important now to evaluate the effect of the contamination of ^{3}He among ^{4}He. It is given in ref.[25] for primary ^{3}He $< n_s > = 4.0±0.1$ and for ^{4}He 4.5±0.1 at the same energy 3.66A GeV. By assuming 100 % ^{3}He in secondary helium fragment, this will lead to decrease not more than 11 %. However, a mixture of equal amount of ^{3}He and ^{4}He will make a decrease of ~ 5.5 %. Keeping in mind that the ratios of ^{3}He to ^{4}He produced from ^{6}Li is $\dfrac{30}{68}$ and ^{7}Li is $\dfrac{72}{22}$ [26] at ~ 4A GeV/c. Thus, the observed decrease for the multiplicity of pions n_s according to the energy conservation could be due to the presence of an unknown neutral source namely bosons.

In Fig. (4) a linear relation is shown to be valid in energy range 3.66 to 200A GeV for the average multiplicity of shower particles produced in the interactions of the primary beams given in table (3). The results have been fitted with the expressions $< n_s > = a E_T^b$ where $\log a = 1.53±0.02$, $b = 0.48±0.01$ and correlation coefficient $r = 0.99895$. Now it is interesting to use these universal linear relations to deduce the average energy corresponding to $< n_s > = 3.6 ±0.1, 6.8±0.1, 13.6±0.2$ and $23.6±1.2$ for interactions of helium daughters 3.66A GeV ^{12}C, 14.6A GeV ^{28}Si, 60A GeV ^{16}O and 200A GeV ^{16}O and ^{32}S beams, respectively. The corresponding energies were found to be equal ≈ 2.5, 11, 40, 140A GeV, respectively, which are less than the parent beam by about $\dfrac{1}{3}$ in Agreement with that mentioned in ref. [23, 27] for 14.5 A GeV ^{28}Si and 200A GeV ^{16}O and ^{32}S. In particular, we know that fragments produced from break up of projectile nuclei at a given energy continue with about the same velocity (and energy) as the parent

nuclei[28]. So it is meaningless to say that He fragments have energy less than parent beam by about $\frac{1}{3}$ (~ 30 %). In particular, the depressions of $< n_s >$ and kinetic energy in the interactions of secondary helium projectile fragments as relative to the corresponding primary ^{4}He beams has been put forward as a possible signature of formation of a neutral boson through a decay of the excited α fragments.

TABLE 3. Average Values of Shower Track Multiplicities for Interactions of Different Beams and Energies in Emulsion.

Beam	Energy GeV/nucleon	$E_{kinetic}$ TeV	$< n_s >$	Ref.
^{4}He	2.000	0.008	3.100±0.100	18
^{4}He	3.660	0.015	4.400±0.100	19, This work
^{12}C	3.660	0.044	7.800±0.200	32
^{4}He	12.000	0.048	8.330±0.270	20
^{16}O	60.000	0.960	34.120±2.300	27
^{16}O	200.000	3.200	57.300±3.100	27
^{32}S	200.000	6.400	79.900±4.100	33

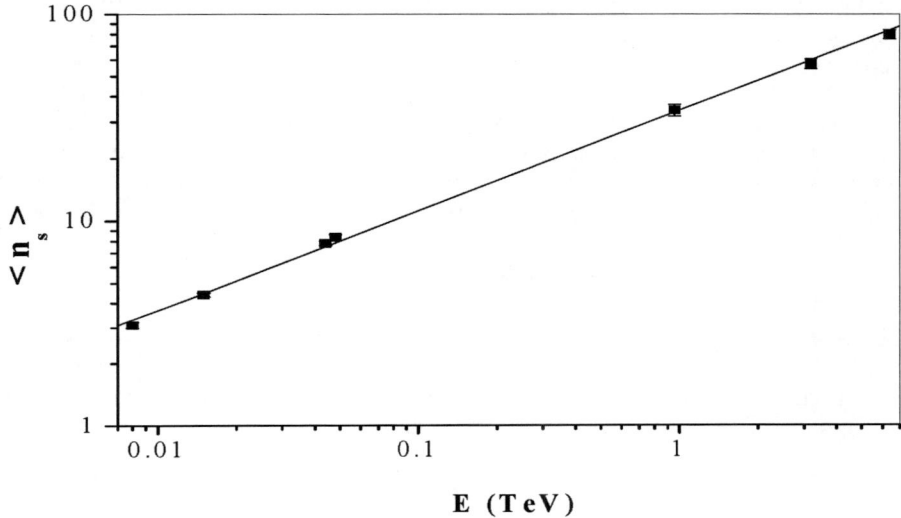

FIGURE 4. Average values of shower tracks $< n_s >$ produced from the interactions of different primary beams as a function of the total energy E_T of the beams given in table (4). The solid line is the best fit to the data points.

Anomalous Behavior of (Z = 2) Fragments

Interestingly, one rather prominent phenomenon may be related to the existence of neutral boson, which is the anomalous short mean free path λ of the projectile fragments at small distance from their emission vertex. Such fragments have greater cross section for interactions. A lot of work has been done for studying such anomalous. It was observed a short mean free path of projectile fragments with charge Z = 2 emitted from 4.5A GeV/c ^{12}C[29] and 200A GeV ^{32}S[30] with λ = 10.95±2.0 cm and 15.4±0.8 cm, respectively. The values are shorter by about $\approx$ 4 standard deviation than that of normal primary α particles ($\lambda_{normal} \approx$ 20 cm). Anomalous behavior was also found for heavier fragments Z $\geq$ 15 from 1.52A GeV ^{84}Kr[31]. Thirteen secondary interactions of projectile fragments with very small change in charge with respect to the primary were observed within about 50μm from their production point. These effects could be attributed to the existence of new neutral boson. The events from secondary interactions are reported[5] to be consistent with large number of (e$^+$ e$^-$ pairs) as a result of decay of a neutral boson with mass $\geq$ 1.5 MeV/c^2 and life time $\approx$ 10^{-15} s, while the pairs induced by primaries are considered as background. These observations can therefore be taken as additional evidence for the presence of particles interacting with anomalously high cross section or short mean free paths among the secondary nuclei emitted from heavy primary interactions which again confirms the production of new neutral boson.

CONCLUSIONS

The results given in the present work extend the confirmation of existence of another source responsible for enhancement of pair production. We would like to stress our ideas to confirm such particles through the following:

1- The data obtained from the interactions of 200A GeV ^{32}S with emulsion nuclei suggest the emission of intermediate neutral bosons of possible average masse 1.51±0.14 and 9.88±2.85 MeV/c^2 and life time τ = 10^{-16} –10^{-15} s. This nicely fall in the clusters "A" and "C" which were previously[5] found in the interactions of fragments with emulsion resulting from collision of ^{12}C and ^{22}Ne at 4.5A GeV/c. The correlation between the emitted projectile fragments and the neutral boson indicate that these neutral bosons could be emitted in the de–excitation of excited fragments ^{4}He, ^{8}Be and ^{12}C produced in ^{32}S fragmentation. A clear example for these excited nuclei is given from analyzing one event includes two visible ^{12}C tracks fragmented from ^{32}S projectile. One is in excited to a short life time of 10.3 MeV (0^+, 0) level which then decays to its g. s. (0^+, 0) emitting a neutral boson decays after 69 μm into e$^+$ e$^-$ pairs. The second carbon fragment is emitted at 7.65 MeV (0^+, 0) level which decays after a detectable distance ($\delta \approx 5$ μm) into its g. s. emitting neutral boson which then decays into (e$^+$ e$^-$ pair) after 50 μm. The 5 μm distance corresponds to 0.05 fem s lifetime of 7.66 MeV level, which is kinematically improbable to be due to Dalitz decay.

2- Afurther interesting feature is that there exist short mean free path for secondary helium fragments as compared to the primary one at similar energy. This, together with the low shower particle multiplicities $< n_s >$ observed and less kinetic energy obtained can be taken as additional evidence for the presence of new neutral boson interacting with anomalously high cross section or short mean free path. In our opinion, $< n_s >$ is introduced where the onset of meson production is important related to or strongly dependent on both the energy and atomic mass number of the projectile beam. So, production of neutral boson would imply a source to that decrease in both $< n_s >$ and kinetic energy in the secondary helium fragments. Again. This confirms the validity of interpretation of production of new neutral boson in heavy ion collisions.

ACKNOWLEDGEMENT

This paper is dedicated to the souls of Professors M. El–Nadi and O. E. Badawy of Cairo University, Egypt. We are pleased to acknowledge the kind help of the CERN authorities for irradiation of photographic plates. We would also like to thank Fulbright commission at Cairo, Alabama and NASA for their support. Thanks are due to Prof. Dr. Lotfia El–Nadi,,Cairo University, for fruitful discussion.

REFERENCES

1. S. Weinberg, *Phys. Rev. Lett.* **40**, 223 (1978).
2. F. Wilczek, *Phys. Rev. Lett.* **40**, 279 (1978).
3. G. 't Hooft, How instantons solve the U(1) problem. *Phys. Reports* **142**, 357 (1986).
4. H. Bokemeyer et al., in Proceedings of the Twenty–Second Recontre de Moriond (Editions Frontiers, Dreux, 1987).
5. M. El–Nadi and O. E. Badawy, Phys. Rev. Lett. **61**, 1271(1988).
6. F. W. N. de Boer and R. van Dantzig, *Phys. Rev. Lett.* **61**, 1274(1988); F. W. N. de Boer, *Phys. Rev. Lett.* **62**, 2639 (1989).
7. JT. Veltman, *Sci Am* **255**, 88 (1986).
8. M. S. El Naschie, *Chaos, Solitons and Fractals* **23**, 683 (2005).
9. M. El–Nadi at al., *IL Nuovo Cimento* **109A**, 1517 (1996).
10. S. Kamel, *Phys. Lett.* **B368**, 291 (1996).
11. W. H. Barkas, *Nuclear Research Emulsion* (Academic Press, New York, 1963, pp. 263; C. F. Powell, P. H. Fowler and D. H. Perkins, *Study of Elementary Particles by the Photographic Method* (Pergamon, London, 1959).
12. A. Bonetti at al., *Nuclear Emulsions* (George Newnes Limited), 1958, pp. 37.
13. B. M. Anand, *Proc. Roy. Soc.* (London) **A220**, 183 (1953).
14. F. Ajzenberg–Selove and T. Lauritsen, *Nucl. Phys.* **A114**, 1 (1968).
15. Table of Isotopes, 7th Edition, Edited by C. Michael Lederer and Virginia S. Shirley, Lawrence Berkeley Laboratory University of California, Berkeley USA(1989).
16. W. Schulz[†] and E. Ganssauge, *J. Phys.* **G18**, 1491 (1992).

17. M. S. El–Nagdy, *Mod. Phys. Lett.* **A16**, 985 (2001).

18. O. E. Badawy, M. El–Nadi, S. A. El–Sharkawy, M. K. Hegab and M. M. Sherif, *Z. Phys.* **A311**, 105 (1983).

19. M. El–Nadi, M. S. El–Nagdy, A. Abdelsalm, E. A. Shaat, N. Ali–Mossa, Z. Abou–Moussa, N. Rashed, S. Kamel and M. A. Fayed, *Egypt. J. Phys.* **33**, 229 (2002).

20. Banaras, Cairo, Chandigarh, Jammu, Lund Collaboration, 7th High Energy Heavy Ion Study, GSI Darmstadt, October 8 – 12(1984).

21. E. M. Friedlander et al., 6th High Energy Heavy Ion Study and 2nd Workshop on Anomalons, LBL, June 28 – 1 July(1983).

22. Abdel–Moniem M. M. Moussa, Ph. D. Thesis Submitted to Faculty of Science, Cairo. University, (1984).

23. P. L. Jain, A. Mukhopadhyay and G. Singh, *Phys. Lett.* **B294**, 27 (1992).

24. B. Judek, *Can. J. Phys.* **46**, 343 (1968); **50**, 2082 (1972).

25. A. Abdelsalam, M. S. El–Nagdy et al., accepted for publication in *Egypt. J. Phys.*, (2005).

26. M. S. El–Nagdy, M. M. Sherif, O. Whaba and F. A. Abdelwahed, *Mod. Phys. Lett.*, **A20**, 18 (2005).

27. P. L. Jain. G. Singh, K. Sengupta and S. N. Kim, *Int. J. Mod. Phys.* **A7**, 1907(1992); *Phys. Rev.* **C43**, R2027 (1991); *Phys. Rev.* **C44**, 844 (1991).

28. H. H. Heckman, D. E. Greiner, P. J. Lindstrom and H. Shwe, *Phys. Rev.* **C17**, 1735 (1978).

29. M. El–Nadi, O. E. Badawy, A. M. Moussa, E. I. Khalil and A. A. El–Hamalawy, *Phys. Rev. Lett.* **52**, 1971 (1984).

30. Z. Abou Moussa and T. Talaat, *Egypt. J. Phys.* **23(1, 2)**, 97 (1992).

31. P. L. Jain, M. M. Aggrawal and K. L. Gomber, *Phys. Rev. Lett.* **52**, 2213 (1984).

32. A. Marin et al., *Sov. J. Nucl. Phys.* **29(1)**, 52 (1979).

33. EMUO1 Collaboration, M. I. Adamovich et al., *Phys. Lett.* **B23**, 369 (1991).

Stakes in plasma physics in the solar system from the sun to the planets

Herve de Feraudy

Centre for studies of the Earth and Planet Environment. CETP/IPSL/CNRS.
University of Versailles-Saint Quentin en Yvelines, FRANCE.

ABSTRACT. In this paper an overview of the main centers of scientific interest for plasma physics in the solar system are presented. The paper is centered successively on the effects of the solar activity on the solar wind, on the structure of the solar wind at the level of the Earth, on the interaction of the solar wind with the earth plasma and magnetic field environment. For the latter the emphasis is put on the interface processes and on the circulation of the plasma from the solar wind to the Earth environment. A section is dedicated to the Space Weather and the impact of large solar events on the low altitude Earth environment and on several aspects of the human activity. Finally the exploration of planets is briefly overviewed.

INTRODUCTION

Until the 1990 years, the Space Physics in the Solar System beyond the Terrestrial environment was mainly exploratory. Since these years, the picture we had of the solar system has been totally renewed in spite of observing instruments which were not always well suited. Today the aim is solving several physical problems that are presently more accurately formulated than 10 years ago getting a quantitative understanding of the main processes that govern the solar system dynamics and the impact of the Solar corona huge events on the planetary environment.

In this paper we shall concentrate on plasma physics problems which the scientific community has identified, today, as important challenges. Accordingly shall present the space programs which are being designed for solving them.

To begin with, problems on the dynamics of the sun corona will be overviewed, including the Corona Mass Ejection events. Then problems related to the transport of the plasma from the solar wind to the Earth environment, in a medium which is strongly structured by the planetary magnetic field, with practically no collisions between particles, will be presented. Emphasis will be put first onto the role of the turbulence for letting the solar wind plasma cross the boundaries of the Earth magnetosphere, then on the instabilities through which matter and energy can be exchanged between the outer and the inner magnetosphere. I will present the problem of the impact of solar large events like impulsive coronal mass ejection, and more generally solar activity, on the low altitude Earth environment including the higher stratosphere, which is so important for the preservation of life on the Earth. This is the challenge of developing a Space Weather forecasting system. It will be done through the analysis of a remarkable solar Coronal Mass ejection event which has been surveyed from the sun corona down to the Earth ionosphere.

Then several aspects of the interaction of the solar wind with the planetary environments will be shortly presented. It will be reminded that quite different situations are encountered according as the planet has or has not an internal magnetic field or has or has not an atmosphere. In this section the interaction of the solar wind with small bodies in the solar system like comets will be briefly evocated.

CP888, *Modern Trends in Physics Research,*
Second International Conference on Modern Trends in Physics Research—MTPR-06,
edited by L. El Nadi
© 2007 American Institute of Physics 978-0-7354-0354-9/07/$23.00

In this paper, most of the figures are provided by the excellent book "Outer Magnetospheric. Boundaries : Cluster Results" [2]

SOLAR PHYSICS, THE SOLAR CORONA

Thanks to the availability of high resolution observations of the sun, provided by several new instruments among which are those of the SOHO space mission, the Hubble Space Telescope and ground-based telescopes dedicated to solar studies like the ESO Themis instrument, our knowledge of the dynamics of the inside of the sun has been strongly improved, putting in evidence a differential rotation of the sun layers, and helping reformulating the energy conversion processes inside the sun and between the inside of the sun and its external layer. Then a better understanding of the origin of the solar cycle has been acquired. On another side, the exploration of the solar wind, thanks to the Ulysses mission and high resolution observations from space, has renewed the picture we had of the structure of the solar corona, showing evidence of a regions with a slow solar wind, surrounded by regions with a fats one, stressing the importance of the corona holes and allowing the discovery of huge Corona Mass Ejection events – CMEs.

This is illustrated on figures 1a and 1b which show a coronal loop which developed after a flare vent and a CME event.

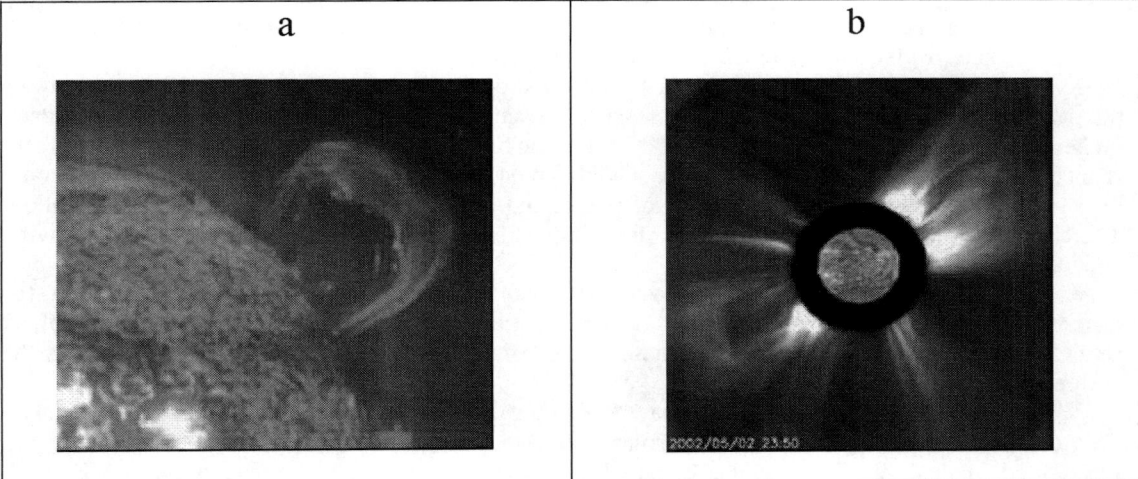

FIGURE 1. a) Coronal loop, observed in H-α; b) Coronal Mass Ejection observed by SOHO in Ly-α in the UV range. The latter is a composite image: the coronograph screens the solar disc (black cisc). An image of the sun has been superposed in the center, with the approximate good size. (Courtesy of ESA).

The availability of high resolution and multi-wavelength observations, from infra-red light to X-rays, provides powerfull tools to analyse underlying processes : Lyman alpha allows the detection of CMEs and of their precursors, Infra Red emissions, corresponding to cyclotron emissions, give trace of electron acceleration processes, γ and X band observations are characteristic of Bremsstrahlung emission and are indicative of electron precipitation into the chromosphere. The polarisations of optical high resolution observations from ground based telescopes (Themis) allow maps of the magnetic field at the solar surface. Lower resolution maps can also be performed through radio-interferometer observations.

Important goals for solar atmosphere and solar system physics are the understanding of the puzzling phenomenon of he corona heating where the corona temperature increases from 30 000 to 10^6 K within few 1000 km in altitude (several processes are suggested: chocks, induced currents ...) and the understanding of the CMEs;

THE SOLAR WIND AT 1 ASTRONOMICAL UNIT.

The morphology of the solar wind is complex, even far from the regions of interaction with planets. It is characterized by turbulence, interplanetary shocks, density depletions, several kinds of discontinuities, current sheets and boundaries. Elucidating their microscopic ant their macroscopic behavior structure is essential for understanding the dynamics and the thermodynamics of the solar wind, and, further the interaction of the solar wind with the planets.

Turbulence is observed as a stationary feature of the solar wind, with a power law spectrum of the magnetic fluctuations. This turbulence is a remnant of the turbulence in the solar corona which has been convected by the solar wind. Is the power law index according to the Kolmogorov theory? This is an open question which can be solved by observations from long lasting spacecraft. The temperature anisotropy of the solar wind and the long range Coulomb interaction force between the solar wind imply that the original turbulence theories, constructed for neutral gazes must be adapted. An observational difficulty is the fast motion of the solar wind across the spacecraft trajectories. This implies that the observed fluctuations are shifted in frequency by the Doppler effect. Restoring the frequency in the solar wind reference frame is a difficult task since the plasma is a dispersive medium with a complex dispersion law. This difficulty has been overcome with the use of multi spacecraft missions as is has been initiated by the CLUSTER four spacecraft mission and as it extended with future projects. They allow independent measurements of the 3 dimensional wave vectors and of the frequencies. This new way of designing space missions will be developed in the next section. Three important issues for the solar wind physics must be underlined:

1. The identification of the nature of the solar wind discontinuities is crucial. In the framework of the Magnetohydrodynamics – MHD – a set of discontinuities has been identified [1]: contact, tangential, rotational discontinuities, shocks. Some do not allow a transfer of matter through them (contact, tangential discontinuities) and should behave as walls. Others, like the rotational discontinuities or shocks, allow a transfer of matter and of energy. Shocks induce energy dissipation. Therefore the role the discontinuities play in the dynamics and the thermodynamics of the solar wind depends critically on their nature.

2. Helisopheric sheets of current have been identified recently in the solar wind, from the observation of the interplanetary magnetic field fluctuations. Several question rise and must be solved through detailed observations: how are these sheets oriented ? What is their thickness? How is the current distributed in these sheets?

3. The structure of the CMEs which have been convected with the solar wind must be elucidated. This essential for the dynamics of the solar wind since large amounts of matter and energy are transported in this way. This is also essential for understanding the interaction of the CMEs with planets. As far as the Earth is concerned, this is a key issue in understanding the effects of solar major events on our environment: this is the scope of Space Weather.

As is has been underlined, solving these questions require observations from multi-spacecraft missions. Only such missions enable one to determine accurately and without ambiguities the three dimensional spatial structure of the solar wind structures. This new type of missions is presented briefly in the next section.

A NEW CONCEPT : MULTI-SPACECRAFT MISSIONS

The observations from a single spacecraft, which travels through plasma structures, varying with space and time, are unable to allow answering those basic questions: do the observations result from time varying phenomena or do they result from the travel of the spacecraft through a purely spatial structure of the plasma, or do they result from both, as we believe being the rule? In the latter case, which part of the fluctuations is to be attributed to time variations and which results form space structures? Besides, when the spacecraft cross any frontier or boundary surfaces in the space environment, what is the local orientation of the surface? Is it moving? What is its velocity?

The discussion above has underlined that those questions can be answered from simultaneous observations from at least 4 spacecraft flying in a cluster.

The first of such missions has been CLUSTER, a program designed by the European Space Agency -ESA, with 4 identical spacecraft, launched in year 2000, with a polar orbit, crossing all the boundaries of the Earth magnetospheric environment. They can be separated by a variable distance, allowing studies with different scale resolution.

Using at least 4 spacecraft provides measurements from the nodes of a tetrahedron. Combined together, these measurements provide a three dimensional picture of the processes, with a space resolution which is the separation distance between the spacecraft. In that way several kind of measurement new can be performed.

1. Vector operators: the 4-point measurement constitutes a 3D finite difference of vector operator quantities. Let illustrate it by four examples :
   a. The estimate of the curl of the magnetic field fluctuations gives access to the determination of current density vectors and, further, to the Lorentz force undergone by the plasma.
   b. The estimate of the curl of the plasma velocity gives access to the vorticity of the flow, which is changed only by dissipative processes.
   c. The divergence of the stress tensor is an element of the momentum balance.
   d. The divergence of the Poynting vector, or of the heat flux tensor, in necessary for determining where the energy flows.
2. The timing of the same event observed by the four spacecraft, a minimum variance analysis of the signals, allows measuring the orientation, the thickness and the velocity of boundaries.
3. Correlation and covariance analysis allows a three dimensional characterization of the turbulence.

The spatial resolution of this new system of observation is the separation between the spacecraft. In the case of CLUSTER it can be adjusted from 100 km at least, 10000 km at most. This allows selecting the scale of the processes that one wants analyzing.

CLUSTER was the pioneer of such missions. The concept is being generalized. Then similar missions, from NASA, will follow: Themis in 2006, with 5 spacecraft, two of them are close, the three other are distant. This allows a multiscale analysis, which is compelling for understanding the behavior of the plasma at the smaller scales, taking properly into account the remote boundary conditions. Later, the NASA MMS mission is made of 4 identical spacecraft, similar to CLUSTER, however the orbit plane will be equatorial. The spacecraft separation could be as small as 1 km, comparable to ion gyroradii, and could reach 1 Earth radius (6370 km), allowing resolving the scale of microscopic processes.

PLASMAS IN THE EARTH ENVIRONMENT

The advent of multi-spacecraft missions with CLUSTER has deeply modified our methods and our requirements in analyzing the plasma processes. The interfaces between plasma domains are presently the center of the scientific interest. We are presently able to get a detailed description of the interfaces themselves, according to, both, a microscopic and a macroscopic approach. Then we can understand in details the processes occurring through the interfaces, leading to the transfer of matter (plasma), energy (electromagnetic and kinetic) and momentum through the interfaces. The stability of the frontier is also a major question which can be presently be approached in renewed way, being able to obtain a detailed description of the boundary conditions in both sides of the interfaces and of the structure of the interface. It is evident today that the turbulence plays a major role among theses processes. Great advances in the understanding of its development are expected from the possibility of multiscale analysis, thanks to suitably designed multispacecraft missions.

Before reviewing these processes, it is necessary to remind briefly the characteristics of the Earth plasma environment.

The main features of the Earth plasma environment

The plasmas forming the Earth environment have two origins: the incoming solar wind and the outflow of the ionospheric plasma.

Similarly, the magnetic structure of the Earth environment has two origins. One is the proper field of the earth. Basically this field is dipolar and its field lines are closed, with their two tips rooted into the Earth surface on both hemispheres. The solar wind magnetic field originates from the solar corona and is convected by the plasma of the solar wind. Its field lines are connected to the solar surface on one side and they expand freely through the interplanetary system with a spiral shape. Since the solar wind plasma cannot be considered as a perfect conductor, the magnetic fields of the two origins, solar and planetary, merge, and the planetary magnetic environment of the planet, instead of forming a closed magnetic "bubble" in the solar wind, is connected to the interplanetary field by his external field lines which are therefore connected to the polar regions of the Earth on one side, and free in the solar wind plasma flow on the other side. The resulting magnetic structure of the Earth is a distorted dipole close to the Earth, compressed in the sunward direction, elongated and forming a magnetic tail in the opposite direction. The tail field lines are connected in some way to the interplanetary magnetic field, the magnetic environment of the Earth is limited by a well defined boundary surface, the magnetopause. The external field lines of the Earth are wrapped on this surface and are also connected to the Interplanetary Magnetic Field - IMF.

Observed from the Earth reference frame, the solar wind flow is supersonic with respect to any characteristic velocity (sound, magnetosonic and Alfvén waves etc.) therefore the magnetopause is preceded by a bow shock in the solar wind. The structure of the shock in complex and its main characteristics are very different according as the IMF field lines are perpendicular to the shock – one speaks of a "parallel" or "quasi-parallel" shock, referring to the orientation the magnetic field with respect to the normal of the shock surface – or parallel to the shock – "perpendicular" or "quasi perpendicular" shock. These structures are illustrated in figure 2.

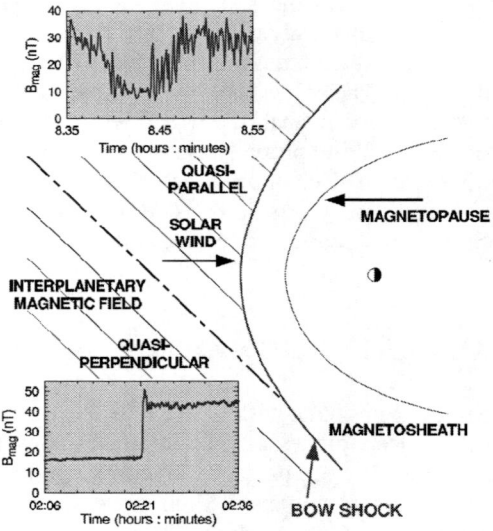

FIGURE 2. Illustration of the magnetic structure of the Earth bow shock in the solar wind. The quite different structures of the quasi-parallel and quasi-perpendicular bow shocks is illustrated by CLUSTER observations (from ref. [2], provided by A. Ballogh)

The magnetic transition region of quasi-perpendicular shocks is very sharp and well defined while quasi-parallel shocks are thick and the shock is preceded by a complex and wavy structure.

Between the bow shock is the magnetopause where the plasma flow pattern and the magnetic structure are less well defined as in the solar wind. Most of the plasma flows along the magnetopause, however the magnetosheath flow and magnetic structure is turbulent. Thanks to this turbulence, to the nature of the magnetopause boundary, to particularities of the Earth magnetic field lines topology and to singular events (reconnection events, Flux Transfer Events or FTEs) some plasma and field energy is transferred through the boundary and penetrate the Earth environment.

The structure of magnetosphere inside the magnetopause is complex. It is illustrated on Figure 3.

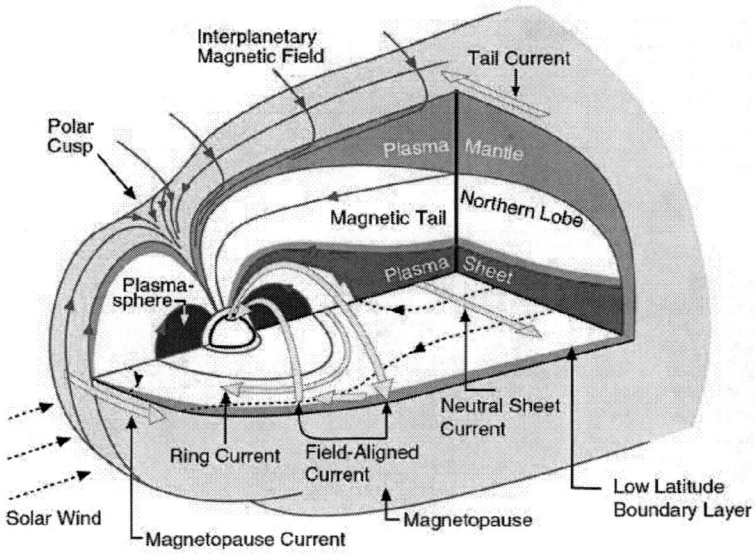

FIGURE 3. A picture of the magnetosphere structure. Yellow arrows are currents. Blue and red are plasma regions with well defined characteristics and parameters (temperature, particle number density, average energy). They are separated by boundary surfaces and boundary layers ("pause" suffix). (From Ref. [2], adapted form Kivelson and Russel, 1995 [3])

Boundary layers study: the Bow Shock

The physics of the Bow Shock is complex since the plasma is collisionless. Therefore the role played by the collisions between particles in neutral gazes is held by the plasma turbulence, resulting itself from the evolution of several instabilities of the plasma up to the saturation level. Turbulence is a key phenomenon for dissipating the solar wind energy and for maintaining the shock as a permanent and more or less stationary structure. It controls the rate of plasma transfer through the shock.

a. Bow Shock surface normal determination.

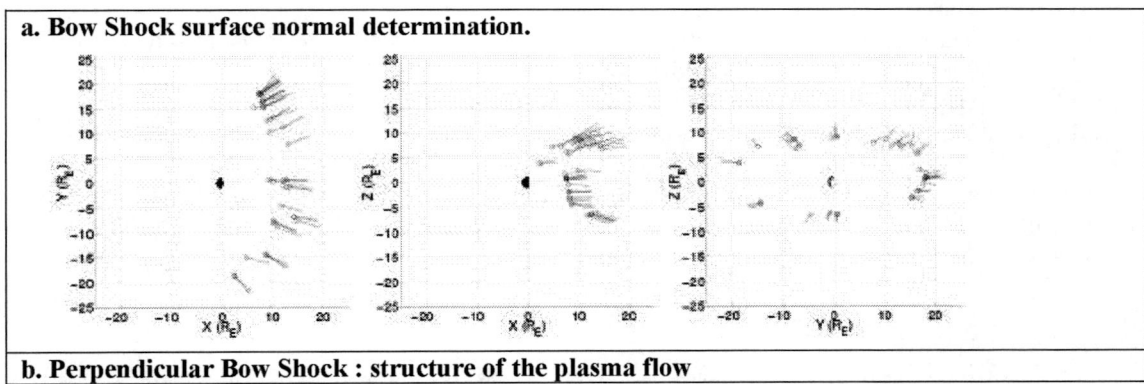

b. Perpendicular Bow Shock : structure of the plasma flow

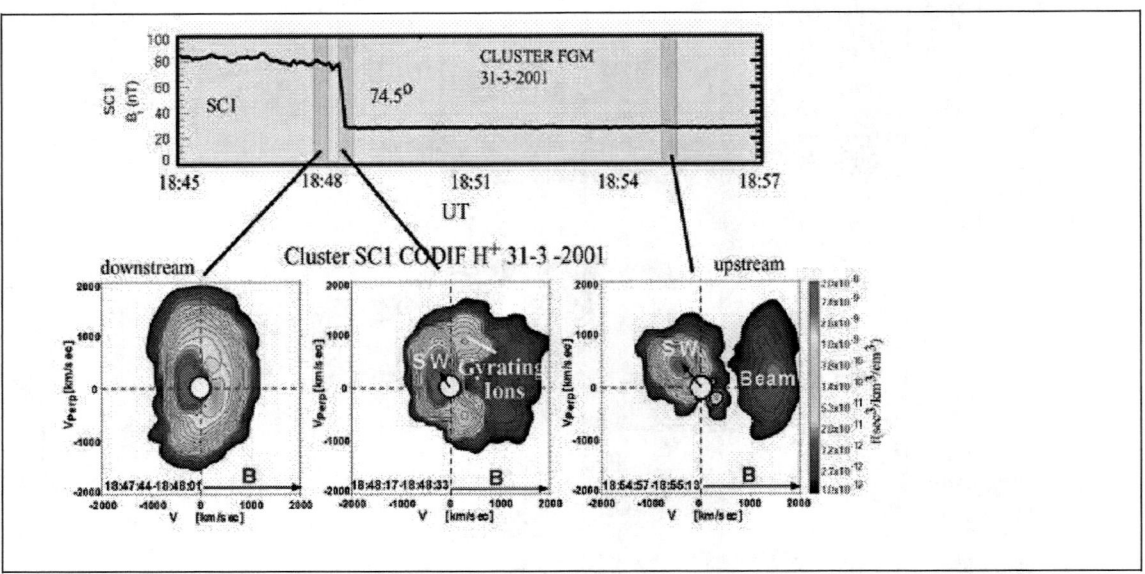

FIGURE 4. a) Measurements of the normal direction to the Bow Shock from Cluster measurements. b) Measurement by particle experiments of CLUSTER of the plasma flow through the shock. The particle velocity probability is represented in a two dimensional velocity space (from Ref. [2], Kucharec et al. et al. [4]).

The dynamics of the shock is also complex. The mechanisms for its persistence are still under debate, and the processes responsible for its motion and dynamics are the center of many studies. This complexity is visible through the quite different morphology of parallel and of perpendicular shocks, which gradually convert into each other as one move along the surface of the shock, thanks to its curvature. The combination of observations and of computer simulation is the privileged tools for these studies. It shows that the charges particles which are reflected by the shock play a significant role. They define a pre-choc forward of the region of quasi-parallel shock.

The main interest, today, focuses on several topics:

- The understanding of the structure of the fore-shock.
- The description of the small scale (microspic) structure of the shocks, particularly of the quasi-parallel shocks.
- The understanding of the dynamics of the shocks, its spatial and temporal variability.
- The identification of the nature and the structure of the flow of the particles reflected by the shock.
- The detailed understanding of the processes of the Fermi acceleration of charged particles by a moving shock.

In order to illustrate the physics that can be expected from multi-spacecraft space missions; Figure 4 show how CLUSTER enables one to determine the orientation of shock normal directions and the structure of the plasma velocity though the shock.

Boundary layers study: the Magnetopause

From the first observations on, the magnetopause is observed as a sharp transition between the Earth magnetic field, which is basically dipolar, and the magnetosheath magnetic field. It amounts to a sharp decrease of the magnetic field intensity, from the inside to the outside, and to a change in its orientation. This is illustrated in figure 5 which is the first observation of the magnetopause.

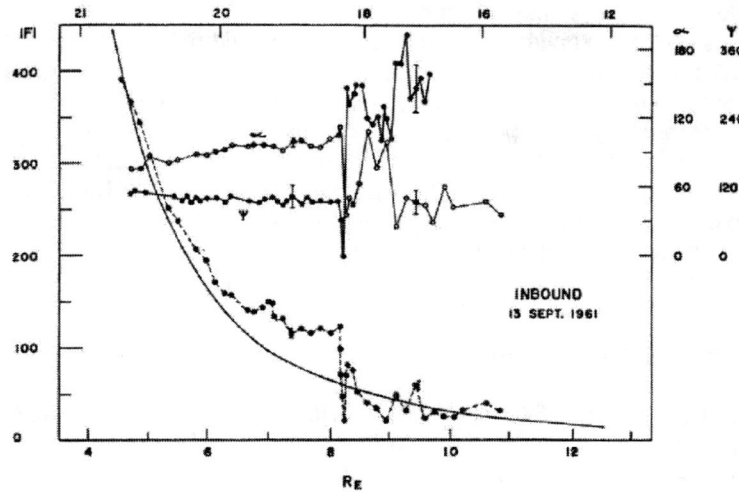

FIGURE 5: first observation of the magnetopause in 1963 [6]. The solid line is the amplitude of a dipolar magnetic field. Dots are observations. Superposed are the changes of the orientation of the field.

The processes of plasma and electromagnetic field transfer through the magnetopause depend strongly on the plasma and field into the boundary layer. Its dynamics is controlled by the plasma and field conditions on both sides of the frontier and on its orientation. Therefore the main topics of interest for its study are:

- The determination of the orientation of the frontier, which can be done in a similar way as it is for the Bow-Shock, using the differences in the timing of the boundary crossing by the four spacecraft. This orientation is a compeling information in understanding how the magnetosphere and the magnetosheath connect to each other, leading or not to the possibility for a plasma transfer. Accurate measurements are made when the boundary is well defined.
- The transfer of the plasma through the magnetopause is controlled by the nature of the discontinuity, which must be identified. According to the ideal magnetohydrodynamics (MHD), a tangential discontinuity does not allow such a mass transfer while a rotational one does.
- The study of the surface motion, inward or outward. It has been observed that the magnetopause reacts rapidly to plasma sheath and solar wind changes. These motions induce oscillations, resonances and instabilities of the global Earth magnetic field.
- One would also want understanding which field and which plasma parameters control the magnetopause thickness. How is maintained the steepness of the plasma gradients. These studies are made difficult, and require a great care, due to the boundary motions.
- The location, the motion and the stability of the magnetopause, when considered from a macroscopic point of view, are controlled by the interaction of the electric currents flowing on its surface (Chapmann currents) and the background magnetic field. Therefore the measurement of these currents is essential for the understanding of the dynamics of the frontier.

We have seen in the introduction of this section that only multi-spacecraft can provide the necessary information for these studies.

Boundary layers study: plasma entry

Were the magnetosphere physic be governed by ideal MHD, no plasma exchange would be allowed through the magnetopause. This is obviously not what is observed. The question of the permeability of the frontier to matter transfer is an open question. Until recently the privileged mechanism was a direct reconnection when, and where, the IMF field lines are anti-parallel to those of the magnetosphere. This

process is only allowed if energy dissipation occurs in some way at the reconnection point. The origin of this dissipation is still not agreed on. More recently localized processes like Flux Transfer Events (FTEs) where a solar wind flux tube merge into the magnetosphere, have been identified. Today one thinks also of a continuous transfer of matter. Quite probably all these processes may operate. Finally the region where topologically the field lines converge, over the North and the South magnetic poles, the Polar Cusps, are the focus of a number of studies. It is believed that a direct entry of the magnetosheath solar wind plasma is possible through this kind of magnetic funnel. Examples of the first and last two processes observed by CLUSTER will be presented here. The analysis of these processes requires both space and remote observations from ground based instruments.

Magnetic Reconnection at the Magnetopause

An example of magnetic reconnection, close to a Polar Cusp is shown on figure 6. The CLUSTER 1 spacecraft was moving from the inside of the magnetosphere (the lobe) to the magnetosheath tailward of the North Polar Cusp. During this time the IMAGE spacecraft was orbiting at a lower altitude through the auroral and Cusp field lines.

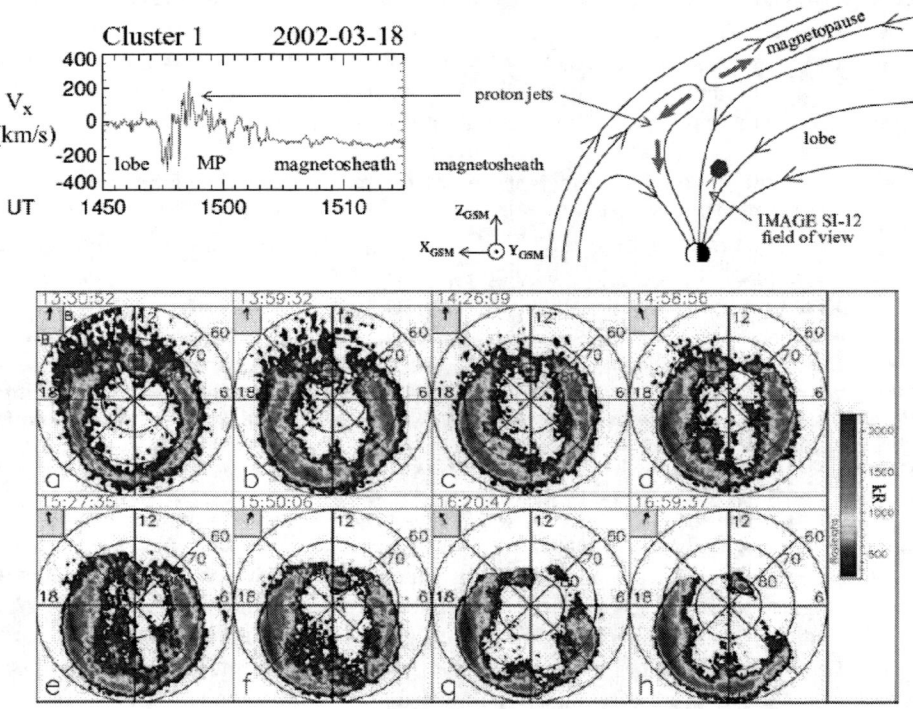

FIGURE 6. Observation by the CLUSTER 1 and IMAGE spacecraft of a reconnection event poleward of the Cusp (from reference [2], adapted from Phan et al. [7] and Frey et al. [8]).

In the upper-left box the plasma velocity measurement by CLUSTER is shown. The identification of the lobe, inside of the magnetopause, with a plasma at rest, the crossing of the magnetopause (MP) with strong fluctuations of the plasma velocity and the magnetosheath with a steady antisunward plasma flow is shown. Upper left is sketched the magnetic field configuration. The IMF is Northward therefore the reconnection is poleward of the Cusp. During this time the IMAGE spacecraft was orbiting in the same latitude range and a similar local time, at a lower altitude. It provided optical images of the auroral structure, downward in the ionosphere. A sequence of these images is shown in the bottom part of the figure. They are separated by approximately 30 min, which means that the whole event lasted for 4 hours. These images show the polar region of the Erth from above. The direction of the sun is upward

(noon – or 12h Magnetic Local Time meridian). The circles are the 60°, 70°, 80° latitude lines. The intensity of the auroras is coded in a colour scale. One clearly observes a corona of auroras and a dinstinct spot in the sun direction, which is produced by electrons precipitating in the cusps.

In this example, characteristic features of reconnection are clearly observed : the reconnection occurs when the interplanetary and planetary (Earth) magnetic fields are anti-parallel. A diffusive process (unknown) is necessary for breaking the field lines and allowing the reconnection. This reconnection is associated to particle precipitation of proton jets, such is sketched by the red arrows of the upper right part of the figure. These jets, when they collide the neutral atoms (O_2, NO) of the atmosphere at the ionospheric altitudes, are responsible for the auroral light spot mentioned above. As it appears, this process can be much localized. The process is therefore controlled by the orientation of the IMF.

Continuous plasma transfer through the magnetpause

We have presented a plasma transfer event well localized in space and in time. At the other extreme are continuous transfers of matter through the frontier. As long as the magnetopause can be considered, not as a tangential, but as a rotational discontinuity, characterized not by a shear of the magnetic field, but by a rotation through the frontier, the plasma is allowed to cross the discontinuity. Using the well known jump relation of plasma and field quantities through the discontinuity, one can predict the jump of the plasma velocity. It is given by the following expression:

$$\Delta \vec{v}_{pred} = \vec{v}_{2t} - \vec{v}_{1t} = \left[\frac{1-\alpha_1}{\mu_0 \rho_1} \right]^{1/2} \left[\left(\frac{1-\alpha_1}{1-\alpha_2} \right) \vec{B}_{2t} - \vec{B}_{1t} \right] \quad ; \quad \alpha = \frac{(p_{//} - p_\perp)\mu_0}{B^2}$$

Where indexes 1 and 2 refer to quantities on both sides of the discontinuity, $\vec{v}_t$ and $\vec{B}_t$ refer to the component of, respectively, the plasma velocity and the magnetic field tangent to the discontinuity surface. ρ is the plasma density, α is a dimensionless parameter describing the anisotropy of the plasma pressure. $p_{//}$ and $p_\perp$ are respectively the pressure of the plasma parallel and perpendicular to the magnetic field.

The involved vector and scalar quantities could be measured from CLUSTER observations and the predicted and observed velocity jumps could be compared. The result is presented on figure 7, where the red lines of the two bottom panels are the theoretical predictions.

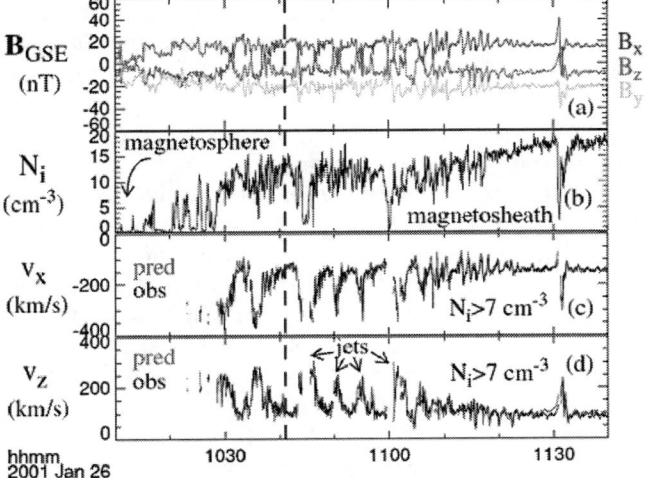

FIGURE 7. Observation of the continuous magnetopause reconnection, equatorward of the Cusp. (from reference [2], Phan et al. [9])

The agreement between observations and predictions is very good. It demonstrate that a steady reconnection is effective and enables one to estimate the rate of transfer of matter through the boundary.

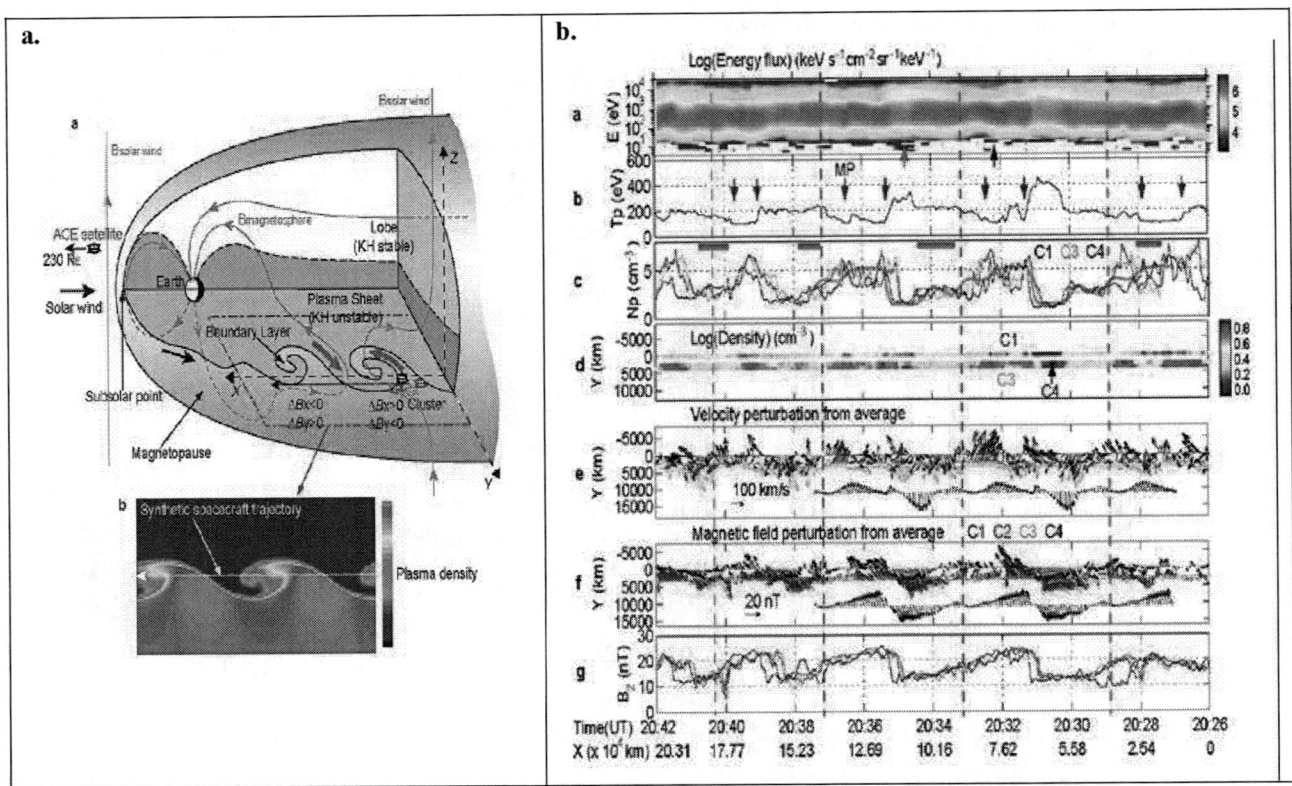

FIGURE 8. a. Sketch of the magnetosheath and tail lobe plasma mixing via Kelvin-Helmotz instability of the boundary. b. observations a such an event by CLUSTER. (from [2], Hasegawa et al. [10])

Plasma entry through instabilities at the magnetopause.

Aside of sporadic reconnection due to a suitable configuration of the magnetic fields and of a continuous reconnection, matter transfers can be produced by the instability of the boundary itself. As a matter of fact the particle and field energetic content of a volume of matter outside and inside the boundary is not the same and the frontier is potentially unstable due to the different flow velocities inside and outside. The most likely instability is the Kelvin-Helmholtz Instability. An illustration is shown on figure 8

The development of a Kelvin-Helmholtz instability at the magnetopause is sketched on the left part of the figure. Vortices form and grow, leading to a mixing of the plasmas of both sides, to the formation of a boundary layer and resulting in an entry of the plasmasheath plasma into the magnetosphere. The trajectory of CLUSTER through the structure is also represented; Corresponding observations by CLUSTER are shown on the right panel. Noticeably the deviations of the magnetic and plasma velocity vector fields along the spacecraft trajectory from their mean values are presented on panels e) and f). They clearly show the wrapping of the vortices. Panels a), b), c) and d) present respectively the energy flux, the proton temperature, the proton and the electron number density. Panel g shows the evolution of the component of the magnetic field perpendicular to the ecliptic plane

The Polar Cusps, a complex region.

The Polar Cusps are key regions for the interaction of the magnetospheric and the solar wind environment. Their great interest lies in several important features:

They are magnetically connected to most of the magnetopause, this means that their boundaries keep trace of most of the reconnection processes.

They are the siege of a strong turbulence activity, of complex shock structures and of intense wave-particle interactions. They allow, although with some difficulties, a direct penetration of magnetosheath plasma.

One reason for the difficulty of their study with particle measurements is the overlapping of the signatures of reconnection and of ordinary energy dispersion produced by their transit time through the cusp region which depends on their energy and their pitch angle. Moreover the boundary of the Cusps moves with changes in the IMF, ant it moves faster than the spacecraft. A sketch of their configuration is given on figure 9. The left panel corresponds to a southward orientation of the IMF. Therefore reconnection points (marked by the label "X-line") are sunward of the cusps. The resulting jets and auroral features are along the low latitude border of the footprint of the cusp region in the ionosphere. The right panel corresponds to a northward IMF. In the auroral ionosphere, the signature of reconnection processes is on the poleward edge of the cusp footprint.

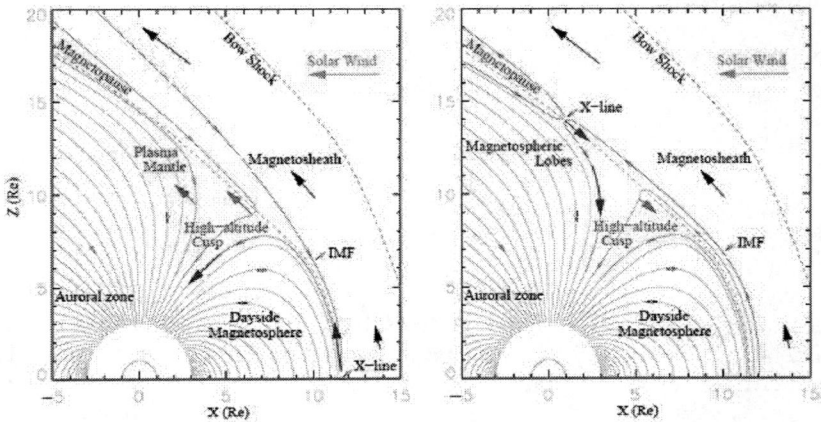

FIGURE 9. A schematic view of the magnetic topology of the cusp in the case of a southward IMF (left) and a northward one (right) (from ref. [2], Lavraud and Cargill [11])

SPACE WEATHER

Today one is able separate, and to some extend to understand, one by one, most of the plasma processes which occur between the solar corona and the planet environment. A next step in to link them in a single causal chain which starts from the solar atmosphere and ends up in the close Earth environment, and to asses the impact on human activity. This is what is named "Space Weather". One operational issue, in the future, is the development of a Space Weather forecast system, with suitable instruments for continuous observations, suitable simulation codes, suitable data bases and a suitable warning system.

For the time being event studies must be accumulated. In order to achieve it, a combination of space and ground based instruments is required. This is illustrated by a sequence of solar events which occurred for May 27 to 29 of year 1998 and were followed down to their impact on the Earth ionosphere. This study has been published in Hanuise et al. [12].

The ground based instrument which were used in this study were coherent radars, from the SuperDARN network, they gave maps of the plasma circulation velocity over the polar caps, EISCAT the incoherent scattering radar which gave the main plasma parameters along his beam (electron and ion density and temperatures, ion composition, plasma velocity), two chains of magnetometers (in Greenland and in Scandinavia), the Themis solar telescope, in the Canaries islands. Space satellites were SOHO for sun observations, ACE at the L1 Lagrangian point for a solar wind survey, CLUSTER and GOES 10 for the magnetosphere and the earth environment.

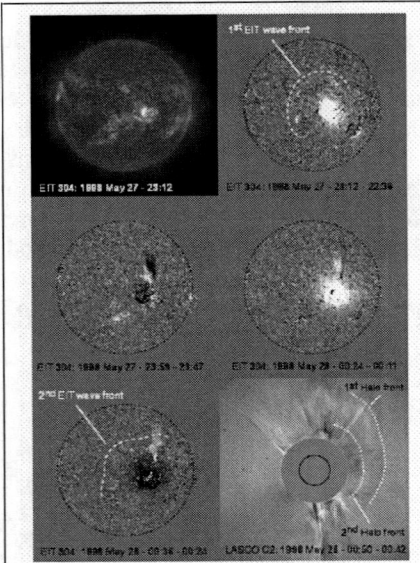 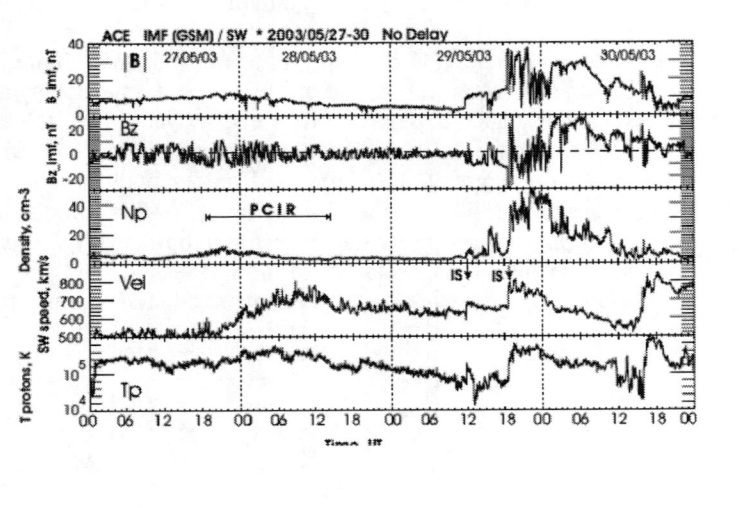

FIGURE 10. Left: SOHO observations from the initial filament and CMEs. For May 27 and 28, 1998 to the ejection of coronal mass. They are issued from the observation of the 30.4 nm HeII line by the EIT instrument. The right part shows plasma and magnetic field observations from the ACE spacecraft, located at the L1 Lagrangian point.

In the beginning a CME was observed in the solar corona by the SOHO spacecraft, located at the Lagrangian point. Several images of this event are shown on the left part of figure 10. The black and white structure of the centre left panel is the signature of the initial ejected filament. Then two wave fronts are observed over the solar disk. They give rise to 2 matter fronts in the solar corona which are observed on the sixth panel of the left part. On the right part of the figure are presented plasma and magnetic field observations from the ACE spacecraft, at the Lagrangian point for the 28-30 of May period. The disturbance is first observed as a "Pseudo Corotating Interaction Region" – PCIR, with a regular increase of the solar wind density followed by a decrease (3rd panel from the top), an increase of the solar wind velocity up to 700 km/s (4th panel) and a solar wind temperature increase. This is believed to be the signature of a coronal hole passing through the ACE location. Then 2 shocks are observed, resulting in a strong enhancement of the solar wind temperature, sharp increases of the solar wind speed and strong enhancement and fluctuations of the IMF.

The signature of the event on CLUSTER observations at the magnetopause level is a delay by 10 hours from model expectations of the boundary crossing, which was interpreted as strong compression of the magnetosphere and a fast inward motion by 5 Earth radii.

The impact on the ionosphere can be estimated from figure 11.

A large scale system of field aligned currents is permanently flowing between the magnetosphere boundaries down to the auroral ionosphere. They form the so-called Region 1 and 2 currents. The main ionospheric impact of the solar event was two fold: the current intensity increased strongly, the whole current pattern drifted equatorward, in agreement with the strong compression of the magnetosphere. Similar observations are made from the magnetometer chains (right part of the figure). The intensity of westward and eastward horizontal electrojets are deduced from the observed magnetic fluctuations. They are shown on the figure. One can observe the intensification of the jets and the south drift of the structure. The shift in timing of the observations from the two chains is caused by the local time shift due to their different longitudes.

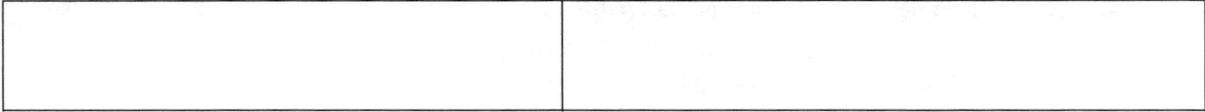

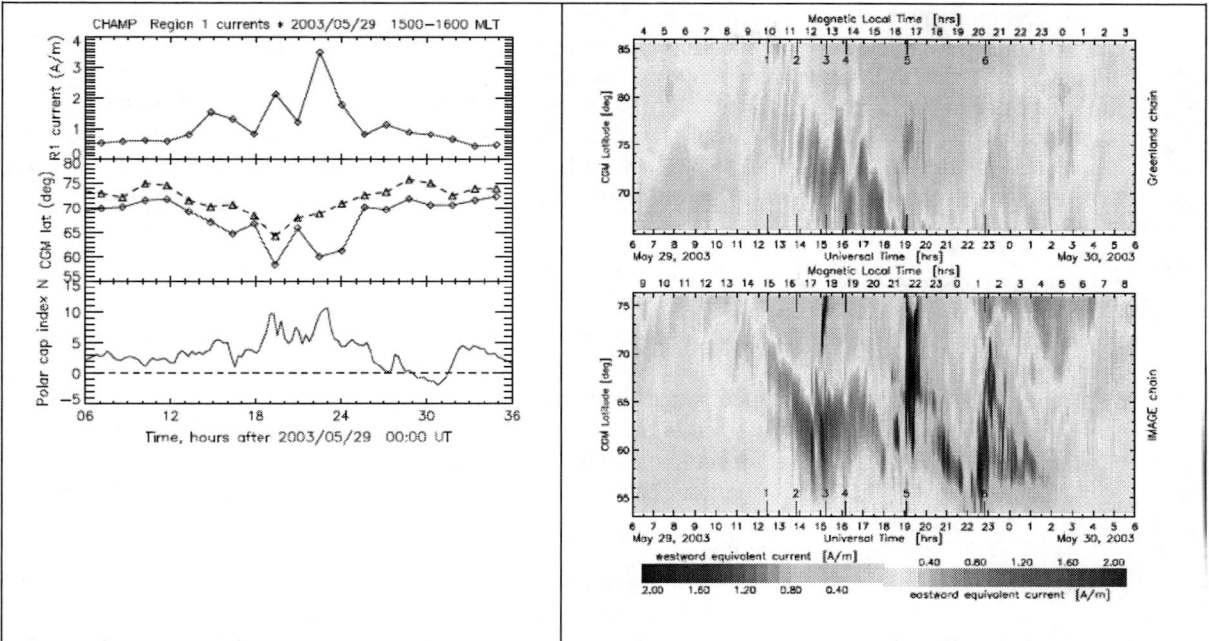

FIGURE 11. Left : Increase of the broad scale field aligned Region 1 currents. Top: increase of the current density, middle: equatorward drift of the low and high latitudes of the boundaries of the current sheets, followed by a recovery, bottom: polar cap magnetic index. Right part: equivalent currents deduced from the two magnetometer chains: top: Greenland, bottom:Scandinavia. In abscissa is platted the local time, in ordinate is the magnitude latitude.

The impact of the event on the thermosphere is shown on figure 12. On the left side one sees the diurnal oscillation of the thermospheric meridional winds. The observations of these winds during the event are shown in red. A strong enhancement of the wind velocity appears clearly.

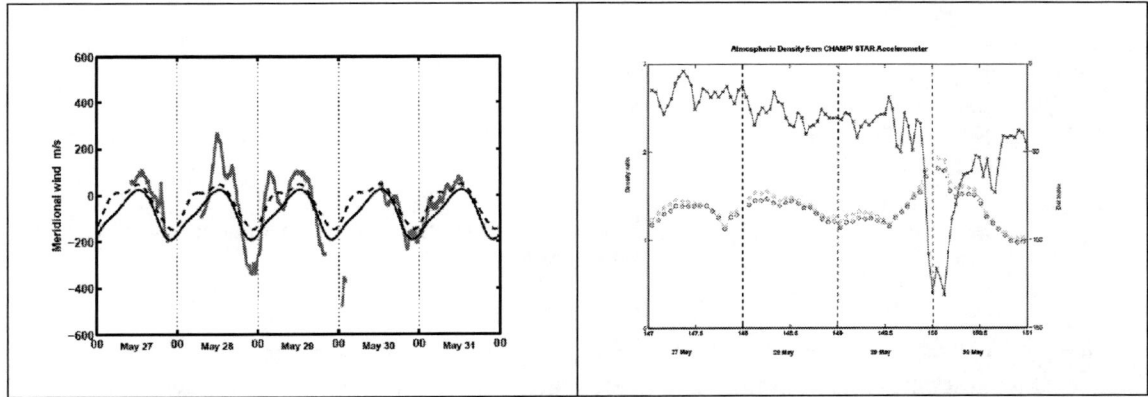

FIGURE 12. Impact of the solar event on the thermosphere. Left : increase of the East-West neutral winds. Right: increase of the thermospheric neutral atmosphrer density compared to two standard models (open dots). Dst magnetic index (blue continuous line).

The right part is a measurement of the thermospheric density deduced from the accelerometer on board the CHAMP satellite. What is plotted is the ratio of the observations and the predictions of two standars models. The enhancement of the density reaches a factor of 2. At the same time the strong depression on the D_{st} index time evolution corresponds to the impact of the event on the ionospheric currents described above. The enhancement of the thermospheric density induces an enhancement of the viscous drag of the atmosphere on the spacecrafts, reducing their life time.

Several consequences for the human activity result. One is the impact on radio communications. It can be induced from a strong attenuation of the radiowaves emitted by the SuperDARN coherent radar, operating at 50 MHz (a frequency much lower than those used for telecommunication). The radar signal dropped close to the noise level for most of the event. It can also be inferred from the strong increase of scintillations on the GPS signal, which were the result of the development of plasma instabilities associated to the increase of the currents and of the plasma convection. The equivalent for telecommunications is an increase of the rate of error on the transmission of the digital signals.

PLASMAS IN THE PLANET ENVIRONMENTS

In this section a mere overview of the center of interest in the plametary exploration will be presented.

The era of a systematic and quantitative exploration of the planets and the planetary systems has started. The enters of the scientific interest, as far as plasma physics is concerned, can be listed as:

- Comparative studies, thank to the diversity of situations which are encountered :
    - o The Earth, Jupiter, Saturn, Neptune have an internal magnetic field and an atmosphere.
    - o Venus has no magnetic field and an atmosphere.
    - o Mars have an atmosphere and remnants of a magnetic field.
    - o Mercury has an internal magnetic field and no atmosphere.
- Studies on the circulation of the charged particles and on the distribution of the plasma in the magnetospheres.
- The identification of the sources and sinks of plasma, which is a key issue when the atmospheres and magnetospheres are loaded with dusts and with planetary rings.
- The interaction of the solar wind with planetary atmospheres, which is particularly important for planets which have no magnetosphere, where the interaction is direct.
- The mechanisms for the atmosphere escape in the interplanetary space and the problem of the atmosphere erosion. A key question for Mars.
- The interaction of the largest planetary satellites with their plasma and magnetic field environment, with the case of Titan which interacts with the solar wind and the Saturn magnetosheath and magnetosphere.

Several space missions are dedicated to these studies :
- Cassini for Saturn and the Huygens probe for its satellite Titan.
- NASA-MSL09 launched in 2009, the ESA Pasteur project, the launch date is not yet settled.
- Venus Express for Venus.
- Mercury Messenger (NASA) and Bepi Colombo (ESA) with both a magnetospheric orbiter (MMO) and a planetary orbiter (MPO) for Mercury.
- Rosetta for cometary studies.

Computer simulations are compelling tools. Several king of simulations are used:
- Test charge codes, in which particle trajectories in a given plasma and field environment are computed.
- Codes simulating the interaction of the solar wind with the planeraty atmosphere. These codes include the photo-ionization, charge exchanges and chemical processes. They can be auto-coherent hybrid codes.
- Particle In Cell - PIC codes, for the study of shocks, plasma acceleration and escape processes.

CONCLUSION

The solar system deserves more than ever its qualification of a system. The key issues are the understanding of the mass and energy ejection from the solar corona, the knowledge of the structure and the dynamics of the solar wind and the understanding of the processes which regulate the exchanges of plasma between the solar wind and the plant environment. The variety of situation met through the solar system and in the environment of the various planets let foresee important progresses not only in the functioning of the solar system, but also in plasma physics.

The interaction of plasmas with neutral atmospheres and with the surfaces of with bodies, like dusts or comets, open a field for interdisciplinary studies involving cold and reactive plasmas and a collaboration with plasma physicists more oriented towards industrial objectives.

As in other fields of physics, fundamental physics leads to operational survey systems, and to the forecast of the impact of natural events, originating in the sun, on the human activity. This is the ultimate aim of Space Weather.

REFERENCES

1. Many textbooks present the MHD discontinuities. I would suggest reading the clear presentation of L. Landau and Lifschitz : Electrodynamics of continuous media.
2. Outer Magnetospherics. Boundaries : Cluster Physics. Space Science Series of ISSI, G. Paschmann, C.P. Escoubet and S. Haaland Eds, Springer, (2005)
3. M.G. Kivelson, and C.T. Russel, Introduction to Space Physics, Cambridge; New York; Canbridge University Press (1995)
4. T.S. Horbury, et al., Four Spacecraft Measurements of the Quasi-perpendicular Terrestrial Bow Shock : Orientation and Motion, J. Geophys. Res., **107** A8, doi. 10.1029/2001JA000273, 2002.
5. H. Kucharec, et al., On the Origin of Field Aligned Beams at the Quasi-perpendicular Shocks: multi-Spacecraft Observations by CLUSTER, Ann. Geophys., 22, 2301-2308 (2004)
6. L.J. Cahill and P.G. Amazeen, The boundary Layer of the Geomagnetic Field, J. Geophys. Res., **85**, 2903 (1963)
7. T.D. Phan et al., Simultaneous CLUSTER and IMAGE observations of cusp reconnection and auroral proton spot for Northward IMF, Geophys. Res. Lett, **30**, 1509, doi. 10.1029/2003GL016885 (2003)
8. H.U. Frey et al., Continuous magnetic reconnection at Earth's magnetopause, Nature, **426**, 533-537 (2003)
9. T.D. Phan et al., Cluster observation of continuous reconnection at the magnetopause under steady interplanetary magnetic field conditions, Ann. Geophys., **22**, 2355-2367 (2004)
10. H. Hasegawa et al., Transport of solar wind into Earth's magnetosphere through rolled-up Kelvin-Helmoltz vortices, Nature, **430**, 755-758 (2004)
11. B. Lavraud and P.J. Cargill, Cluster reveals magnetospheric cusps, Asronomy and Geophysics, (2006).
12. C. Hanuise et al., From the Sun o the Earth: impact of the 27-28 May 2003 solar event on the magnetosphere, ionosphere and thermosphere, Ann. Geophys. **24**, 129-151 (2006)

The Role of Free Falling Hypersurfaces

Ayub Faridi[1] Amjad Pervez[2], Haris Rashid[1], Fazal-e-Aleem[1]

[1] Centre for High Energy Physics, University of the Punjab, Lahore- 54590 Pakistan
[2] Institute of Chemical Engineering and Technology, University of the Punjab, Lahore-54590, Pakistan

Abstract. In General Relativity, free falling hypersurfaces play an important role. Special significance is attached to the frame of observer falling freely from infinity, starting at rest. In such a frame, the gravitational force deduced for a Schwarzschild source would be just the Newtonian force. In this paper, it has been shown that complete foliation of Penrose diagram by free falling hypersurfaces is possible in the Schwarzschild geometry.

INTRODUCTION

Linchnerowiez in his pioneering work beautifully spelled out the significance of maximal and constant mean curvature hypersurfaces [1] which now have a key role in the various aspects of general relativity [2-9]. There have been many insights obtained by using various foliation approaches [10-17]. A physically preferred frame is the Pseudo Newtonian (ψN) frame which corresponds to an observer falling freely from infinity. Significant progress has been reported in the earlier work in this area of research [14, 15]. Here we will further explore various aspects of these developments.

THE FOLIATION PROCEDURE

In order to obtain free falling hypersurfaces that foliate the Schwarzschild spacetime, the world lines of observers falling freely from infinity, starting at rest, must be orthogonal to such surfaces [18-20]. This enables one to write down the tangent vectors to the hypersurfaces themselves. This procedure works well and gives the required foliation of Schwarzschild spacetime [10-15]. We will briefly explain the procedure first and then give our most recent preliminary results. To be able to display the foliation more conveniently, we use Kruskal-Szekeres coordinates appropriate for the Carter-Penrose diagram [18-22]. Writing the unit tangent vector to the world-line of the freely falling observer as t^μ and the unit tangent vector to the flat hypersurface as T^μ, it is required that [20]

$$t^\mu t_\mu = -T^\mu T_\mu = 1, \ T^\mu t_\mu = 0 \tag{1}$$

The geodesic equation for the extreme path between two points can be expressed as [20]

$$\ddot{x}^\mu + \Gamma^\mu_{\nu\sigma} \dot{x}^\nu \dot{x}^\sigma = 0 \tag{2}$$

where $\Gamma^\mu_{\nu\sigma}$ is the Christofell symbols and the dot represents the derivative with respect to the arc length parameter, s. The tangent vector to the hypersurfaces can be found [18-20] by solving Eq. (2) using Eq. (1) for a given spacetime. The components of the tangent vector can be used to obtain an expression for the hypersurfaces

CP888, *Modern Trends in Physics Research,*
Second International Conference on Modern Trends in Physics Research—MTPR-06,
edited by L. El Nadi
© 2007 American Institute of Physics 978-0-7354-0354-9/07/$23.00

[18-20]. The Schwarzschild spacetime is completely foliated by this procedure [20]. The Schwarzschild spacetime metric in spherical polar coordinates [22] is defined as

$$ds^2 = -\left(1 - \frac{2M}{r}\right)dt^2 + \left(1 - \frac{2M}{r}\right)^{-1} dr^2 + r^2 d\theta^2 + r^2 \sin^2 \theta d\phi^2 \qquad (3)$$

It is required that the spacetime be viewed by the class of observers referred to above, i.e. those in the freely falling rest-frame at the appropriate place and time. It has been established [20] that the spacelike hypersurfaces orthogonal to the world lines of these observers then pass through the horizon smoothly. In Schwarzschild coordinate they are given by

$$t = 4mA - 2m \ln \left| \frac{\sqrt{r/2m - 1}}{\sqrt{r/2m + 1}} \right| \qquad (4)$$

where

$$A = \frac{t_0}{4m} - \sqrt{\frac{r}{2m}}$$

It is further observed [18-20] that in order to remove the coordinate singularity, the hypersurfaces in Compactified Kruskal-Szekeres coordinates are written as

$$\psi = \tan^{-1}(B) - \tan^{-1}(C) \qquad (5)$$
$$\xi = \tan^{-1}(B) + \tan^{-1}(C) \qquad (6)$$

where

$$B = (1 + \sqrt{r/2m})\exp(r/4m + A)$$

and

$$C = (1 - \sqrt{r/2m})\exp(r/4m - A)$$

It is clear that these hypersurfaces start at $r = 0$ with

$$\frac{d\xi}{d\psi} = 0$$

and end up at $r = \infty$ with

$$\frac{d\xi}{d\psi} = 1$$

In other words, these hypersurfaces come from spatial infinity and hit the singularity without going through the throat of the Einstein- Rosen Bridge. Free Falling Hypersurfaces have zero intrinsic curvature tensors and the mean extrinsic curvature of the hypersurfaces is given by [20]

$$K = \frac{1}{4m} - \frac{3}{2r}, \qquad (6)$$

The foliation as undertaken in present work is shown in Figure 2 that represents free falling hypersurfaces for certain values of time coordinates in upper-half of Penrose diagram.

RESULTS & DISCUSSION

The remarkable feature of the sequence of free falling hypersurfaces is that they provide a complete foliation of the upper as well as lower half of the Penrose diagram. The foliation as undertaken in our work is depicted in Figure 2. It represents the foliation of free falling hypersurfaces for different values of time coordinates in upper-half of Penrose diagram. The regions III and IV are foliated by the compactified coordinates by changing ξ to $-\xi$ in Eq 6. The foliation process shows that the upper left corner, D, of the Penrose diagram as shown in Figure 2 is

simultaneous with the lower right corner, A, in this particular frame. Since the geometry does not depend upon the choice of frames, we would argue that D should be labeled as I^- and not $\mathscr{I}^+$. This is also reasonable from another point of view. Inside the horizon t is a spacelike coordinate and r is a timelike coordinate. Thus $r = 0$ is a spacelike singularity whose points are labeled by different values of t. Since C is at $t = \infty$, D should be at $t = -\infty$. Hence it should be I^-. The edge joining D and E should be $\mathscr{I}^-$ in the maximal extension. Here A is a past time like infinity I^-, C is a future time like infinity I^+ and B is a spacelike infinity I^0. The compactification of these two sequences of hypersurfaces foliates the complete Penrose diagram.

TABLE 1. Compactified Coordinates ξ and ψ for different values of t_0 at $\frac{d\xi}{d\psi} = 0$

t_0	ξ	ψ
-8	1.59	- 1.53
-7	1.61	- 1.50
-6	1.63	- 1.47
-5	1.67	- 1.41
-4	1.74	- 1.31
-3	1.85	- 1.15
-2	2.01	- 0.939
-1	2.27	- 0.669
0	2.56	- 0.388
1	2.77	- 0.178
2	2.98	- 0.0653
3	3.06	- 0.0164
4	3.11	0.00170
5	3.13	0.00775
6	3.13	0.0283
7	3.13	0.0483
8	3.13	0.0386

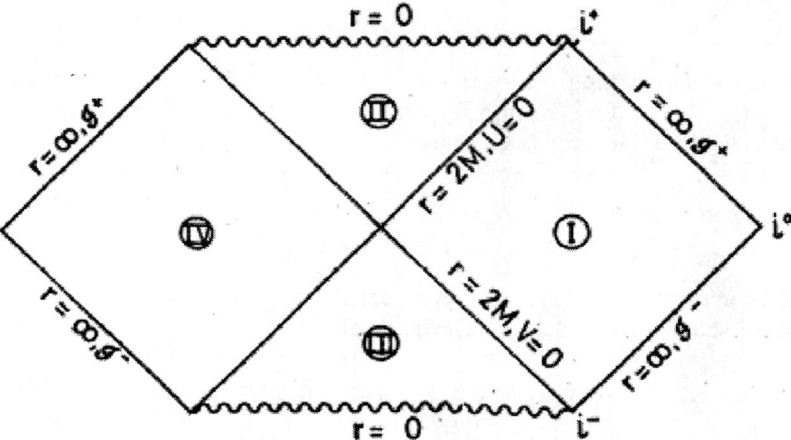

FIGURE 1. The Penrose- Carte diagram of the Schwarzschild solution [21-22]

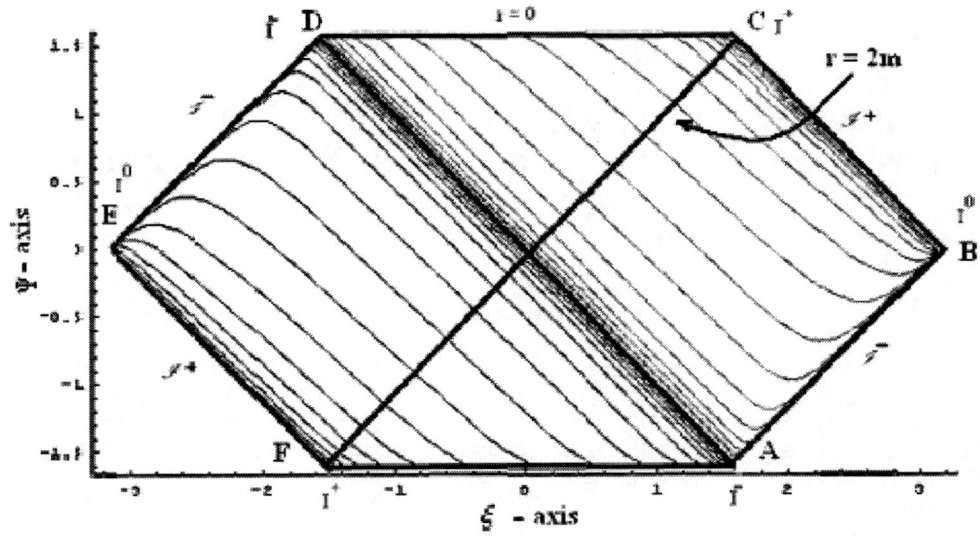

FIGURE 2. Union of the two sequences of Free Falling Hypersurfaces for the Schwarzschild geometry which results in complete foliation of Penrose diagram.

REFERENCES

1. A. Lichnerowicz, J. Math. Pure Appl. **23**(1944) 37
2. J.E.Marsden and F.J.Tipler, Phys. Reports **66**(1980) 109
3. J. Guven and N. Ó Murchadha Phys. Rev. D **60** (1999)104015
4. Rainer Burghardt " Free Falling Observers" ARG-2004-03
5. Qadir and A. A. Siddiqui, "Foliation of the Schwarzschild and Reissner-Nordstrom spacetimes by flat spacelike hypersurfaces," preprint Dec. 2000; K. Martel and E. Poisson, Am. J. Phys. 69, 476 (2001).
6. D.R.Brill, J.M.Cavallo and J.A.Isenberg, J.Math. Phys. **21**(1980) 2789;
7. J.York, Phys.Rev. Letters, 28(1971) 1656;
8. F.Estabrook, H.Wahlquist, S.Christensen, B.DeWitt, L.Smarr and E.Tsiang, Phys.Rev.**D7** (1973) 2814;
9. A.Qadir, Relativity: *An Introduction to the Special Theory* (World Scientific 989);
10. A. Pervez, A. Qadir and A. Siddiqui, Phys. Rev. D**51** (1995) 4598.
11. A. Qadir and Azad A. Siddiqui, J. Math. Phys. **40** (1996) 5883
12. A. Pervez, A. Qadir, *Proceedings of the Mini workshop on Astrophysics, Relativity & Cosmology,* eds. A. *Qadir and M. Sharif, QAU, 1992.*
13. A. Pervez, A. Qadir, *Proceedings of the Fourth Regional conference on Math. Phys., eds.* F Ardalan, H. Arfaei and S. Rouhani 1993.
14. A. Qadir and P.M. Valanju, Nuovo Cimento **B65** (1981) 404; A. Qadir and J. Quamar, *Proceedings of the Third Marcel Grossmann Meeting,* ed. Hu Ning, North Holland Publishing Co. (1983) 189
15. A. Qadir, Physics Letters, **A99** (1983) 419; Qadir, Europhysics Letters, **2** (1986) 426
16. A. Qadir, *Proceedings of the Second Chittagong conference Math. Phys., ed. J.N. Islam*
17. A. Qadir and I. Zafarullah, Nuovo Cimento **B111** (1996) 79; A. Qadir and A. Siddiqui Nuovo Cimento **B117** (1996)
18. A. Pervez, Ph.D. thesis, Quaid-I-Azam University, (1994) ; Azad A. Siddiqui Ph.D. thesis, Quaid-I-Azam University (2000) and references given therein..
19. Ayub Faridi, M.Phil Thesis University of the Punjab Lahore **(2002)** and references given therein.
20. Ayub Faridi et al., CHEP- Preprint 8/2006 and references given therein.
21. G. W Gibbons and S. W. Hawking, Phys. Rev. D **15** (1977) 2738.
22. C. W. Misner, K. S. Thorne, and J. A. Wheeler, Gravitation (Freeman, San Francisco,(1973).

Neutron Inelastic Scattering Mechanism and Measurement of Neutron Asymmetry Using Time of Flight Technique

AL AZZAWE, A. J. M.

Nuclear Energy Center, Argentina

Abstract. Inelastic scattering is an essential reaction for other nuclear reactions to detect the optical model and compound nucleus formation within the range of (0.4– 5.0) MeV neutron incident energy by using time of flight technique. The time of flight system (TOFS) installed on the horizontal channel reactor RRA has been used to measure the asymmetry of scattered fast neutrons, when data acquisition and system control were recorded event by event by HP – computer via CAMAC system. Eight NE 213 neutron counters were used in order to detect neutron inelastic scattering in the forward direction (4 neutron counters at 0° angle) and in the backward direction (4 neutron counters at 180° angle) to measure the asymmetry of fast neutron. Each neutron counter was 50cm in length and 8cm in diameter, viewed by two (58 – DVP) photomultiplier tubes. The contribution of direct interaction to the compound nucleus formation was deduced from the asymmetry in the neutron detection at the same direction of these eight neutron counters. A time resolution of 8.2 ns between the eight neutron counters and one of the two Ge(Li) detectors has been obtained.

INTRODUCTION

The majority of studies and investigations of neutron reactions with nucleons are very important for all advanced stages of nuclear physics investigations.

Neutron inelastic scattering occurs like other neutron reactions by absorbing the incident neutron and forming the compound nucleus in an excited state, or a result of neutron collision directly with any nucleon emission with reduced energy to excite the nucleus target a higher level without compound nucleus formation as intermediate stage (FIG. 1) [1]. Compound nucleus age is characterized with large time of 10^{-15} ns compared with nuclear transient time (neutron time required to pass the diameter of the nucleus = $(d / \upsilon_n) = 10^{-21}$ ns, where d is the nucleus diameter and υ_n is the incident neutron velocity). The compound nucleus may be forget its formation method when stay in the excited state for large time of 10^{-6} ns [2], after that begin to decay and reach to the stable state by means of the neutron emission, any nucleons or α – particles with γ – ray decay [3].

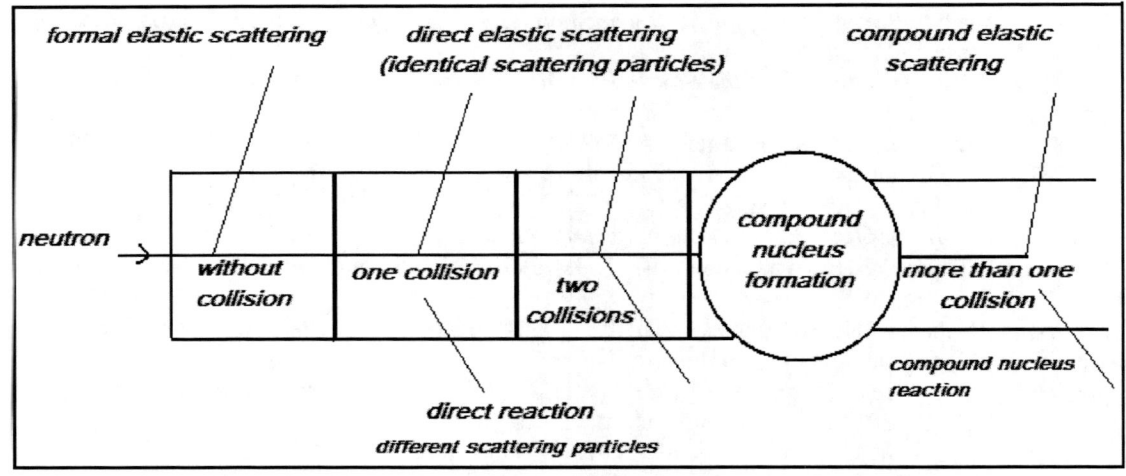

FIGURE 1. Direct reaction position relative to other nuclear reactions.

$$A_x + n \rightarrow (A_x)^* + n' \rightarrow A_x + \gamma + n', \quad E_n = E_n' + E_\gamma \qquad (1)$$

CP888, *Modern Trends in Physics Research,*
Second International Conference on Modern Trends in Physics Research—MTPR-06,
edited by L. El Nadi

Scattering neutrons from the target nucleus has been detected by time of flight measurements (time of scattering neutron from the target nucleus to any of eight neutron counter) [4].

$$TOF\ (ns) = (72.3\ D) / [E_n{'}(MeV)]^{1/2},\ E_{n'} = \tfrac{1}{2}\ m_n\ \upsilon_n^{2} \qquad (2)$$

Where:
D: distance from target nucleus to neutron counter in (m).
υ_n: neutron velocity in (m/ns).
m_n: neutron mass.

Two Ge (Li) detectors were used to detect the γ − ray emission from the excited levels after inelastic scattering. The distance between any one of these two Ge (Li) detectors and the target nucleus was 16cm within large solid angle respective to the target.

Eight neutron organic oscillation counters were used also to detect the scattering neutrons and limit the flight path of neutron distinguished from γ − ray and neutrons within low energies to obtain good energy resolution [5]. These eight neutron counters were used to calculate the scattering neutrons from Yb^{174} - levels and measure the neutron asymmetry. The flight path was limited with 100cm from neutron counters (NE − 213) and target nucleus.

Optical model produced by incident particle reaction with nucleus depending on compound potential method of individual particle model when the imagine part give the inelastic processes together with elastic scattering.
The selection of optical model depending on many parameters [6]:

$$V_{(r)} = - (V_v + iW_v)\ F_v\ (r) + 4ia_sW_s\ [dF\ (r)\ /dr] + V_{so}\ (\lambda_\pi^2\ /\ r)\ [dF_{so}(r)\ /\ dr]\ (\sigma.l) \qquad (3)$$

$$F_i\ (r) = [1 + exp.\ \{(r - r_i\ A)^{1/3}\ /\ a_i\}]^{-1} \qquad (4)$$

Where:
V_v: real depth of optical potential.
W_s, W_v: surface and volume imaginary optical potential depth.
V_{so}: spin and orbital potential components.
λ_π: wave length.
F_i (r): distribution function related with WOOD-SAXON potential:

EXPERIMENTAL PROCEDURE

A. Mechanical Setup

Scattering fast neutrons were obtained by using many types of materials inside the reactor horizontal channel of RRA, Pb collimator with another three of iron were put inside the channel. An cylindrical fragment of depleted uranium − 238 (5cm length and 4cm diameter) was put inside the first iron collimator near to reactor core to attenuate the emission of γ − rays coming together with the neutron beam. A Cd − layer of 0.5mm thickness was put to remove the thermal neutrons. A B_4C fragment was used to limit the resonance neutrons. Six collimators pairs of lead and iron were arranged on the channel gate (outside of reactor core), the length of each one equal to 15cm to limit the neutron beam (1cm diameter at the channel gate and 4cm at target nucleus) (see FIG. 2) [7].

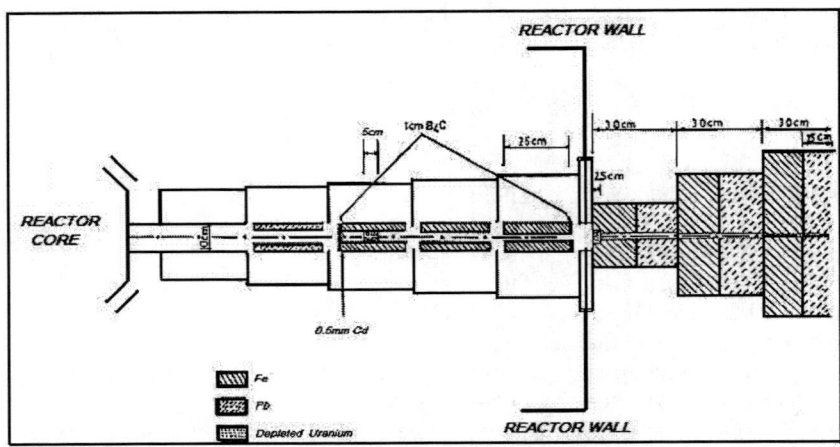

FIGURE 2. Schematic diagram of horizontal reactor channel of RRA.

During the calibration experiments, were observed background rays coming together with the neutron beam and were removed by using another U^{238} – depleted that was put at the channel gate (2.5cm length and 4cm diameter).

By using the neutron activation for many foils, the neutron emission flux has been measured $\approx 1.25 \times 10^8$ n/s. Eight similar neutron counters were used to detect the scattering neutrons from Yb^{174} target nucleus, each one 50cm length and 8cm of diameter fabricated from glass material of 3mm thickness; viewed by two photomultiplier tubes (58 – DVP). Each neutron counter was filled with organic scintillation liquid (NE – 213) to separate between neutron and γ – ray. The O_2 dissolved within NE – 213 was removed by using pure nitrogen corresponding to mean pressure on the counter tube [8].

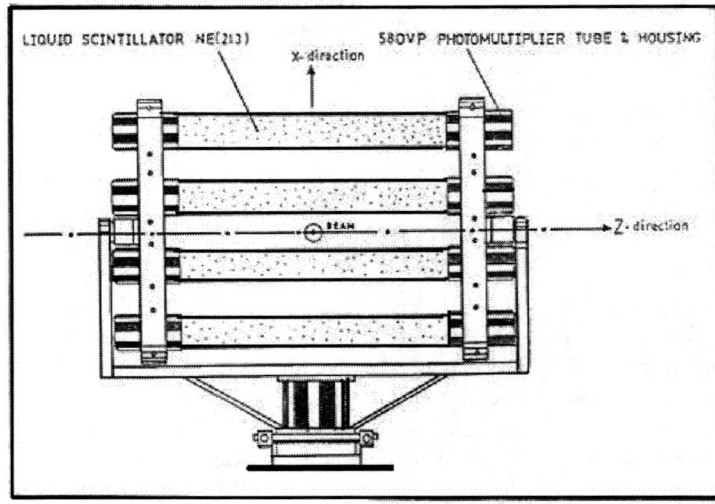

FIGURE 3. Forward hodoscopes of neutron scintillation counters.

The eight neutron counters were arranged in two groups, 4 neutron counters in the forward direction ($\sum F$ hodoscopes at 0° angle) as shown in FIG. 3 and 4 neutron counters in the backward direction ($\sum B$ hodoscopes at 180° angle) corresponding to neutron incident beam. The distance between neutrons counter center and other in the same hodoscopes equal to 25cm.

The two hodoscopes have rotation ability around target nucleus and can move toward the target nucleus within (50 – 120) cm, also have ability to interchange between them to measure the asymmetry of fast neutrons. For this investigation, the distance between any hodoscopes and target nucleus has been limited as 100cm (shown in FIG. 4).

Fast scattering neutrons that do not interact with target nucleus have been removed by means of neutron beam catcher (3m distance from the target nucleus) to minimize their velocities to change a thermal neutrons and absorbed through beam catcher materials.

Two Ge (Li) detectors have been used to detect the γ – rays emitted from the nucleus excited levels after the inelastic scattering reaction, the crystal of each one having 66mm length and 47mm in diameter with 20% efficiency corresponding to NaI(Tl) 3"x3" detector efficiency. The energy resolution of two Ge (Li) detectors was measured as (1.9KeV) at (1332.3KeV) excited level energy and the time resolution of 8.2ns was measured between the two neutron hodoscopes and one of the γ – ray detector.

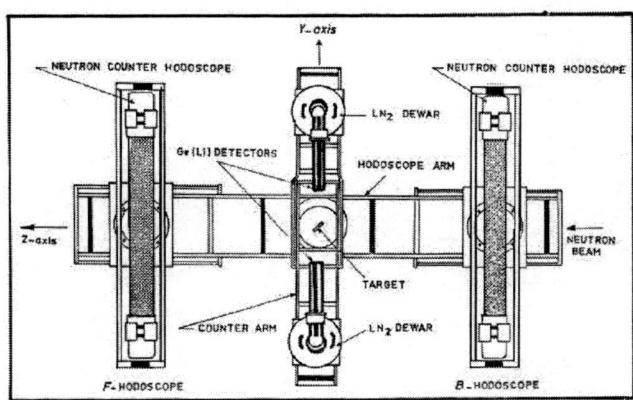

FIGURE 4. Schematic diagram of neutron counters with two Ge (Li) detectors.

A Pb shield of 10cm length and 5cm thickness was used to protect them from background rays, Lithium – Polyethylene plate of 5cm thickness was put brow the each Ge (Li) crystal to reduce the incident neutron effect. Each γ – ray detector has the ability to move toward the target nucleus within (15 – 30) cm and for this experiment is 16cm perpendicular to neutron beam.

B. Electronic Design

Various electronic instruments (see FIG. 5) were used to obtain the pulses related with the time of flight of scattering particles inside neutron counters when its shape limit the scattering particle type from excited state of target nucleus. The instruments groups provide also the pulse width of two Ge (Li) detectors related with γ – ray energies after inelastic scattering processes.

The select of (γ – n) signal was obtained by double coincidence between any Ge (Li) detectors signal and any one of eight neutron counters signals to record four parameters to every event from this coincidence related with time of flight of neutrons reach to left and right sides of each neutron counters through (58 – DVP) photomultiplier tubes and two measurements of charge produced from the integration process of anode currents of the two (58 – DVP).

The produced signal from any 16 photomultiplier tubes inter through an amplifier and after that to discriminator unit to cut the noise pulses depending on the threshold voltage to obtain logical signal (width time 50ns). The two discriminator signals of the photomultipliers of any neutron counter inter to delay time unit, another signal was take from the discriminator after an delay time of 150ns inter to analog to digital converter to represent the final signal of time of flight. Eight neutron counter signals were take to the two assembly units (A, B) to get the two out pulses ($\sum$ F) and ($\sum$ B) respectively and inter every one to coincidence units (a, b, c, d) with the two signals of (γ_1) and (γ_2) from the two Ge (Li) detectors to represent (γ_1.n) as (γ_1. $\sum$ F, γ_1. $\sum$ B), or (γ_2. n) as (γ_2. $\sum$ F, γ_2. $\sum$ B).
The start signal of time of flight is the coincidence time signal between scattering neutrons and the γ – ray release at time of γ – ray signal.

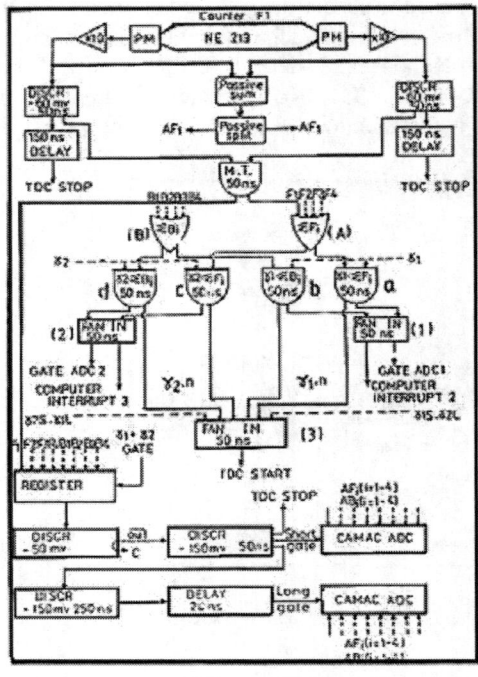

FIGURE 5. Schematic electronic diagram of one neutron counter.

The (γ_1. γ_2) signal was select from double coincidences between the two pulses of the two Ge (Li) detectors to represent the start signal when measuring time of flight spectra, recording four parameters of (γ_1. γ_2) and their times to obtain pulses width by means of the two analog to digital converter of (8192) channels.

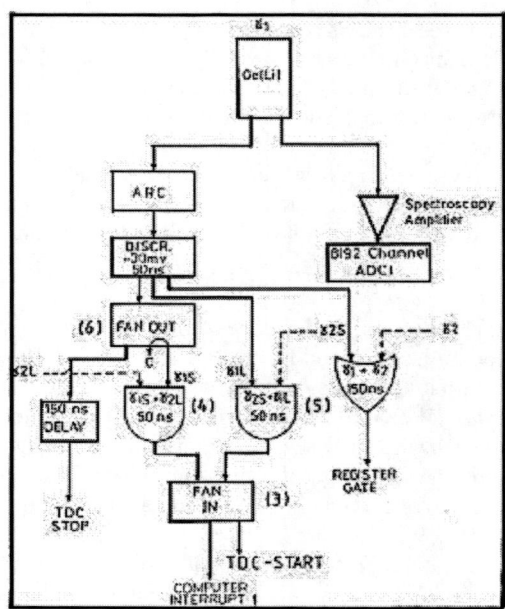

FIGURE 6. Schematic electronic diagram of one γ – detector.

Short time signal (γ_1S) from the first Ge (Li) detector was taken from discharge unit after clip process and long time signal (γ_2L) from the second Ge (Li) detector was taken after an delay of (50ns) to coincidence unit. In the same method, the two time signals (γ_1L, γ_2S) were taken to another coincidence unit. All signals inter to FAN IN / FAN OUT unit to represent the start signal in time to digital converter (TDC) unit.

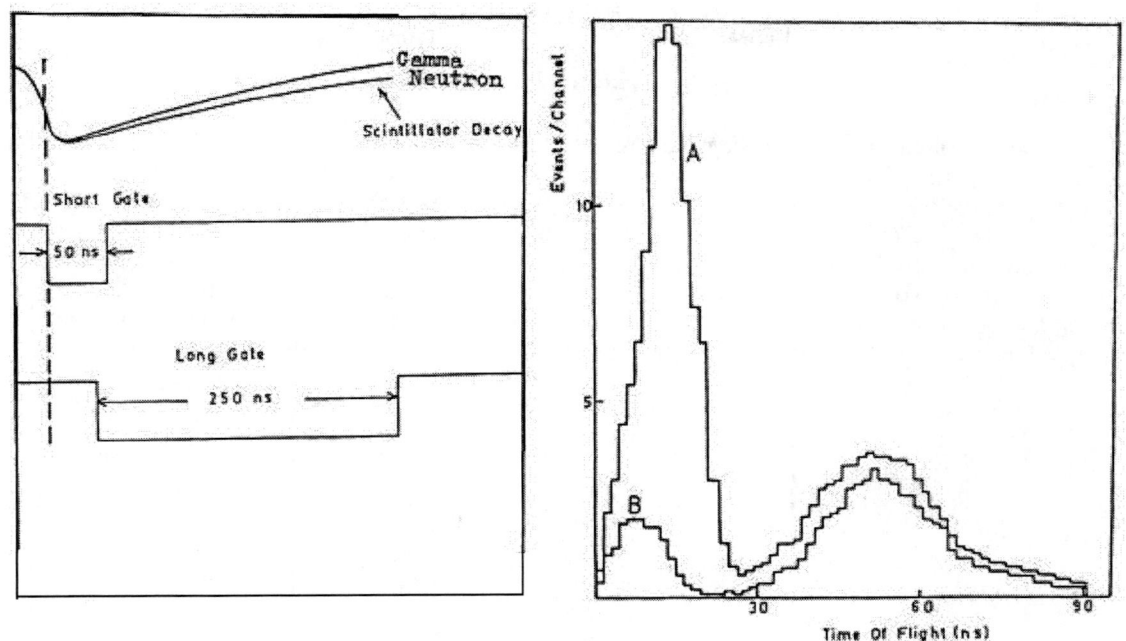

FIGURE 7. Short and long gate time of organic scintillation pulse.

FIGURE 8. Diagram time distribution:
(A): Before separation process.
(B): Charge comparison condition.

Charge comparison method (FIG. 8) was used to separate between scattering neutron and γ – ray depending on charge integration of low and fast component of neutron scintillation counter and comparison between two charge quantities after convert them a digital data by used two analog to digital converters. A Cf^{252} have been used for charge comparison method to reduced the γ – ray for 7 times with (12%) neutron events lost.

Stop signal of time of flight was taken directly from discharge unit after a delay of (150ns) inter in time to digital converter (TDC) of CAMAC system which control automatically all experimental measurements to transport the information through an magnetic tape by means of electronic terminal computer (Hewlett Packard 21 MX / E) with (64K x 16 bits) memory where K = 1024 words connecting with CAMAC Multi – Crate Interference unit as shown in FIG. 9 below.

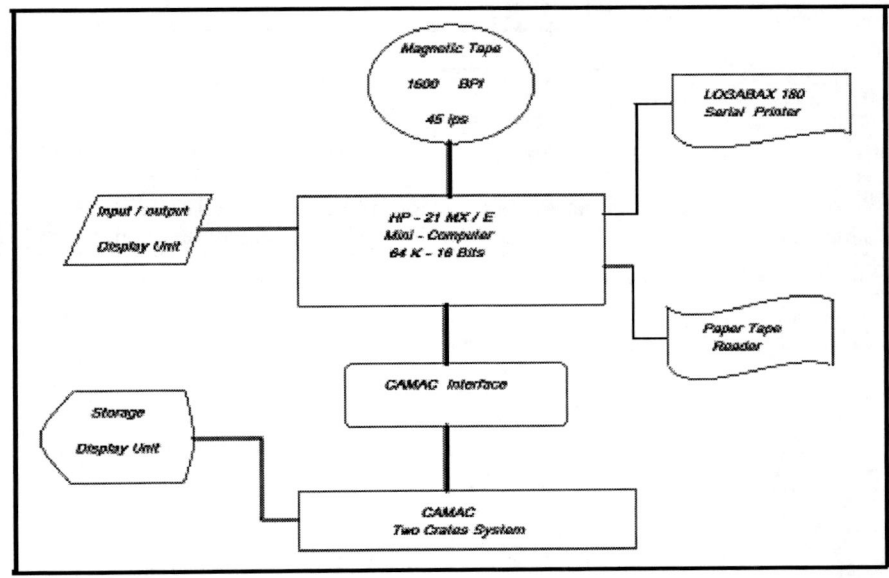

FIGURE 9. Schematic diagram of data acquisition system (DAQ).

283

EXPERIMENTAL RESULTS

A. Primary Experiments

Varied measurements and primary experiments were done by using Cf^{252} and $Am^{241} - Be^9$ neutron sources by prepare high efficiency nuclear instruments with low background.

An Am^{241} source was used to limit the threshold energy of the reaction by selecting 0.3 MeV equivalent to γ – ray energy of (59 KeV) of this source by means of measurements the record number of neutron counter ends, produced signal was taken from the discriminator (CFD) to timer unit with and without source. Threshold voltage value of neutron counter end represented in the FIG. 10, the straight line cutting the (1) value relative to coaxial ratio when neutron counter does not detect neutrons with energy equal or low than 0.3 Mev energy and γ – ray with energy equal or low than 59 KeV.

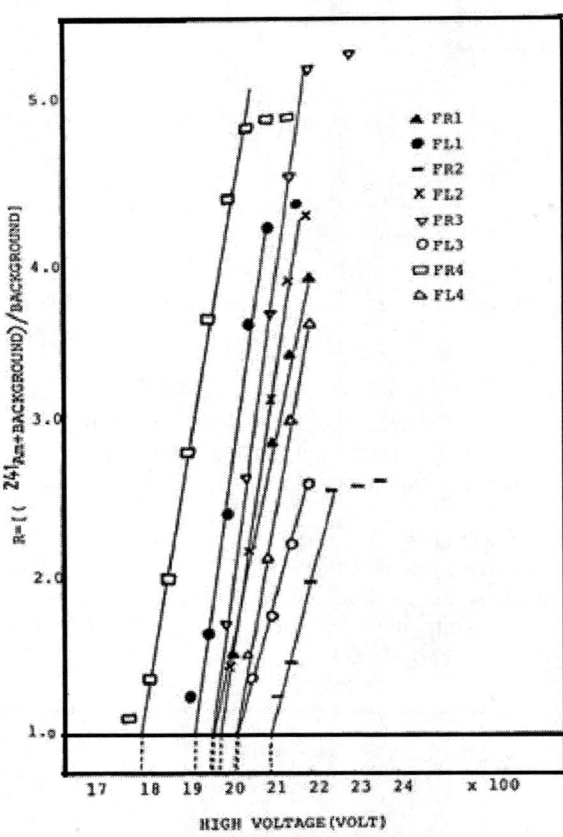

FIGURE 10. Relation between the ratio R and feeding voltage of the ends of forward neutron counter group.

A Eu^{152} γ – source was put at target position [16cm from the two Ge (Li) detectors] to record γ – ray spectra emission (see FIG. 11) from the source by using two analog to digital converters (ADC'S) with MINUIT program [9].

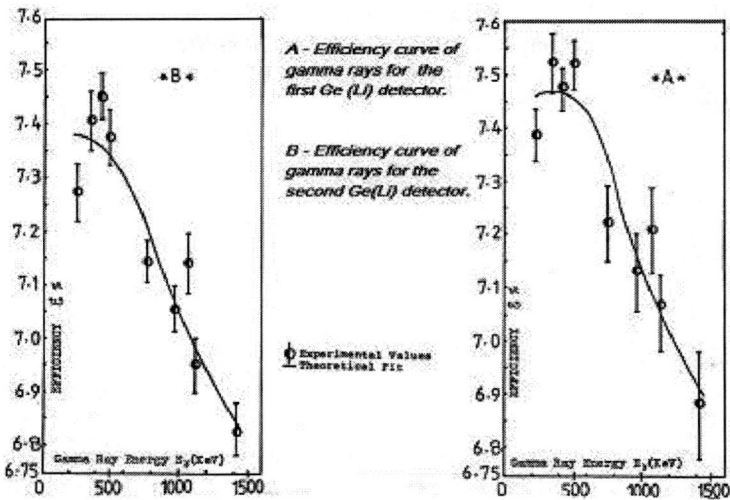

FIGURE 11. Efficiency curve of the two Ge (Li) detectors.

Varied experimental measurements were done on the Fe^{56} isotope, specially the selected first excited level 847 (2^+) KeV with distinguished emission of γ – ray (847) KeV illustrated through FIG. 12 and comparing the results with Yb^{174} excited levels to measure the asymmetry of inelastic fast neutrons.

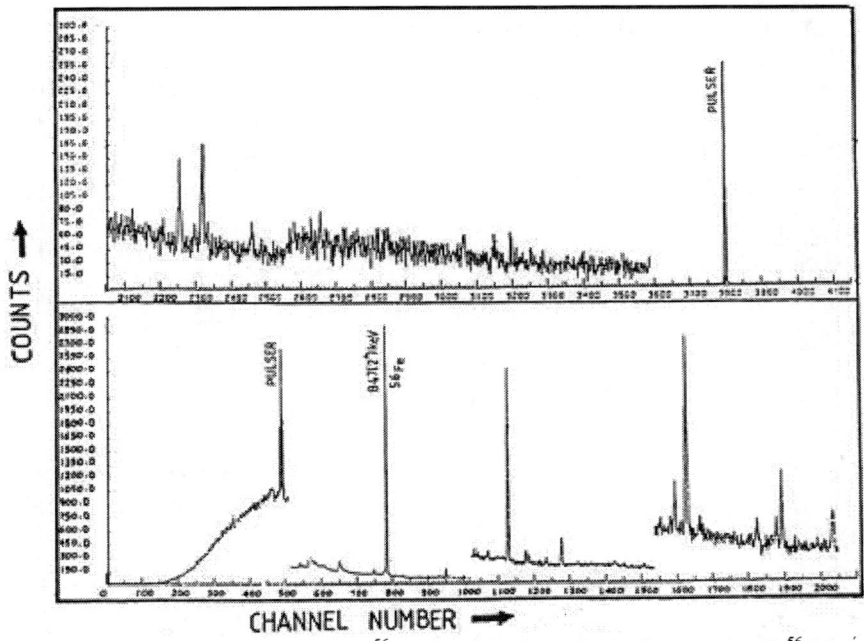

FIGURE 12. Emission of γ – ray spectrum from Fe^{56} isotope levels after inelastic scattering Fe^{56} (n, n'γ) of fast reactor neutrons.

B. Asymmetry Measurements

The direct reaction contribution in the inelastic scattering of fast neutrons reactor has been calculated through measurement of asymmetry as the best method to indicate the comparison between the compound nucleus formation (the reaction is distinguished as homogenous in all directions controlled by the statistical contribution and the direct reaction (scattering neutrons take resonance peaks in the forward direction to excited limiting levels).

The total number of scattering neutrons from excited nucleus target of Yb^{174} and Fe^{56} – levels was calculated respected to the direction of the two neutron counter hodoscopes for two different positions:

(1) Ordinary position(forward neutron counter hodoscopes $\sum$ F near to beam neutron catcher at $0°$ angle and backward neutron counter hodoscopes $\sum$ B near to channel gate at $180°$ angle) where these two angles represents the neutron counters position to neutron incident beam.

Asymmetry equation for this position:

$$\prod (ORDINARY) = [\, N_1(F)\, \varepsilon_F - N_1(B)\, \varepsilon_F\, \alpha\,] \,/\, [N_1(F)\,/\,\alpha + N_1(B)] \qquad (5)$$

(2) Rotated position (backward neutron counters hodoscopes $\sum$ B near to beam neutron catcher at $0°$ angle and forward neutron counter hodoscopes $\sum$ F near to channel gate at $180°$ angle)

Asymmetry equation for this position:

$$\prod (ROTATED) = [N_2(B)\, \varepsilon_B\, \alpha - N_2(F)\, \varepsilon_B] \,/\, [N_2(B) + N_2(F)\,/\,\alpha] \qquad (6)$$

Where:

$N_1(F)$, $N_2(F)$: Neutron number detected by forward neutron counters $\sum$ F in ordinary and rotated position.
$N_1(B)$, $N_2(B)$: Neutron number detected by backward neutron counters $\sum$ B in ordinary and rotated position.
α: Neutron counter efficiency ratio = $(\varepsilon_F\,/\,\varepsilon_B)$, equal a unity when all efficiencies are the same and near to unity when they are different.

The solution of these two equations has been done to obtain the final asymmetry equation:

$$\prod (TRUE) = [N(F) - N(B)] \,/\, [N(F) + N(B)] \qquad (7)$$

Where:

$$N(F) = N_1(F) + N_2(B) \quad \text{and} \quad N(B) = N_1(B) + N_2(F)$$

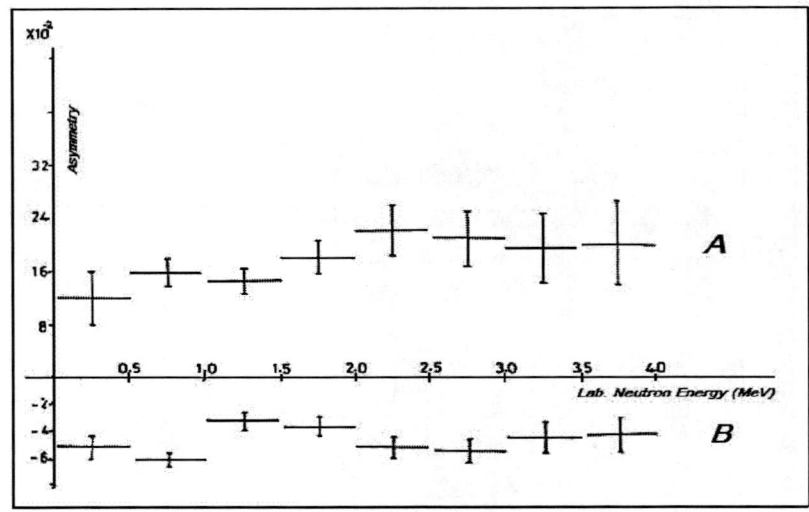

FIGURE 13. A – Asymmetry of inelastic scattering neutrons of 874 (2^+) KeV level of Fe^{56}.
B – Asymmetry in time of flight system using Cf^{252} neutron source.

The part (A) of the FIG. 13 illustrate the asymmetry of inelastic scattering of fast reactor neutrons from $847(2^+)$ KeV level in Fe^{56} isotope after taking in acount the time coincidence events detected by the two neutron counter hodoscopes for the two positions. Correction method for these events was done by using Cf^{252} neutron source (have the same neutrons emission for all directions) which be shown in the part (B) of the same figure.

Error ratio for asymmetry values represents the statistical error in the experiment, it is noted that the quantum quantity of direct interaction contribution in front of compound nucleus formation in the inelastic reaction for this level and observed an increasing of asymmetry from (12%) at low energies to (20%) for several MeV neutron energy.

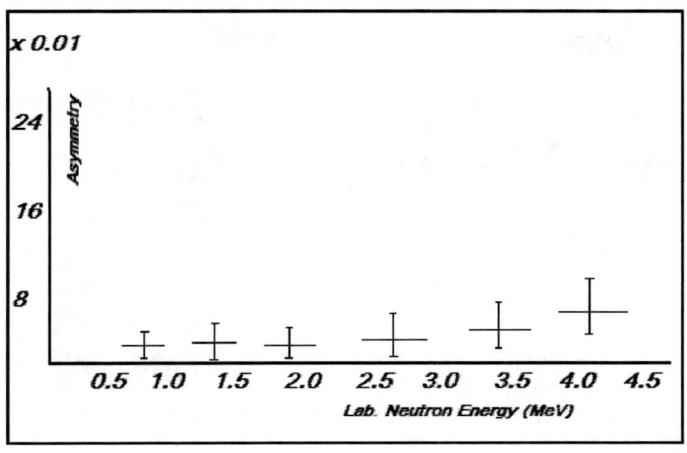

FIGURE 14. Asymmetry of inelastic scattering neutrons from 889.7 (8$^+$) KeV level of Yb^{174} isotope.

The same experiment was done by using equation (7) on Yb^{174} isotope levels when the asymmetry have not contribution in high energy levels and appear with (3%) at high incident neutron energy in one of the low energy levels 889.7 (8$^+$) KeV as shown in Fig 14.

The final results of asymmetry from Yb^{174} levels indicate no direct interaction share and the inelastic scattering reaction occurred through compound nucleus formation take in count the low statistical fluctuation.

REFERENCES

1. J. R. Lamarch; Introduction to Nuclear Reactor Theory – New York, 1966, Chap. 2.
2. D. F. Jackson; Nuclear Reactions – London, 1970, Chap. 5.
3. P. E. Hodgson, Nuclear Reactions & Nuclear Structure – London, 1971, Chap. 3, 13.
4. J. B. Marion & J. L. Fowler; Fast Neutron Physics – New York, 1960, Part I, P. xiv.
5. J. B. Birks; The Theory & Practice of Scintillation Counter , Pergamon, London, New edition, 2003.
6. V. M. Bychkov et al. ; INDC (CCP) – 217 / LI (1983).
7. A. J. Al Azzawe; Measurement of Inelastic Scattering Cross – Sections for Reactor Fast Neutrons from Yb^{174} using Time of Flight Technique, Unpublished, NUPPAC – CAIRO (2005).
8. F. D. Brooks; Nucl. Instr. & Meth. , **162** (1979).
9. F. James & M. Roos; MINUIT – Computer Physics Commun. 210 (2004).

Effect of Gamma Irradiation on The Dielectric Properties of La Doped CuZn Ferrite

M.A.Ahmed[*], A.A.Ramadan[**], M. A.El-Ahdal[**], M.M .Kamal[*] and A. yousef[**]

[*] *Physics Department, Faculty of Science, Cairo University, Giza, Egypt.*
[**]*Atomic Energy Authority, National Center for Nuclear Safety, Nasr City, Cairo Egypt.*

Abstract. The electrical Properties of Lanthanum doped Cu-Zn ferrite of the general formula $Cu_{0.5}Zn_{0.5}La_{0.35}Fe_{1.65}O_4$ was carried out at different temperatures as a function of the applied frequency. The studied samples were prepared by double sintering ceramic technique. X-ray diffraction analysis was performed to assure the formation of the sample in single phases. The effect of gamma irradiation on the investigated samples was also studied. Two types of charge carriers were found in the form of two types of conduction, n-type and p- type as it was found by Seebeck measurements. Hopping mechanisms was the most predominant one that exists in the conduction processes. The variation of dielectric loss with gamma irradiation dose illustrates the different responses which enhances the change of magnetic ordering from ferromagnetic to paramagnetic state.

INTRODUCTION

Ferrite materials have many interesting physical and technological applications specially those containing rare earth metals ions [1, 2]. The rare earth ferrites have found important applications in modern telecommunication and electronic devices. For this reason, engineers and scientists are keenly interested in determining their characterizations [3, 4]. Because of low electrical conductivity of ferrites and rare earth ferrite in comparison with those of magnetic materials they play a useful role in many magnetic applications. The electrical properties of ferrite materials have been found to be altered depending on the substitution of different valence cations as well as the preparation conditions such as sintering temperature, sintering time and rate of heating and cooling [5, 6]. So that the electron exchange interaction between Fe^{2+} and Fe^{3+} results in a local displacement of the electrons during the sintering of ferrites [7].

During the last few years, many kinds of oxide ceramics have been investigated as materials for humidity sensors used in automatic humidity controlling systems. Basically, ceramics can detect humidity on the basis of the change of the surface electrical conductivity by vapor adsorption. Therefore the ferrites possess a dielectric constant as high as 10^5 and are useful in designing good microwave devices such as isolators, circulators.

In certain materials a permanent change in electrical conductivity may be produced by radiation damage to the crystal [8]. In these materials the change is a function of the total dose absorbed in the material. When radiant energy acts on materials its electrical properties may change and new electrical phenomena may develop in it; furthermore, profound change may occur in the structure of the materials, its mechanical strength, optical properties, etc. The chief advantage of solid- state conductivity devices is that their high – density and low ionization potentials enable dosimeters of small volume to be fabricated. The large dose of gamma rays permanently alters the electrical conductivity [9]. This change in conductivity can be measured as a function of the adsorbed dose in the detectors, so the system then becomes an integrating dosimeter. Ferrites are being considered to be suitable for this application.

The present work reports an attempt to throw light on the effect of ionizing radiation, temperature and frequency on the dielectric constant, dielectric loss, ac conductivity and Seebeck coefficient of Cu–Zn-La substituted ferrite. This will open a new era for using irradiation in optimizing the electrical properties

EXPERIMENTAL TECHNIQUES

The Ferrite sample $Cu_{0.5}Zn_{0.5}La_{0.35}Fe_{1.65}O_4$ was prepared by the standard ceramic technique from pure Analar oxides (BDH) CuO, ZnO, (Fe_2O_3), and (La_2O_3) these oxides were mixed perfectly using agate ball – mill and pressed into pellet form using uniaxial press of pressure of 8×10^5 N/m^2. The samples pre sintered

CP888, *Modern Trends in Physics Research,*
Second International Conference on Modern Trends in Physics Research—MTPR-06,
edited by L. El Nadi

at 800°C for 10 hours and final sintering at 950° /30h using rate of heating 4°C/min .The two surfaces of each sample were polished, coated with silver paste and checked for good conduction. The ac conductivity and dielectric constant are measured using RLC bridge model 3531 (Japan) self calibrated. The temperature of the sample was measured using T-type thermocouple with accuracy better than ±1°C where its junction in conductor with the sample.

RESULTS AND DISCUSSION

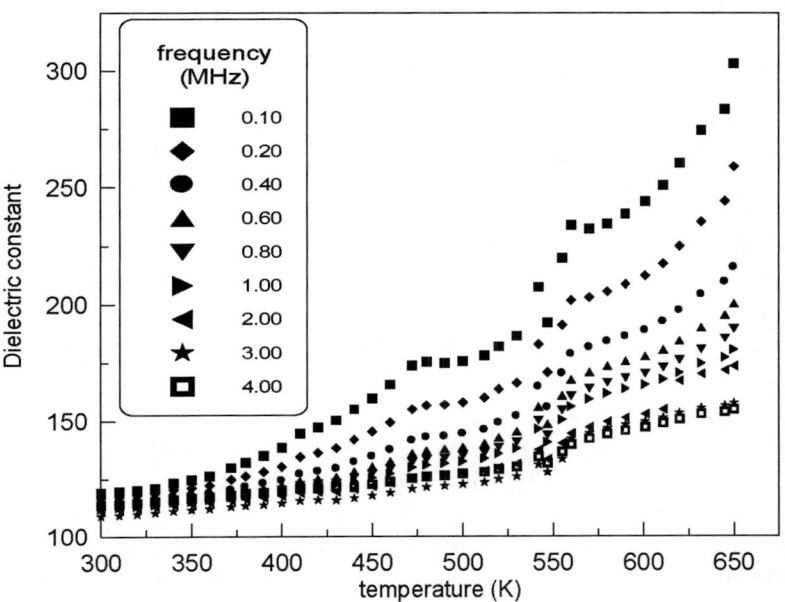

FIGURE1. The variation of the dielectric constant with temperature for Different frequencies.

Analysis of X-ray diffraction patterns reveals that all the samples were found   in a single phase spinel ($Cu_{0.5}Zn_{0.5}Fe_2O_4$) and secondary phase ($Cu_{0.5}Zn_{0.5}La_{0.35}Fe_{1.65}O_4$).

Fig. (1) Shows the   dielectric constant variation with temperature at Different frequencies for the unirradiated Cu-Zn-La ferrite samples ($Cu_{0.5}Zn_{0.5}La_{0.35}Fe_{1.65}O_4$). Figure shows an increase in έ with temperature giving raise a step transition at ≈ 475K. Another kink appeared at ≈565 K.

According to references [10, 11], the abnormal dielectric behavior of some ferrites is due to collective contribution of two types of carriers in p-type carriers and n-type to the polarization. The appearance of p-type carriers is due to the reduction tendency of Cu^{2+} to Cu^{1+} ion, which means that in the present sample the two types of conduction exist, i.e. $Cu^{2+} \leftrightarrow Cu^{1+} +e^+$ resulting in p-type conduction and $Fe^{2+} \leftrightarrow Fe^{3+} +e^-$ in the n-type conduction. It is well known that the local displacement of p-type carriers takes part in the polarization in the opposite direction to that of the external field. In addition since the mobility of p-type carrier is lower than that of n-type carrier, their contribution to polarization decreases more rapidly at lower frequency. Therefore, the position of the peak depends on whether the majority in p-type or n -type carriers in the sample. It is obvious, as the iron content is increased resulting in more $Fe^{2+} \leftrightarrow Fe^{3+}+e^-$ transitions. Therefore, the increase in n -type carrier with increasing Fe content results in a shift of peak temperature towards higher temperature side with increase in frequency.

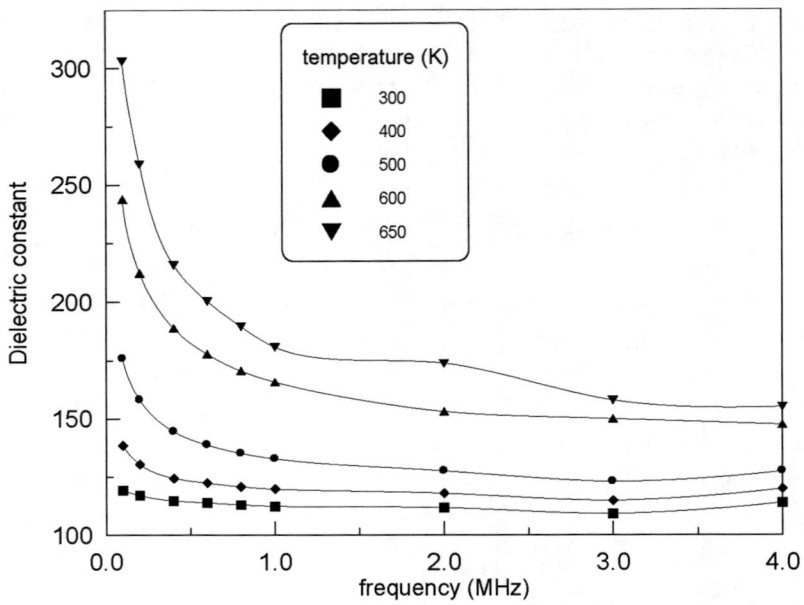

FIGURE 2. The variation of the dielectric constant with frequency for Different temperatures.

The variation of dielectric constant έ with frequency at different temperatures is shown in Fig. (2). From graph it can be seen that the dielectric constant decreases rapidly with the electric field frequency. The maximum drop in έ appears at 200 kHz after which έ gradually decreases until reaching a saturation value at around 4MHz for all temperatures. The dielectric behavior for this sample can be explained on the basis that the mechanism of polarization process in ferrites is similar to that of conduction. The electron exchange $Fe^{2+} \leftrightarrow Fe^{3+} + e^-$ gives local displacement of electrons in the direction of an applied electric field, which induces polarization in ferrites.

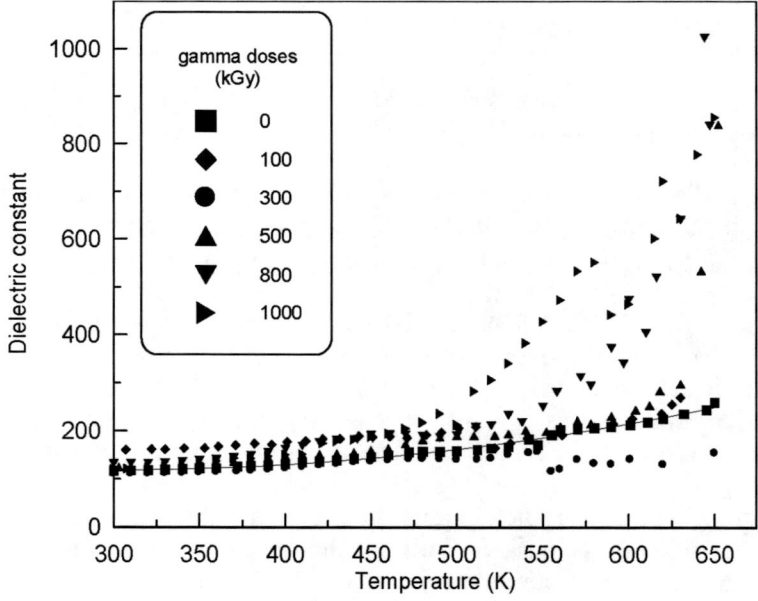

FIGURE 3. The dielectric constant change with temperature for different Gamma doses at 200 kHz.

Variation of the dielectric constant with temperature for different gamma radiation doses (Fig.3) illustarates the increase of it with gamma radiation doses.

The measurement of Seebeck coefficient was carried out for the unirradiated and irradiated samples Fig (4). The positive and negative values of Seebeck coefficient, for all samples, over the investigated

temperature range denote the presence of p-type and n-type charge carriers The presence of the two types of charge carriers define the two regions of temperatures in which the sample changed from p-type to n-type and vise versa.

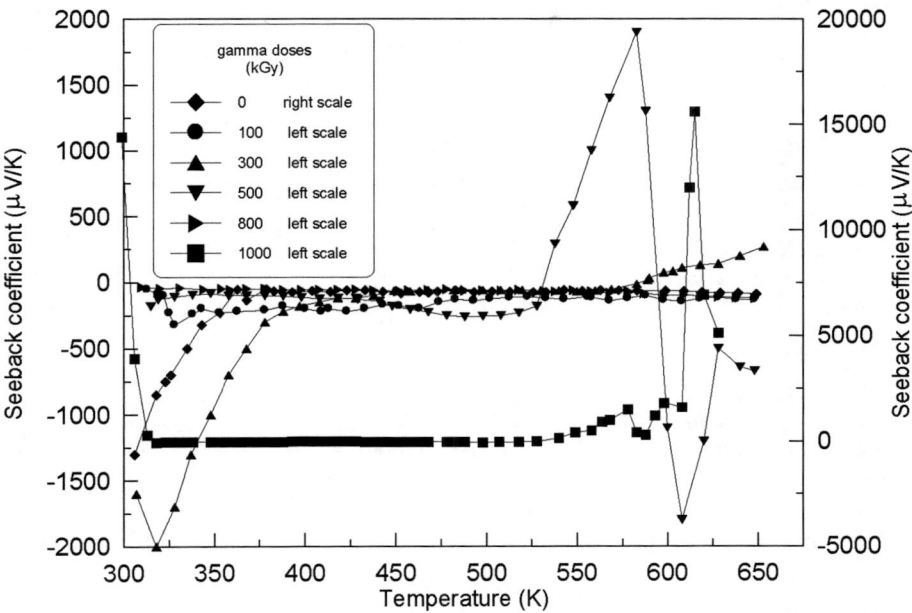

FIGURE 4. Seeback coefficient change with temperature for different Gamma doses.

Some of the energy of the applied field frequency consumed in the orientation of the dipoles and other part of this energy enhanced the electrical conduction.

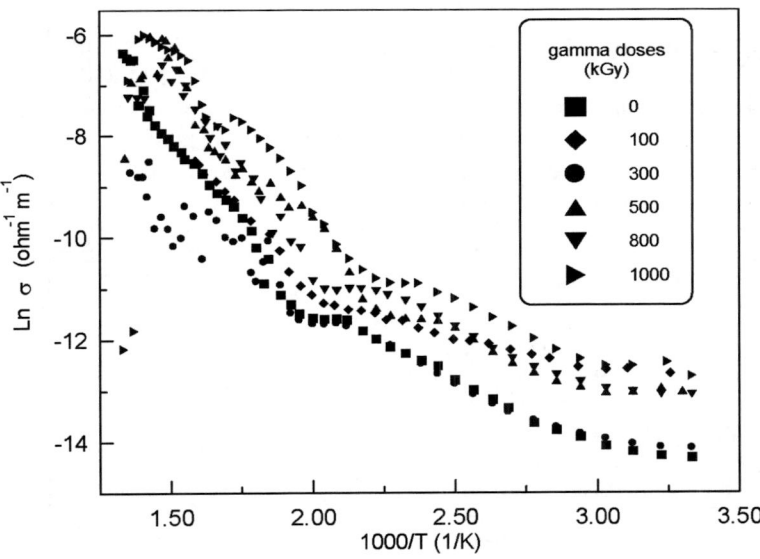

FIGURE 5. The logarithm of ac conductivity change with the reciprocal of temperature for different Gamma doses at 200 kHz.

A relation between the logarithms of the electrical conductivity (Ln σ) versus 1000/T for different gamma doses, at 200 kHz, is shown in Fig (5). Four regions of temperature with different values of activation energy are obtained (300-350K), (350-450K), (450-500K) and (500-650K). The different values of activation energy indicate the different conduction mechanisms. The activation energy in the high temperature range is larger than that in the lower temperature range [12]. The increase in conductivity with increasing temperature from one region to another, which yields a decrease of the activation energy means

that the charge carriers are pumped to higher localized states near the conduction band and then participate in the conduction process [8]. It was found that the observed increase in activation energy at different temperatures depends on the gamma irradiation doses.

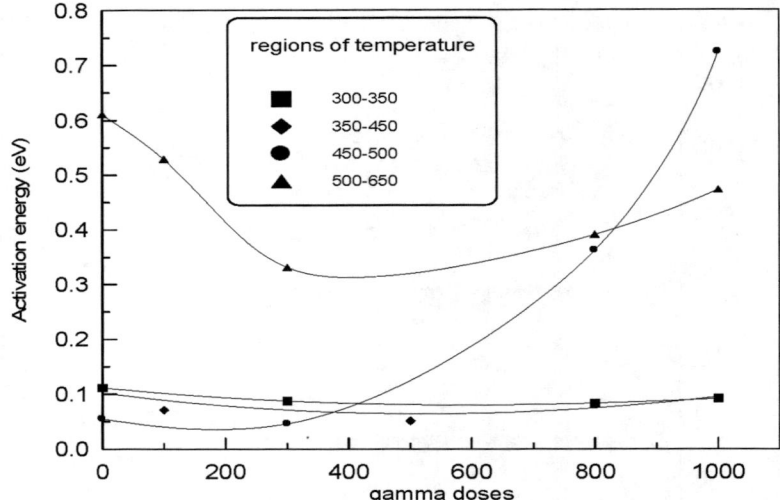

FIGURE 6. The variation of the activation energy with gamma doses for Different temperature regions at 200 kHz.

The activation energies deduced from Fig. (5) Are plotted against the gamma radiation doses in the different temperature regions are shown in Fig. (6). the data shows that the activation energy decreses up to 300 kGy and then increses. The increase of the activation energy with gamma doses is the result of initiating defects in the samples, which increases the resistivity. The decrease of activation energy with gamma doses may be due to either initiating holes from varying the valance of Cu ions or forming a small polarons and the high thermal energy was sufficient to move them to participating in conduction process.

On the other hand, the dielectric loss (D) gives an indication about some of the electric field energy consumed as heat loss. Figure (7) gives an impression about the change of D with temperature for different gamma radiation doses at frequency equal to 200 kHz. The dielectric loss changes in growing steps with temperature. This result agrees well with the result of increasing the activation energy with temperature. Variation of D with gamma radiation doses illustrates different responses, which ensures the change of magnetic ordering from ferromagnetic to paramagnetic state.

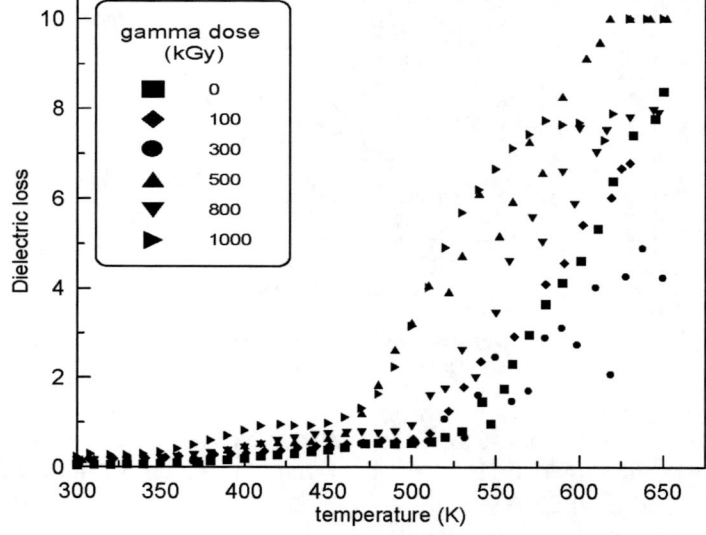

FIGURE 7. The variation of the dielectric loss with temperature for Different gamma radiation doses at 200 kHz.

CONCLUSION

Wide applications of Cu-Zn ferrites in different fields were attributed to their prime properties. $Cu_{0.5}Zn_{0.5}La_{0.35}Fe_{1.65}O_4$ was subjected to an external ac electric field. The frequency of this field ranged from 100 kHz up to 4MHz. Some of electrical energy of this field was oriented to mobile the dipole inside the ferrite. The rest of the energy was dissipated as a heat leading to change in the sample conductivity. Drop in the dielectric constant value occurred at 200 kHz for all samples. Response of these materials (at 200 kHz) to high gamma radiation doses (100 -1000kGy) showed different effects on the conductivity depending on the temperature of the samples. Four regions of conductions (300-350 K), (350-450 K), (450-500 K) and (500-650K) were obtained. Both electrons and holes participate in conduction mechanism.

REFERENCES

1. E.Rezlescu, phys. Stat. sol.**17**, 139 (1973).
2. V.R.Kulkarni,A.S.Vaingankar,J. Mater. Sci. **22**, 4087 (1987).
3. M. Ristic, S.Popvice, S.Music, J.Mater Sci.Letters **9**, 872 (1990).
4. L.M.Lelyuk, G.I.Zhuravlen, Chemistry and Technology of ferrites, Khimiya, Leningrad, 1983, p.256 (in Russian).
5. V.R.Kulkarni, M.M. Todkaar, A.S.Vaingankar, Indian, J.Pure Appl. Phys. **24**, 294 (1986).
6. N.Rezlescu, E.Rezlescu, Phys. Stat. Sol. (a) **23**, 575 (1974).
7. P.V. Reddy, T.S. Rao, J.Less.Conmon Met. **86,** 255 (1982).
8. O.M. Hemeda, Phase Transition **51**, 87 (1994).
9. F.H. Attrix, W.C.Roesch, Radiation Dosimetry, **293**, Second Edition, Academic press, NY, San Francisco, London, 1996.
10. A.A. Sattar, Egypt. J. Sol.,**26**, No.2 (2003).
11.S.A. Mazen, Materials Chemistry and Physics **62**, 131 (2000).
12. A Npatil, M G Patil, K K Patankar, V L Mthe, R P Mahajan and S A Patil, Bull. Mater. Sci., **23**, 447 (2000).

Study on the Enhanced Contribution in Noble Metals from Positron Annihilation

A. S. Hamid, M. M. Ahmed, M.S. M. Abu-Elmagd, and R. A. M. Rizk

Department of Physics, Faculty of Science, Helwan University, Helwan Cairo, Egypt

Abstract. Our motivations in the current work were to inspect various enhancement formulae along with to classify the electronic structure of noble metals Cu, Ag and Au. The measurements were performed via the two dimensional angular correlation of annihilation radiation 2D-ACAR apparatus. The electron density in momentum space $\rho(p)$ was reconstructed and it displayed the following features. Initially, the reciprocal lattice points underscored the calculations, and they revealed Fermi surface features. Additionally, enhanced anisotropy exposed nearby Fermi momentum. They attributed to enhancement of the electronic wave function at the position of the positron. Finally, the high momentum contributions, due to interaction of positron with core-like-state, conducted the electronic structure of the metals under investigations. From another viewpoint, the features of Fermi surface of Cu, Ag and Au showed an expected analogous behavior as multiply connected sphere inside the first Brillouin zone.

INTRODUCTION

The measurement of angular correlation of positron annihilation radiation has established to be a useful technique for absorbing information of solids. Given that, the enhanced electronic wave function at the position of the positron increases the annihilation probability. This effect has proven to be regulator in understanding the electronic structure of the materials. Early, S. Kahana [1] has accomplished his first conclusive try in this field to discover a correlation function for a positron-electron pair within a metal. The results have indicated that the distribution of the total momentum of the annihilating pair was very close to that of the Fermi distribution of valence electrons. Since his calculation [2] has predicted a roughly constant enhancement factor across the Fermi Sea, in agreement with the angular correlation data [3,4]. However, Kahana's approximation has received much attention. This has attributed to the subsequent measurements of alkali metals that have confirmed Kahana's prediction and its dependence on density parameter r_s and the enhancement parameters b/a and c/a. Incongruously, Boronski et al [5] have suggested that Kahana's approximation formula needs a logarithmic singularity treatment nearby Fermi boundaries p_F. Consequently, new values of the enhancement factors and total annihilation rates were computed. Notwithstanding, in another manuscript [6] of the same group they have pointed out that the enhancement problem is still under question. Contemporary, newborn theory has been applied by J. Arponen et al [7], in which an electron gas was considered as a system of interacting collective excitation. The conduction electrons were treated in a jellium model with additional effects due to core annihilation and lattice periodicity. Their results have revealed that the ratios b/a and c/a decrease as r_s increases, which, confirm quite contradiction to the results of Kahana. Scrutinizing, A. Rubaszek et al [8] has compared the experimental data on positron annihilation in alkali metals to the results of different theoretical approaches to this problem within the Kahana formalism. They emphasized that the necessity of abandoning the biparabolic formula, of Kahana, in order to made progress in the investigation of momentum-dependent enhancement factors. In their significant manuscript, they verified a very important principal that Kahana formalism was well satisfied for momenta $p \leq 0.8 p_F$. However this is corresponding only to about 50% of the electronic states. G. Kontrym-Sznajd et al [9], in series of publications, have presented a method of extracting a full shape of the electron-positron on the momentum distribution of annihilation quanta [10]. Furthermore, they compared their results to the experimental angular correlation of positron annihilation radiation curves. Their method was more reliable verification of the form of electron-positron enhancement factors near the Fermi surface in simple metals. In addition, they studied the influence of the positron distribution and the electron-positron

CP888, *Modern Trends in Physics Research,*
Second International Conference on Modern Trends in Physics Research—MTPR-06,
edited by L. El Nadi

interactions on the momentum density $\rho(p)$ of annihilation quanta in real metals Mg, Zn and Cd [11]. The motivation of deliberating the current work is; firstly, there is no comparison work has been done yet concerning the enhancement effect on noble metals; Secondly, as regards, anomalous work have been done concerning the enhancement effect. Nevertheless, they did not answer exactly which hypothesis gave proper results of the enhancement effect in the central Fermi surface and in the reciprocal lattice point. This makes the work in this aspect still open until now. In the current work, the electron density in the momentum space $\rho(p)$ reveals enhanced contribution as approaching Fermi momentum. This contribution is scrutinizing in terms of Kahana, Boronski and Rubaszek enhanced formulae. From another viewpoint, the features of the Fermi surfaces of the current metals are obtained by subtracting the isotropic part from the spectra. It revealed multiply connected surface in the 1st Brillouin zone.

THEORETICAL ASPECTS

The positron perturbs the many-electron system of the solid. It attracts a cloud of electrons, which screens its charge. This influences the annihilation process in two ways. Firstly, it changes the mean-field potential felt by the positron when it diverts through the crystal by adding an extra potential V_{ep} describing the electron-positron $e^{-} - e^{+}$ correlation to the Coulomb-Hartree potential V_{c} (with inverted sign) due to the ions and the electrons.

$$V^{+}(r) = -V_{c}(r) + V_{ep}(r) \tag{1}$$

Secondly, the pile-up of the electronic wave functions at the position of the positron increases the annihilation probability to a value much larger than that obtained via the overlap of the independent-particle wave functions. The theory of Kahana [2] is the basis for many of the calculations for a positron in a homogeneous electron gas. However, there is a major problem in the fact that the Kahana approach gives a divergence in the positron annihilation rate for $r_{s} > 4$. The Kahana approach uses electron-positron pair wavefunctions

$$
\begin{aligned}
\psi_{p}(x_{e}, x_{p}) &= \frac{e^{ip.x_{e}}}{V} + \frac{1}{V} \sum_{k>1} \chi(p,k) e^{ik.x_{e}} e^{i(p-k).x_{p}} \\
&= \frac{e^{ip.x_{p}}}{\sqrt{V}} \left(\frac{e^{ip.r}}{\sqrt{V}} + \sum_{k} \chi(p,k) e^{ik.r} \right) \\
&= \frac{e^{ip.x_{p}}}{\sqrt{V}} \phi_{p}(r)
\end{aligned} \tag{2}
$$

Where x_{e} and x_{p} are the electron and positron coordinates and $r = x_{e} - x_{p}$. The first term on the right is the electron wavefunction of momentum p approaching the positron, and the second term is a spherical wave describing the spherical scattering of an electron with momentum p of the positron. The Fourier coefficients $\chi(p,k)$ are found from a Bethe-Goldstone type equation,

$$\chi(p,k) = \frac{Av(|k-p|)}{k^{2} + (k-p)^{2} - p^{2}} + \frac{A}{k^{2} + (k-p)^{2} - p^{2}} \int_{q>1} v(|k-q|) \chi(p,q) d^{3}q \tag{3}$$

Where $v(q)$ is the Fourier transform of the effective electron-positron interaction potential, p is the initial electron momentum, V is the volume of the sample and

$$A = \frac{1}{4\pi^{3} k_{F} a_{o}} \tag{4}$$

with all momenta in units of k_F (the Fermi wave vector). However, Kahana parameterized his theory to be applicable as, the so called, biparabolic formula,

$$\varepsilon(p) = a + b\gamma^2 + c\gamma^4, \qquad\qquad \gamma = \frac{p}{p_F} \qquad\qquad (5)$$

This theory appeared to be in good accordance with early 1D ACAR [12] and later 2D ACAR experiments [13] on the largely free electron–like alkali metals. However, it can't be applied in a straightforward manner to materials with a more complex electronic structure involving $d-$ or $f-$ bands. This formalism postulated that no high-momentum component, the enhancement factor in the region $(p > p_F)$ is given by the previous equation and the total annihilation rate and the enhancement factor diverge at high r_s.

E. Boronski et al [5] suggested a logarithmic singularity in $\varepsilon(p)$ in the limit $p \to p_F$. Thus their equation becomes

$$\varepsilon(p) = a + b\gamma^2 + c\gamma^4 + d\ln(1 - \gamma) \qquad\qquad (6)$$

From another view point, A. Rubaszek et al [8] proposed analysis of experimental data allows more reliable verification of the form of electron-positron enhancement factors nearby Fermi surface in simple metals and the corresponding factor may be written as

$$\varepsilon(p) = \varepsilon_b(p) + \varepsilon_c(p)\theta(0.8p_f - p) \qquad\qquad (7)$$

Where $\varepsilon_b(p)$ is Kahana equation, θ is the unit-step function and $\varepsilon_c(p)$ is the correction function.

$$\varepsilon_c(p) = \frac{d}{[1.06 - \gamma^2]} \qquad\qquad (8)$$

Finally, G. Kontrym-Sznajd et al [10] have improved the Kahana approach by using a better effective electron-positron interaction potential. By determining the interaction potential self-consistently and taking account of the electron-electron correlation they were able to remove the divergence in the annihilation rate.

RESULTS AND DISCUSSIONS

2D-ACAR measurements have been performed on Cu, Ag and Au single crystals. This experiment is discussed elsewhere [14,15]. Subsequently, the electron densities in momentum space $\rho(p)$, of the metals under study, are obtained from the measured results, by applying the reconstruction technique based on Fourier transformation. This method includes interpolation in the Fourier space, and inverse Fourier transformation following basic Fourier projection theorem. The estimation of the error in $\rho(p)$ of this technique is about 0.02 a. u. in all directions [16]. Figure 1 shows $\rho(p)$ spectra in (100) plane as a contour map for Cu, Ag and Au, respectively. The contours are presented as 10% from the maximum height of the peak. The Brillouin zone of FCC structure metals is symbolized in the figures and the reciprocal lattice points are characterized in the spectra as dark circles. These figures show the following features:

Initially, they reveal the contributions of the reciprocal lattice points on the high momentum components HMCs, which are attributed to Umklapp process. Specifically, the nearest neighbor reciprocal lattice points G_{111}, in Cu, Ag and Au are located at $p_{[001]}$= 6.7, 5.9 and 5.9 mrad, $p_{[110]}$ = 9.5, 8.4 and 8.4 mrad and at $p_{[111]}$ =11.5, 10.0 and 10.1 mrad, respectively. Almost, these values are larger than Fermi momentum P_F of Cu (=5.308 mrad), Ag (=4.686 mrad) and Au (=4.687 mrad). As a result, the contribution to Fermi surface is mainly due to G_o (G=0; G is the reciprocal lattice points vector). Accordingly, Fermi surface features in the first Brillouin zone grow to be the strongest and they are weak in the higher zones. This exhibits clearly that the noble metals are influenced by $s-$ signal, which predominated in the first Brillouin zone. From another viewpoint, around 7 mrad for Cu, 6 mrad for Ag and Au (about 1.35 p/p_F) the momentum density displays the discontinuity between the Fermi momentum and the momentum $P_F + G_o$, see Fig. 2. As it is well known, the radii of the Fermi surfaces of the present metals are shorter than any vector joining the center to the surface of the Brillouin zone. Consequently, this Fermi surface is completely within the first Brillouin zone in extended zone scheme. Specifically, it is entirely contained in the 1st

band in the more conventional reduced zone scheme. This is sustaining the fact that, these metals contain only one valence electron, Cu ($3d^{10}4s^1$), Ag ($4d^{10}5s^1$) and Au ($5d^{10}6s^1$), and hence a partially filled conduction band.

Secondly, Fig 1 reveals, also, an enhanced contribution nearby Fermi momentum, which is attributed to Kahana-like enhancement effect [2]. To clarify this feature, $\rho(p)$ is represented through [100] direction for Cu, Ag and Au, as shown in Fig 2. The results of different enhancement formulae are presented in the same figure, (a) Kahana-like enhancement (Khconv), (b) Bronski (Brconv) and (c) Rabzak (Rzconv). These spectra are convoluted by the resolution function of the experiment. Subsequently, they are fitted to the experimental spectra via the least square fitting analysis. This is after adding the core contributions, which estimated to be Gaussian function, see Fig 3. In respect of the electron gas calculations, $\rho(p)$ increases as p approaches Fermi momentum p_F. This is in view of the following fact; according to the electron positron attraction, an electronic charge screens the positron and it increases as the number of valence electrons increases, i.e., as the unit cell volume increases. These electrons are mainly in a state close to p_F. The above-mentioned positron screening cloud in such system is built up by scattering in the states close to Fermi surface. Accordingly, an additional electron positron correlation potential is formed. Table 1 declares the enhancement parameters $(b+c)/a$ of different enhancement formulae in [001], [100] and [110] direction, for Cu, Ag and Au, respectively. The square difference between the experimental and theoretical value is represented as x^2. It is obvious that the χ^2 shows small value for the formulae of both Kahana and Rabzak. However, as mentioned above Kahana formula represents only about 50% of the electronic states. From another viewpoint, the obtained values using the additional term of Rubaszek [8] show more reality choice of the enhancement factor. Regrettably, the additional logarithmic term of Boronski [5] shows singularity in the integral convolution and it is fairly weak. Accordingly, it could not be experimentally identified.

Finally figure 3 presents the high momentum component HMC of the metals under study in [001] plane. It reveals that the core contribution increases from Cu, Ag to Au. This is due to the interaction of the positron with the core electrons. Seeing as, the localized core states 5d of Au is larger than that of localized 4d-state of Ag as well the localized 3d-state of Cu.

From another perspective, Figure 4 a, b and c represents the result of subtraction the mean value from different direction of symmetry of $\rho(P)$, and lines passing through [001] direction are considered. The solid and dotted lines represent the p_F and the border of the Brillouin zone, respectively. Anisotropy is entirely apparent by inquisitive the spectra. In this way the isotropic contributions are subtracted and Fermi surface features are enhanced. Cu, Ag and Au presented nearly the same behavior, i.e. multiply connected surface in the 1st Brillouin zone. For momentum larger than Fermi momentum, the high momentum component is emerged. This is due to the above-mentioned discontinuity between the Fermi momentum and the momentum $p_F + G_o$. It is obvious that, Fermi surfaces of the above Cu, Ag and Au are deviated from the free electron spherically symmetric. Even though it is considered as a good first approximation to describe the electronic structure of the noble metals.

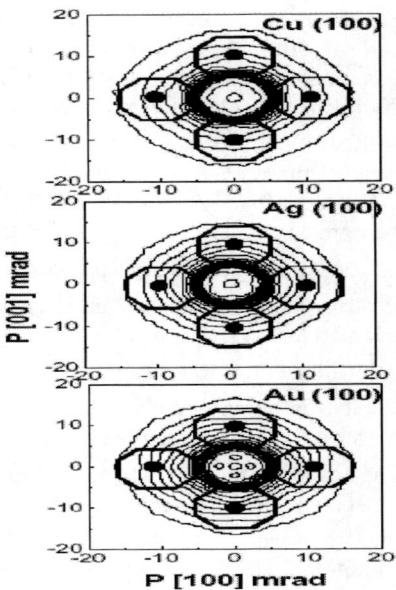

FIGURE 1. The electron momentum density $\rho(p)$ in (ΓMK) plane for Cu, Ag and Au, respectively. The distribution is shown as a contour map in a step of 10% of the peak height. The dark circles show the positions of the nearest reciprocal lattice points.

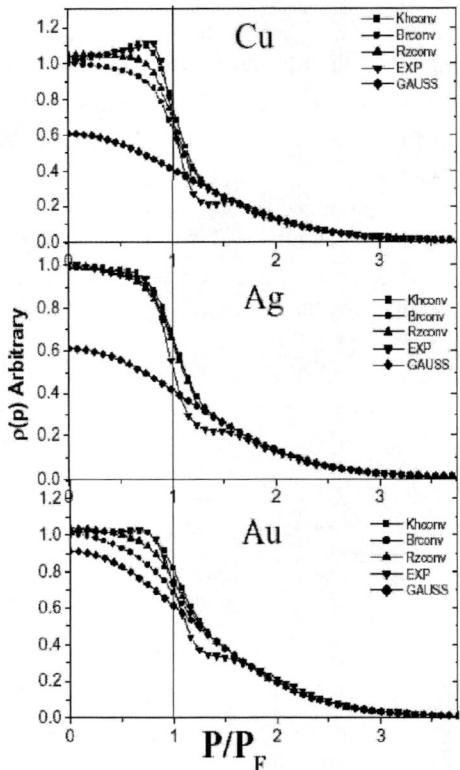

FIGURE 2. The electron momentum density of Cu, Ag and Au in [001]direction, respectively, with that convoluted fitting data using Kahana [2] (Khconv), Boronski [5] (Brconv) and Rubaszek [8] (Rzconv) enhancement equation. Gauss distribution is fitted to all the spectra then it added to the convoluted results.

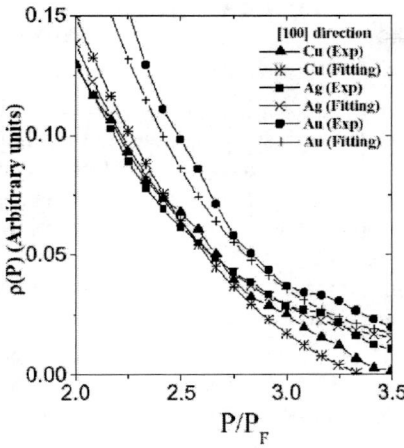

FIGURE 3. The high momentum component HMC of the metals under study in [100] plane with the Gaussian fitting results.

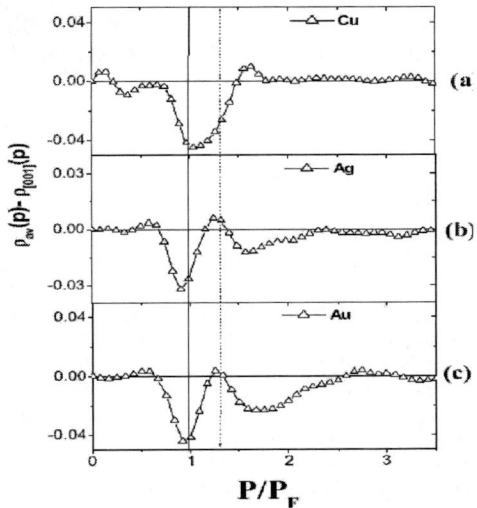

FIGURE 4. The anisotropy due to subtraction of mean value from different direction of symmetry of $\rho(P)$ for Cu, Ag and Au in [001] and [100]. The dotted lines show the Brillouin zone boundary.

TABLE 1. The parameter (b+c)/a from Kahana [2], Bronski [5] and Rabszak [8] enhancement formula from Cu, Ag and Au in [001], [100] and [110] directions, respectively. The square differences between the experimental and theoretical values x^2 is presented in the table.

	Direction	Kahana [2]		Bronski [5]		Rabaszk [8]	
		(b+c/a)	x^2	(b+c/a)	x^2	(b+c/a)	x^2
	[001]	0.26	0.009	0.001	0.16	0.43	0.052
Cu	[100]	0.30	0.011	0.001	0.18	0.43	0.058
	[110]	0.28	0.009	0.001	0.15	0.43	0.051
	[001]	0.11	0.035	0.001	0.19	0.11	0.034
Ag	[100]	0.11	0.036	0.001	0.20	0.12	0.035
	[110]	0.11	0.087	0.001	0.40	0.13	0.058
	[001]	0.08	0.010	0.001	0.26	0.26	0.062
Au	[100]	0.06	0.009	0.001	0.24	0.24	0.060
	[110]	0.01	0.024	0.001	0.40	0.20	0.085

CONCLUSIONS

The electron densities in momentum space $\rho(P)$ of Cu, Ag and Au are obtained using 2D-ACAR experimental set up. Some significant notifications are observed. Initially, the reciprocal lattice points are located far from Fermi momentum. This reveals the strong features of Fermi surface in the first Brillouin zone. Therefore, the Noble metals influenced by the s − signal, which are preponderated in the first Brillouin zone. Additionally, enhanced contribution, nearby Fermi momentum, are attributed to Kahana-like enhancement effect. The additional term of Rabzack shows good choice of the enhancement factor for the Noble metals. Finally, the contribution due to the interaction of the positrons with core state reveals that the core contribution increases from Cu, Ag and Au. This is attributed to the interaction of the positron with the core electrons. Concerning to the Fermi surface of the current metals, its features are enhanced by inquisitive the spectra. Cu, Ag and Au presented nearly the same Fermi surface behavior i.e. multiply connected surface in the 1st Brillouin zone. As a conclusion, Fermi surfaces of Cu, Ag and Au are distorted from the spherically symmetric behavior of the free electron model. However, as a good first approximation the free electron model could describe the electronic structure of the noble metals.

ACKNOWLEDGMENTS

The experimental work has been done at the Institute of Applied Physics, Graduate School of Pure and Applied Sciences, University of Tsukuba, Tsukuba, Ibaraki 305-8573, Japan. The first author would like to thank Prof. A Uedono for his help and support.

REFERENCES

1. S. Kahana, Phys. Rev., **117**, 123 (1960).
2. S. Kahana, Phys. Rev., **129** (1963).
3. J. J. Donaghy and A. T. Stewart, Phys. Rev., **164** (1967).
4. J. A. Arise-Limonta and P. G. Varlashkin, Phys. Rev. B**1**, 142 (1970).
5. E. Boronski, Z. Szotek and H. Stachowiak, unpublished manuscript (1978).
6. H. Stachowiak and E. Boroński, Polish Nucl. Soc., **42**, 225 (1979)
7. J. Arponen and E. Pajanne, J. Phys. F **9**, 2359 (1979)
8. A. Rubaszek and H. Stachowiak, J. Phys. F**15**, 231(1985).
9. G. Kontrym-Sznajd and A. Rubaszek, Acta Physica Polonica A., **83**, 339 (1993)
10. G. Kontrym-Sznajd and A. Rubaszek, Phys. Rev. B**47** , 6950 (1993)
11. G. Kontrym-Sznajd, H. Sormann, Acta Physica Polonica A., **88**, 177 (1994)
12. J. J. Donaghy and A. T. Stewart, Phys. Rev., **164** ,396 (1967)
13. L. Oberli, A. A. Manuel, R. Sachot. P. Descouts, and M. Peter, Phys. Rev. B**31**, 1147 (1985)
14. H. Kondo, T. Kubota, H. Nakashima and S. Tanigawa, Materials Science Forum **105** , 110 (1992).
15. R. Suzuki and M. Osawa, S.Tanigawa, M. Matsumoto, and N. Shiotani, J. Phys. Soc. Jpn., **58**, 3251 (1989).
16. A S. Abdul Hamid, S. Tanigawa, Z Q. Mao, and Y. Maeno, Phys. Stat. Sol. (b), **231**, 149 (2002).

GEANT4: Applications in High Energy Physics

Tariq Mahmood, Abrar Ahmed Zafar, Talib Hussain and Haris Rashid

Centre for High Energy Physics
University of the Punjab, Lahore-54590, Pakistan

Abstract.   GEANT4 is a detector simulation toolkit aimed at studying, mainly experimental high energy physics. In this paper we will give an overview of this software with special reference to its applications in high energy physics experiments. A brief of process methods is given. Object-oriented nature of the simulation toolkit is highlighted.

Keywords: Object-oriented technology, Simulation

PACS: 07.05.Tp; 13; 23

INTRODUCTION

GEANT4 [1] (**GE**ometry **AN**d **T**racking) is a software package written in object-oriented C++ programming language and is an upgraded version of GEANT3 [2]. The software fully exploits the functionality of OOP (**O**bject **O**riented **P**rogramming) technology. Primarily designed for detector simulations, it is meant to study different physical phenomena which take place when particles pass through matter. Besides high energy physics experiments, it has vast applications in other fields of physics such as medical physics, nuclear physics, space physics, etc [1]. In view of its versatility, it allows to simulate and study variety of particle interactions at energies which will be available at Large Hadron Collider (LHC) and cosmic ray experiments [3]. In view of its versatility of application we will give a short overview of this important toolkit specially for the benefit of our young participants.

In this paper we will briefly describe essential components to study various processes in GEANT4. We also give a short review of its application in currently ongoing high energy physics experiments where physicists are working at the frontiers of knowledge. These experiments, for example: BaBar (SLAC), ATLAS (LHC), OSCAR (LHC), GAUSS (LHC), ALICE (LHC), and future experiments planned at ILC will unearth several aspects of new physics. These experiments are also a source of new and cutting edge technologies. Simulation of various phases of experimentation will therefore provide good insight of the events which are to take place when the experiment is actually performed.

Process Methods in GEANT4

In order to understand GEANT4, it is necessary to know the hierarchy of its C++ classes, their data items and functions they provide. Once the structure of GEANT4 classes is known, its functionality can be enhanced according to application needs. It is done by modifying an already existing physical process or by adding a new process in GEANT4. Data variables and functions of existing classes can be used though inheritance. For any process, the general purpose functions of the abstract class 'G4VProcess' can be used. There are many other classes defined in GEANT4 to represent different physical processes - particles, their properties and so on [4].

As described in the Users Guide [4], "All physical processes are characterized by two kinds of methods called ***GetPhysicalInteractionLength*** and ***DoIt***. *GetPhysicalInteractionLength* is used to provide the length (in space or time) before which the particle undertakes the interaction. *DoIt* invokes the interaction of the physics process and determines the final results of it". Each of these two methods prototyped in abstract class G4VProcess has three types. Every process of GEANT4 will have at least one of these three methods, at least one from *DoIt* triplet and one from *GetPhysicalInteractionLength* triplet [4].   Order of the process invocations is very important and is provided by *ProcessManger* of the particle. A user may change the order of invoking a process. Three types of *GetPhysicalInteractionLength* and of *DoIt* therefore play an important role in characterizing and tracking in a process [4-7]. More details are available on the webpage of GEANT 4 in General User's Guide.

CP888, *Modern Trends in Physics Research,*
Second International Conference on Modern Trends in Physics Research—MTPR-06,
edited by L. El Nadi
© 2007 American Institute of Physics 978-0-7354-0354-9/07/$23.00

Applications in High Energy Physics

A list of physical processes developed in GEANT4 is available in references 1-7. Users can select a physics process of their choice from the list and then suitably modify according to the needs of their application. Given below are some of the ongoing and future experiments which utilized the available resources of this simulation toolkit.

BaBar (SLAC)

At BaBar, all the simulation activities are handled by a GEANT4 oriented software package called BOGUS (**B**aBar **O**riented **G**EANT4-based **U**nified **S**imulation). This experiment uses its own customized functionality of GEANT4 like transportation etc [8].

ATLAS (LHC)

During the initial ten years of experimentation, ATLAS has been using GEANT3 for detector simulations. In 2004 it has been replaced by GEANT4. GEANT3 played important role in refinig the detector simulation and geometry during the course of its use. The simulation started with the help of event generators present in the 'simulation software suit' at ATLAS [9-11].

OSCAR (LHC)

In this experiment, GEANT4 has played an important role in specifying geometry of the CMS detector, its sub-detectors, tracking of the events, etc. OSCAR group [12-13] adopted a unique geometrical description of CMS detector completely independent of different views that different applications may need. Figure 1 gives a schematic description of the use of GEANT 4 by OSCAR group.

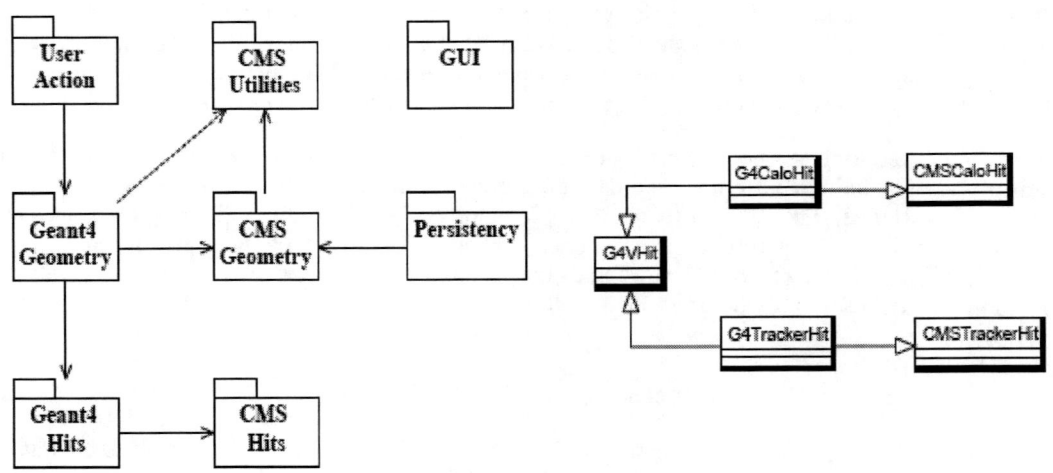

Figure 1: OSCAR category diagram (left) and hits class diagram (right) [12]

GAUSS (LHC)

GAUSS is also an object oriented simulation application for LHCb. It is based upon Gaudi framework and GEANT4 interfaced through GiGa. The development of this interface started in 1999. Its first prototype became available in 2000. A complete LHCb simulation under this software application was started in 2001[14, 15].

ALICE (LHC)

The simulation code of ALICE experiment is based upon VMC (Virtual Monte Carlo) event generator. This code can be run along with GEANT4. Implementation of VMC in GEANT4 has been completed for complete simulation in complex geometries [16, 17].

ILC

ILC (International Linear Collider) is using LCDG4, a general purpose simulation package which is based upon GEANT4. The simulation package has an easy to use XML-based interface for displaying geometry representations specified by desired parameters [18].

Conclusion

In order to use GEANT4, one should have sufficient knowledge of object-oriented C++ language. One can then easily modify or add new processes in already available repository. In view of the expertise involved in the development of the software, it provides the functionality of performing detector simulation very near to the reality. It can be used in all fields of physics involving different particle interactions. The domain of applications of GEANT4 is increasing very rapidly. With the advent of every new addition, functionality of the software becomes vaster. It allows many easy to use interactive user friendly interfaces in order to perform the simulations.

Note: The material in this article is not original and has primarily been taken from the scholarly work cited in references. In view of the large amount of resource available in literature, many details have not been included which are available on hep archives. Authors apologize to all whose scholarly work could not be included or get the due share. Two of us (AAZ & TH) acknowledge the enabling role of the Higher Education Commission Islamabad, Pakistan and appreciate its financial support.

References

1. S.Agostinelli et al., "Geant4 - A Simulation Toolkit", Nucl. Instrum. Meth. A, **506**, 250-303, **2003;** Also see http://wwwinfo.cern.ch/asd/geant4/geant4.html.

2. Application Software Group, CN Div., "GEANT Detector Description and Simulation Tool (Version 3.21)", CERN Program Library W5013.

3. J. Allison, K. Amako et al. "Geant4 Developments and Applications" IEEE Transactions on Nuclear Science, Vol. **53**, No. **1**, 270-278, February **2006**.

4. Physics reference Manual of Geant4 and Toolkit developers Manual: http://geant4.web.cern.ch/geant4/G4UsersDocuments/UsersGuides/

5. K.Murakami, K.Amako et al. "Development of an Interface for Using EGS4 Physics Processes in Geant4" Proceeding of Computing in High Energy and Nuclear Physics, 24-28 March **2003**, La Jolla, California

6. http://geant4.web.cern.ch/geant4/support/ about.shtml

7. http://geant4.web.cern.ch/geant4/applications/hepapp.shtml

8. http://www.slac.stanford.edu/BFROOT/www/Computing/Offline/Simulation/web/simover/simover_main.html

9. *http://atlas.web.cern.ch/Atlas/GROUPS/SOFTWARE/OO/architecture/General/Documentation/AthenaDeveloperGuide-8.0.0-draft.pdf*

10. ATL-SOFT-2003-013 Strategy for the transition from GEANT3 to GEANT4 in ATLAS. by:Barberis, D.; Polesello, G.; Rimoldi, A.; Geneva : CERN, 13 Nov 2003

11. A.Dell'Acqua CERN EP/SFT, Geneva, CH, "Status of the physics validation studies using GEANT4 in ATLAS", Computing in High Energy and Nuclear Physics, 24-28 March 2003, La Jolla, California

12. S. Banerjee, I. Gonz᾽alez, V.Lef᾽ebure "CMS simulation software using GEANT4", CMS Note 1999/072, 10 December 1999.

13. P. Arce1, et al. Use of GEANT4 in CMS The OSCAR project , chep2000.pd. infn.it/paper/pap-a335.pdf

14. http://www.trigconsulting.co.uk/gauss/man_intro.html

15. http://www.ruf.rice.edu/~bwbwn/econ400_files/UserGuide.pdf

16. I. Gon᾽alez Caballero et. al. "ALICE experience with GEANT4", Computing in High Energy and Nuclear Physics, 24-28 March 2003, La Jolla, California

17. http://root.cern.ch/root/vmc/VirtualMC.html]

18. http://www.linearcollider.org/cms/ ; http://www.symmetrymagazine.org/cms/?pid=1000221.

Inelastic Interactions of Proton with Emulsion Nuclei without Shower Particle Creation

A. Abdelsalam[1], M .S. El-Nagdy[2], N. Rashed[3],and B. M. Badawy[4]

1   Physics Department, Faculty of science, Cairo University, Giza, Egypt.
2   Physics Department, Faculty of science, Helwan University, Helwan, Egypt.
3   Physics Department, Faculty of science, Fayoum University, Fayoum, Egypt.
4   Reactor Physics Department, Nuclear Research Center, Atomic Energy Authority, Egypt.
E-maill: naglaarashed@hotmail.com

ABSTRACT. This paper presents exhaustively the general characteristics of the inelastic interactions of P, ^{4}He and ^{7}Li with emulsion nuclei distinguished without relativistic hadrons (n_s = 0) in Lab. system. The dependence of these interactions on the projectile and target sizes is presented. It is found that, the probability of the events having (n_s = 0) is dependent on projectile size and incident energy. The average no. of grey particles $<N_g>$ and black particles $<N_b>$ as well as the ratio $<N_g>$ / $<n_s>$ are displayed for different target size. The multiplicity distribution of different target fragments for the events having (n_s = 0), $n_s \geq 0$ and those of complete destruction ($N_h \geq 28$) are presented.

INTRODUCTION

The study of projectile and target fragmentation processes gives every indication of being a rich source of information on nuclear structure. One is free to study the fragmentation process of either target or projectile, the choice being one of experimental technique. That is, projectile fragments will tend to be fast, forward-going fragments. Whereas, target fragments are generally low energy in the laboratory (target) frame, except for high energy tails which are known to exist. If we consider the process of projectile fragmentation, we see that there is a number of advantages provided by doing the experiment at high energy. The fragments are emitted at energies typically near the beam velocity and are collimated into a narrow forward cone (at 2.1 A GeV subtended an angle of about 3° in the laboratory frame).

Now we want to transform to the target frame and investigate reaction products emitted in nucleon-nucleus (N-A) and nucleus-nucleus (A-A) collision. At high energies, where the rapidity interval between projectile and target are well separated, the physics of the two fragmentation regions must be similar: However, the experimental techniques are often considerably different. The projectile fragmentation peaks which appear near the projectile velocity, when transformed to the target frame correspond to near zero energy fragments. The fragmentation process itself has been attended throughout many experimental and theoretical study [1-11] along the last decade.

In the present paper, we were interested in studying the interactions of p (3.7 GeV), ^{4}He (2.1 A GeV) and ^{7}Li (2.2 A GeV) with emulsion regarding to the target fragmentation occurring accompanying such interaction. This process is one of our Lab. Group program concerning the break up process in high energy interactions

EXPERIMENTAL DETAILS

Three stacks of NIKFI-BR2 nuclear emulsion with dimensions of 10 cm x 20 cm x 600 μm were irradiated to P (3.7GeV),^{4}He (2.1 A GeV) and ^{7}Li (2.2 A GeV) beams at Dubna /JINR Synchrophasotron. Along the track double scanning, fast in the forward direction and slow in the backward one, was carried out using a total magnification of about 1500X. Data studied in the present work consist of 2649 , 2066 and 1003 total interactions of P, ,^{4}He and ^{7}Li projectiles with emulsion nuclei, respectively. The events due to the electromagnetic dissociation and elastic scattering are rejected from the used samples. The remaining unbiased samples of inelastic interactions, indicate mean free paths (λ_{Inel}) equal 30.20 ± 0.70 cm, 20.76 ± 0.46 cm, and 15.30 ± 0.48 cm for inelastic interactions of P +Em , ^{4}He+Em,   and ^{7}Li + Em respectively.

The tracks of the emitted secondary charged particles in each event were classified according to the traditional emulsion criteria as follows:

CP888, *Modern Trends in Physics Research,*
Second International Conference on Modern Trends in Physics Research—MTPR-06,
edited by L. El Nadi
© 2007 American Institute of Physics 978-0-7354-0354-9/07/$23.00

a) Shower tracks (s)

are singly charged relativistic particles with relative ionization $I/I_o \leq 1.4$ where I is the track ionization and I_o is the plateau ionization for singly charged minimum ionizing particles. Their multiplicity is denoted by n_s.

b) Grey tracks (g)

have range $L \geq 3mm$ in emulsion and relative ionization $I/I_o \geq 1.4$. These tracks are mostly due to protons kinetic energy in the range 26-400 MeV. Their multiplicity is denoted by N_g.

c) Black tracks (b)

are those having range of L<3mm, corresponding to protons with kinetic energy ≤ 26 MeV. They are mainly due to evaporated target fragments. Their multiplicity is denoted by N_b. In each event, the black and grey tracks together are called highly ionizing tracks. Their multiplicity is denoted by ($N_h = N_g + N_b$).

d) PF's (projectile fragments) emitted with an angle $\theta_{Lab} \leq 3°$, with respect to the incident direction and characterized by no change in their ionization for at least 2cm from the interaction point. In the present case, they are singly and doubly charged particles, spectator (stripped) fragments of the projectile nucleus, having velocity $v \approx 0.97c$. The total charge of the stripped fragments in the forward cone per event is denoted by Q. The single charged fragments emitted in the fragmentation cone were subjected to rigorous multiple scattering measurements for the momentum determination in order to separate the produced pions from the singly charged fragments.

RESULTS AND DISCUSSION

A study of general characteristics of the events not accompanied by the emission of relativistic charged particles (in the laboratory system) has been carried out for interactions of proton (2.2 GeV [12] , 3.7 GeV) , d (3.7 A GeV) [13], ^{4}He (2.1 A GeV) and ^{7}Li (2.2 A GeV) with emulsion nuclei . The dependence of the probability of the interactions not accompanied by relativistic hadrons on the projectile size is presented in table (1). One can observe that, this probability is nearly dependent on the projectile size and the incident energy.

TABLE 1. The percentage of events for (2.2 GeV, 3.7 GeV) proton, (3.7 A GeV) d, (2.1 A G e V) ^{4}He and (2.2 A GeV) ^{7}Li without relativistic hadrons (n_s=0) and also of events with complete destruction of heavy target ($N_h \geq 28$)..

Projectile	Energy A GeV	Total No. of Events	P(n_s=0) %	P($N_h \geq 28$) %
P	2.20	709	31.45±2.11	0.00±0.00
P	3.70	2649	9.85±0.61	0.00±0.00
d	3.70	1644	2.60±0.40	1.58±0.31
^{4}He	2.10	2066	7.12±0.59	2.86±0.37
^{7}Li	2.20	1003	10.37±1.02	3.69±0.61

The dependence on the incident energy is glow for the two incident beams of protons. The probability at energy 2.2 GeV is nearly three times its value at energy 3.7 GeV. In the case of hadron nucleus interaction (proton) indicating the pionization absence in our protons interactions. Of course the excess of the energy gives a chance for pionization to be more probable.

The complete destructions of heavy target (Ag nuclei) is very interesting, this interest is form the fact that most of these interactions are due to very central collisions. In table (1) we give the probability for

complete destruction ($N_h \geq 28$) for various projectile nuclei. For protons, the destruction is not achieved because the energy is not enough to reach the region of the central collision. For the other used projectiles, the probability begins to increase with increasing the mass of the projectiles. This result may be explained by the fact that, at high energies, the inelastic cross section increases as the projectile mass increases [14]. This increase has a limiting behavior where it was shown [15] that this probability begins to saturate at A=12.

The dependence of the mean number of emitted shower particles, $<n_s>$, on N_h for different projectiles (P, d, ^{4}He and ^{7}Li) are plotted in figure (1). One can see that $<n_s>$ seems to change linearly with N_h ($<n_s>= a N_h + b$) and the line slope increases with increasing the mass number of the projectile or the number of interacting nucleons from the projectile (see table 2).

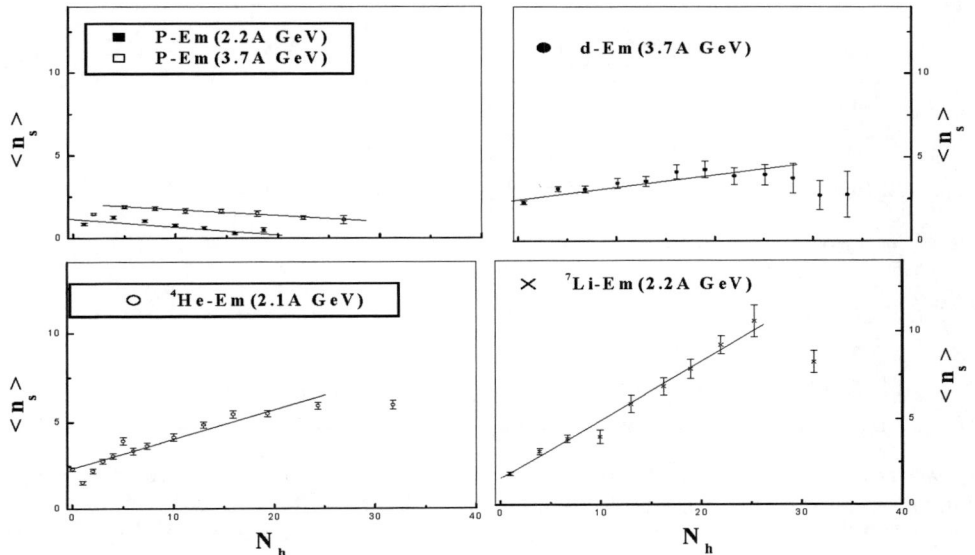

FIGURE 1. Dependence of $<n_s>$ on N_h in the interactions of p, d, ^{4}He and ^{7}Li with emulsion nuclei. The line represents the best fit of the experimental data.

TABLE 2. The values of the slope (a) and the intercept (b) determined by the fitting of the experimental data illustrated by the equation $<n_s>= a N_h + b$ for the interaction of p, d, ^{4}He and ^{7}Li projectiles with emulsion nuclei.

Projectile	a	B
P (2.2 GeV)	-0.047 ± 0.0135	1.137 ± 0.164
P (3.7 GeV)	-0.010 ± 0.011	1.645 ± 0.120
d (3.7 A GeV)	0.074 ± 0.016	2.440 ± 0.135
^{4}He (2.1 A GeV)	0.166 ± 0.016	2.381 ± 0.182
^{7}Li (2.2 A GeV)	0.336 ± 0.017	1.560 ± 0.115

Table (3) shows that for the total sample of events ($n_s \geq 0$) the average number of shower particles $<n_s>$ is strongly dependant on the projectile mass and energy. For the target fragmentation behavior indicated by $<N_g>$ and $<N_b>$, their ratio $<N_g>/<N_b>$ is nearly constant for ^{4}He and ^{7}Li interaction with emulsion nuclei. For protons and deuterons interaction as a result of the absence of pionization the incident energy may be dissipated in the fragmentation of the target fragments, so their ratios $<N_g>/<N_b>$ have no constancy.

case of p-Em, a weak decrease of $<n_s>$ with the increasing number of heavily ionizing particles attracted our attention. This indicates the absence of visible meson formation in the secondary processes as will as knocking out the relativistic particles with the increase of the impact parameter after the scattering [16].

TABLE 3. The values of average multiplicities of different emitted particles for the total sample ($n_s \geq 0$) and for sample without relativistic hadrons ($n_s = 0$).

Projectile	Total Sample				$n_s = 0$		
	$<n_s>$	$<N_g>$	$<N_b>$	$<N_g>/<N_b>$	$<N_g>$	$<N_b>$	$<N_g>/<N_b>$
P (2.2GeV)	1.27±0.05	2.42±0.12	4.45±0.22	0.54±0.03			
P (3.7 GeV)	2.10±0.03	1.76±.03	6.86±0.09	0.26±0.00	1.59±0.09	6.45±0.25	0.25±0.01
d (3.7 GeV)	2.53±0.07	3.90±0.10	4.60±0.20	0.85±0.04			
^{4}He (2.1AGeV)	3.64±0.05	2.60±0.07	5.82±0.13	0.45±0.01	0.74±0.07	1.67±0.15	0.44±0.05
^{7}Li (2.2AGeV)	4.15±0.12	2.42±0.09	5.68±0.19	0.43±0.02	0.60±0.08	2.22±0.24	0.27±0.04

For samples not accompanied by the relativistic charged particles ($n_s=0$) the ratio $<N_g>/<N_b>$ corresponding to proton interaction is constant for events have ($n_s=0$) compared with that for the total sample ($n_s \geq 0$). The impact parameter in this case is nearly constant, hence the target fragmentation behavior must be similar. For ^{4}He the ratios $<N_g>/<N_b>$ are the same for events have ($n_s=0$) and ($n_s \geq 0$) where the effect of the impact parameter still insufficient to show the fragmentation process. Meanwhile, such effect begins to appear for ^{7}Li reflected on the ratios. The ratio at ($n_s=0$) is nearly half its value at ($n_s \geq 0$).

Figure (2) confirms our vision discussed in table (3). The distributions in the proton shows the same behavior for the proton at ($n_s=0$) and ($n_s \geq 0$). The deuteron has nearly similar behavior regarding the excess in N_h tale at events with ($n_s \geq 0$). In case of ^{4}He and ^{7}Li projectiles, the behavior same to differ at the two criteria of n_s.

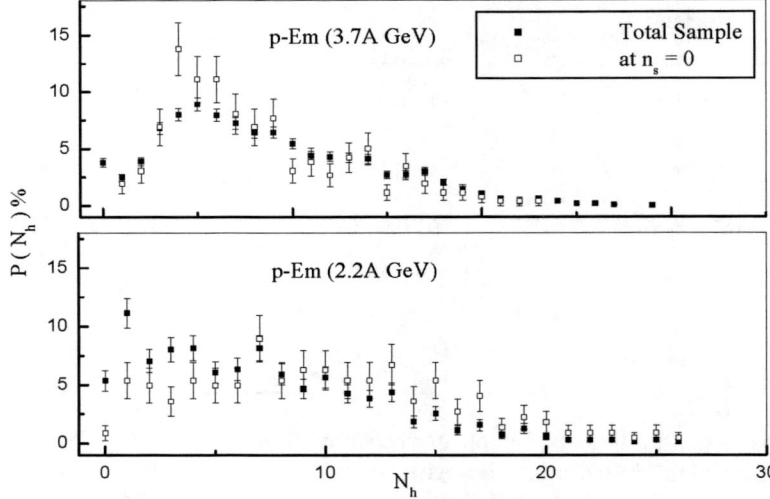

FIGURE 2-a. The multiplicity distributions of heavily ionizing particles (N_h)) emitted from protons for total samples($n_s \geq 0$) and for events emitted without relativistic hadrons ($n_s=0$))

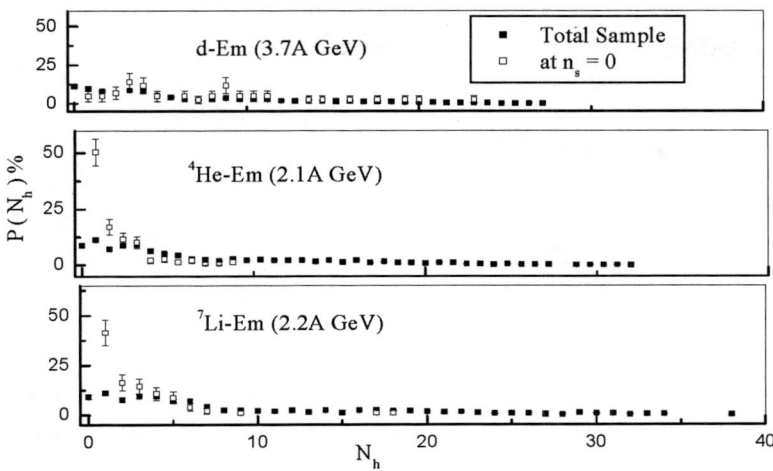

FIGURE 2-b. The multiplicity distributions of heavily ionizing particle (N_h)emitted from d, ^{4}He and ^{7}Li for total samples $(n \geq 0)$ and for events without relativistic hadrons $(n_s=0)$

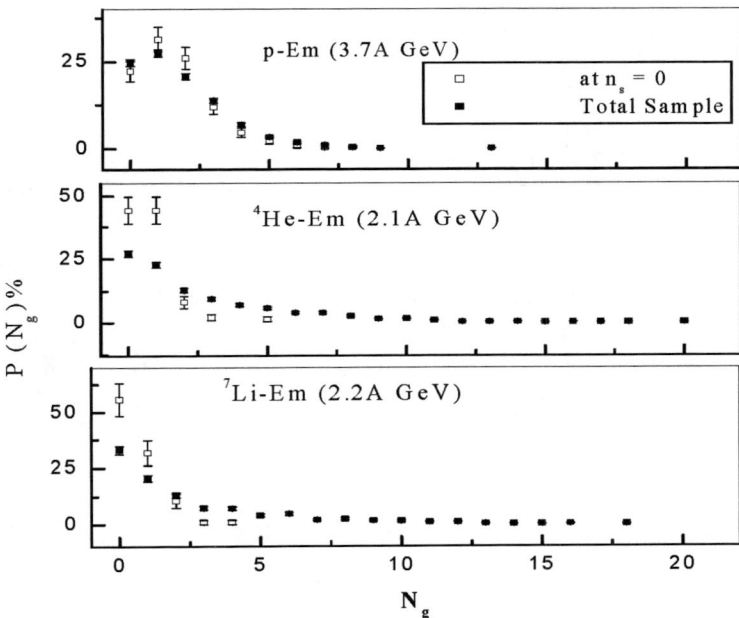

FIGURE 3. The multiplicity distributions of gray particles (N_g)) emitted from p, ^{4}He and ^{7}Li for total sample $(n_s \geq 0)$ and for events without relativistic hadrons (ns=0).

Now, the target fragmentation behavior is more clear in figures (3) and (4) presenting the N_g distribution and N_b distribution. For proton, the N_g and N_b distributions are the same at the two criteria of n_s. For ^{4}He and ^{7}Li, it's clear that the behavior is completely different at the two criteria of n_s, further more, the obvious difference between the two projectile distributions. This trend is clearer to be observed in N_g distribution than N_b distribution. From figure (5), the dependence of $<N_g>$ on n_s is seen to be weak for proton interaction. For ^{4}He and ^{7}Li the dependence seem to be strong. Although the correlation between them is thought to be weaker than observed such behavior may be attributed to the emission of some of pions with energy similar to that of gray particles. This group of pions are not suffering successive collisions in the overlapping region. The dependence of $<N_g>/<n_s>$ on N_h for different projectiles is plotted in figure (6). One can see that $<N_g>/<n_s>$ seems to change linearly with N_h, and the line slope decreases with increasing the mass number of the projectile or the number of interacting nucleons from the projectile (see table 4) .

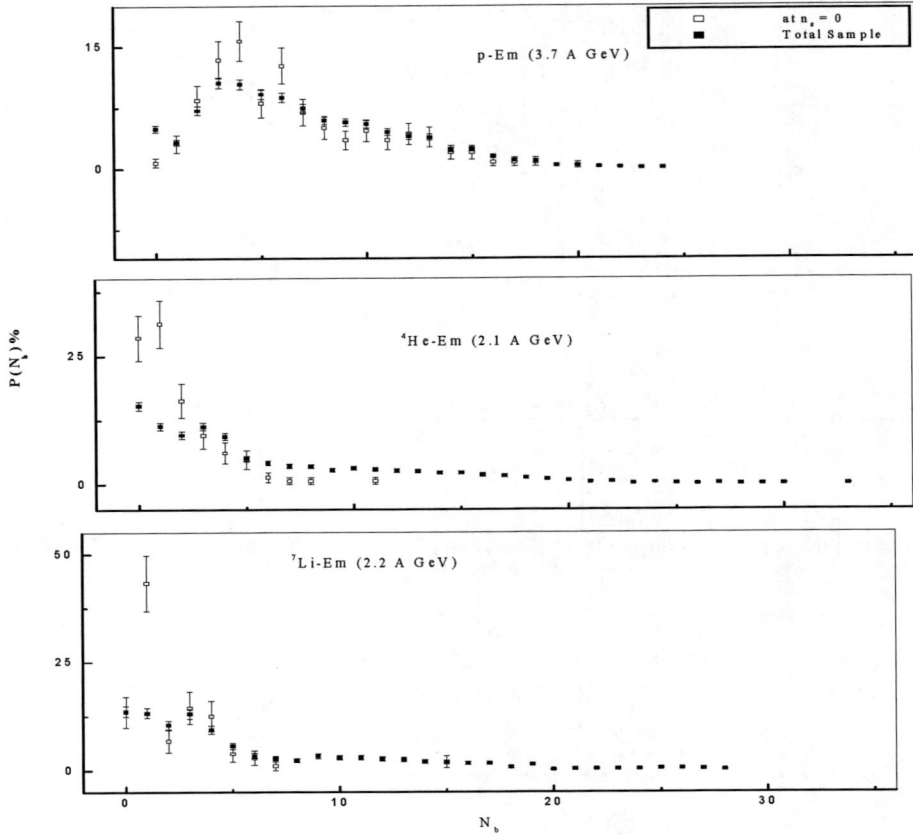

FIGURE 4. The multiplicity distributions of black particles (N_b) emitted from p, ^{4}He and ^{7}Li for total sample ($n_{s \geq} 0$) and for events without relativistic shower hadrons (ns=0).

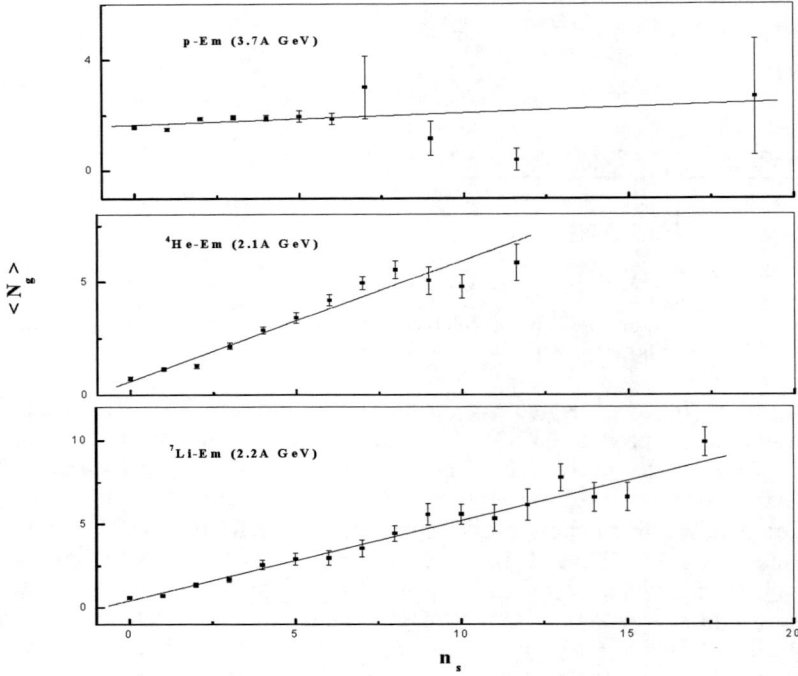

FIGURE 5. The dependence of <Ng > on n_s in the interactions of p, ^{4}He and ^{7}Li with emulsion nuclei. The line represents the best fit of the experimental data.

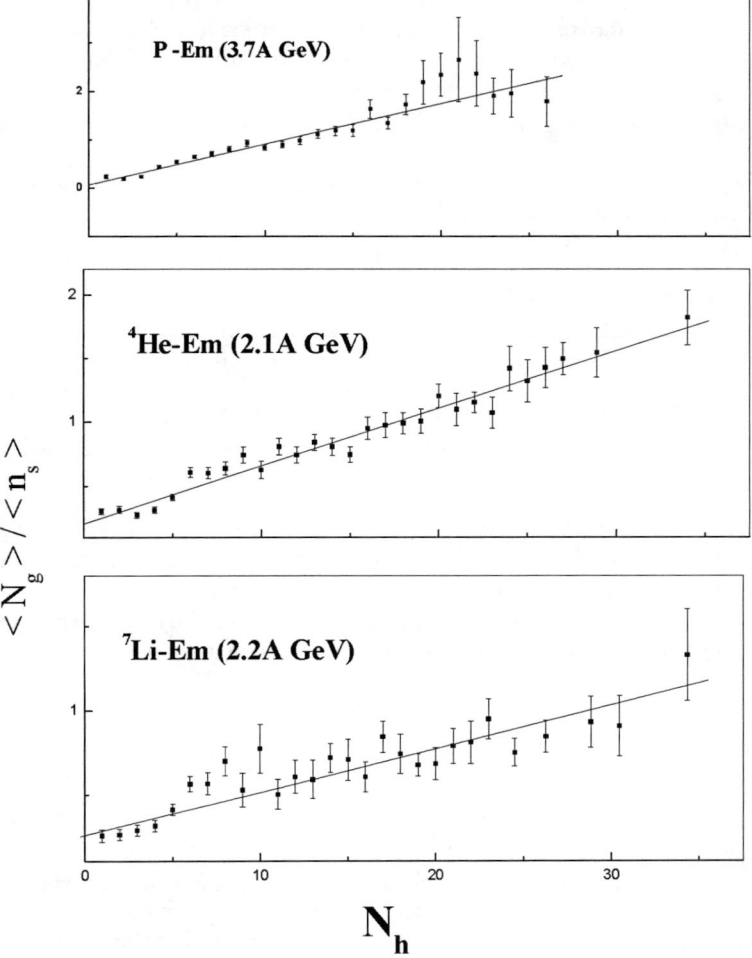

FIGURE 6. The dependence of the ratio $<N_g>/<n_s>$ on N_h in the interactions of p, ^{4}He and ^{7}Li with emulsion nuclei. The line represents the best fit of the experimental data.

TABLE 4. The values of the slope (α) and the intercept (β) determined by the fitting of the experimental data illustrated by the equation ($<N_g>/<n_s> = \alpha N_h + \beta$) for the interaction of p, d, ^{4}He and ^{7}Li projectiles with emulsion nuclei.

projectile	α	β
d (3.7 A GeV)	0.084± 0.004	0.065 ± 0.027
^{4}He (2.1 A GeV)	0.045± 0.002	0.205 ± 0. 0.019
^{7}Li (2.2 A GeV)	0.026± 0.002	0.249 ± 0.023

It is widely believed that, the inclusive nucleus – nucleus data can be explained within the frame work of the superposition models[17-19]. In order to investigate this model , we considered the data of the created shower particles in case of d - E m , ^{4}He – E m .and ^{7}Li – E m. $<n_s>_{A-A}$ at 3.7 A GeV , 2.2 A GeV , and 2.2 A GeV respectively . Then we compared each interaction with the single collisions taken from P-Em .interactions $<n_s>_{P-A}$ at similar incident energy. The values of $<n_s>_{A-A} / <n_s>_{P-A}$ are given in table (5) and compared with the average number of interacting nucleus $<N_{int.}>$.

From this table, one may regard that, the value of $< N_{int.} >$ and $<n_s>_{A-A} / <n_s>_{P-A}$ within the experimental error are nearly the same for d, ^{4}He and ^{7}Li projectiles. One can conclude that, the average number of interacting nucleons can be calculated from the relative shower particles multiplicities. This analysis leads to the conclusion that, the nucleus – nucleus interactions can be viewed as a superposition of nucleon – nucleus collisions at Dubna energies.

TABLE 5. The ratio $<n_s>_{A.A}/<n_s>_{P.A}$ and the experimental average number of interacting nucleons $< N_{int.}>$ in addition to the ratio of the dispersions $D/<n_s>$ for d, ^{4}He and ^{7}Li projectiles.

Projectile	$<n_s>_{A.A} /<n_s>_{P.A}$	$< N_{int.} >$	$D/ <n_s>$
d (3.7 A G e V)	1.55 ± 0.10	1.60 ± 0.10	0.60 ± 0.10
^{4}He (2.1 A G e V)	2.87 ± 0.17	2.77 ± 0.13	0.68 ± 0.15
^{7}Li (2.2 A G e V)	3.79 ± 0.24	3.92 ± 0.12	0.83 ± 0.15

The ratios of the dispersion and the average value of shower particles $D / <n_s>$, are listed also in table (4) for the projectiles d , ^{4}He and ^{7}Li projectiles. It is clear that these ratios are nearly independent on the average number of the interacting nucleus from each projectile, (i.e independent on the projectile mass number). These results agree with those found in ref [20-26]. This conclusion also agrees with the prediction of the superposition models [17-19] considering the nucleus – nucleus as superposition of nucleon – nucleus collisions.

CONCLUSION

From the above data analysis, it can be concluded that:

The probability of the interactions not accompanied by relativistic hadrons depends on the projectile size and the incident energy. The complete destruction of Ag target by proton is not achieved but for the other used projectiles the probability begins to increase with increasing the mass of the projectiles. The mean number of the emitted shower particles $< n s >$ increase with increasing the mass of the projectiles and the target size. In case of proton, this value decrease with increasing the target size.
Nucleus – nucleus interactions can be interpreted as a superposition of nucleon – nucleus interactions in this incident energy. The effect of the impact parameter on the fragmentation process begins to appear from 7Li projectile. For proton and deuteron interaction (hadron -nucleus interactions), the major part of incident energy may be dissipated in the fragmentation of the target .

REFERENCES

1. W. R. Webber, J. C. Kish, and D. A. Schirier, Phys. Rev. C **41**, 520 (1990); **41**, 533 (1990); **41**, 547 (1990); and **41**, 566 (1990).
2. F. Deák et al., phys. Rev. C **42**, 1029 (1990).
3. C. Stephán et al., Phys. Lett. B **262**, 6 (1991).
4. J. Dreute, W. Heinrich, G. Rusch, and B. Weigel, Phys. Rev. C **44**, 1057 (1991).
5. G. A. Souliotis et al., Phys. Rev. C **46**, 1383 (1992).
6. K. H> Schmidt et al., Phys. Lett. B **300**, 313 (1993).
7. G. Bertsch, H. Esbensen, A. Sustitch, Phys. Rev. C **42**, 758 (1990).
8. J. J. Gaimard and K. H. Schmidt, Nucl. Phys. A **531**, 709 (1991).
9. G. Lanzanó et al., Z. **43**, 429 (1992).
10. b. V. Carlson, M. S. Hussein, R. C. Mastroleo, Phys. Rev. C **46**, R30 (1992).
11. F. Khan and L. W. Townsend, Phys. Rev. C **48**, R513 (1993).
12. M. Bagdanski et al., Helv. Phys. Acta, vol. **42** , 485 (1969).
13. V. G. Bogadanov et al., Sol. J. Nucl. Phys. **38**, 909 (1993).
14. P. Singh, M. S. Khan and H. Khushancead. Can. J. Phys. **76**, 559 (1998).
15. N. N. Abdullah, Phys. Seripta, **47**, 501 (1993).
16. B. P. Bannik et al., Gzech. J. Phys. B. **31**, 490 (1981).
17. M.K.Hegab and J. Hufner Nucl. Phys, A**384**, 353 (1982).
18. Bials, M. Bleszynki and W . Czyz, Nucl. Phys.B**111**, 461 (1976).

19. M. M. Sherif , M. K. Hegab, S. A. El-Sharkawy and A. M. Tawfik, Egypt. J. Phys. **23** , 55 (1992).

20. M. M. Sherif , M. K. Hegab, A. Abd El –Salam, S. A. El-Sharkawy and A. M. Tawfik, International Journal of modern physics E, **2**, 835 (1993).

21. M. I. Adamovich et al., JINR. Preprint, El-10838, Dubna (1977) . BWDKLMT Collab., Sov. J. Nucl. Phys . **29** 52 (1979).

22. S. A. El-Sharkawy, M. K. Hegab, O. M. Osman and M.A.Jilany, Physica Scripta. **47**, 512 (1993).

23. M.A.Jilany, Nucl. Phys., A **579**, 627 (1994).

24. Bannik, B. P., et al., Czech. J. Phys. B**31** 490 (1981).

25. A. Abd El –Salam, JINR, E1-**81**, 623 (1981) and references therein.

26. E.O.Abdrakhmanov et al. , Z.Phys. C.5.(1980)1, N.Angelov et al., Sov.J.Nucl. Phys.**33**, 552 (1981) and references therein.

Fragmentation Patterns Of Argon At ~ 1 And 2A GeV In Collisions With Emulsion

M. S. El–Nagdy[1], A. Abdelsalam[2], E. A. Shaat[2], B. M. Badawy[3] and E. M. Khashaba[2]

1 *Phys. Dpt., Faculty of Science, Helwan University, Helwan, Egypt.*
2 *Phys. Dpt., Faculty of Science, Cairo University, Giza, Egypt.*
3 *Reactor Phys. Dpt., Nucl. Research Center, Atomic Energy Authority, Egypt.*
E. Mail: badawyfathalla@hotmail.com

ABSTRACT. This paper combines the results due to interactions of ^{40}Ar at two energies 1.00 and 1.88A GeV in emulsion targets. We discuss and identify the projectile fragments with charge $2 \leq Z \leq 18$. The identification of projectile fragments is based on two measuring methods δ–ray and gap density. The charges tend to appear in groups $Z(2 - 8)$, $(9 - 13)$ and $Z \geq 14$ with a positive long–range correlation. A polynomial function form describes well the δ–ray multiplicity distributions correlating the two used ^{40}Ar beams with each other. A correlation between the average values of δ–ray and gap densities is depicted. We can correlate the values of δ–ray density per fragment and the corresponding gap densities with the fragment charge.

INTRODUCTION

Nuclear fragmentation in heavy ion collisions has been measured with extremely high precision since many years. It stills a subject of great current interest. The study of the properties of nuclear matter within fragmentation process has come into focus of recent research [1–5] activities. They provide essential tools for our understanding of the reaction mechanism. Most investigations [6,7] are concentrated on the detection of nucleons produced particles and very light fragments in full acceptance experiments. Many important results have been obtained by detecting such light nuclei.

In this paper we do an analysis of nuclear fragmentation produced in emulsion from ^{40}Ar projectiles at 1.00 and 1.88A GeV. We report projectile fragments of different charges (2 – 18) emitted from interactions of two Argon beams. The identification of these fragments is based on measurements of δ–ray and gap density methods [8, 9].

EXPERIMENTAL DETAILS

Emulsion Stack

In this experiment, we use two stacks of low sensitivity Ilford G.5 nuclear research emulsion pellicles 7 grain/100 µm for singly charged relativistic particles, 600 µm thick, and 10X10 cm dimensions. These stacks were exposed to ^{40}Ar beam parallel to the surface of emulsion pellicles at ~ 1.00 and 1.88 GeV/nucleon at Lawrence Berkeley laboratory (LBL).

Scanning Technique And Event Classification

Each beam trajectory is scanned along the track under high magnification. The inelastic interactions with nuclear emulsion observed for each of the two used projectiles detected about 500 and 2300 events for ^{40}Ar projectile with energies of 1.00 and 1.88A GeV, respectively. Recorded for each interaction were event type, relativistic secondaries, the charges of the secondary fragments (PF's) determined by a technique discussed later. The secondary

CP888, *Modern Trends in Physics Research,*
Second International Conference on Modern Trends in Physics Research—MTPR-06,
edited by L. El Nadi
© 2007 American Institute of Physics 978-0-7354-0354-9/07/$23.00

tracks resulting from each inelastic interactions are classified into three types on the basis of their grain density (g) measured [10]: Shower tracks those with $g \leq 1.5g_p$; gray tracks for which $1.5g_p < g < 4.5g_p$; black tracks with $g > 4.5g_p$. Here g_p corresponds to the grain density of a minimum ionizing track.

In each event, the charges $Z \geq 2$ of individual PF's are determined by the combination of several methods, which include grain, gap and δ–ray densities [11].

RESULTS AND DISCUSSIONS

The projectile fragments PF's essentially travel with the same speed as that of the parent beam nucleus, so the energy of the produced PF's is high enough to distinguish them easily from the target fragments. All PF's are emitted in a very narrow forward direction ($\theta_{Lab} \leq 3°$) within an angle given by the Fermi momentum. More details are given in [12]. The projectile fragments of charge $Z = 1$ are not included in the following analysis due to the low sensitivity of the used emulsion. It is possible to determine, by eye, the charge of the fragments with $Z = 2$ (mainly He nuclei). The confusion between the He PF's and other PF's with $Z = 3$ or 4 has only a small effect due to the rare production of these latter fragments. Therefore, systematic charge measurements are performed only on fragments with $Z \geq 3$. However, a sample of fragments with $Z = 2$ is also measured.

The charge identification of PF's is based on δ–ray (low energy electron) densities which produces a track containing four or more grains having energy > 15 KeV, and by gap density.

The linear density of δ–rays for a particle with $\beta \approx 1$ is $\dfrac{dn}{dx} \propto Z^2$ in nuclear emulsion[8]. The primary beam with the nuclear charge $Z = 18$ is used in calibration. According to the above proportionality, the number of δ–rays N_δ / mm for ^{40}Ar ($Z = 18$) is proportional to Z^2. Using that proportionality one can obtain the expected N_δ for any PF. Hence,

$$\frac{N_\delta \, for \, ^{40}Ar}{N_\delta \, for PF} = \frac{Z^2 \, (^{40}Ar)}{Z^2 \, for PF} \tag{1}$$

Applying equ. (1), Fig. (1) can give a linear relation expressed by

$$Z^2 = A + BN_\delta \tag{2}$$

The fitting parameters A and B are given in table (1) characterizing the δ–rays calibration of PF's due to the interactions of ^{40}Ar (1.00 and 1.88A GeV) with emulsion. From Fig. (1), it can be noted that, the calibration of the two Argon beams is similar, where their fitting parameters are nearly the same. This reflects the accuracy in counting δ–rays for the two beams.

TABLE 1. The fitting parameters of the calibration curves identifying the PF's due to ^{40}Ar–Em interactions at 1.00 and 1.88A GeV.

Fitting Parameters	^{40}Ar (1.00A GeV)	^{40}Ar (1.88A GeV)
A	$1.95 \times 10^{-6} \pm 5.83 \times 10^{-6}$	$-2.45 \times 10^{-7} \pm 5.11 \times 10^{-6}$
B	$5.40 \pm 1.92 \times 10^{-7}$	$5.25 \pm 1.63 \times 10^{-7}$

Figure (2) displays typical charge spectra, obtained by counting δ–ray densities on primary and secondary tracks. For PF's having charge $Z = 2 - 18$ emitted from ^{40}Ar (1.00 and 1.88A GeV) projectiles by following each track for at least 1 cm track length, the corresponding distributions are shown in this figure. The $Z = 2$ histogram represents the δ–rays measurements for tracks produced for helium PF. On the other hand, the $Z = 18$ histogram indicates the results of the measurements for a sample of produced tracks of ^{40}Ar beam in addition to projectile tracks having $Z = 18$. A series of histograms are observed belonging to a certain value of Z. Each of these histograms can be fitted by a Gaussian distribution with a peak agrees with the calibrated value of N_δ.

Table (2), displays the results of the present investigation of the average number of δ–ray per mm < $N_δ$ > of the fragments at each charge Z (2 – 18) produced from ^{40}Ar ions at 1.00 and 1.88A GeV. Each value of < $N_δ$ > is in agreement with the peak value of each corresponding histogram in Fig. (2). A good agreement with both beams is obtained.

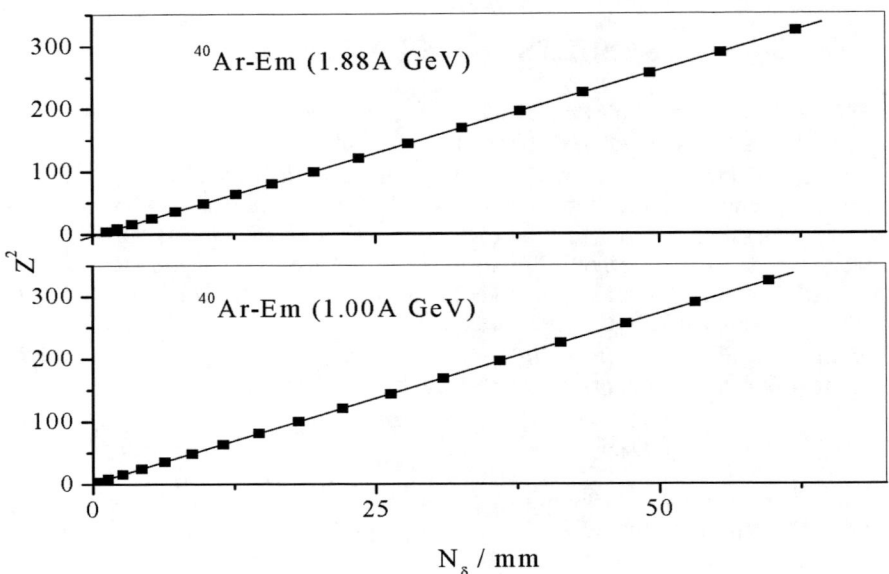

FIGURE 1. The Calibration Lines Due to Counting δ–Ray Density / mm of Primary Argon Tracks.

Each δ–ray spectrum belonging to a certain charge in Fig. (2) shows a dispersion agreement with the measured values of < $N_δ$ >. The dispersion $D_δ$ of each δ–ray spectrum is defined as,

$$D_δ^2 = < N_δ^2 > - < N_δ >^2 \tag{3}$$

This dispersion can be correlated with the average no. of δ–ray / mm < $N_δ$ > corresponding to a certain charge.

In Fig. (3), the ratios of $\dfrac{D_δ}{< N_δ >}$ for the δ–ray spectrum belonging to each charged projectile fragment resulted through the interactions of ^{40}Ar (1.00 and 1.88A GeV) with emulsion are correlated with < $N_δ$ > in hyperbolic shapes. The two curves in Fig. (3) are the fitting of the data resulted from the two ^{40}Ar beams. They are determined by 2nd order exponential decay of the form

$$\frac{D_δ}{< N_δ >} = a_1 e^{-<N_δ>/t_1} + a_2 e^{-<N_δ>/t_2} \tag{4}$$

The fitting parameters a_1, a_2, t_1 and t_2 are given in table (3). The behavior of the curves illustrated in Fig. (3) suggests that, there is a positive long-range correlation in emitting different charges. The ratio $\dfrac{D_δ}{< N_δ >}$ decreases strongly with < $N_δ$ > at the small values of < $N_δ$ >, which correspond to He, Li or Be nuclei. This may be due to some confusion in the identification of such light nuclei resulting in inconsistency between the dispersion and its corresponding average value. At larger values of < $N_δ$ > corresponding to Z > 4 that ratio tend saturate linearly which is a reasonable behavior.

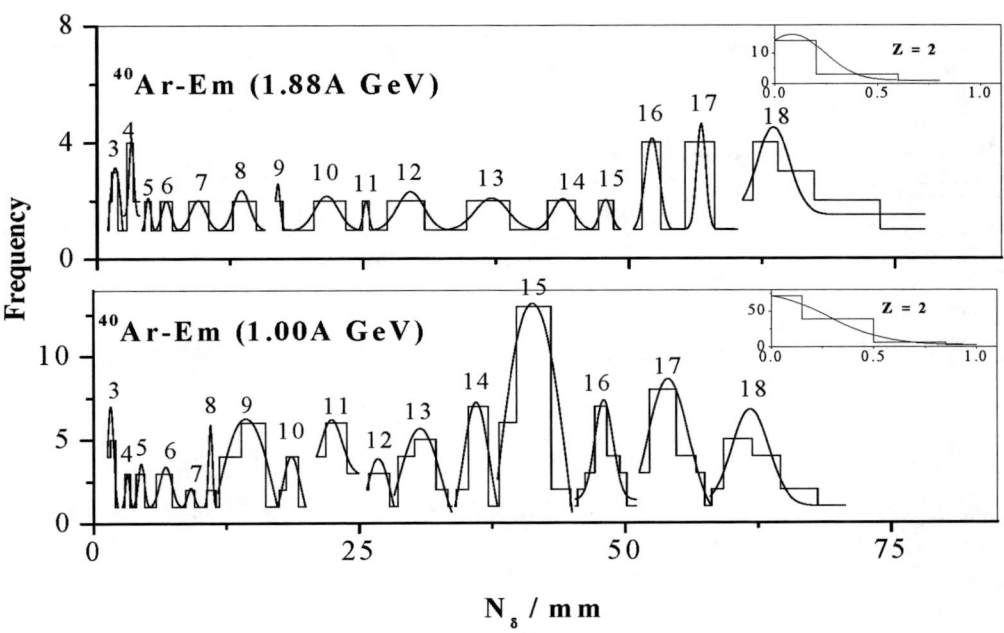

FIGURE 2. The Charge Distributions of δ–Ray Density / mm for Secondary Projectile Fragments Due to ^{40}Ar (1.00 and 1.88A GeV) Interactions with Emulsion (Histograms) Fitted by typical Gaussian Shapes (Solid Curves).

TABLE 2. The Average Values of δ–Ray per mm Characterizing Each Fragmented Track of Charge Z Through the Interactions of ^{40}Ar (1.00 and 1.88A GeV) with Emulsion.

Charge Z	^{40}Ar–Em (1.00A GeV) $< N_\delta >$	^{40}Ar–Em (1.88A GeV) $< N_\delta >$
2	0.16±0.01	0.13±0.03
3	1.69±0.53	1.48±0.56
4	3.20±1.43	3.36±1.12
5	4.38±1.96	4.75±2.38
6	6.60±2.95	6.75±3.38
7	9.08±4.54	9.55±4.78
8	11.08±5.54	13.90±6.95
9	13.68±3.42	17.60±8.80
10	18.41±6.51	21.90±10.95
11	22.73±6.30	25.45±12.73
12	26.73±8.91	29.05±14.53
13	30.88±8.56	36.80±18.40
14	35.88±10.82	43.05±21.53
15	40.89±8.53	47.85±23.93
16	48.06±10.49	52.07±21.26
17	54.46±12.49	57.14±21.60
18	62.15±16.05	65.42±18.89

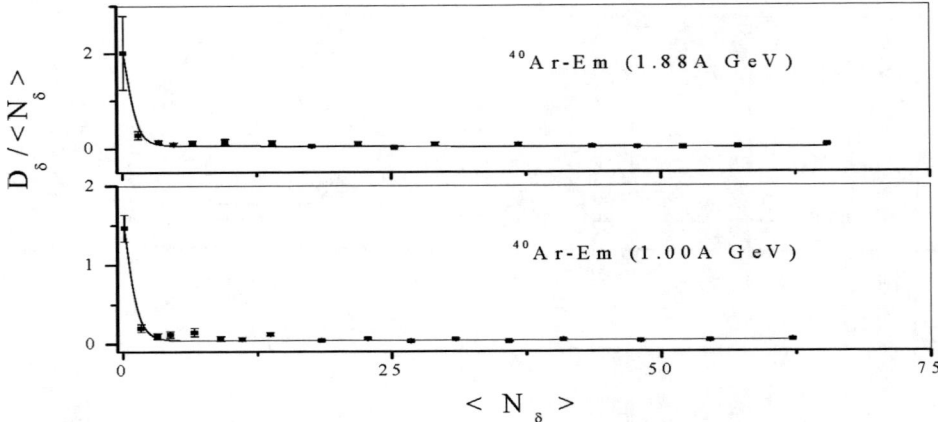

FIGURE 3. The Ratio $\dfrac{D_\delta}{<N_\delta>}$ for the Interactions of ^{40}Ar (1.00 and 1.88A GeV) with Emulsion Against the $<N_\delta>$ Belonging to Each Charge δ–Ray Spectrum, Fitted by Hyperbolic Shapes, (Solid Curves).

TABLE 3. The Fitting Parameters of the Correlation Between $\dfrac{D_\delta}{<N_\delta>}$ Versus $<N_\delta>$ Through the Interactions of ^{40}Ar (1.00 and 1.88A GeV) with Emulsion.

Projectile	^{40}Ar (1.00A GeV)	^{40}Ar (1.88A GeV)
a_1	0.12±0.02	0.13±0.02
a_2	1.79±0.07	2.30±0.05
t_1	38.99±11.90	41.62±12.19
t_2	0.56±0.07	0.55±0.05

The behavior of fragmentation may evidence that it varies according to the charges of fragments for each beam of ^{40}Ar into three groups. These groups can be categorized from the δ–ray normalized distributions given in Fig. (4). It is obvious from the figure that, the 1st group has $N_\delta \leq 13$. This group is corresponding to $2 \leq Z \leq 8$. The 2nd group having $13 < N_\delta \leq 34$ corresponds to $9 \leq Z \leq 13$. The 3rd group having $N_\delta > 34$ corresponds to $Z \geq 14$. The average values of δ–ray according to each group are given in table (4), where they are similar for the two ^{40}Ar beams within experimental error. The data in Fig. (4) are fitted well using, polynomial regression of the form,

$$P(N_\delta) = \sum_{i=0}^{7} b_i N_\delta^i \tag{5}$$

The fitting is illustrated in Fig. (4) by the solid curves. The fitting parameters according to each ^{40}Ar beam are given in table (5). These fitting parameters are presented diagrammatically giving a parabolic correlation between both ^{40}Ar beams through Fig. (5). The data in Fig. (5) can be fitted by the equation of the form,

$$b_i(1.88A\ GeV) = \alpha + \beta b_i(1.00A\ GeV) + \gamma[b_i(1.00A\ GeV)]^2 \tag{6}$$

where, $\alpha = 0.18\pm0.18$, $\beta = 2.82\pm0.04$, $\gamma = -0.02\pm6.86\times10^{-4}$ and $i = 0 \rightarrow 7$. Fig. (5) can correlate the two ^{40}Ar irrespective of their energies. This indicates that the number of δ–ray can shift systematically between the two beams

in a parabolic increments. Hence, using equations 5 and 6 one can predict the fragmented charges due to ^{40}Ar beams at the used range of energy.

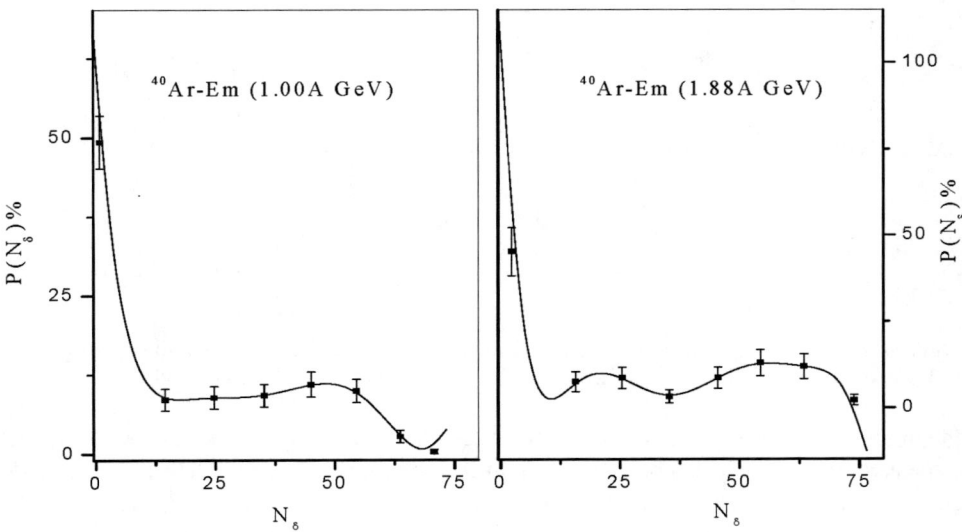

FIGURE 4. The normalized multiplicity distributions of the available values of δ–ray produced on the charged projectile fragments due to the interactions of ^{40}Ar (1.00 and 1.88A GeV) with emulsion, fitted by polynomial shapes, (Solid Curves).

TABLE 4. The Average Values of δ–Ray Density According to Each Interval of Z.

^{40}Ar–Em (1.00A GeV)	^{40}Ar-Em (1.88A GeV)
$N_\delta \leq 13$ $2 \leq Z \leq 8$ $< N_\delta > = 3.05 \pm 0.33$	$N_\delta \leq 13$ $2 \leq Z \leq 8$ $< N_\delta > = 4.16 \pm 0.74$
$13 < N_\delta \leq 34$ $9 \leq Z \leq 13$ $< N_\delta > = 23.90 \pm 3.45$	$13 < N_\delta \leq 34$ $9 \leq Z \leq 13$ $< N_\delta > = 23.68 \pm 6.12$
$N_\delta > 34$ $Z \geq 14$ $< N_\delta > = 48.45 \pm 5.17$	$N_\delta > 34$ $Z \geq 14$ $< N_\delta > = 55.90 \pm 9.60$

TABLE 5. The Fitting Parameters for the δ–Ray Multiplicity Distributions Characterizing the Fragmentation of ^{40}Ar (1.00 and 1.88A GeV).

Fitting Parameters	^{40}Ar (1.00A GeV)	^{40}Ar (1.88A GeV)
b_0	57.42	103.52
b_1	−10.82	−32.37
b_2	0.95	3.76
b_3	−0.04	−0.21
b_4	0.00	0.01
b_5	-1.39×10^{-5}	-9.91×10^{-5}
b_6	8.96×10^{-8}	8.29×10^{-7}
b_7	-2.04×10^{-10}	-2.82×10^{-9}

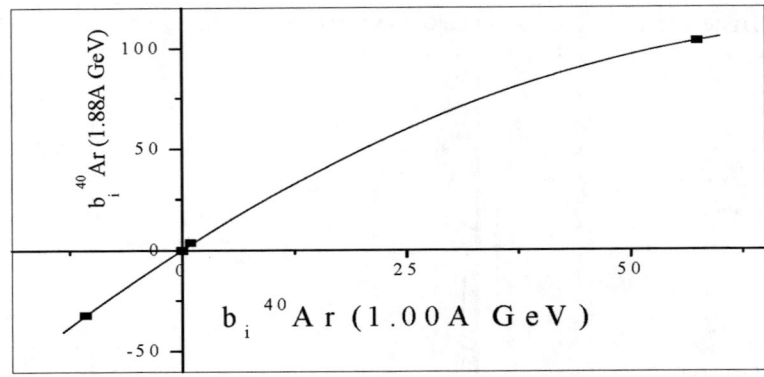

FIGURE 5. The correlation between the fitting parameters characterizing the fragmented tracks through the interactions of ^{40}Ar (1.00A GeV) with emulsion and that of ^{40}Ar (1.88A GeV) fitted by a parabolic relation (solid curves).

Now, the charge identification of relativistic multiplicity charged fragments is made also, by measuring the δ–ray gap densities [13] (the gaps on the δ–ray with length > 3 μm). In the present investigation, a sample of gap densities for both ^{40}Ar beams fragmentation is measured. This sample has been measured before, using δ–ray densities.

Figure (6) depicts a correlation between the average δ–ray densities < N_δ > belonging to a certain charge Z and the corresponding average gap densities < gap > for the fragmentation of the two ^{40}Ar beams. The data in Fig. (6) are fitted by a 2nd order exponential decay of the form,

$$< gap >= A_1 e^{-<N_\delta>/t_1} + A_2 e^{-<N_\delta>/t_2} \qquad (7)$$

where, [A_1 = 208.26±18.39 and 146.25±44.74, A_2 = 3.83±8.38 and 41.79±30.56, t_1 = 9.98± 1.55 and 4.61±2.65, t_2 = 3266.54±393488.90 and 18.15±4.46] for ^{40}Ar at 1.00 and 1.88A GeV respectively.

In order to combine the three parameters < N_δ >, < gap > and Z characterizing each fragmented track in a unique relation, Fig. (7) displays the ratio < N_δ > / < gap > versus charge Z. The ratios for the two ^{40}Ar beams are similar and constant up to Z ≈ 12. Then it seems to grow exponentially with Z where < N_δ > seem to be more than the corresponding value of < gap >. Hence, it is more accurate and preferable to identify the fragments having Z > 12 using gap density method where their densities are lower than δ–ray densities. The energy may be an effective parameter in the fragmentation process to produce heavier fragments with excess of δ–ray density than at lower energy. The fitting of the data is illustrated by the dashed and solid curves through the form,

$$\frac{<N_\delta>}{<gap>} = A e^{Z/t} \qquad (8)$$

where, [A = 2.34x10^{-4}±3.20x10^{-5} and 1.91x10^{-4}±7.47x10^{-5}, t = 1.59±0.03 and 1.42±0.05] for ^{40}Ar at 1.00 and 1.88A GeV respectively.

From figures (6 and 7) it can be noticed that, there is a clear evidence of a critical behavior of the fragmentation process with universal features using both or one δ–ray or gap density measuring method.

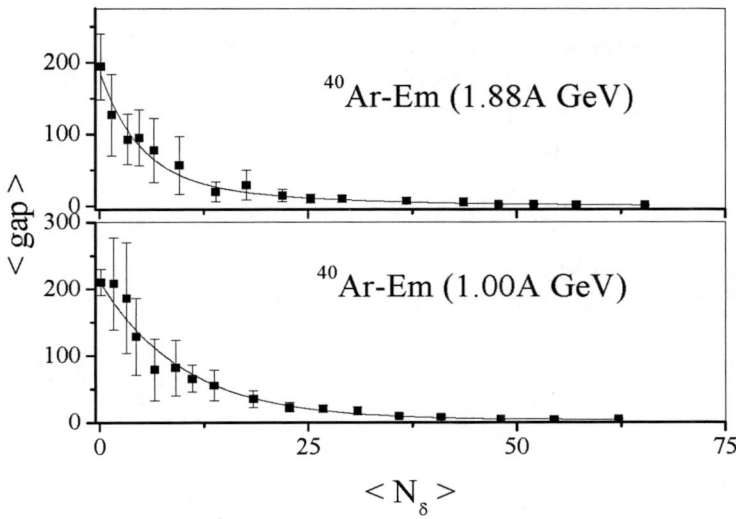

FIGURE 6. The change of average gap density with average δ–ray density belonging to each charge of the fragmented tracks through the interactions of ^{40}Ar (1.00 and 1.88A GeV) with emulsion fitted by 2nd order exponential decay (solid curves).

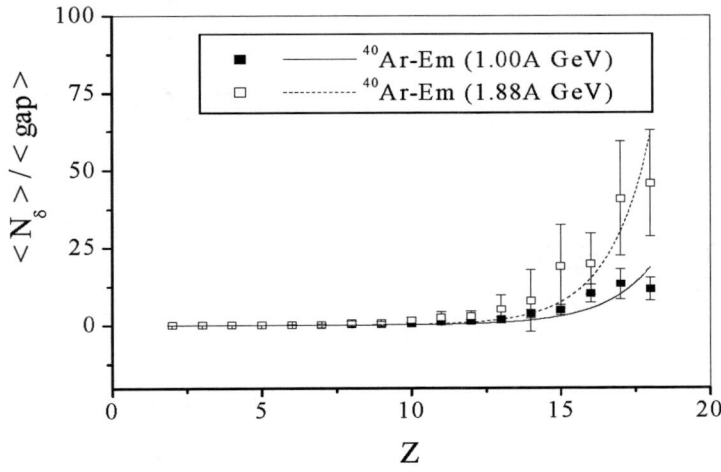

FIGURE 7. The relation correlating the δ–ray and the gap density with the corresponding charge Z characterizing the fragmented tracks through the interactions of ^{40}Ar (1.00 and 1.88A GeV) with emulsion fitted by exponential growth (solid curves).

SUMMURY

We have studied the interactions of ^{40}Ar at two different energies 1.00 and 1.88A GeV in emulsion targets. A systematic comparison using the data of each beam fragment is made. The experimental data on the nuclear fragmentation of each projectile beam are analyzed. The charge of each produced fragment is identified using δ–ray and gap density measurements. A positive long-range correlation in emitting these charged fragments is present. From the δ–ray distributions, it is found that the charges tend to appear in groups with charges Z (2 – 8), (9 – 13) and Z ≥ 14. The δ–ray production spectrum may be convenient to be described in a polynomial shape equation. A clear evidence of a critical behavior discriminating the fragmentation process with universal features is observed throughout using the identification methods.

ACKNOWLEDGEMENTS

We are thankful to Prof. P. L. Jain at the New York state university, Buffalo, U. S. A. for supplying the emulsion plates of 1.00A GeV ^{40}Ar. The authors are grateful to Prof. I. Otterlund at university of Lund, Sweden for providing us the emulsion plates of 1.88A GeV ^{40}Ar.

REFERENCES

1. A. Stolz, et al., *Phys. Rev.* **C65**, 064603 (2002).
2. S. Cecchini, G. Giacomelli, M. Giorgini, G. Mandrioli, L. Patrizii, V. Papa, P. Serra, G. Sirri and M. Spurio, *Nucl. Phys.* **A707**, 513 (2002).
3. J. Miller, *Physica Medica*, Vol. **XVII**, Supplement **1**, 45 (2001).
4. M. El–Nadi, M. S. El–Nagdy, N. Ali–Mossa, A. Abdelsalam, A. M. Abdalla and A. A. Hamed, *J. Phys.* **G25**, 1169 (1999).
5. M. L. Cherry, et al., *Eur. Phys. J.* **C5**, 641 (1998).
6. A. Kelic et al., GSI Report 0174–0814 P47(2004–1).
7. M. El–Nadi, M. S. El–Nagdy, N. Ali–Mossa, A. Abdelsalam, A. M. Abdalla and S. M. Abdel–Halim, *J. Phys.* **G28**, 1251 (2002).
8. W. H. Barkas, *Nuclear Research Emulsion*, vol. **1**, Academic Press, New York (1963).
9. C. F. Powell, P. H. Fowler and D. H. Perkins, *The Study of Elementary Particles by The Photographic Method*, Pergamon, New York (1959).
10. M. S. El–Nagdy, A. Abdelsalam, N. Ali–Mossa, A. M. Abdalla, S. M. Abdel–Halim and Khaled Abdel–Waged, *Nucl. Phys.* **A730**, 419 (2004).
11. A. Z. M. Ismail, M. S. El–Nagdy, K. L. Gomber, M. M. Aggrawal and P. L. Jain, *Phys. Rev. Lett.* **52**, 1280 (1984).
12. M. S. El–Nagdy, *Mod. Phys. Lett.* **A16**, **15**, 985 (2001).
13. B. Jakobsson and R. Kulberg, Preprint, LUIP–CR–75–14 Lund, (October 1975).

Multiplicities and Angular Distributions of Charged Particles in Interactions of Helium Isotopes with Nuclear Emulsion at 3.7A GeV

A. Abdelsalam[1], M. S. El-Nagdy[2], N. Ali-Mossa[3], M. M. Ahmed[2] and R. Elkholy[2]

1) Physics Department, Faculty of Science, Cairo University - Giza, Egypt
2) Physics Department, Faculty of Science, Helwan University - Cairo, Egypt
3) Basic Science Dept., Faculty of Engineering Shoubra, Banha Branch Zagazig University – Cairo, Shoubra, Egypt

ABSTRACT. In inelastic collisions of ^{3}He and ^{4}He with photoemulsion nuclei at 3.7A GeV, experimental data have been obtained on the multiplicities of charged shower, N_s, grey, N_g, and black, N_b, particles and analyzed. The multiplicity correlations between various kinds of produced particles are presented. Angular distributions for grey and black track particles emitted from collision between Helium isotopes and different emulsion nuclei are presented and analyzed in term of the statistical model. Comparison between pseudorapidity distributions of shower particles for Helium isotopes has been made and fitted with Gaussian distribution.

INTRODUCTION

The study of heavy ion interactions at high energies is an important field of modern particle and nuclear physics. Under present experimental conditions, high energy heavy ion interactions are achieved via fixed target experiments. By serving as a fixed target and detector, nuclear emulsion can capture an interaction event in its entirety and with a high resolution of particle tracks, and also the use of heavy nuclei as targets in relativistic-particle collisions enables scientists to study the hadronic-production mechanism.

Many experimental data have been obtained from measurements of star-produced events. This allows one to investigate nuclear matter under unique conditions of strong compression and high temperature.

Isotopes are of high interest in nuclear reaction studies since the observations can be related to the change of only the neutron number, for this reason this paper is devoted for studying the interactions of two helium isotopes (^{3}He, and ^{4}He) with emulsion nuclei at energy 3.7A GeV.

The main process taking place in relativistic heavy ion collisions is, just as in hadron-hadron interactions at high energies, the multiple productions of particles. In the present paper we discuss the basic characteristics of this process, such as multiplicity of charged secondaries, and multiplicity correlations, as well as their dependence on the number of interacting projectile nucleons and the atomic number of a target in inelastic collisions of the two helium isotopes.

In nuclear emulsion experiments [1-5], the emission angles, θ, of shower particles can be measured with great precision, and usually they are presented in terms of the pseudorapidity variable $\eta = -\ln\tan(\theta/2)$. The resulting distribution can be often be described by a single Gaussian distribution. We also present the angular distributions of grey and black particles produced from the interactions between Helium isotopes and emulsion nuclei and the Maxwell-Boltzmann distribution is used to fit those distributions [6].

EXPERIMENTAL DETAILS AND SELECTION RULES

Two NIKFI Br-2 nuclear emulsion stacks were exposed to two helium isotopes (^{3}He, and ^{4}He) beams at incident energy 3.7A GeV at Dubna/JINR Synchrophasatron, Russia. The dimensions of each pellicle are 20 X 10 cm^2 and 600 μm thick with sensitivity of about 30 grains per 100 μm for the minimum ionizing particles. The chemical compositions of this emulsion [7] are hydrogen (39.54%), carbon (17.70%), nitrogen (4.96%), oxygen (12.00%),

CP888, Modern Trends in Physics Research,
Second International Conference on Modern Trends in Physics Research—MTPR-06,
edited by L. El Nadi

bromine (12.90%), and silver (12.90%). The percentage given after each element is the weight of nuclei in the used emulsion. By along the track double scanning technique (fast in the forward direction and slow in the backward one), 1685 and 1092 inelastic interactions were picked up for ^{3}He and ^{4}He projectiles, respectively, after excluding the electromagnetic dissociation and elastic interactions using criteria in ref. [8]. The main characteristics of these interactions, selection rules and other details concerning the experimental procedure have been published elsewhere [9-13].

The secondary charged particles from the above events were divided according to the usual photoemulsion criteria; (a) Shower tracks (s-particles) with specific ionization $g/g_o < 1.4$, g_o being the plateau grain density for singly charged relativistic particle, and velocities β (=v/c) > 0.7; (b) Grey tracks (g-particles) with ionization $1.4< g/g_o<10$ and the range R $\geq$ 3mm, and with velocities $0.3< \beta <0.7$; and (c) Black tracks (b-particles) having $g/g_o \geq 10$, R < 3mm, and $\beta <0.2$. Black plus grey particles are called heavily ionizing particles (h-particles), i.e. $N_h = N_g+N_b$.

Each event was qualitatively classified according to the different types of collisions, i.e. peripheral, quasicentral or central. From a geometrical point of view, the type of collision is defined according to the value of the impact parameter, b, of collision. To classify the collisions with respect to the value of the impact parameter, the quantity Q (the total charge of noninteracting projectile fragment "P_{fs}") was used. The parameter Q represents the combined charge of the spectator fragments of the incident nucleus in a given interaction, i.e. $Q = \Sigma Z_{pfs}$. The value of Qis taken to estimate "N_{int}", the number of nucleons of the incident nucleus that take part in the interaction with target nuclei, i.e., $<N_{int}> = A_p - (A_p/Z_p)Q$, where A_p and Z_p are the mass and atomic numbers of the incident nucleus, respectively. Thus, the following categories can be defined: (i) Q = 0 events that exhibit no visible forward cone fragments P_{fs}. These are called central events. (ii) Q = 1 events with one projectile fragment emitted in a forward cone with charge Z = 1,and (iii) Q = 2 events with total noninteracting outgoing charges equal to 2, such events have either one projectile fragment emitted in a forward cone with charge Z = 2 or two P_{fs} each with Z = 1.

The multiple-scattering measurements were done for the singly charged fragments with $\theta \leq 3°$ in order to reject the projectile fragments from the multiparticle creation (mainly pions). The emitted angles θ (space angle) for each emitted secondary from the different interactions of ^{3}He and ^{4}He nuclei with different emulsion nuclei were measured.

TABLE 1. Experimental values of mean free path (λ_{exp}) in addition to that calculated according to EMUO1 collaboration [23], in 3He, and 4He in comparison with different projectile beams at similar energy per nucleon (3.7 GeV).

projectile	(λ_{exp}) cm	(λ_{th}) cm	Ref.
P	30.0 ± 0.7	35.15	[14]
^{2}H	26.9 ± 0.6	23.74	[15]
^{3}He	19.4 ± 0.5	21.04	present work
^{4}He	19.9 ± 0.6	20.60	present work
^{6}Li	14.5 ± 0.5	17.13	[16]
^{7}Li	15.2 ± 0.5	16.28	[16]
^{12}C	13.7 ± 0.1	13.49	[17]
^{22}Ne	9.9 ± 0.3	10.71	[18]
^{24}Mg	9.6 ± 0.2	10.35	[19]
^{28}Si	8.8 ± 0.3	9.72	[20]
^{32}S	8.3 ± 0.1	8.90	[21]
^{54}Fe	7.3 ± 1.4	6.80	[22]

DATA ANALYSIS AND RESULTS

Reaction Cross-Section and Mean Free Path

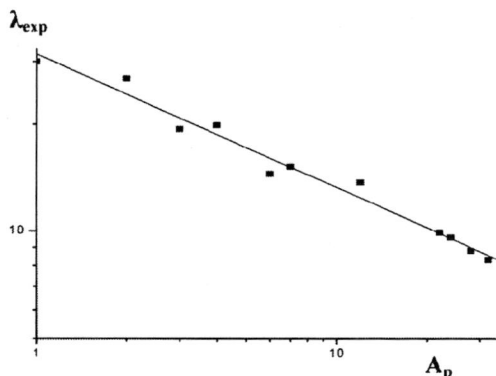

FIGURE 1. The Dependence of the Values of λ_{exp} on the Projectile Mass Number A_p. The Line Represents the Best Fit to the Experimental Points From Table1, it is Clear that the Mean Free Paths Increase with Decreasing the Projectile Mass Number.

A total of 1685 and 1092 inelastic interactions were collected in a total scanned length of 332.6 m and 217.6 m for [3]He and [4]He interactions with emulsion, respectively, after excluding the electromagnetic dissociation and elastic events. The experimental mean free paths in emulsion, λ_{exp}, were determined and were found to be 19.4 ± 0.48 cm and 19.93 ± 0.60 cm for [3]He and [4]He, respectively. The values of λ_{exp} for different projectiles [14-22], presented in Table 1, where compared with the corresponding values calculated from the inelastic reaction cross-section according to the equation $\sigma_i = 109.2\,(A_p^{0.29} + A_T^{0.29} - 1.3)^2$ mb, obtained by the EMUO1 collaboration [23].

Figure 1, illustrates the dependence of the values of λ_{exp} on the projectile mass number A_p, and the fitting relation to the experimental points is

$$\lambda_{exp} = 31.52\,A_p^{-0.37611}$$

Multiplicity of Charged Secondaries

It is well known that in nuclear collisions the dependence of various characteristics on the atomic number of the target A_T, can shed light on the production mechanism. Therefore, we have selected from our samples, according to the necessary criteria of selecting events for emulsion, inelastic interactions of [3]He and [4]He on the free emulsion hydrogen. Thereon the remaining events were statistically divided into collisions with light (CNO) and heavy (AgBr) emulsion in accordance with the method described in details in ref.[13].

TABLE 2. Average multiplicities of shower ($<N_s>$), grey ($<N_g>$), and black ($<N_b>$) particles produced from the interactions of [3]He, and [4]He beams with CNO, Em, and AgBr nuclei.

Projectile	(a) CNO			(b) Em			(c) AgBr		
	$<N_s>$	$<N_g>$	$<N_b>$	$<N_s>$	$<N_g>$	$<N_b>$	$<N_s>$	$<N_g>$	$<N_b>$
[3]He	3.14	1.02	2.36	3.96±0.1	2.53±0.1	5.85±0.1	4.71	3.45	7.98
[4]He	3.62	1.08	2.32	4.54±0.1	2.58±0.1	5.07±0.2	5.40	3.41	6.64

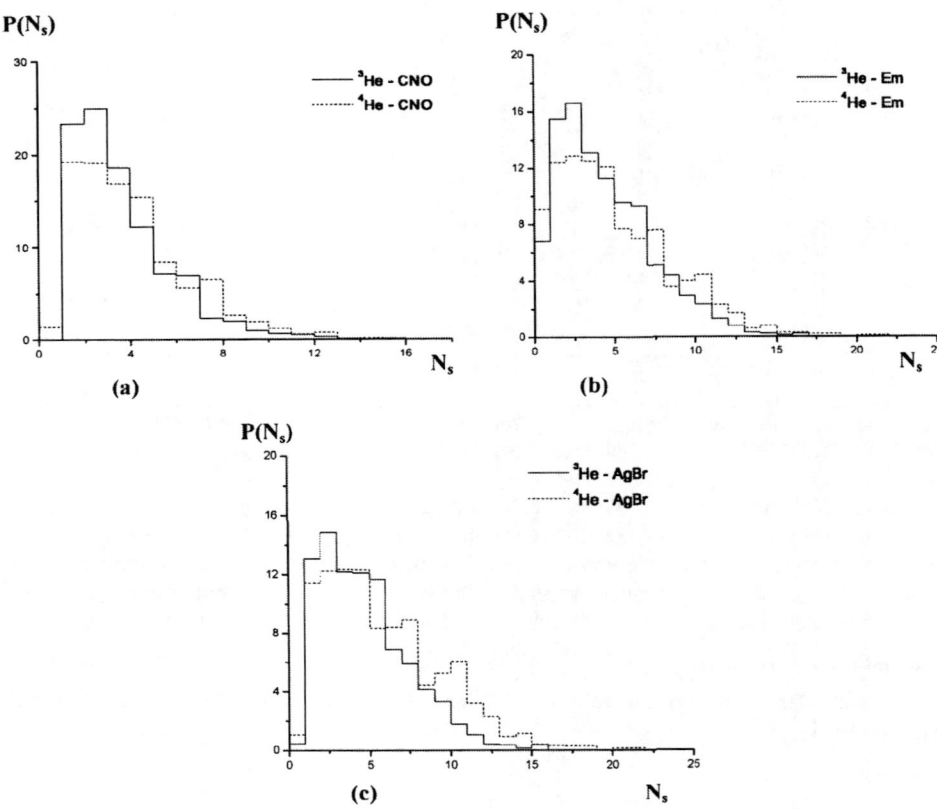

FIGURE 2. Normalized Shower Track Particles "N_s" Multiplicity Distribution of ^{3}He (Solid Histogram) Compared with that Obtained for ^{4}He (Dashed Histogram) Interactions with **(a)** CNO, **(b)** Em, and **(c)** AgBr Nuclei at the Same Energy 3.7A GeV.

In Figures 2(a, b, and c), we show the normalized multiplicity distributions of shower particles, N_s, produced in the inelastic interactions of 3.7A GeV ^{3}He and ^{4}He beams with light nuclei (CNO), emulsion nuclei, and heavy target nuclei (AgBr), respectively. Table 2, presents the mean values of the shower multiplicity, $<N_s>$, for each of the studied interactions. From Figure 2, it can be seen clearly that for the three targets (CNO, Em, and AgBr), ^{4}He exhibits much broader N_s distributions than those for ^{3}He. However, the crests of the N_s distributions of ^{3}He are somewhat higher than those of ^{4}He for small N_s values while this is reversed for higher values of N_s. This means that, the contribution from small values of N_s decreases as the projectile mass number increases, while the contribution from large values of N_s increases as the projectile mass number increases.

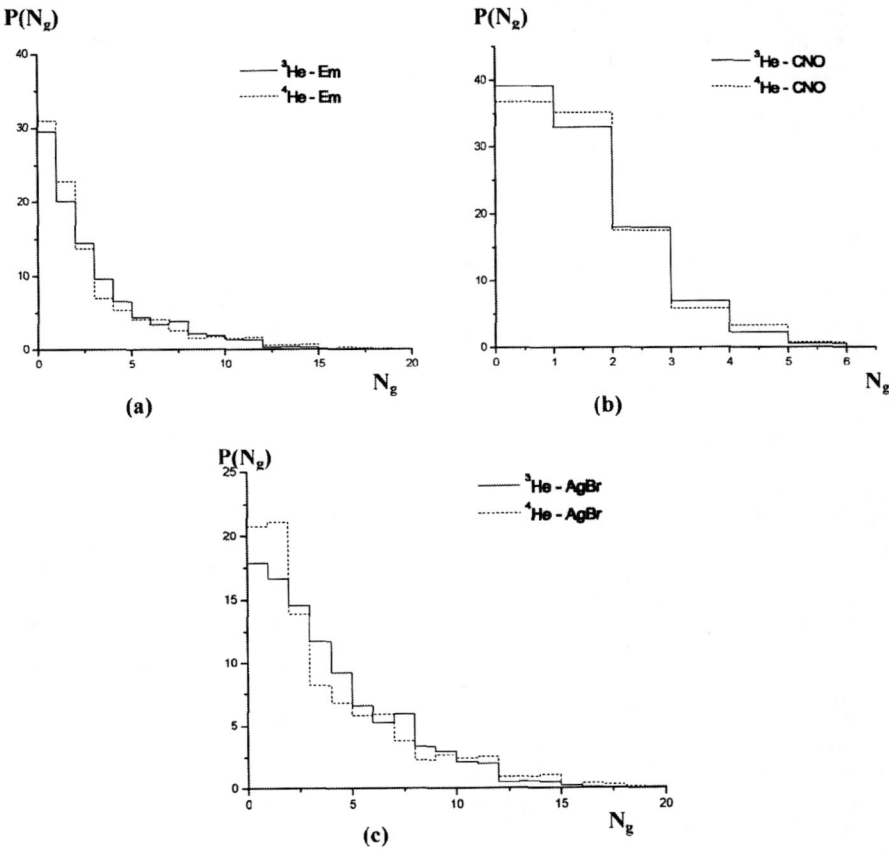

FIGURE 3. Normalized Grey Track Particles "N_g" Multiplicity Distribution of ^{3}He (Solid Histogram) Compared with that Obtained for ^{4}He (Dashed Histogram) Interactions with **(a)** CNO, **(b)** Em, and **(c)** AgBr Nuclei at the Same Energy 3.7A GeV.

Figures 3(a, b, and c) show the normalized multiplicity distributions for the grey particles, N_g, due to the interactions with CNO, emulsion, and AgBr nuclei, respectively. Table 2, presents their corresponding mean values, $<N_g>$. From Figure 3, it is obvious that all distributions follow the same trend. Table 2, confirms the previous results where it illustrates that for the interactions with CNO, Em, and AgBr, the values of $<N_g>$ for the two beams (^{3}He, and ^{4}He) seems to be unchangeable, but $<N_g>$ increases as the mass number of the target increase. Thus, one can conclude that, the N_g distribution and mean value depend on the mass number of the target and independent of the mass number of the projectile.

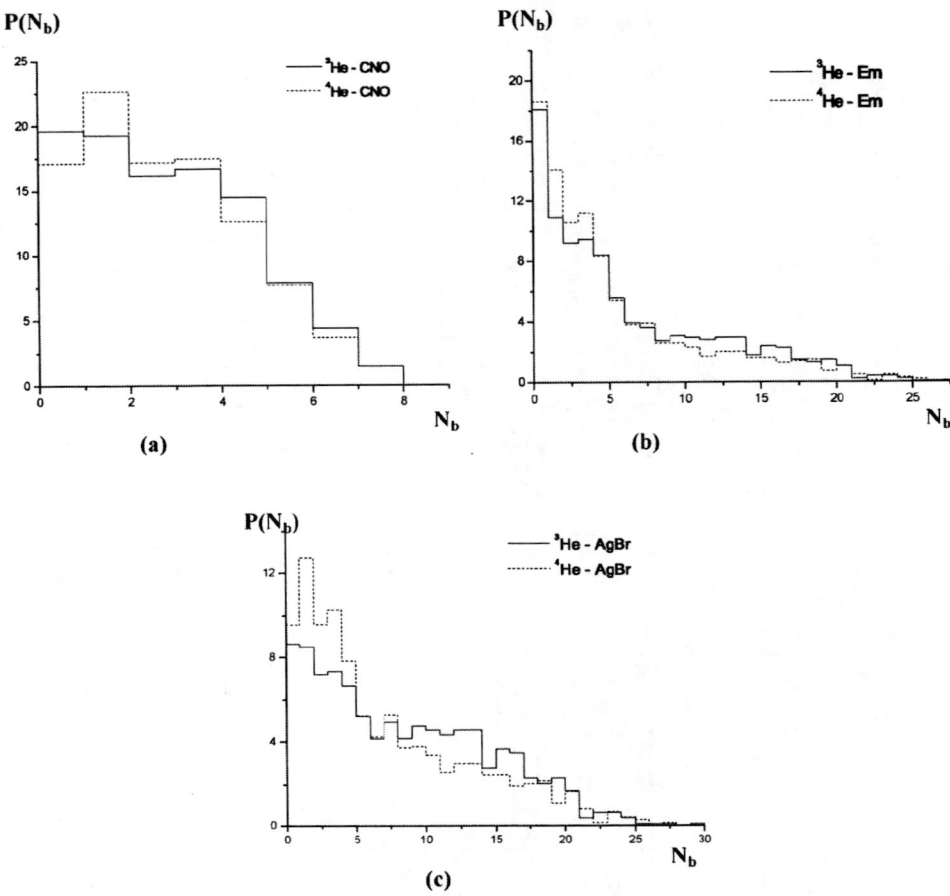

FIGURE 4. Normalized Black Track Particles "N_b" Multiplicity Distribution of ^{3}He (Solid Histogram) Compared with that Obtained for ^{4}He (Dashed Histogram) Interactions with (a) CNO, (b) Em, and (c) AgBr Nuclei at the Same Energy 3.7A GeV.

Figures 4(a, b, and c) illustrate the normalized multiplicity distributions of the black particles, N_b, due to the interactions of ^{3}He and ^{4}He beams with CNO, Em, and AgBr, respectively. This figure shows that the three distributions for the two projectiles (^{3}He and ^{4}He) exhibit the same behavior. Table 2, includes the average multiplicity values, $<N_b>$, obtained for each interaction.

From Figure 4, and Table 2, we conclude that Nb mean value and distribution is strongly dependent only on the mass number of the target nucleus, A_T, whereas it does not depend on the projectile mass number, A_p.

Multiplicity Correlations

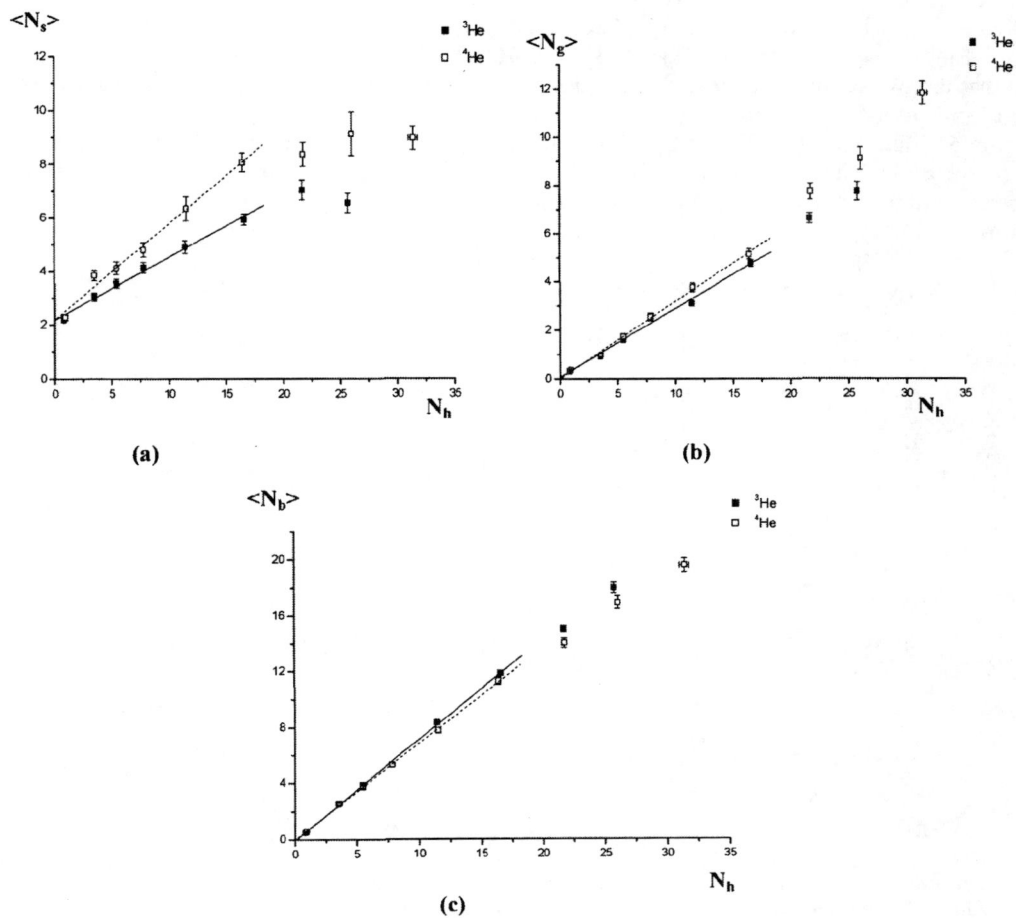

FIGURE 5. Correlations between the Multiplicity of Heavy Ionizing Particles, N_h, and Mean Multiplicities of **(a)** Shower tracks, $\langle N_s \rangle$, **(b)** Grey tracks, $\langle N_g \rangle$, and **(c)** Black tracks, $\langle N_b \rangle$, and the Least Square Fit to the Points from ^{3}He-Em (Solid line) and ^{4}He-Em (Dashed line) Collisions at Energy 3.7A GeV.

Figures 5(a, b, and c) show the correlations between the multiplicity of heavy ionizing particles "N_h" and mean multiplicities of others *i.e.* shower, $\langle N_s \rangle$, grey, $\langle N_g \rangle$, and black, $\langle N_b \rangle$, respectively, from ^{3}He-Em and ^{4}He-Em collisions at the same incident energy per nucleon (3.7 GeV).

From Figure 5a, that shows the variation of $\langle N_s \rangle$ with the variations of N_h for ^{3}He-Em and ^{4}He-Em interactions, it is clear that, the dependence is positive (i.e., $\langle N_s \rangle$ increases as N_h increases), and is also linear over the range of N_h values from 0 to 18 only. The relation between $\langle N_s \rangle$ and N_h can be determined by the least square fit of the experimental data, and it was found to be

for ^{3}He-Em $\qquad \langle N_s \rangle = (2.167 \pm 0.1) + (0.234 \pm 0.01)\, N_h$

and for ^{4}He-Em $\qquad \langle N_s \rangle = (2.167 \pm 0.2) + (0.358 \pm 0.02)\, N_h$

From these two equations, it can be easily noticed that from the higher slope in the equation for ^{4}He-Em interaction, $\langle N_s \rangle$ increases more rapidly with increasing N_h for than for ^{3}He-Em interaction. However, the value of $\langle N_s \rangle$ when N_h is equal to 0is the same for both projectiles. And this confirms our previous results in section 3.2.1, that $\langle N_s \rangle$ depends on both the mass numbers of the projectile, A_p, and target, A_T.

Figure 5b illustrates the variation of $<N_g>$ with N_h for ^{3}He and ^{4}He projectiles in collisions with emulsion nuclei, here the relation is positive and linear only in the range of N_h values from 0 to 18. The least square fit of the experimental data gave us the following equations

for ^{3}He-Em $\qquad <N_g> = (0.065\pm 0.12) + (0.282 \pm 0.01) N_h$

and for ^{4}He-Em $\qquad <N_g> = (0.002\pm 0.08) + (0.317 \pm 0.01) N_h$

It is obvious that the two equations have nearly equal slopes, which indicates that $<N_g>$ is independent of the mass number of projectile as indicated before in section 3.1.2.

From Figure 5c illustrating the relation between $<N_b>$ and N_h for both projectiles (^{3}He and ^{4}He), it is clear that the relation between $<N_b>$ and N_h is linear only for N_h values in the range from 0 to 18. Using the least square fit with the experimental data e obtained the following relations

for ^{3}He-Em $\qquad <N_b> = (-0.0650\pm 0.12) + (0.718 \pm 0.01) N_h$

and for ^{4}He-Em $\qquad <N_b> = (-0.0002\pm 0.08) + (0.683 \pm 0.01) N_h$

It is clear from these equations that $<N_b>$ is independent of the mass number of the projectile.

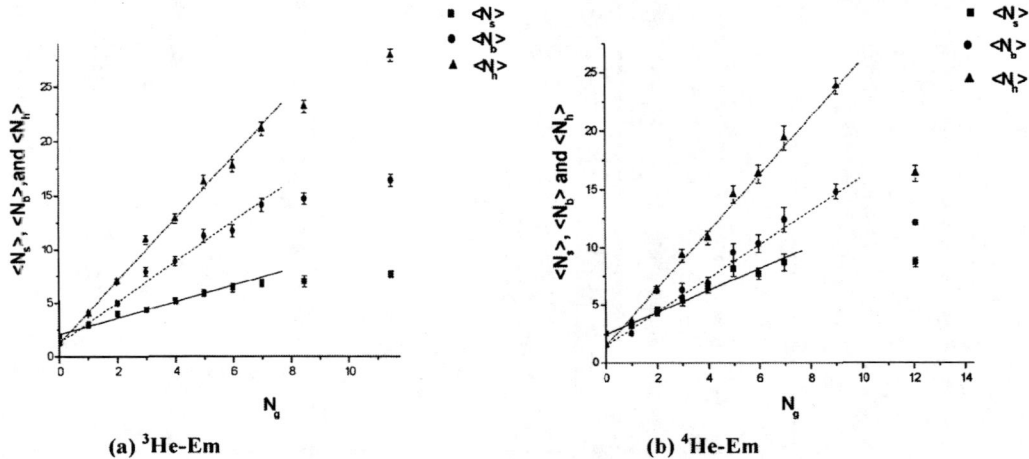

(a) ^{3}He-Em (b) ^{4}He-Em

FIGURE 6. Correlations between the Multiplicity of Grey Particles, N_g, and Mean Multiplicities of Shower, $<N_s>$, Black, $<N_b>$, and Heavy ionizing, $<N_h>$, Particles for **(a)** ^{3}He-Em, and **(b)** ^{4}He-Em Interactions at 3.7A GeV.

Figures 6(a and b) show the mean values of shower, $<N_s>$, black, $<N_b>$, and heavy ionizing, $<N_h>$, particles for ^{3}He-Em, and ^{4}He-Em, respectively interactions at 3.7A GeV. From this figure, it is clear that the dependence of $<N_s>$, $<N_b>$, and $<N_h>$ on N_g value for ^{3}He-Em, and ^{4}He-Em collisions is positive and linear over the range of N_h values from 0 to 7. For ^{3}He-Em collisions, the least square fit to the experimental data in Figure 6a gave the following equations

$<N_s> = (2.069 \pm 0.07) + (0.759 \pm 0.03) N_g$

$<N_b> = (1.269 \pm 0.08) + (1.883 \pm 0.05) N_g$

$<N_h> = (1.269 \pm 0.08) + (2.883 \pm 0.05) N_g$

and for ^{4}He-Em collisions

$<N_s> = (2.477 \pm 0.1) + (0.942 \pm 0.05) N_g$

$<N_b> = (1.47 \pm 0.09) + (1.457 \pm 0.05) N_g$

$<N_h> = (1.47 \pm 0.09) + (2.457 \pm 0.05) N_g$

Angular Characteristics of Charged Secondary Particles

For Shower Particles

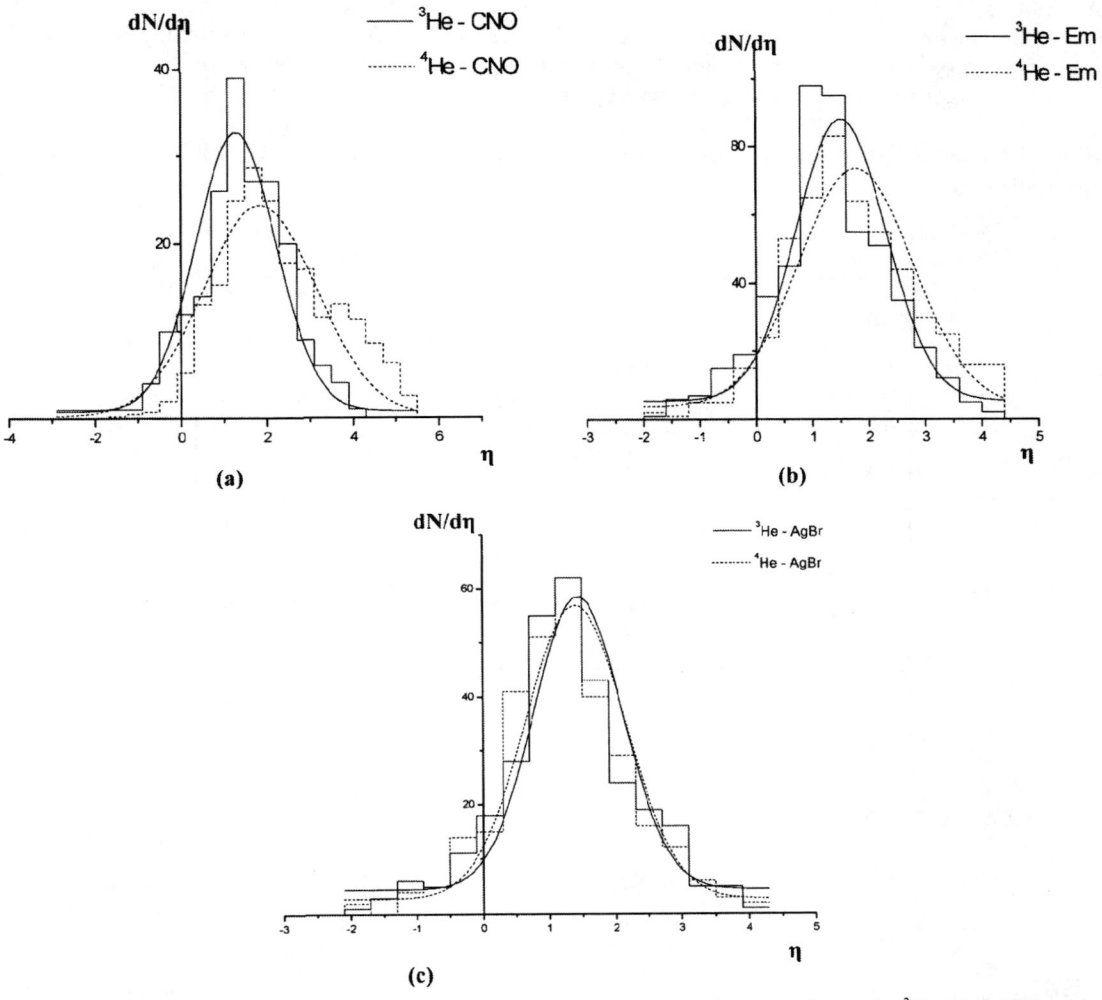

FIGURE 7. The Pseudorapidity Distribution of the Shower Particles in the Laboratory Frame in ^{3}He (Solid Histogram) and in ^{4}He (Dashed Histogram) Interactions with **(a)** CNO, **(b)** Em, and **(c)** AgBr, Nuclei. The Curves are the Gaussian Fit to the Data.

The angular distributions of the emitted shower particles are plotted in terms of the pseudorapidity variable $\eta = - \ln \tan(\theta/2)$ (where θ is the space angle of the emitted particle relative to the incident primary beam direction). Figure 7(a, b, and c) presents a comparison between the pseudorapidity of the shower tracks from the interactions of ^{3}He and ^{4}He with emulsion, CNO, AgBr nuclei, respectively, at the same energy 3.7A GeV. The experimental histograms in Figure 7, are fitted by a Gaussian distribution of the form

$$dN / d\eta = \frac{1}{\sigma \sqrt{2 \pi}} \exp[(\eta - <\eta>)^2 / 2\sigma^2]$$

The average pseudorapidity <η> and the dispersion σ are calculated for each distribution in Figure 7, and are listed in Table 3. It also includes the corresponding values given by EMUO-1 [23] in a central sample for different projectiles.

From Figure 7(a, b, and c) and Table 3, one can notice that, the pseudorapidity distribution of the produced shower particles for different colliding systems at incident energies 3.7A GeV could be reasonably represented by the Gaussian distribution.

From Table 3, it is clear that, the values of the dispersion σ are independent of the projectile mass number and are found to be close to unity. However, the values of the average Pseudorapidity <η> slightly increase with the increasing of the projectile mass number, and for each type of the projectile the values of <η> decrease with the increasing of the target size (or decreasing the impact parameter).

TABLE 4. The dispersion, σ, and the average pseudorapidity, <η>, for the different types of collisions at the same incident energy per nucleon (3.7 GeV).

Interaction	<η>	σ	References
^{3}He + CNO	1.29	0.93	present work
^{3}He + Em	1.51	0.79	present work
^{3}He + AgBr	1.44	0.67	present work
^{4}He + CNO	1.84	1.31	present work
^{4}He + Em	1.77	1.01	present work
^{4}He + AgBr	1.40	0.75	present work
^{7}Li + CNO	1.70	1.00	[24]
^{7}Li + Em	1.52	1.00	[24]
^{7}Li + AgBr	1.50	1.00	[24]
^{12}C + CNO	1.93	1.00	[6]
^{12}C + Em	1.65	1.03	[6]
^{12}C + AgBr	1.71	1.00	[6]
^{16}O + AgBr	1.60	0.80	[25]
^{22}Ne + Em	1.71	1.00	[26]
^{28}Si + Em	1.80	1.00	[26]

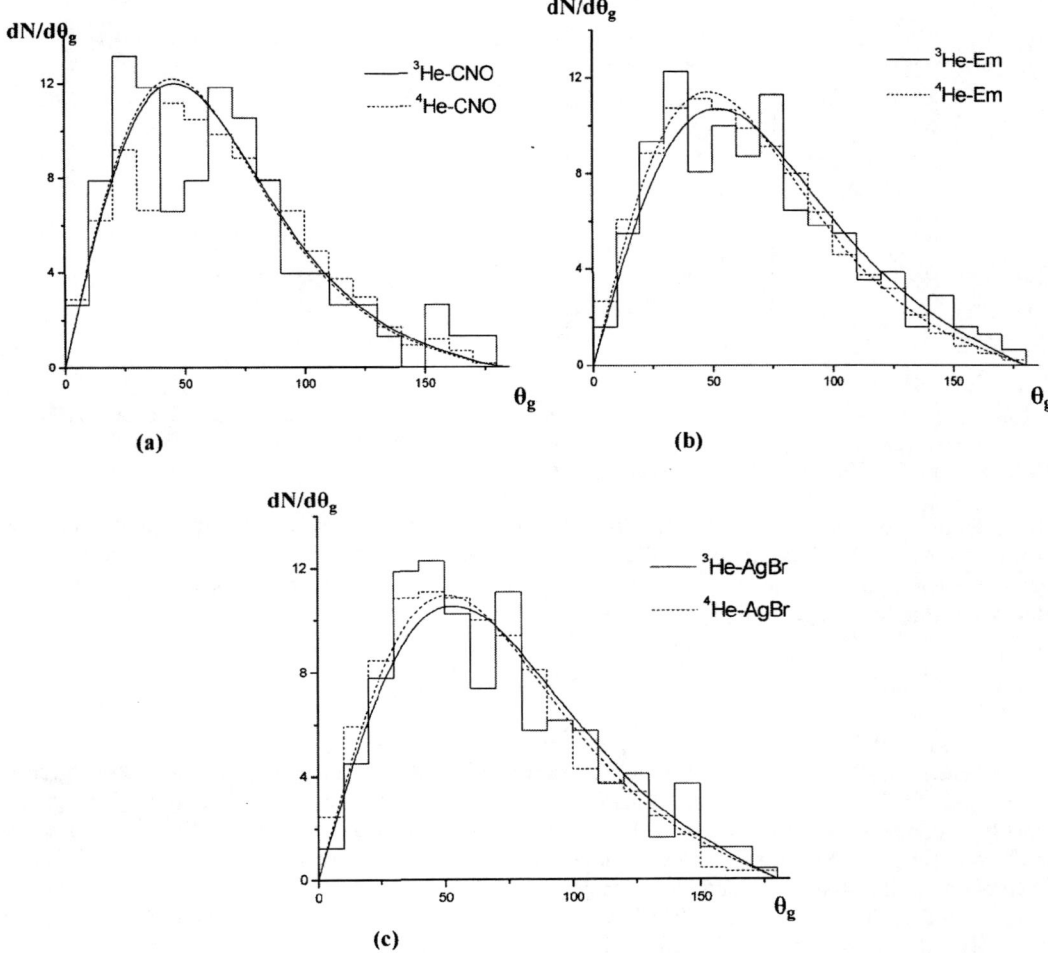

FIGURE 8. The Angular Distributions of Grey Tracks Emitted from the Interactions of ^{3}He (Solid Histogram) and ^{4}He (Dashed Histogram) Interactions with **(a)** CNO, **(b)** Em, **(c)** AgBr Nuclei at Energy 3.7A GeV. The Experimental Histogram is Fitted by the Maxwell-Boltzamann Distribution (Solid Curves for ^{3}He and Dashed Curves for ^{4}He).

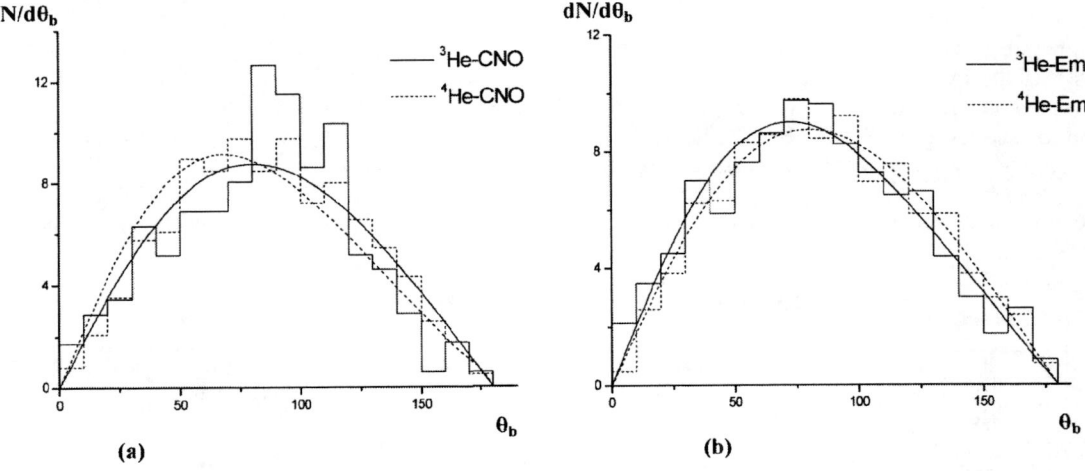

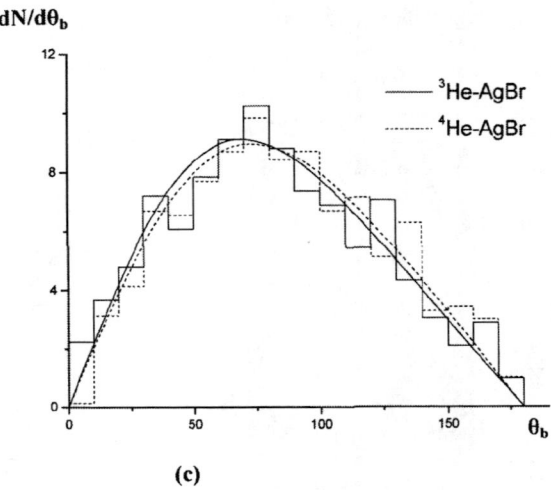

dN/dθ_b

³He-AgBr
⁴He-AgBr

(c)

FIGURE 9. The Angular Distributions of Black Tracks Emitted from the Interactions of ³He (Solid Histogram) and ⁴He (Dashed Histogram) Interactions with **(a)** CNO, **(b)** Em, **(c)** AgBr Nuclei at Energy 3.7A GeV. The Experimental Histogram is Fitted by the Maxwell-Boltzamann Distribution (Solid Curves for ³He and Dashed Curves for ⁴He).

Figure 8 (a, b, and c) presents the angular distribution of the emitted grey particles in the interaction of ³He (solid histogram) and ⁴He (dashed histogram) with CNO, emulsion, and AgBr nuclei, respectively. The smooth curves (solid curves for ³He and dashed ones for ⁴He) presented in Figure 8, based on the statistical model [6, 23, 28] and are calculated according to the equation

$$dN / d\theta_g \approx \sin \theta_g \, (F / B)_g^{\cos \theta_g}$$

where $(F/B)_g$ is the ratio of forward-to-backward emitted grey particles.

From Figure 8, within the statistical errors the experimental histograms for both helium isotopes agree with the calculated curves, and it is seen as has noted in many papers [23,29,30] the angular distribution of the grey-particle tracks is weakly dependent on the variation of the target size. Table 4, lists the average emission angle, $<\theta_g>$ and $(F/B)_g$ ratios. From Table 4, one can notice that, although the values of $<\theta_g>$ and $(F/B)_g$ ratios are nearly independent of the incident energy, they weakly depend on the projectile mass number.

Figure 9 (a, b, and c) shows the angular distributions of the emitted black particles in the interaction of ³He (solid histogram) and ⁴He (dashed histogram) with CNO, emulsion, and AgBr nuclei, respectively.

The experimental histograms are fitted by the calculated curves by using the formula

$$dN / d\theta_b \approx \sin \theta_b \, (F / B)_b^{\cos \theta_b}$$

A close agreement with the experimental data is present in Figure 9 (a, b, and c) for both helium isotopes. Table 4, also lists the average emission angle, $<\theta_b>$ and $(F/B)_b$ ratios. It is clear from Table 4 that, the values of $(F/B)_b$ for different projectiles are greater than the one which reflects the possibility of evaporation from the moving system as suggested in Ref.[6,28]. The values of the $(F/B)_b$ ratios and $<\theta_b>$ are nearly independent of the variation of projectile mass number and of the target size as well as the incident energy.

TABLE 4. The values of the forward-to-backward ratio and the mean angle of emission of grey and black particles at incident energy 3.7A GeV.

Beam	Grey particles		Black particles		References
	$(F/B)_g$	$<\theta_g>$	$(F/B)_b$	$<\theta_b>$	
P	3.80 ± 0.2	67.80 ± 1.2	1.28 ± 0.1	85.30 ± 1.9	[6]
³He	2.73	68.61	1.42	82.30	present work
⁴He	3.38	64.4	1.2	86.5	Present work
⁶Li	2.90 ± 0.2	64.90 ± 2.2	1.30 ± 0.2	82.20 ± 2.6	[23]
⁷Li	3.82	63.50 ± 2.0	1.28	81.30 ± 2.3	[24]
¹²C	3.51 ± 0.2	64.00 ± 1.9	1.27 ± 0.1	79.50 ± 2.4	[6]
²²Ne	3.50 ± 0.2	63.70	-------	-------	[27]
²⁸Si	3.52 ± 0.2	63.30	-------	-------	[27]

CONCLUSION

From the study of the characteristics of the interactions of Helium isotopes (^{3}He and ^{4}He) with different emulsion nuclei at similar energy 3.7A GeV, we reached at the following conclusions:

- The mean free paths increase with decreasing the projectile mass number. The fitting relation to the experimental points is

$$\lambda_{exp} = 31.52 \ A_P^{-0.37611}$$

- The contribution from small values of N_s decreases as the projectile mass number increases, while the contribution from large values of N_s increases as the projectile mass number increases.
- The N_g distribution and mean value depends on the mass number of the target and independent of the mass number of the projectile.
- N_b mean value and distribution is strongly dependent only on the mass number of the target nucleus, A_T, whereas it does not depend on the projectile mass number, A_p.
- The dependence of $<N_s>$, $<N_g>$, and $<N_b>$ on N_h for ^{3}He-Em, and ^{4}He-Em collisions is positive (*i.e.* $<N_s>$, $<N_g>$, and $<N_b>$ increase as N_h increases), and is also linear over the range of N_h values from 0 to 18 only.
- The dependence of $<N_s>$, $<N_b>$, and $<N_h>$ on N_g value for ^{3}He-Em, and ^{4}He-Em collisions is positive and linear over the range of N_h values from 0 to 7.
- The pseudorapidity distribution of the produced shower particles for different colliding systems at incident energies 3.7A GeV could be reasonably represented by the Gaussian distribution.
- The values of the dispersion σ are independent of the projectile mass number and are found to be close to unity. However, the values of the average pseudorapidity $<\eta>$ slightly increase with the increasing of the projectile mass number, and for each type of the projectile the values of $<\eta>$ decrease with the increasing of the target size.
- The angular distribution of the produced grey and black particles for ^{3}He-Em and ^{4}He-Em collisions at incident energy 3.7A GeV could be represented by the Maxwell-Boltzmann distribution.

ACKNOWLEDGMENT

We would like to thank Dubna/JINR Synchrophasatron in Russia, for supplying us with the irradiated emulsion plates. Special thanks to Prof. A.I. Malakhov, Prof. A.D. Kovalenko and Prof. P.I. Zarubin.

REFERENCES

1. M.I. Adamovich et al., EMUO1 Collaboration, *Phys. Lett.* **B 262**, 371 (1991).
2. P.L. Jain et al., *Phys. Rev.* **C 44**, 844 (1991).
3. M. El-Nadi et al., *Nuovo Cimento* **A 108**, 809 (1995).
4. D. Ghosh et al., *Nuovo Cimento* **A 103**, 423 (1990).
5. D. Ghosh et al., *Nuovo Cimento* **A 110**, 565 (1997).
6. A. Abdelsalam, *Phys. Scr.* **47**, 124 (1993).
7. M. El-Nadi, A. Abdelsalam, N. Ali-Mossa, Z. Abou-Moussa, S. Kamel, Kh. Abdel-waged, W. Osman and B.M. Badawy, *Eur. Phys. J.* **A 3**, 183 (1998).
8. M. El-Nadi et al., *Journal of Phys. G. Nucl. Part* **28**, 241 (2002).
9. A-BDDKLMTU-B Collaboration, *Cech. J. Phys.* **B 31**, 490 (1981).
10. Alma – Ala – Bucharest – Dubna – Dushambe – Kishinev – Kosice – Leningrad – Moscow – Tashkent – Ulaan – Batoor – Collaboration, *Z. Phys.* **A 302**, 133 (1981).
11. A-BDDKLMTU-B, Collaboration, *JINR*-**E1**-10838, Dubna (1977).
12. A. Abdelsalam, M. Sumbera, and S. Vokal, JINR-Preprint Dubna. E1-82-509 (1982); Proceedings of the International Conference of Nucleus-Nucleus Collision, East-Lansing, Michigin (1982).
13. A. Abdelsalam, *JINR* **E1**-81-623 Dubna (1981).
14. V.I. Bubnov et al., *Z. Phys.* **A 303**, 133 (1981).
15. M.I. admovich et al., *Preprint JINER*, **E1** 10838, Dubna (1977); A. El-Naghy and V.D. Toneev, *Z. Phyzik* **A 298**, 55 (1980).
16. M. El-Nadi et al., International School of Cosmic Ray Astrophysics 10th course-Erice, Italy, June 16-23, 1996.
17. M. El-Nadi, O.E. Badawy, A. Moussa, E. Khalil and A. Hamalawy, *Phys. Rev. Lett.* **52**, 1971 (1984).
18. B. P. Bannik ct al., *Sov. J. Nucl. Phys.* **52**, 982 (1984).
19. S. El-Sharkawy, M.K. Hegab, O.M. Osman and M.A. Jilany, *Phys. Scrripta* **47**, 512 (1993).

20. M.A. Jilany, *Nucl. Phys.* **A 579**, 627 (1994).
21. N.N. Abd-Allah, *Can. J. Physics* **78**, 915 (2000).
22. L.K. Magnotra and I.K. Daftari, *Nuovo Ceminto Soc. Ital. Fis.* **A 87**, 279 (1985).
23. EMUO1-Collaboration, M.I. Adamovich et al., *Lund Univ. Report* Nos. LUIP 8904, 8906, 8907 (1989).
24. M. El-Nadi et al., *Nuovo Cimento* **A 107**, 31 (1993).
25. EMUO1-Collaboration, M.I. Adamovich et al., *Lund Univ. Report* Nos. LUIP 9202, 9203 (1992).
26. M. El-Nadi, M.M. Sherif, M.S. El-Nagdy, A. Abdelsalam, M.N. Yasin, M.A. Jilany and A. Bakr, Egyptian-German Spring School and Conference on Particle and Nuclear Physics, Cairo, April 11-19, 1992.
27. M. El-Nadi, N. Ali-Moussa and A. Abdelsalam, *IL- Nuovo Ceminto* **110**, 1255 (1997).
28. H.H. Heckman, D.E. Greiner, P.L. Lindstrom and II. Shewe, *Phys. Rev.* **C 17**, 1651 (1978).
29. B. Andersson et al., *Phys. Lett.* **B 73**, (1978), M.M. Aggarwal et al., *Phys. Scr.* **26**, 262 (1982).
30. M.M. Sherif et al., *J. Mod. Phys.* **E** (1993).

Analysis of Central Events in the Interactions of Relativistic Heavy Ions with Emulsion Nuclei at 118.4 GeV

E.EL-Falaky

Physics Department, Faculty of Education, Suez Canal University,
El-Suez, Egypt

Abstract : Data on the multiplicity of the secondary produced particles in the central events from the interactions of ^{32}S with AgBr nuclei at 118.4 GeV. A different selection criteria of the central collision in heavy ion interactions was investigated. The multiplicity distributions of the different produced shower particles (mainly pions) in the central events for each criteria was studied. The multiplicity distributions of the target fragments emitted in the central events was fitted by a Gaussian distribution. The target analysis of the experimental data shows agreement with the limiting fragmentation hypothesis.

Introduction

Many currently theoretical and experimental activities [1-8] deal with analyzing the data of high – energy interactions in terms of selecting central events. In studying the inelastic collisions of relativistic particles with AgBr nuclei using the nuclear emulsion as target and detector, the investigators are attracted by events with complete disintegration (i.e. without an appreciable residual nucleus). The characteristics of such events are expected to be more sensitive to the choice of interaction model than are the characteristics of the other general events i.e., without special selection of the disintegration degree of the target. The selected events have been regarded as a potentially useful events source of information about the underlying production processes as well as the behavior of nuclear matter in extremes states.

From a geometrical point of view, the impact parameter in an asymmetrical central collision is less than or equal to the absolute value of the difference between the radii of the interacting nuclei. Due to the fact that in emulsion experiments the impact parameter can not experimentally measured, the development of selection criteria for central collisions becomes very important. At present a several criteria have been proposed to select central collisions [5,6].

In the present paper, central events induced by the interactions of 118.4 GeV ^{32}S ions with the heaviest nuclei presented in nuclear emulsion (i.e., AgBr) are systematically selected and analyzed. The different selection criteria based on either high degree of target destruction or high multiplicity of outgoing secondary charged particles. These criteria are examined with and without the appearance of the non-interacting projectile fragments. The data sets are described by Gaussian distributions.

Experimental technique

This work was performed using stack of BR-2 type. The emulsion pellicles of dimension of 10 x 20 x 0.06 cm^3 . The stack was horizontally exposed to 118.4 GeV ^{32}S ions in Dubna Synchrotron.

The grain density of singly charged relativistic particles is about 30-35 grains per 100 μm at minimum ionization. The beam tracks of ^{32}S projectile at 118.4 GeV were followed up to $\approx$ 7.0 cm potential path length from the beam entrance. The total scanned length about 74.99 meters of the primary sulpher ^{32}S tracks gave (785) events were attributed to inelastic interactions of ^{32}S ions with the emulsion nuclei. The corresponding experimental value of the average mean free path (λ_{exp}) of ^{32}S ions in the emulsion is found to be equal = 9.55±0.34 cm.

For each detected events, the charged secondaries are classified as follows:

a) Relativistic shower particles (N_s), with velocity $\upsilon \geq 0.7c$. These particles are mostly produced pions.

b)Grey particles (N_g), with velocity $0.2c < v < 0.7c$ and range in emulsion $R > 3mm$ .This particles are mainly protons with kinetic energy $26 < E_k \leq 400 MeV$.

c) Black particles (N_b), having velocity $v < 0.2c$ and range $R < 3mm$, corresponding

CP888, *Modern Trends in Physics Research,*
Second International Conference on Modern Trends in Physics Research—MTPR-06,
edited by L. El Nadi
© 2007 American Institute of Physics 978-0-7354-0354-9/07/$23.00

to $E_k < 20 MeV$. Grey and black particles together are referred to as The total number of these charged secondaries per events (i.e. $N_s + N_g + N_b$) are denoted by N_{ch}. The projectile fragments (PFs) for each interaction are also observed. Such PF's are emitted within a fragmentation forward cone defined by a critical value of 52 mrad for 118.4 GeV ^{32}S . The total charge (or the sum of the charges) of these PF's is denoted by Q. More details about these PF's are given in ref. [9].

Concerning the criteria for selecting a central collision, it has been shown in ref. [10] that the present of high multiplicity of fragment and pions at large angles and with intermediate energies, may be used as a distinctive feature which allows one to select near-central collisions of relativistic nuclei. From the geometrical concept, an events characterized by such a feature occurs as the result of a small impact collision parameter b within the range $0 \leq b \leq |R_1 - R_2|$. where R_1 and R_2 are the radii of target and projectile nuclei, respectively. Heckman et. al. [1] defined central events as interactions that exhibit the absence of projectile fragmentation (i.e. $Q = 0$). However many other criteria to select central events were used concerning high N_{ch} multiplicity [5,11] and high degree of target destruction $N_h \geq 28$ [6,12]. In a previous work [13], the central collisions of ^{12}C-Em. at 3.7 A GeV were selected as those with $N_h \geq 28$ with and without forward cone fragments (i.e., with $Q \neq 0$ and $Q = 0$).

In this work from 785 unbiased inelastic interactions of ^{32}S (118.4 GeV) with emulsion, 307 events were found to be $N_h > 7$. Such events are due to collisions with AgBr nuclei and are thought to be created by violent destruction of projectile and target nuclei at small impact parameter.

It was observed that the beam energy has an effect on the chosen criteria according to which centrality can be studied for the used sample. The events of ^{32}S- AgBr are categorized according to the following criteria:

1) $Q = 0$, i.e., events characterized by the absence of PF's in the fragmentation cone.

2) $N_h \geq 28$,

3) $N_h \geq 28$ and $Q = 0$,

4) $N_{ch} \geq 45$,

5) and $Q = 0$.

The mean values of interacting nucleons $\langle N_{int} \rangle$ are experimentally determined from the formula

$$\langle N_{int} \rangle = A_{proj} - \left(A_{proj} / Z_{proj} \right) \sum Z_f.$$

where A_{proj}, Z_{proj} are the masses and charge number of inelastic beam and $\sum Z_f$ is the total charge of the non-interacting fragments from the projectile in each event.

Results and discussion

Central collisions are generally defined as those characterized by small impact parameter (b) between the interacting nuclei. Here we shall stick to the unique theoretical definition for central collisions as those having impact parameters corresponding to complete overlap of the projectile and target, i.e, $0 \leq b \leq R_t - R_P$, where $R_t (R_p)$ is the radius of the target (projectile). In such collisions, the energy and momentum deposited to the target nucleus will be increased, then the probability for interactions would be increased, and as consequence of this , a large number of particles will be emitted, so the large multiplicities of the emitted particles from any events is one possible signature for central collisions.

Also nuclear emulsions are adequate for studying central collisions, due to their high spatial resolution and possibility of investigating event by event in a 4π geometry.

Since, the impact parameter is an unobserved quantity, several experimental selection criteria for central collisions have been proposed to cope with the above definition. The most often used are summarized in the following:-

1) Small total stripped charge of the projectile fragments (Q) [1,14-16] and specially $Q = 0$ [Fig. (1)] events. Such these events easily separated experimentally which

characterized by the absences of the projectile fragments in the forward cone ($\theta \leq 3^\circ$) .

2) Observation of the degree of target destruction. The multiplicity of the heavily ionizing tracks, N_h, is usually taken as the measure for the degree of destruction. In the hadron interactions with the heavy (AgBr) group of emulsion, events having $N_h \geq 28$ are often classified as central or complete interactions [17-19] . Since, only the target destruction is consider then the limit $N_h \geq 28$ will be the same for nuclear interactions in emulsion. Of course, the probability or cross-section of these interactions will depend on the projectile.

3) High multiplicity of all charged secondaries [11,20-22] (excluding fragments).

$$N_{ch} = n_s + N_g + N_b = n_s + N_h$$

The authors in Ref. (23) taken the central events characterized by

$$N_{ch} \geq 2 < N_{ch} >$$

4) A large number of interacting protons, N_p, from both projectile and target nuclei [24].

5) High total transverse energy E_t [24,25] . The limiting values for the criteria (3-5) were determine in Ref. (26) for the ^{12}C- ^{181}Ta interactions at 4.2 A GeV/C.

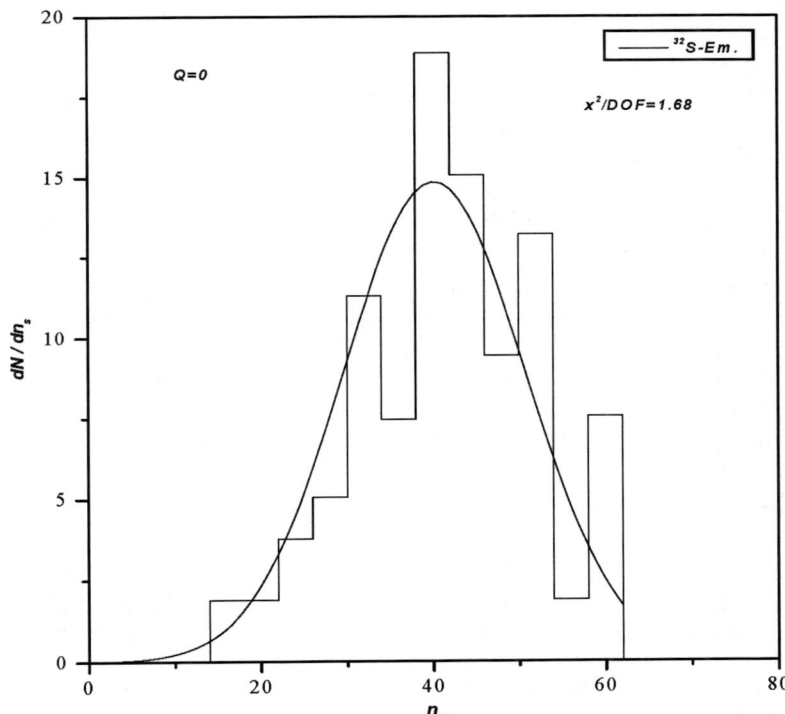

Fig. (1):The n_s-multiplicity distributiom of the shower particles emitted from central collision of ^{32}S with the AgBr nuclei at 1184 GeV according to the criteria Q=0 , The smooth curve represented the Gaussian distributions fitted the experimental histogram.

In Ref. (27) the authors studied the interactions of ^{4}He and ^{12}C ions with emulsion nuclei at 3.7 A GeV . They suggested that, a suitable criteria for selecting events due to central collisions as which have no projectile spectators ($Q = 0$) and $N_h \geq 28$. [Fig.(2) part (a) and (b)].

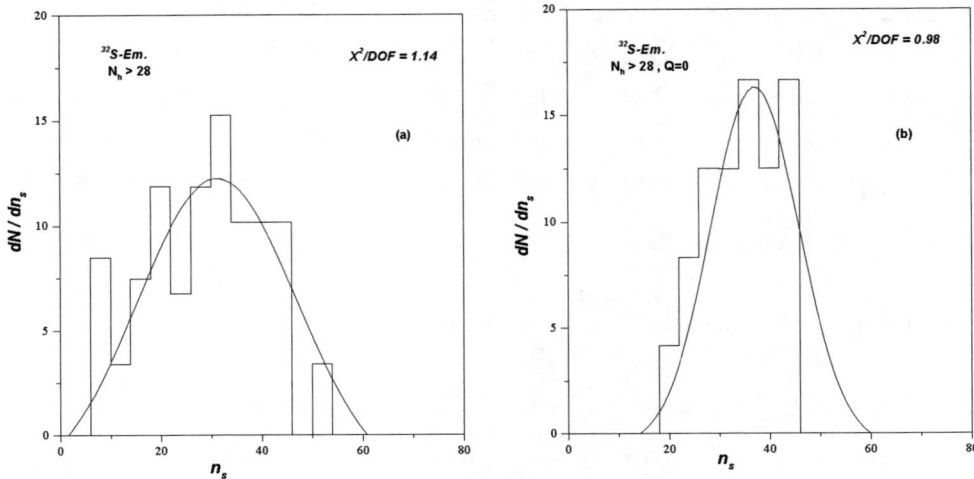

Fig. (2):The n_s-multiplicity distributiom of the shower particles emitted from central collision of ^{32}S with the AgBr nuclei at 1184 GeV according to the criteria $N_h > 28$ [part(a)] and $N_h > 28$, Q=0 [part (b)],(The smooth curve represented the Gaussian distributions fitted the experimental histogram.)

In the work under investigation all that we can do at present is multiplicity measurements. Thus criteria (5) above will be excluded from the investigation. Also criteria (4) needs an exact determination of the number of interacting protons which is not available especially for all the target part. Therefore, we tested here the above criteria's (1-3) in the case of the interactions of ^{32}S with (AgBr) nuclei at 118.4 GeV.

Table (1) present the number of inelastic interactions of ^{32}S ion with emulsion nuclei at 118.4 GeV. Which attributed to the central collisions according to the above criteria's (1-3).The table (1) also shows a comparison between the average numbers of interacting nucleons from the ^{32}S projectile $\overline{N}_{int.}$ and the average multiplicities for different secondaries belong the different group of events, selected according to the criteria (1-3). From the above table one can conclude that :-

(I) The probability of the central interactions of ^{32}S ions with emulsion nuclei according the above criteries (1-2) nearly equals within statistical errors and of order (10%).

(II) The average numbers of interacting projectile nucleons from the events which selected according the above criteria as central interactions of ^{32}S ions with emulsion nuclei are nearly equal within experimental error to the projectile mass number. (i.e. $A_p \approx \overline{N}_{int.}$).

(III) The values of the ratio of $< n_s > / \overline{N}_{int.}$ from the selected events approximately constant which represent the number of generated shower particles per projectile interacting nucleon. These values are compared with the corresponding ones for proton – AgBr interactions [15] at the same incident energy per nucleon which equal to $\approx$ (1.4) .

These results are consisted with those of Refs. (1,27,28) and they given evidence for the validity of considering the nucleus-nucleus interaction as a superposition of nucleon-nucleus collisions.

(IV) The values of the ratio $< n_s > / < N_g >$ for the different group of events selected as central interactions of ^{32}S ions with emulsion nuclei are nearly constant.

(V) The values of $< N_h >$ in the groups of events characterized by $Q = 0$ [criteria (1)] nearly equal = 25.13 $\pm$ 1.05 . This means that the central collisions occur mainly with (AgBr) group of nuclei in the emulsion. The same conclusion are observed in the criteria $N_{ch} \geq 45$. [Fig. (3) part (a) and (b)].

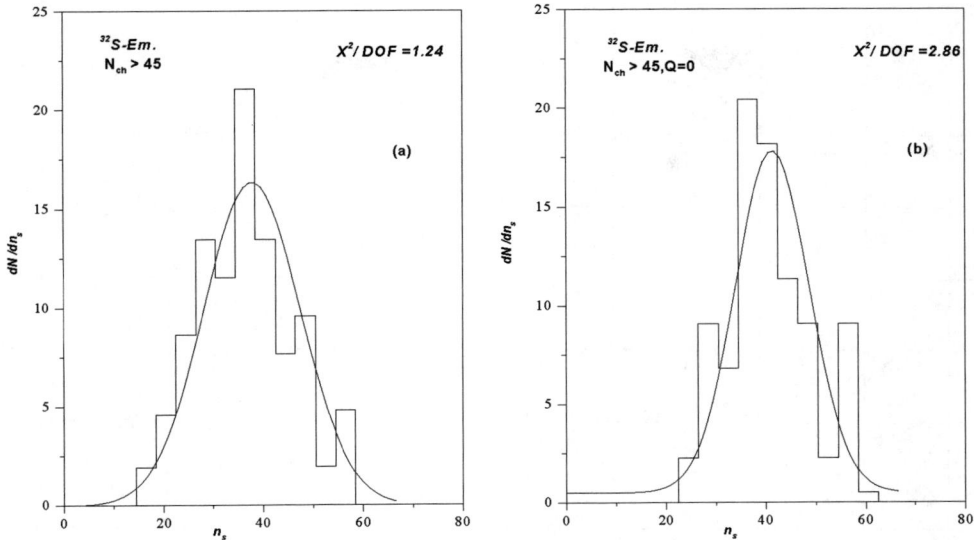

Fig. (3):The n_s-multiplicity distributiom of the shower particles emitted from central collision of ^{32}S with the AgBr nuclei at 1184 GeV according to the criteria $N_{ch} > 45$ [part(a)] and $N_{ch} > 45$, Q=0 [part (b)],(The smooth curve represented the Gaussian distributions fitted the experimental histogram.)

Fig. (4) shows the multiplicity distributions for the target associated particles. N_h, from the events which attributed as a central interactions of ^{32}S with (AgBr) nuclei according to criteria (1-3).

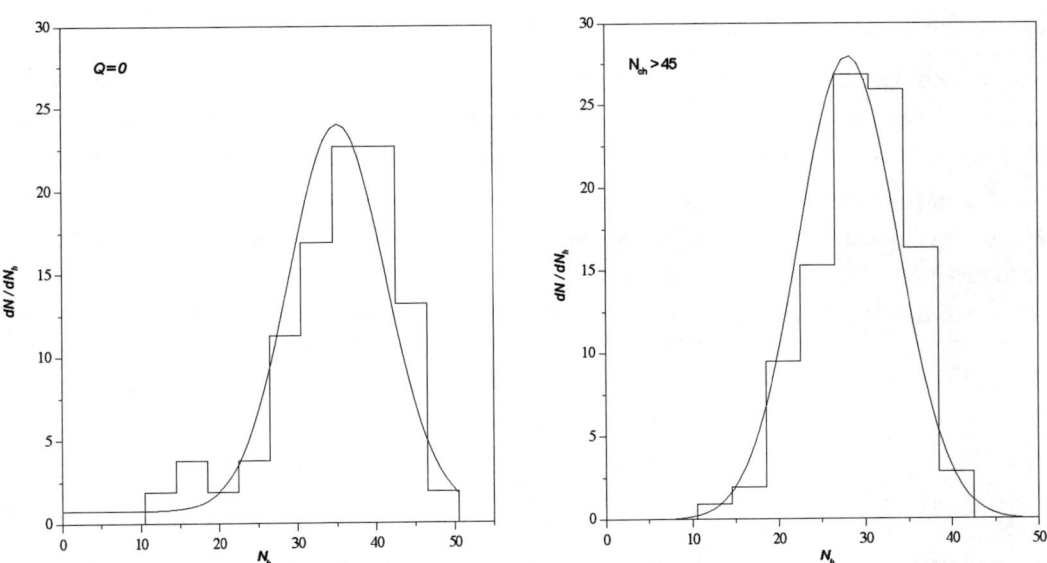

Fig. (4):The N_h-multiplicity distributiom for the target associated particles from events which attributed as a central interactions of ^{32}S with AgBr nuclei according to the criteria Q = 0 [part(a)] and $N_{ch} > 45$ [part (b)],(The smooth curve represented the Gaussian distributions fitted the experimental histogram.)

Selected Criteria	Number of events	Probability	$<n_s>$	$\dfrac{D(n_s)}{\sqrt{<n_s>}}$	$\overline{N}_{int.}$	$\dfrac{<n_s>}{\overline{N}_{int.}}$	$<N_g>$	$<N_b>$	$\dfrac{<n_s>}{<N_g>}$	$\dfrac{<N_g>}{\overline{N}_{int.}}$
$Q=0$	68	8.66%	35.52 ± 1.58	1.91	32.00 ±0.00	1.11	9.47 ± 0.64	15.66 ± 0.73	3.75	0.29
$N_h \geq 28$	72	9.17%	30.89 ± 1.60	2.20	26.98 ±0.88	1.14	11.91 ± 0.50	19.61 ± 0.46	2.95	0.44
$N_h \geq 28$ & $Q=0$	24	3.90%	38.45 ± 2.34	1.84	32.00 ±0.00	1.20	12.95 ± 0.83	18.33 ± 0.82	2.96	0.40
$N_{ch} \geq 45$	131	16.68 %	34.79 ± 0.97	1.66	24.42 ±0.44	1.42	10.27 ± 0.37	17.23 ± 0.41	3.38	0.42
$N_{ch} \geq 45$ & $Q=0$	54	6.62%	39.61 ± 1.38	1.44	32.00 ±0.00	1.23	10.47 ± 0.65	16.52 ± 0.65	3.78	0.32

Table (1): The Comparison between the average number of interacting nucleons and the average multiplicies for different emitted secondaries belong the different groups of events for ^{32}S-Em. At 118.4 GeV.

In the case of $Q=0$, there are some events have $N_h \leq 8$ (6%), this events have a target spectators of the order of the CNO group of nuclei . The smooth curve in Fig. (4) represent the Gaussian distribution fitted the experimental data. In the events characterized with $N_{ch} \geq 45$, the N_h- distribution observed no significant events that have $N_h \leq 8$. Thus we can conclude that the event which selected according to criteria (3) as a central collisions in case of ^{32}S interactions with emulsion nuclei occur mainly with (AgBr) nuclei. The two distributions in Fig. (4) can be represented by a Gaussian form and peaked nearly at $N_h \approx 28$. In the case of the event selected according to the second criteria ($N_h \geq 28$), there is a projectile spectator nucleons [$2\overline{Q} = 5$] while in the case of the events which selected according to criteria (3), the average number of spectator nucleon from the projectile = 3.7 , therefore, we take the criteria (2-3) [$N_h \geq 28$ and $N_{ch} \geq 45$] without spectator nucleon fragments from the projectile ($Q=0$).

Table (2) gives the probability of the central interactions of different projectiles with emulsion nuclei selected according to the criteria ($Q=0$, $N_h \geq 28$ and $N_{ch} \geq 45$).

Fig. (3) show the multiplicity distributions of the shower particles emitted from the events which selected as a central collisions of ^{32}S with (AgBr) nuclei according to different criteria. The smooth curve is a Gaussian distribution of the form :

$$\frac{dN}{dn_s} = \frac{1}{\sigma\sqrt{2\pi}} e^{-\frac{(n_s-<n_s>)^2}{2\sigma^2}}$$

Projectile	$Q=0$	$N_h \geq 28$	$N_{ch} \geq 45$	Ref.
^{12}C	14.00 %	9.60 %	-	28
^{22}Ne	14.48 %	10.10 %	-	29
^{24}Mg	9.60 %	18.50 %	-	30
^{28}Si	15.30 %	11.20 %	-	29
^{32}S	8.66 %	9.17 %	16.68 %	Present Work

Table (2) : The Probability of the central interactions of different projectile with emulsion nuclei selected according to different criteria's .

Where $<n_s>$ is the average number of shower particles in each group and σ is the dispersion of the distribution. Within statistical errors the experimental histograms agree with the Gaussian distribution. According to the values of the Chi-squares per degree of freedom (χ^2/DOF), the criteria (1) is a weak one for centrality. The above n_s - distribution for the events which selected as central interactions of ^{32}S-with (AgBr) can not represented by a Piosson distribution.

This is an impact result, since, the Piosson distribution was is expected for fixed (nearly zero) impact parameter nucleus-nucleus collision and tested for projectiles with smaller mass number ($A_p \leq 12$).

In Ref. (27,28), as in case of ^{32}S at zero impact parameter the projectile is an extended object rather than nearly point as in case of projectile with small mass number. From table (1) , we notice that, the values of $D(n_s)$ far from the values of $\sqrt{<n_s>}$ for all n_s - distribution selected according to the different criteria. Therefore, the distribution can not represented by the Piosson distribution in our experiment. This an important conclusion from the our data for ^{32}S-Em. interaction in compared with the interactions of light projectiles ($A_p \geq 12$) with emulsion nuclei [27,28] at 118.4 GeV .

Conclusions

The study of the various selecting criteria for classifying central interactions of ^{32}S with AgBr nuclei in the emulsion at 118.4 GeV, we can drow the following conclusion:

1)The events characterized by $N_{ch} \geq 45$ and $Q = 0$ for ^{32}S possess the highest values of $\langle N_S \rangle$ and $\langle N_{int} \rangle$ therefore can be considered as the most central ones. These events are characterized by the smallest impact parameter (around zero), representing complete overlap of the projectile and target nuclei.

2)The n_s - distributions of these events are also found to give the minimum values of χ^2/DOF when compared with the suitable mathematical predictions (NB or Poisson distribution.

3)The net charged yield distribution of the non-interacted charged projectile fragment with different target size reflected the type of temperature of the interactions [gentle or violent processes] of different colliding systems.
A selection criteria for central events are investigated here in case of ^{32}S-Em. interactions. The multiplicity distributions of shower particles emitted from the events which selected as a central interactions according to different criteria can be represented by Gaussian formula instead of the Piosson distribution which fitted the shower particles multiplicity distribution in case of the interactions of light projectile $A_p \leq 12$.

4)The number of the projectile nucleons which take part in an interaction, decreases with the increase of the projectile energy. This reflects that the transparency of the target for a projectile becomes more pronounced as the incident energy increases.

Acknowledgments

We would like to thank the staff of the Joint Institute of Nuclear Research (JINR, Russia) for providing us with the irradiated emulsion plates. I wish to express may deep thanks to prof. A. Abdelsalam who help me to perform this work.

Reference

1. H.H. Heckman et al., *Phys. Rev. C*17, (5) (1978) 1651.
2. H.W. Barz et al., *Nucl. Phys. A* 548, 427 (1992).
3. U. Lynen et al., *Nucl. Phys. A* 545, 329c (1992).
4. F. Schussler, H. Nifenecker, B. Jakobsson, V. Kopljar, K. Soder Strom, S. Leray, C. Ngo, S. Souza, J.B. Bondorf, K. Sneppen, *Nucl. Phys. A* 584, 704 (1995).
5. M.M. Sherif, M.A. Jilany, M.N. Yasih, S.M. Abd-Elhalim, *Physical Scripta*, 51, 431 (1995).
6. B.K. Singh, S.K. Tuli, *Nucl. Phys. A* 602, 487 (1996).
7. P. Deines-Jores et al. (KLMM Collaboartion) *Phys. Rev. C* 53, 3044 (1996).
8. H. Sako et al. (E802 Collaboration), *Nucl. Phys. A* 63, 427c (1998).
9. M. EL-Nadi, M.S. EL-Nagdy, A. Abdslsalam, E.A. Shaat, N. Ali Mousa, Z. Abou Mousa, Kh. Abdel-Waged, W. Osman and F. Abd-EL-Wahed, *J. Phys. G.* 24, 2265 (1998).
10. J. Gosset, H.H. Gutbord, W.G. Mater, A.M. Posksnzer, A. Sandoval, R. Stock and G.D. Westall, *Phys. Rev. C* 16, 629 (1977).
11. R. Ihara, Phys. Lett. B 106, 179 (1981).
12. B.P. Bannik, A. EL-Naghy, R. Ibatov, J.A. Salamov, G.S. Shabratov, M. Sherif and K.D. Tolstov, *Z. Phys. A*284, 283 (1978).
13. M.S. EL-Nagdy, Phys. Rev. C47, 346(1993) and references therein.
14. A.Kh. Abdurakhimov et al., *Nucl. Phys.* 33 (1981) 552.
15. S.Y. Fung et al., *Phys Rev. Lett.* 40 (1978) 292.
16. A. Sandoval et al., *Phys. Rev. Lett.* 45 (1980) 874.
17. A. Abdelsalam et al., *Czech. J. Phys.*, B34 (1984), 1196.
18. O. Akhrorov et al., *JINR Report P1-9963 Dubna* (1976).
19. B.P. Bannik et al., *JINR Report P1-*10762 *Dubna* (1977); *Z. Phys. A*284 (1978) 283.
20. B. Jakobsson et al., *Lund University Preprint Lund, LUIP*, 7809 (1978).
21. R. Stock et al., *Phys. Rev. Lett.* 44 (1980) 1243.
22. S. Nagamiya et al., *Phys. Rev. Lett.* 45 (1980) 602.
23. M.A. Jilany, M.Sc. thesis. *Faculty of Science. Assiut University.*
24. V. Boldea et al., Sov. J. *Nucl. Phys.* 44 (1) (1988) 94.
25. R.E. Renfort et al., *Darmstalt Preprient, GSI-*84-35, Darmstadt (1984).
26. Shapiro, Stiller and O.Dell, Bull Am. *Phys. Soc.*, 1,319 (1956).
27. N.Ali-Mousa *J. of the Faculty of Education* No. 17 (1992) 361.
28. A.EL-Naghy *IL NUOVO Cimento*, 71 A (1982) 245 ,K.D.Z.Phys. A 301 (1981) 339.
29. M.EL-Nadi, A.Abdelsalam and N. Ali-Moussa, *Int. J. Mod. Phys. E*, 3 (1994) 811.
30. S. EL-Sharkawy, M.K. Hegab, O.M. Osman and M.A. Jilany, *Physical Scripta*. Vol. 47, (1993) 512.

The production cross-sections for the process
$$e^-(p_1)e^+(p_2) \rightarrow \tilde{\chi}_i^0(p_3)\breve{\chi}_j^0(p_4)H_\ell^0(p_5)$$

M.A.Kamel , M.H.Nous , M.M.Ahmed* , Doaa.M.H

Faculty of Education , Ain-Shams university , Cairo , Egypt
* Faculty of Science , Helwan university , Cairo , Egypt

Abstract. The cross-sections for the process: $e^-(p_1)e^+(p_2) \rightarrow \tilde{\chi}_i^0(p_3)\breve{\chi}_j^0(p_4)H_\ell^0(p_5)$ Have been calculated for all different situations, which are (1520) situations. Three different groups of Feynman diagrams are taken into account a- Production of H_ℓ^0 from different propagators (from 1-752 Feynman diagrams). b- Production of H_ℓ^0 from different legs when Z^0 is the propagator (from 753-944 Feynman diagrams). c- Production of H_ℓ^0 from different legs when H^0 is the propagator (from 945-1520 Feynman diagrams). The values of the cross-sections have been calculated for different incident energy (S) which ranges from 500 to 2000 GeV and for different Higgs'masses. The most probable mechanisms for such reaction are determined.

INTRODUCTION

One of the outstanding questions in particle physics[1] is that of electroweak symmetry breaking and the origin of mass. The electroweak Standard Model[2] is a gauge theory with SU(2)×U(1) symmetry, which is spontaneously broken down to the electromagnetic U(1); electroweak symmetry breaking is the so-called Higgs mechanism, which produces particle masses.The Standard Model[1] requires one Higgs field doublet and predicts a single neutral Higgs boson of unknown mass.

Supersymmetry (SUSY)extensions of the Standard Model[1] are of interest since they provide a consistent framework for the unification of the gauge interactions at a high energy scale and for the stability of the universe at the electroweak scale. Moreover, their predictions are compatible with existing high-precision data.

The Minimal Supersymmetric Standard Model(MSSM)[3] is the SUSY extension with minimal new particle content. It introduces two Higgs field doublets, and predicts the existence of three neutral and two charged Higgs bosons.

The MSSM contains four neutralinos[4] the mass eigen states of the Photino, Zino and neutral Higgsino, and two charginos being mixtures of Wino and charged Hggsino. The neutralinos / charginos sector [5]depends at tree level on four parameters: the U(1) and SU(2) gaugino masses M_1 and M_2 ,the higgsino mass parameter μ , and the ratio $\tan\beta$ of the vacuum expectation values of the higgs fields. In this work the Cross-Sections of the electron-positron interaction producing Higgs boson with the neutralinos will be studied. The Cross-Sections for such process will be calculated for different propagators. The most propable mechanism will be determined.

CP888, *Modern Trends in Physics Research,*
Second International Conference on Modern Trends in Physics Research—MTPR-06,
edited by L. El Nadi
© 2007 American Institute of Physics 978-0-7354-0354-9/07/$23.00

FEYNMAN DIAGRMS FOR THE PROCESS:

$$e^- \, e^+ \rightarrow \tilde{\chi}_i^0 \, \chi_j^0 \, H_\ell^0$$

a-Production via different propagator:

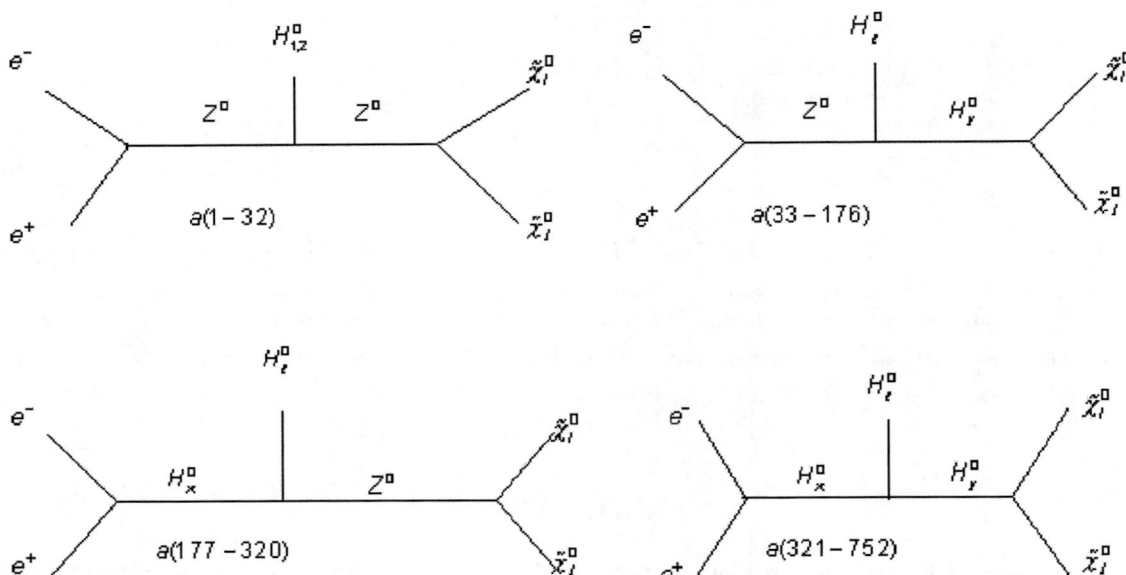

FIGURE 1. Feynman diagrams for the process via Z and Higgs boson exchange (where x and $y = 1, 2,$). There are 752 diagrams .

The matrix elements for a(1-32) are

$$M_{a(1-32)} = \frac{-ig^3 \, M_Z}{8Cos^3\theta_W} \, a_{1,2} \, \overline{V}(P_2) \, \gamma_k \, C \, U(P_1) \, (S^2 - M_Z^2)^{-1} \, (\sigma^2 - M_Z^2)^{-1} \, V(P_4) \, \gamma_k \, [O_{ij}^{//L} \, P_L + O_{ij}^{//R} \, P_R] \, \overline{U}(P_3)$$

where:

$$P_1 + P_2 = P_3 + P_4 + P_5$$
$$S \quad = \quad \sigma + P_5$$

$a_{1,2} = Cos\,(\beta - \alpha)\,, Sin(\beta - \alpha)$ Are the Feynman rules for $H_\ell^0 \, Z^0 \, Z^0$ vertices ($\ell =1$, 2). The tri linear $H_3^0 \, Z^0 \, Z^0$ vertex vanish at the tree level [6],

$$C = 4\operatorname{Sin}^2\theta_W + \gamma_5 - 1 \quad , \quad P_{R(L)} = 1 \pm \gamma_5$$

$$O_{ij}^{//L} = -\frac{1}{2}N_{i3}\,N_{j3}^* + \frac{1}{2}N_{i4}\,N_{j4}^* \quad , \quad O_{ij}^{//R} = \frac{1}{2}N_{i3}^*\,N_{j3} - \frac{1}{2}N_{i4}^*\,N_{j4}$$

Are the Feynman rules for the interaction of a gauge boson with neutralinos [7] Where N is the (4x4) matrices diagonalizing the neutralino mass matrix.

The matrix element for a (33-176) are

$$M_{a(33-176)} = \frac{-ig^3}{8\operatorname{Cos}^2\theta_W \operatorname{Sin}\beta}\, b_{\ell y}\, \overline{V}(P_2)\,\gamma_\mu\, C\, U(P_1)\,(S^2 - M_Z^2)^{-1}(P_5 + \sigma)_\mu (\sigma^2 - m_{H_y^0}^2)^{-1}\, V(P_4)\, A_y$$

$$\overline{U}(P_3)$$

For $b_{\ell y}$ we have ($b_{13} = b_{31} = \operatorname{Sin}(\alpha - \beta)$, $b_{23} = b_{32} = \operatorname{Cos}(\alpha - \beta)$)

Where CP- invariance forbids b_{11} , b_{22} , b_{33} , b_{12} , b_{21}

and A_y $(y = 1,2,3)$ are the Feynman rules for $\widetilde{\chi}_i^0\,\chi_j^0\,H_\ell^0$ vertices [7] given as:

$$A_1 = \frac{\widetilde{M}_i^{(0)}\,\delta_{ij}\,\operatorname{Sin}\alpha}{M_W} + [Q_{ij}^{//*}\operatorname{Sin}(\beta - \alpha) - R_{ij}^{//*}Sin\alpha](1 - \gamma_5) + [Q_{ij}^{//}\operatorname{Sin}(\beta - \alpha) - R_{ij}^{//}Sin\alpha](1 + \gamma_5)$$

$$A_2 = \frac{\widetilde{M}_i^{(0)}\,\delta_{ij}\,\operatorname{Cos}\alpha}{M_W} - [Q_{ij}^{//*}\operatorname{Cos}(\beta - \alpha) - R_{ij}^{//*}Cos\alpha](1 - \gamma_5) - [Q_{ij}^{//}\operatorname{Cos}(\beta - \alpha) - R_{ij}^{//}Cos\alpha](1 + \gamma_5)$$

$$A_3 = i(\frac{\widetilde{M}_i^{(0)}\,\delta_{ij}\,\operatorname{Cos}\beta}{M_W}\,\gamma_5 - [Q_{ij}^{//*}\operatorname{Cos}2\beta + R_{ij}^{//*}Cos\beta](1 - \gamma_5) + [Q_{ij}^{//}\operatorname{Cos}2\beta + R_{ij}^{//}Cos\alpha](1 + \gamma_5))$$

Where $\widetilde{M}_i^{(0)}$ are the neutralinos mass (i=1, 2, 3, 4) and m_{H^0} is the mass of Higgs boson.

The matrix element for a (177-320) are

$$M_{a(177-320)} = \frac{g^3\, m_e}{8\operatorname{Cos}^3\theta_W\, M_Z}\, b_{\ell x}\, \overline{V}(P_2)\, B_x\, U(P_1)\,(S^2 - m_{H_x^0}^2)^{-1}(S + P_5)_k(\sigma^2 - M_Z^2)^{-1}\, V(P_4)\,\gamma_k\, [O_{ij}^{//L}\, P_L + O_{ij}^{//R}\, P_R]\,\overline{U}(P_3)$$

Where m_e is the electron mass, for B_x (x=1, 2, 3) are

$$B_1 = \frac{\operatorname{Cos}\beta}{\operatorname{Cos}\alpha} \quad , \quad B_2 = \frac{\operatorname{Sin}\beta}{\operatorname{Cos}\alpha} \quad , \quad B_3 = \gamma_5\tan\alpha$$

Are the Feynman rules for $e^-\,e^+\,H_\ell^0$ vertices [6], and $b_{\ell x} = b_{\ell y}$ in the previous matrix.

The matrix element for a (321-752) are

$$M_{a(321-752)} = \frac{-ig^3\, m_e}{8\operatorname{Cos}^3\theta_W\, \operatorname{Sin}\beta}\, \overline{V}(P_2)\, B_x\, U(P_1)\,(S^2 - m_{H_x^0}^2)^{-1}(\sigma^2 - m_{H_y^0}^2)^{-1}\, V(P_4)\, F_{xy\ell}\, A_y\, \overline{U}(P_3)$$

Where $F_{xy\ell}$ is given as:

$F_{111} = -Cos2\alpha\ Cos(\beta + \alpha)$

$F_{222} = -Cos2\alpha\ Sin(\beta + \alpha)$

$F_{112} = F_{121} = F_{211} = 2Sin2\alpha\ Cos(\beta + \alpha) + Sin(\beta + \alpha)\ Cos2\alpha$

$F_{122} = F_{212} = F_{221} = -2Sin2\alpha\ Sin(\beta + \alpha) + Cos(\beta + \alpha)\ Cos2\alpha$

$F_{133} = F_{331} = F_{313} = Cos2\beta\ Cos(\beta + \alpha)$

$F_{323} = F_{233} = F_{332} = -Cos2\beta\ Sin(\beta + \alpha)$

CROSS SECTION CALCULATIONS

In this work we have 3-body final states with momentum P_3, P_4, P_5 and the initial states have momentum P_1, P_2.

In general, the cross section for the process $e^- e^+ \rightarrow \tilde{\chi}_i^0\ \tilde{\chi}_j^0\ H_\ell^0$ can be written in the form

$$\sigma = \int \Pi^2 |M|^2\ \frac{dx\ dy\ d\sigma^2}{\Lambda(S, m_1, m_2)\ \Lambda(S, \sigma, m_5)}$$

where M is the matrix element previously mentioned, the integration is performed using a simple approximation obtained by an improved Weizsacker-Williamson procedure where

$$\Lambda(x, y, z) = [x^4 + y^4 + z^4 - 2x^2 y^2 - 2x^2 z^2 - 2y^2 z^2]^{1/2}$$

The limit of integration is given as follows

$$x_\pm = \frac{1}{4S^2}[(S^2 + m_1^2 - m_2^2)(S^2 - \sigma^2 + m_5^2) \pm \Lambda(S, m_1, m_2)\ \Lambda(S, \sigma, m_5)]$$

$$y_\pm = \frac{1}{4\sigma^2}[(\sigma^2 + m_3^2 - m_4^2)(S^2 - \sigma^2 + m_5^2) \pm \Lambda(\sigma, m_3, m_4)\ \Lambda(S, \sigma, m_5)]$$

$$(m_{\tilde{\chi}_i^0} + m_{\tilde{\chi}_j^0})^2 \le \sigma^2 \le S^2 - 2m_5 S + m_5^2$$

In all our calculations, we assume the following values for vector-boson masses suggested by recent collider runs[5]:

$M_W = 80\ GeV$

$M_Z = 100\ GeV$

$m_{H_1^0} = 700\ GeV$, $m_{H_2^0} = 800\ GeV$, $m_{H_3^0} = 900\ GeV$

$m_{\tilde{\chi}_1^0} = 700\ GeV$, $m_{\tilde{\chi}_2^0} = 800\ GeV$, $m_{\tilde{\chi}_3^0} = 900\ GeV$, $m_{\tilde{\chi}_4^0} = 1000\ GeV$

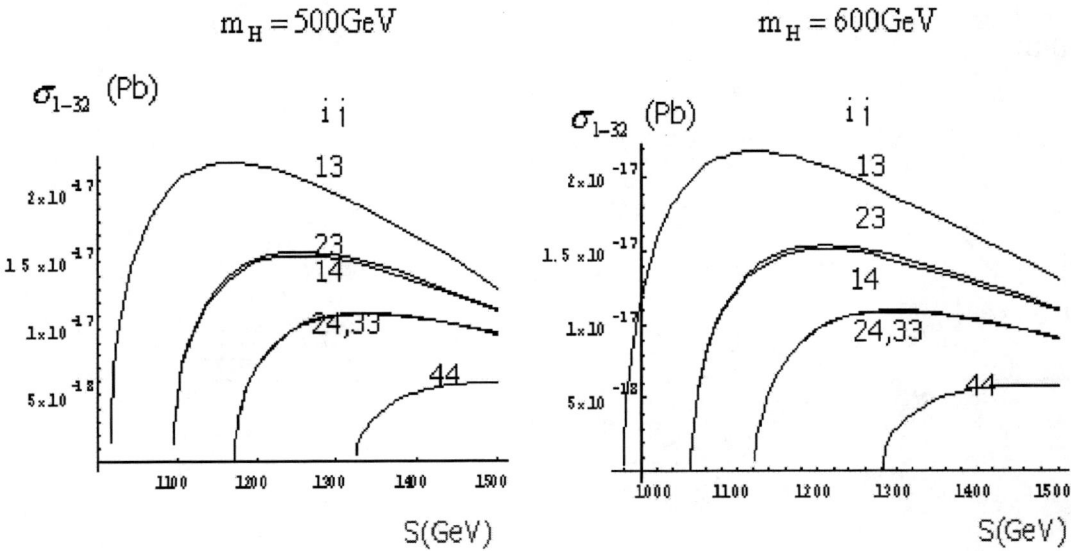

Figure 2. The cross section for a (1-32).

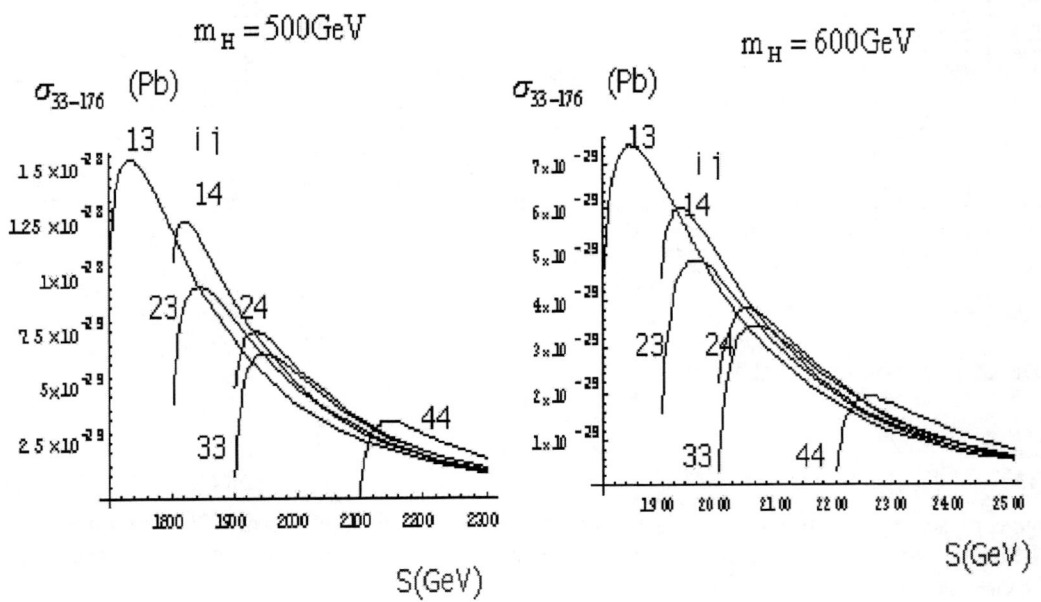

FIGURE 3. The cross section for a (33-176).

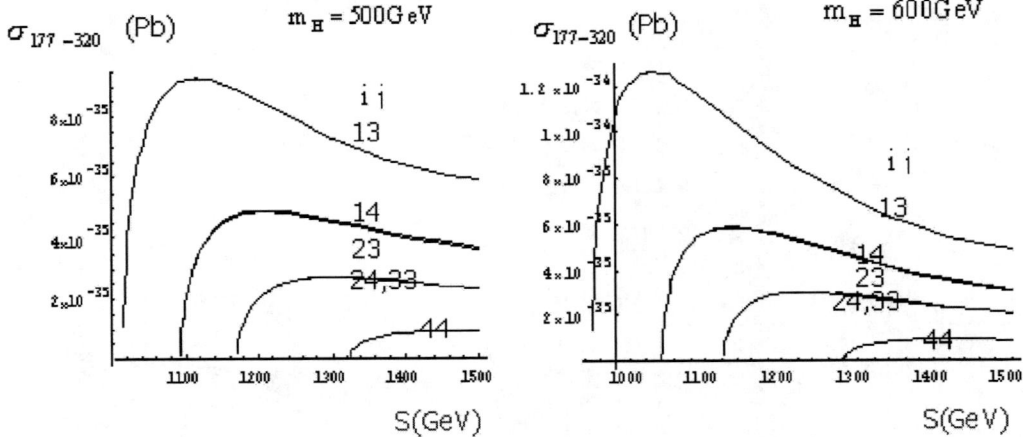

FIGURE 4. The cross section for a (177-320).

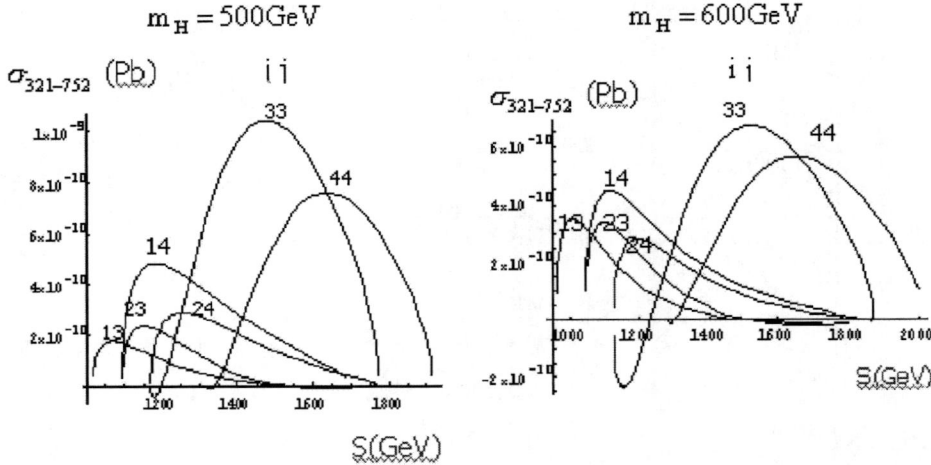

FIGURE 5. The cross section for a (321-752).

RESULTS

The results of the calculated cross sections for the process $e^- \; e^+ \rightarrow \widetilde{\chi}_i^0 \; \chi_j^0 \; H_\ell^0$ via Z and Higgs Boson exchange is presented from Fig. (2) To Fig. (5) As a function of the incident energy (S) We found that: At S increase from 1000 to 1700 we have maximum value for the cross-section. As S increase - σ decrease. The value of σ various from 10^{-10} to 10^{-34}.

b-Production via Z boson Exchange

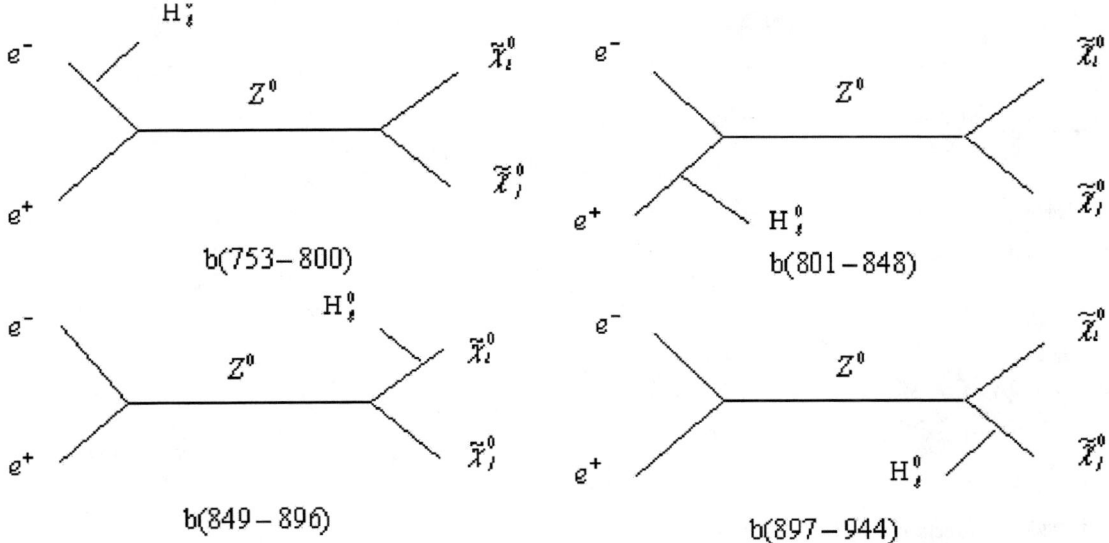

FIGURE 6. Feynman diagrams for the process   mediated by Z boson propagator. There are 192 diagrams.

The matrix elements for b (753-800) are:

$$M_{b(753-800)} = \frac{-ig^3 \, m_e}{8 Cos^3 \theta_W \, M_Z} \, \overline{V}(P_2) \gamma_\mu \, C \frac{\gamma_k (P_1 - P_5)_k + m_e}{(P_1 - P_5)^2 - m_e^2} \, B_\ell \, U(P_1)(\sigma^2 - M_Z^2)^{-1} V(P_4) \gamma_\mu [O_{ij}^{//L} \, P_L + O_{ij}^{//R} \, P_R] \overline{U}(P_3)$$

The matrix elements for b (801-848) are:

$$M_{b(801-848)} = \frac{-ig^3 \, m_e}{8 Cos^3 \theta_W \, M_Z} \, \overline{V}(P_2) \, B_\ell \, \frac{\gamma_k (P_2 - P_5)_k + m_e}{(P_2 - P_5)^2 - m_e^2} \, \gamma_\mu \, C \, U(P_1)(\sigma^2 - M_Z^2)^{-1} V(P_4) \gamma_\mu [O_{ij}^{//L} \, P_L + O_{ij}^{//R} \, P_R] \overline{U}(P_3)$$

The matrix elements for b (849-896) are:

$$M_{b(849-896)} = \frac{-ig^3}{8 Cos^2 \theta_W \, Sin\beta} \, \overline{V}(P_2) \gamma_\mu \, C U(P_1)(S^2 - M_Z^2)^{-1} V(P_4) \gamma_\mu [O_{ij}^{//L} \, P_L + O_{ij}^{//R} \, P_R] \frac{\gamma_k (P_3 - P_5)_k + m_{\tilde{\chi}_i^0}}{(P_3 - P_5)^2 - m_{\tilde{\chi}_i^0}^2} \, A_\ell \, \overline{U}(P_3)$$

The matrix elements for b (897-944) are:

$$M_{b(897-944)} = \frac{-ig^3}{8 Cos^2 \theta_W \, Sin\beta} \, \overline{V}(P_2) \gamma_\mu \, C U(P_1)(S^2 - M_Z^2)^{-1} V(P_4) A_\ell \, \frac{\gamma_k (P_4 - P_5)_k + m_{\tilde{\chi}_j^0}}{(P_4 - P_5)^2 - m_{\tilde{\chi}_{ji}^0}^2} \, \gamma_\mu$$

$$[O_{ij}^{//L} \, P_L + O_{ij}^{//R} \, P_R] \overline{U}(P_3)$$

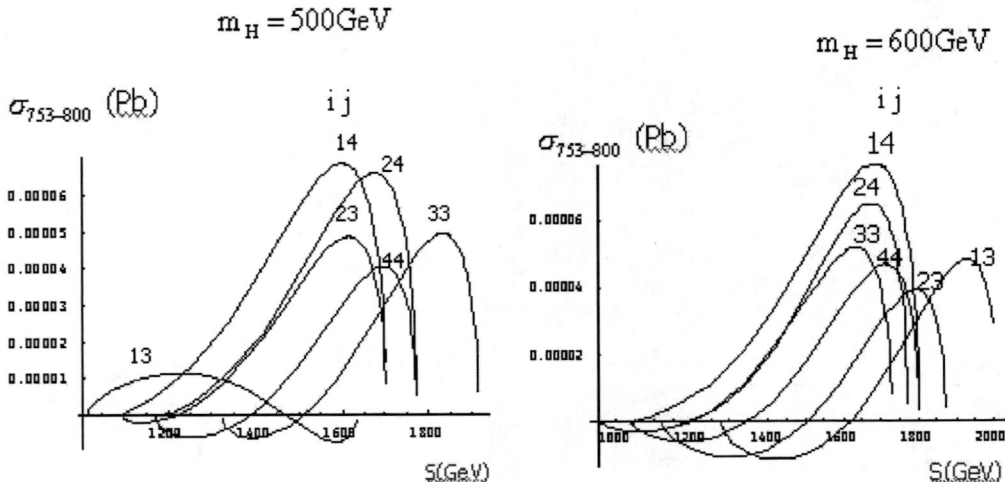

FIGURE 7. The cross section for b (753-800).

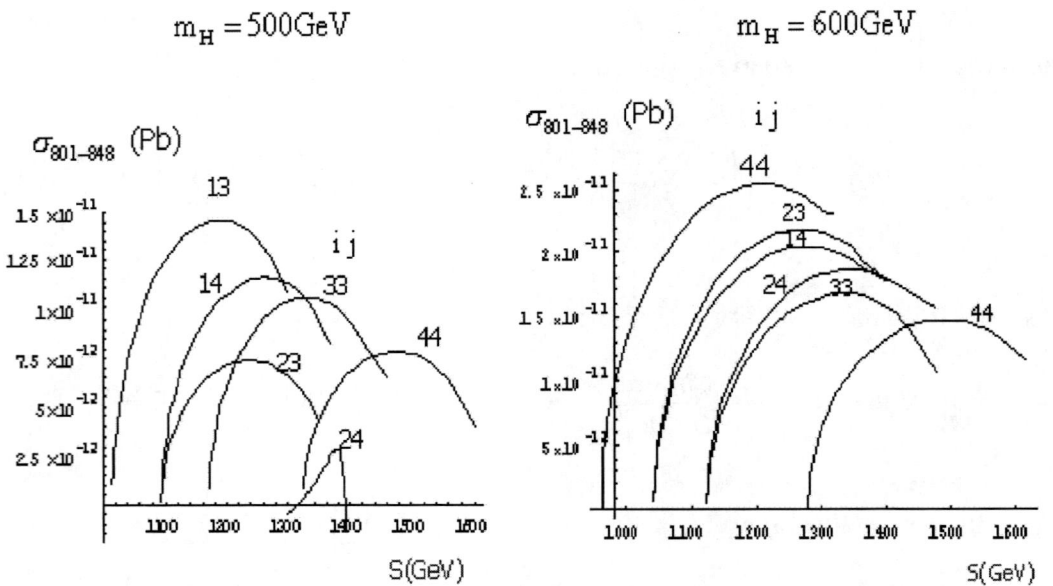

FIGURE 8. the cross section for b (801-848).

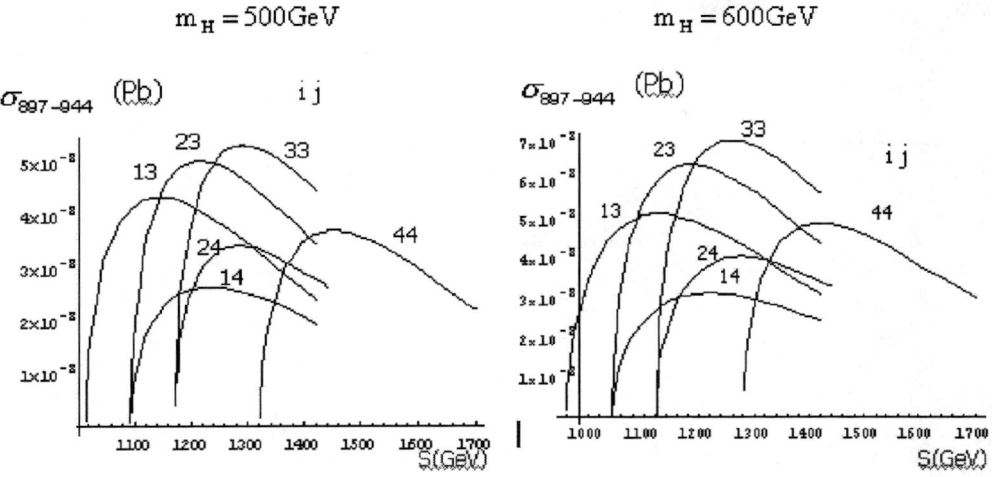

FIGURE 9. The cross section for b (849-896).

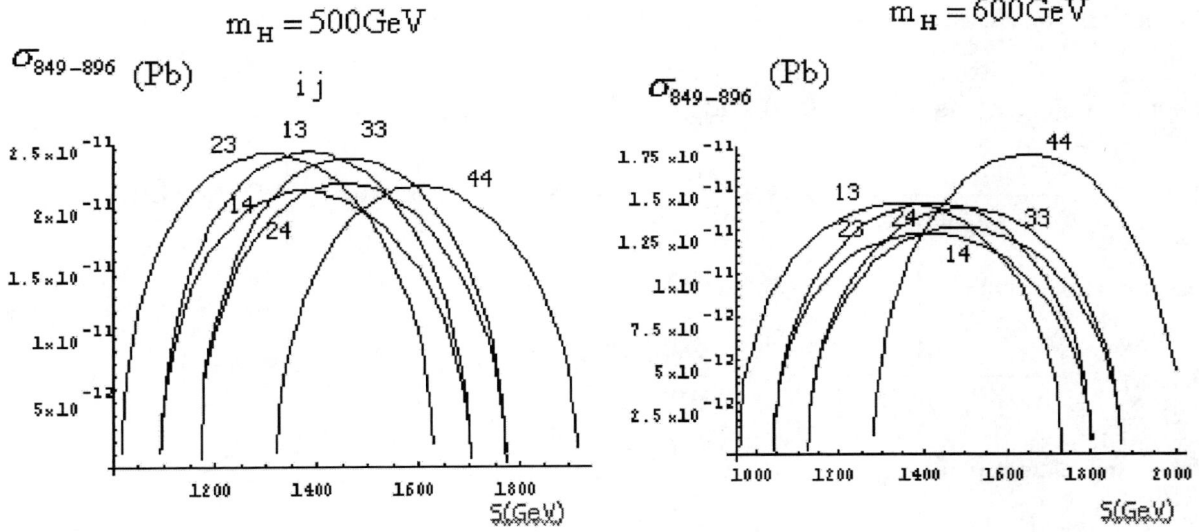

FIGURE 10. the cross section for b (897-944).

RESULTS

Figures from 7 to 10 show the cross-sections for the process as a function of (S) (Higgs boson is emitted from the initial or final legs through Z boson exchange). At S increase from 1100 to 1900 GeV we have maximum value for the cross-section, the value of σ various from 10^{-5} to 10^{-12}.

C-Production via Higgs boson exchange:

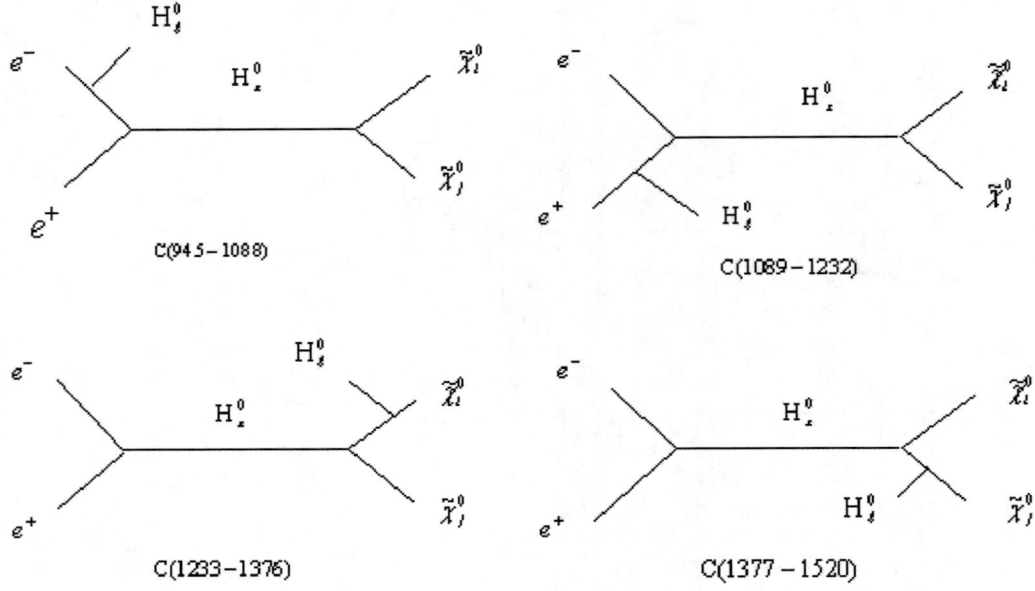

FIGURE 11. Feynman diagrams for the process mediated by Higgs boson propagator. there are576 diagrams.

The matrix elements for c (945-1088) are:

$$M_{C(954-1088)} = \frac{-ig^3 m_e^2}{8Cos^2\theta_W \, Sin\beta \, M_Z} \overline{V}(P_2) B_x \frac{\gamma_k (P_1 - P_5)_k + m_e}{(P_1 - P_5)^2 - m_e^2} B_\ell \, U(P_1)(\sigma^2 - m_{H_x^0}^2)^{-1} V(P_4) A_x \, \overline{U}(P_3)$$

The matrix elements for c (1089-1232) are:

$$M_{C(1089-1232)} = \frac{-ig^3 m_e^2}{8Cos^2\theta_W \, Sin\beta \, M_Z} \overline{V}(P_2) B_\ell \frac{\gamma_k (P_2 - P_5)_k + m_e}{(P_2 - P_5)^2 - m_e^2} B_x \, U(P_1)(\sigma^2 - m_{H_x^0}^2)^{-1} V(P_4) A_x \, \overline{U}(P_3)$$

The matrix elements for c (1233-1376) are:

$$M_{C(1233-1376)} = \frac{-ig^3 m_e}{8Cos\theta_W \, Sin^2\beta \, M_Z} \overline{V}(P_2) B_x \, U(P_1)(S^2 - m_{H_x^0}^2)^{-1} V(P_4) A_x \frac{\gamma_k (P_3 - P_5)_k + m_{\tilde{\chi}_i^0}}{(P_3 - P_5)^2 - m_{\tilde{\chi}_i^0}^2}$$

$$A_\ell \, \overline{U}(P_3)$$

The matrix elements for c (1377-1520) are:

$$M_{C(1377-1520)} = \frac{-ig^3 m_e}{8Cos\theta_W \, Sin^2\beta \, M_Z} \overline{V}(P_2) B_x \, U(P_1)(S^2 - m_{H_x^0}^2)^{-1} V(P_4) A_\ell \frac{\gamma_k (P_4 - P_5)_k + m_{\tilde{\chi}_j^0}}{(P_4 - P_5)^2 - m_{\tilde{\chi}_j^0}^2} A_x \, \overline{U}(P_3)$$

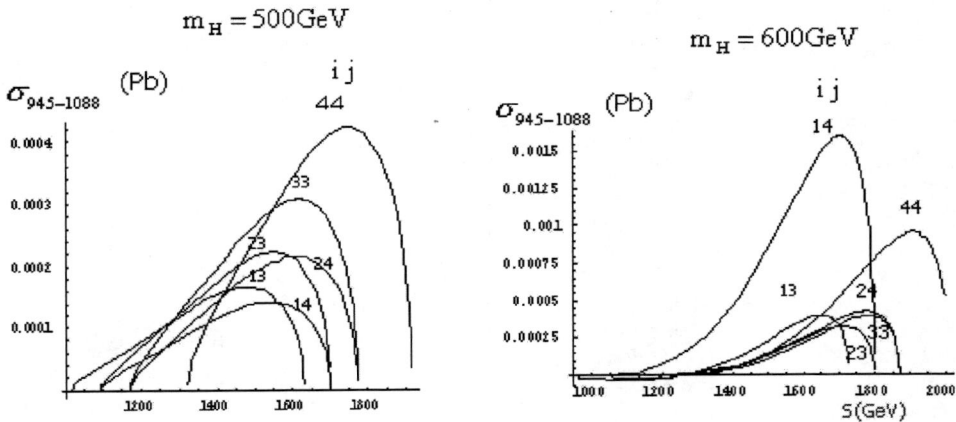

FIGURE 12. The cross section for c (945-1088).

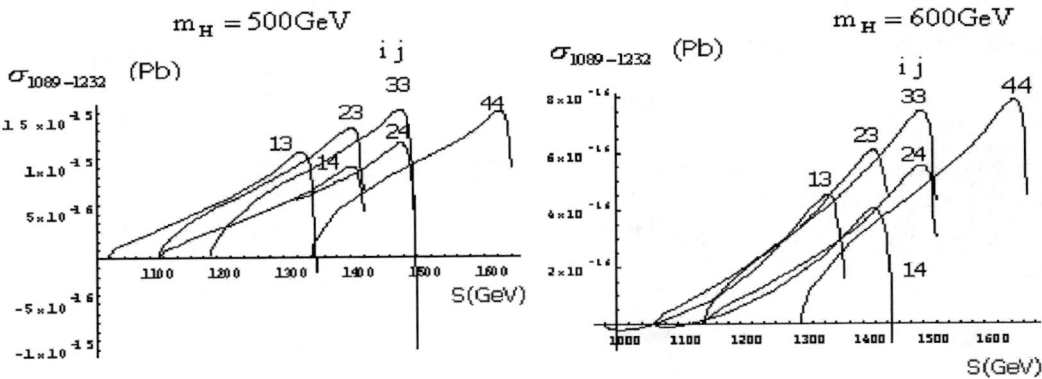

FIGURE 13. The cross section for c (1089-1232).

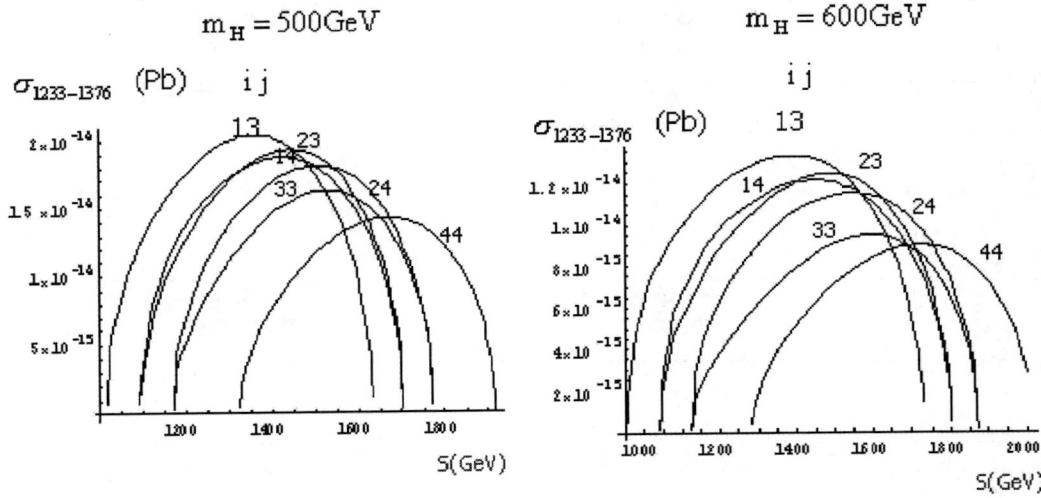

Figure 14.   the cross section for c (1233-1376).

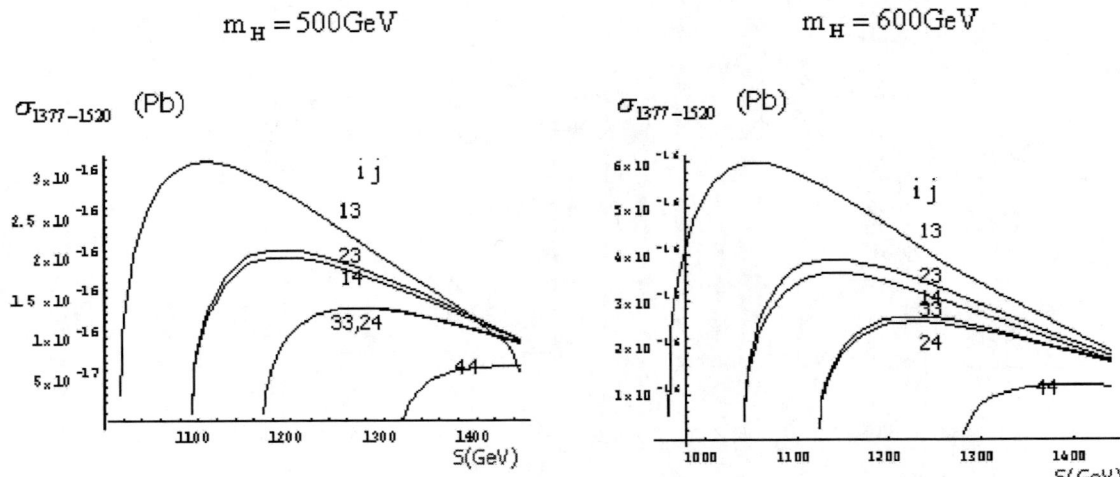

Figure 15. The cross section for c (1377-1520).

RESULTS

Figures from 12 to 15 show the cross-sections for the process as a function of (S) (Higgs boson is emitted from the initial or final legs through Higgs boson exchange). At S increase from 1000 to 1900 GeV we have maximum value for the cross-section, the value of σ various from 10^{-4} to 10^{-16}.

CONCLUSION

Comparing the values for Cross-sections obtained through different propagators for the three studied categories, it was found that:

TABLE 1. First category.

Figure	Maximum Cross-Sections Higgs mass 500GeV	Maximum Cross-Sections Higgs mass 600GeV
a(1-32)	2.2×10^{-17}	2.2×10^{-17}
a(33-176)	1.5×10^{-28}	7.4×10^{-29}
a(177-320)	9×10^{-35}	1.3×10^{-34}
a(321-752)	1×10^{-9}	6.5×10^{-10}

TABLE 2. Second category.

Figure	Maximum Cross-Sections Higgs mass 500GeV	Maximum Cross-Sections Higgs mass 600GeV
b(753-800)	0.00007	0.00008
b(801-848)	1.5×10^{-11}	2.5×10^{-11}
b(849-896)	2.5×10^{-11}	1.75×10^{-11}
b(897-944)	5.2×10^{-8}	7×10^{-8}

TABLE 3. Third category.

Figure	Maximum Cross-Sections Higgs mass 500GeV	Maximum Cross-Sections Higgs mass 600GeV
c(945-1088)	0.0004	0.0015
c(1089-1232)	1.6×10^{-15}	8×10^{-16}
c(1233-1376)	2×10^{-14}	1.35×10^{-14}
c(1377-1520)	3.1×10^{-16}	6×10^{-16}

So it can be noticied that the highest cross-sections are found in figures c(945-1088) and b(753-800) in which H_ℓ^0 is emitted from electron through Higgs boson and Z^0 exchange respectively.

The mechanism for figs. c(945-1088)

$$e^-\ (P_1) \to e^-\ (P_1 - P_5) + H_\ell^0 (P_5),$$
$$e^-\ (P_1 - P_5) + e^+(P_2) \to\ H^0 (P_3 + P_4) \to \tilde{\chi}_i^0 (P_3)\ \tilde{\chi}_j^0 (P_4)$$

The mechanism for figs. b(753-800)

$$e^-\ (P_1) \to e^-\ (P_1 - P_5) + H_\ell^0 (P_5),$$
$$e^-\ (P_1 - P_5) + e^+(P_2) \to\ Z^0 (P_3 + P_4) \to \tilde{\chi}_i^0 (P_3)\ \tilde{\chi}_j^0 (P_4)$$

Finally the most propable mechanism for the reaction is

$$e^-\ (P_1) \to e^-\ (P_1 - P_5) + H_\ell^0 (P_5),$$
$$e^-\ (P_1 - P_5) + e^+(P_2) \to\ H^0 (P_3 + P_4) \to \tilde{\chi}_i^0 (P_3)\ \tilde{\chi}_j^0 (P_4)$$
$$e^-\ (P_1 - P_5) + e^+(P_2) \to\ Z^0 (P_3 + P_4) \to \tilde{\chi}_i^0 (P_3)\ \tilde{\chi}_j^0 (P_4)$$

REFERENCES

1. A.Salam,Elementary Particle Theory , ed.N.Svartholm (Almquist and Wiksells,Stock holm , **367** (1968).
2. D. L. Davids, "Search for Neutral Higgs Bosons", Ph.D. Thesis, Humboldt- Universitat zu Berlin, 2002.
3. H.E.Haber,Supersymmetry,Phys.Rev.D**66**,010001-895(2002)
4. J.KALINOWSKI ,Journal of Physics **55**, 259-264 (2000).
5. G.Moortgat,S.Hesselbach,F.Frank and H.Fraas,"Distinguishing between MSSM and NMSSM by combined LHC and ILC analyses" ,Report from Uppsala Universitet,February 2005.
6. Howard E.Haber, Nuclear Physics B **272**, 1-76 (1986).
7. H.E.Haber and G.L.Kane, the search for supersymmetry, Physics Letters**117**, 225-226(1985).

Positron Scattering by Kr Atoms At The Low Energy

M. M. Abdel-Mageed, M. Abdel-Aziz And H. S. Zaghloul

Physics Department, Faculty of Science, Ain Shams University, Cairo, Egypt

Abstract. The scattering of positron from Kr atom is investigated with the least square variational method (LSVM). Elastic cross sections and scattering length, in the low-energy region, as well as the annihilation parameter Z_{eff} are presented. The bound state wave-functions of the target orbitals are obtained from Cowan computer code, using Hartree-Fock-Slater method with Hartree plus statistical exchange approximated potential. The computed results are consistent with some other theoretical, semi-empirical model and experimental results.

INTRODUCTION

In recent years there has been much interest in the study of positrons scattering by atomic and molecular gases [1]. Both theoretically and experimentally calculations on the collisions of slow positrons with atoms have been of interest because they provide a means of exploring various theoretical methods without the complication of electron exchange [2]. The calculation of positron-atom scattering is simpler than electron-atom scattering. There is no exchange interaction between the positron and target electrons. But in every other respect, the theoretical treatment of positron-atom scattering is a more difficult proposition than electron-atom scattering. The reason of this lies in the attractive nature of the positron-electron interaction. The interaction of a low-energy positron with a many-electron atom is characterized by strong correlation effects. A part from the dynamic polarization of the electron cloud by the field of the positron, the positron can also form positronium (Ps), by picking up one of the atomic electrons. If the kinetic energy of the positron is less than $I - 6.8$ eV, the positronium formation threshold and I is the ionization energy of the atom with 6.8 eV being the binding energy of the positronium atom ground state [3] , the only elastic scattering or direct annihilation are possible. As the incident positron kinetic energy increases, various inelastic channels become accessible, including positronium formation, target excitation and ionization [4]. For positron-krypton collision, for example, the elastic scattering remains the only open channel for energies below the positronium formation threshold of 13.998-6.8 eV. Furthermore, the total wave function is fully guided by Pauli's principle in the case of electron scattering, but in the case of positron-atom scattering is still incomplete for small separations.

The annihilation of positrons in atomic and molecular gases has been a topic of interest, various interesting phenomena associated with the positron-annihilation process such as very large annihilation rates [5-7], high sensitivity of the rates to small changes in molecular structure [8], and rapid increase of fragmentation and annihilation rats at small temperatures [9,10].

There were two different mechanisms for positron-annihilation [11], these were direct annihilation and resonant annihilation. Direct annihilation describes the annihilation of the positron with the target electrons and the direct annihilation rate was strongly correlated with the size of the elastic cross section. Resonant annihilation was mainly important for large molecules with closely spaced vibrational levels. In resonant annihilation, the positron is trapped in a Feshbach resonance associated with a vibrationally excited state. The resonant-annihilation process was suggested to be the mechanism responsible for the large annihilation rates seen for some molecules [10,11]. In the present work, we concentrate ourselves on the direct annihilation process.

In the present paper the least-squares variation method (LSVM) is applied to determine the s-wave scattering wave-function of a slow positron-atom interaction below positronium (Ps) formation threshold. This scattering wave-function will be used to calculate the effective charge (annihilation parameter) Z_{eff} (the effective number of electrons at the position of positron) at each orbit of the target atoms. A recent version of Cowan computer code [12,13] (program RCN32) is used to calculate the electrons orbital wave-functions of the target atoms.

CP888, *Modern Trends in Physics Research,*
Second International Conference on Modern Trends in Physics Research—MTPR-06,
edited by L. El Nadi

DETAILS OF CALCULATION

A. The Scattering Hamiltonian

In non-relativistic time-independent quantum mechanics, Schrodinger's equation is equivalent in form to the conventional eigenvalue problem:

$$(H - E)\,|\,\Psi\,\rangle = |\,0\,\rangle \tag{1}$$

where H and E are the total Hamiltonian and energy, respectively, of a quantum mechanical system described by the vector $|\,\Psi\,\rangle$. The interaction Hamiltonian for positron-Neon atom is written as

$$H = H_T - \nabla_X^2 + V_{int}(x,r) + V_{pi}(r) \tag{2}$$

where H_T being the Hamiltonian for the target atom, ∇_X^2 is the kinetic energy operator for the incident positron, $V_{int}(r,x)$ stands for the interaction potential between the positron and the target and is given by:

$$V_{int}(r,x) = \frac{2Z}{x} - \sum_i^N \frac{2}{|x-r_i|} \tag{3}$$

where x and r_i are the position vectors of positron and i^{th} electrons respectively. The polarization potential $V_{pl}(r)$ is given in the form [14]

$$V_{pl}(r) = -\alpha_d(1 - \exp(-r^6/r_c^{\,6}))\,/\,2r^4 \tag{4}$$

The dipole polarizability $\alpha_d = 16.70$ [15], and the cut off parameter $r_c = 2.83$ [14].
The total energy E of the system may be written as

$$E = E_T + k_p^2 , \tag{5}$$

where E_T and k_p^2 are the energy of the target and the kinetic energy of the incident positrons, respectively.

B. The Least-Squares Variational Method And Trial Wave-function

For s-wave scattering process, the variational treatment [16] starts by defining a trial wavefunction $|\,\Psi_t^n\,(x, r_N; k)\,\rangle$. It consists of two multiplicative wave-functions

$$|\,\Psi_t^n\,(x, r_N; k)\,\rangle = |\,\Phi_T(r_N)\,\rangle\,|\,\Psi_P^n(x;k)\,\rangle, \tag{6}$$

where $|\,\Phi_T(r_N)\,\rangle$ represents the target in its ground-state and $|\,\Psi_P^n(x;k)\,\rangle$ is the positron scattering wavefunction which is composed of the angular part ($Y_{0,0} = \sqrt{1/4\pi}$) multiplied by the radial part ($R_P^n(x;k)$).

We have:

$$|\,R_P^n(x;k)\,\rangle = a^n\,|\,\hat{S}(x;k)\,\rangle + b^n\,|\,\hat{C}(x;k)\,\rangle + \sum_{i=1}^n d_i \chi_i(x), \tag{7a}$$

n refers to the dimension of the Hilbert-space part of the trial wavefunction representing all possible virtual states of quantum mechanical system composed of the positron and the target. $\left| \hat{S}(x;k) \right\rangle$ and $\left| \hat{C}(x;k) \right\rangle$ are the regular and irregular parts of the scattering wavefunction, respectively. $\left| \hat{C}(x;k) \right\rangle$ should contain a cut-off function to avoid the singularity at the origin. This cut-off function will tend to zero at the origin and to unity at infinity. $\left| R_P^n(x;k) \right\rangle$ has to satisfy the boundary condition

$$\left| R_P^n(0) \right\rangle = \left| 0 \right\rangle$$

$$\left| R_P^n(x;k) \right\rangle \xrightarrow{\;x\to\infty\;} a^n \left| \hat{S}(x;k) \right\rangle + b^n \left| \hat{C}(x;k) \right\rangle \tag{7b}$$

where $\left| \chi_i(x) \right\rangle$ is a quadratic integrable wavefunction. a^n, b^n and d_i are variational parameters.

In this case the reactance matrix K contains a single element which is identical with the tangent of the s-wave scattering phase shift (η_0) and is calculated by:

$$K_{11} = \tan\eta_0 = b^n/a^n \tag{8}$$

The s-wave elastic scattering trial wavefunction for the system may be written in abbreviated form as:

$$\left| \Psi_t^n \right\rangle = \left| S \right\rangle + K_{11} \left| C \right\rangle + \left| \phi_n \right\rangle \tag{9}$$

where S is the regular part ;

$$S = \hat{S}.\Phi_T(r_N) = \frac{1}{\sqrt{4\pi}} \; Sinc\,\beta \,.\, \Phi_T(r_N) \; , \tag{10}$$

($Sinc\,\beta = \dfrac{Sin\,\beta}{\beta}$, $\beta = (kx)$ where k is the momentum of the incident positron) and $K_{11} = \tan\eta_0$. The function C, consists of a cut-off function and the irregular part of the asymptotic solution. It has the form:

$$C = \hat{C}.\Phi_T(r_N) = \frac{1}{\sqrt{4\pi}} \; (1 - e^{-\alpha x})(Cosc\,\beta)\,\Phi_T(r_N) \; , \tag{11}$$

where $Cosc\,\beta = \dfrac{Cos\,\beta}{\beta}$ and α is an adjustable (free) parameter which is selected from the values that give a plateau of K_{11} (see ref. [16] p.73). $\Phi_T(r_N)$ is the target ground state wavefunction (see ref. [16] Appendix). The Hilbert-space part $\left| \phi_n(x,r_N) \right\rangle$ possesses the form

$$\left| \phi_n(x,r_N) \right\rangle = \Phi_T(r_N).\sum_{i=1}^{n} d_i \left| \chi_i(x) \right\rangle = \sum_{i=1}^{n} d_i \left| \phi_i \right\rangle \tag{12}$$

where $\qquad \chi_i = x^i \, e^{-\alpha x}$ and $\phi_i = \chi_i \, \Phi_T$ \hfill (13)

The next step in the variational treatment is to select a proper test-wave function $\left| \phi_s \right\rangle$ and which satisfies the following:

$$\left\langle \phi_s \left| H - E \right| \Psi_t^n \right\rangle = V \tag{14}$$

The linear variational parameters K_{11} and d_i are chosen according to the following variational principle:

$$\partial V^2 = 0 \tag{15}$$

Thus, they are chosen following to a least-squares variational principle in which all projections of the vector $(H - E) \left| \Psi_t^n \right\rangle$ on $\left| \phi_S \right\rangle$ are minimum. The test wavefunction $\left| \phi_S \right\rangle$ is constructed [17,18] by:

$$\left| \phi_S \right\rangle = \left\{ \left| S \right\rangle, \left| C \right\rangle, \left| \phi_j \right\rangle; \; j = 1, 2, \ldots n \right\}. \tag{16}$$

In this case we have the system of projections

$$\left(S \mid S \right) + K_{11} \left(S \mid C \right) + \sum_{i=1}^{n} d_i \left(S \mid \phi_i \right) = V_1$$

$$\left(C \mid S \right) + K_{11} \left(C \mid C \right) + \sum_{i=1}^{n} d_i \left(C \mid \phi_i \right) = V_2 \tag{17}$$

$$\left(\phi_j \mid S \right) + K_{11} \left(\phi_j \mid C \right) + \sum_{i=1}^{n} d_i \left(\phi_j \mid \phi_i \right) = V_{j+2} \; ; \; j = 1, 2, \ldots n$$

where the matrix elements $\left(S \mid S \right)$, $\left(S \mid C \right)$, $\left(S \mid \phi_i \right)$, $\left(C \mid S \right)$, $\left(C \mid C \right)$, $\left(C \mid \phi_i \right)$, $\left(\phi_j \mid S \right)$, $\left(\phi_j \mid C \right)$, and $\left(\phi_j \mid \phi_i \right)$ have the general form [16-18]:

$$\left(g \mid f \right) = \left\langle g \mid E - H \mid f \right\rangle = \left\langle g \mid \hat{H} \mid f \right\rangle = \left\langle g \mid \hat{f} \right\rangle$$

$$= \int_0^{\pi} Sin\theta \, d\theta \int_0^{2\pi} d\varphi \int_0^{\infty} x^2 g \, \hat{f} \, dx \tag{18}$$

where x is the position vector of the positron, θ is the angle between x and the Z-axis and φ is the azimuthal angle. The operator $\hat{H}$ possesses the form:

$$\hat{H} = \left(E - H \right),$$

and the LSVM implies :

$$\partial \sum_{j=1}^{n+2} V_j^2 = 0 \quad , \tag{19}$$

which means that the sum of the projections of $\left(H - E \right) \left| \Psi_t^n \right\rangle$ on the test function space ϕ_S are minimum. The

minimization of $\sum_{j=1}^{n+2} V_j^2 = 0$ guarantees that the vector $\left(H - E \right) \left| \Psi_t^n \right\rangle$ has a minimum length. The variational

parameters are obtained by applying this variational principle.

C. Analytic Expressions For Z_{eff} And Elastic Scattering Cross Section σ_{el}

In the collision of positrons with target atoms, the positrons may annihilate according to one of the following processes:

In the first process which is called direct annihilation, the incident positron with energy below the Ps formation threshold, annihilate with one of the atomic electrons of the neutral target atom and the annihilation rate is calculated using the electron charge density (Z_{eff}) at the positron position.

In the second process the incident positron with energy above the Ps formation threshold may pick up an electron to form positronium and after that annihilates. The positron in the third process may be captured to the atom to form bound system and the photon annihilation then occurs within the positron-many-electron complex system.

In the present work, we focus on the first process. In positron-atom scattering, λ, the rate of annihilation of an incoming positron and an atomic electron with the emission of two gamma rays, is given by the expression [19]

$$\lambda = \pi \, r_0^2 \, c \, \rho \, Z_{eff}(k) \tag{20}$$

where r_0 is the classical radius of the electron, c is the velocity of light, ρ is the density of electrons per atom available to the positron for annihilation and k is the positron wave number. $Z_{eff}(k)$ is the effective number of electrons per atom available to the positron for annihilation. It depends on specific properties of the e^+-Atom system under consideration and is equal to Z, the number of atomic electrons, if the interaction potential between the positron and the atom is set to zero. The annihilation parameter $Z_{eff}(k)$ can be calculated by using scattering wavefunction obtained using the least-squares variational method. The annihilation parameter is given by [20]:

$$Z_{eff}(k) = \left\langle \Psi\left(x, r_N; k\right) \right| \sum_{i=1}^{N} \delta\left(r_i - x\right) \left| \Psi\left(x, r_N; k\right) \right\rangle \tag{21}$$

It is related to the probability of an electron and a positron to be found in the same position. The wave-function $\Psi\left(x, r_N; k\right)$ is the total scattering wavefunction, including all partial waves, for the system made up of the incident positron with wave vector k and the target atom. x and r_N stand for the position vectors of the positron and the target (composed of N electrons), respectively.

Z_{eff} can be determined experimentally to a high degree of accuracy and thus, the calculation of this parameter is a criterion for the goodness of the employed wavefunction Ψ_t which represents the system of a low energy positron moving in the field of the atom .

The s-wave elastic scattering cross section (in πa_0^2 units) is related to the phase shift [16] by the following relation

$$\sigma_{el} = \frac{4}{k^2} \, Sin^2\left(\eta_0\right). \tag{22}$$

Also the determination of the phase shift is useful to calculate the s-wave scattering length [16] which is defined as:

$$A_o = \lim_{k \to 0} \left\{ \frac{-\tan \eta_0}{k} \right\} \tag{23}$$

RESULTS AND DISCUSSION

The computation of the annihilation rates was started by the calculation of the orbital wave-functions and energies of the target atoms using Cowan computer code (program RCN32). These wave-functions which generated from Cowan code are used for the calculation of the positron-atom potential. After the construction of the matrix elements $(S|S)$, $(S|C)$, $(S|\phi_i)$, $(C|S)$, $(C|C)$, $(C|\phi_i)$, $(\phi_j|S)$, $(\phi_j|C)$, and $(\phi_j|\phi_i)$, using LSVM program (LLSQ) with starting value of the free parameter (non linear parameter) α and certain value of n (the dimension of the Hilbert-space part of the trial wavefunction) and changing α and increasing n until we obtain a convergence in the results of K_{11}. We have found that this corresponds to $\alpha = 1.4$ and n = 3 .

Our computational process of the annihilation rates was proceeded by the calculations for the positron annihilation rates in case of krypton atoms where there exists some experimental data [22-24] as shown in Figure (1). This figure shows that our theoretical calculations have an agreement with the available experimental data Z_{eff} = 90.1 by UCL group [22,23] and Z_{eff} = 65.7 by San Diego group [8,24] at k = 0.05 a_o^{-1}. The curve possess the same behavior as a developed two-parameter semi-empirical theory, of positron scattering and annihilation [1], and the polarized-orbital approximation model [21].

The annihilation rate rapidly decreased as the wave number increase at the low positron kinetic energy, while after k = 0.1 a_o^{-1} the annihilation rate is slightly change and then starts to increase as the positron kinetic energy increases towards the Ps threshold energy $E = 7.198$ eV.

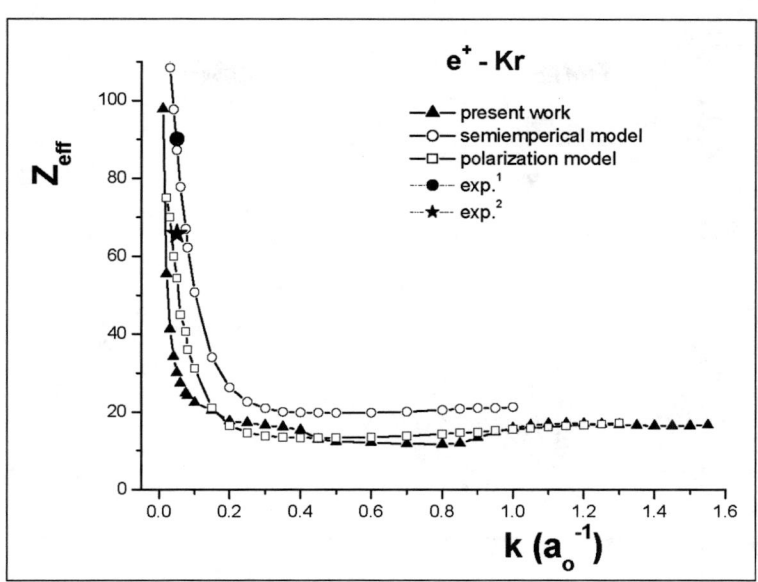

FIGURE 1. The annihilation parameter Z_{eff} for e^+- Kr scattering as a function of k (in a_o^{-1} units) for momenta below the Ps-formation threshold. Our present work results with semi-empirical results [1], polarized-orbital approximation model [21], exp.[1] measurements of UCL group[22,23] and exp.[2] measurements of San Diego group [8,24].

Figure 2 shows a comparison between our calculated s-wave elastic scattering cross-section for krypton atom, as a function of the positron wave number, and some theoretical results from polarized-orbital approximation method [21] and semi-empirical model results [1].

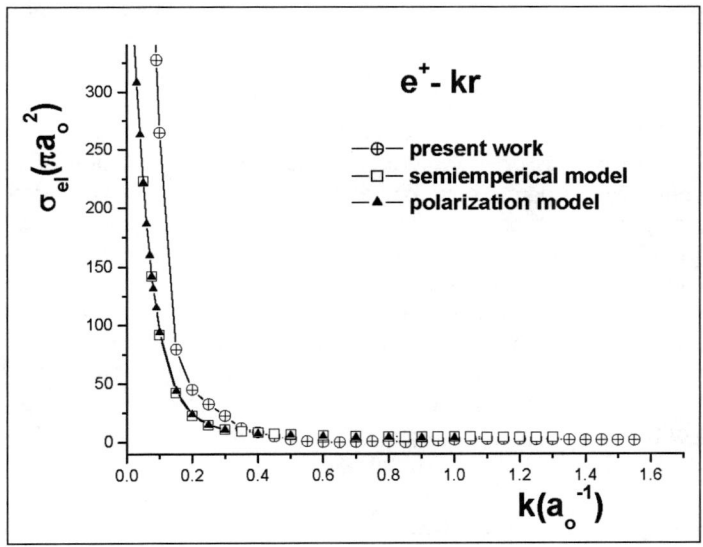

FIGURE 2. A comparison of elastic scattering cross-section for krypton as a function of k (in a_o^{-1} units)

The elastic scattering cross-section has a large value at very small wave number and rapidly decrease to 25 πa_o^2 at $k = 0.2$ a_o^{-1}. There is a dip in the elastic cross-section and the three curves coincide together at that value of k.
We have calculated the scattering length of positron-krypton , A_o = - 11.73 at $k = 0.15$ a_o^{-1}, such as the krypton atom has an ionization potential I = 13.998 eV which is twice the Ps ionization potential (6.8 eV) but the negative sign of the scattering length means that the positron-krypton bound state system is impossible.

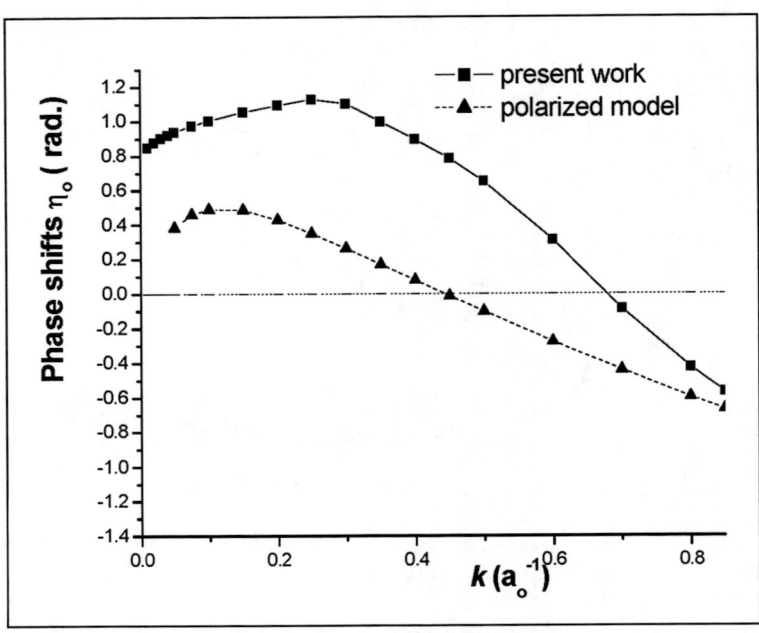

FIGURE 3. The elastic scattering phase shifts η_o for e^+- Kr as a function of k (in a_o^{-1} units). Our calculated results (solid line) in comparison with the polarized -orbital approximation model [21] (dot line).

The s-wave positron scattering phase shifts are plotted in Figure 3. When the scattering length is negative, the s-wave phase shift increases from zero until it reaches a maximum in the phase shift, the maxima cannot exceed π / 2 [1], then begin to decrease. The figure shows the typical behavior, while it has been noticed in the polarized-orbital approximation model calculation of McEachran et. al.[21] that the phase shift converge faster than our results.

CONCLUSION

The least square variational method (LSVM) has been applied to the calculation of the elastic scattering cross-section, phase shifts and annihilation parameters for positron scattering from krypton atoms. A recent version of Cowan computer code, using Hartree-Fock-Slater method with Hartree plus statistical exchange approximated potential, has been used to calculate the electrons orbital wave-functions of the target krypton atoms.

The annihilation parameters agree with those of other theoretical calculations, semi-empirical results and experimental results. Our calculated results demonstrate that the positron-krypton atom annihilation with the inner shell electrons can be studied in a quantitative manner in isolated atomic systems.

The rapid variation in the annihilation parameter as the positron energy approaches threshold is largely driven by kinematical factors related to the ability of the positron to tunnel into the repulsive Coulomb barrier.

The scattering length of positron-krypton atom has been calculated and it was found it has negative value which means that the positron-krypton atom bound state is impossible.

REFERENCES

1. J. Mitroy and I. A. Ivanov, *Phys.Rev.***A65**, 042705 (2002).
2. H. S. W. Massey, *Positron annihilation*, edited by A. T. Stewart and L. O. Roelig"Academic Press", New York, 1967, pp. 113.
3. M. W. Bromley, J. Mitroy and G. G. Ryzhikh, *J.Phys.***B31**, 4449 (1998).
4. M. R. Nikolic and A. R. Tancic, *Phys. Chim. And Tech.***3**, 141 (2005).
5. D. A. L. Paul and L. Saint-Pierre, *Phys.Rev.Lett.***11**, 493 (1963) .
6. G. R. Heyland, M. Charlton, T. C. Griffith and G. L. Wrigt, Can.*J.Phys.***60**, 503 (1982).
7. C. M. Surko, A. Passner, M. Leventhal and F. J. Wysocki, *Phys.Rev.Lett.***61**, 183 (1988).
8. I. Iwata, R. G. Greaves, T. J. Murphy, M. D. Tinkle and C. M. Surko, *Phys.Rev.***A51** (1995) 473.
9. Jun Xu, L. D. Hulett, T. A. Lewis, D. L. Donohue, S. A. McLuckey and O. H. Crawford, *Phys.Rev.***A49** , R3151 (1994).

10. I. Iwata, G. F. Gribakin, R. G. Greaves, C. Kurz, M. D. Tinkle and C. M. Surko, *Phys. Rev.* **A61** , 022719 (2000) .

11. G. F. Gribakin, *Phys.Rev.***A61**, 022720 (2000).

12. Robert D. Cowan, " The Theory of Atomic Structure and Spectra", Univ. of California Press (1981); *Phys.Rev.***163**, 54 (1967).

13. F. Herman and S. Skillman, " Atomic Structure Calculation", Prentice-Hall (1963).

14. J. Mitroy and M. W. J. Bromley A **67**,034502 (2003)

15. A. Kumar and W. J. Meath, Can. J. Chem , **36**, 1616 .,(1985).

16. M. A. Abdel-Raouf, *Phys.Rep.***108,** 164 (1984).

17. E. W. Schmid and K. H. Hoffmann, *Nucl.Phys.***A175**, 443 (1971).

18. E. W. Schmid, *Nucl.Phys.***A180** (1972) 434; Nuovo Cim. **A18**, 171 (1973).

19. R. Ferrell, *Review of Modern Physics* **28**, 308 (1956) ; P. A. Fraser, *Adv. At. Mol. Phys.***4**, 63-107 (1968).

20. R. E. Montgomery and R. W. Labahn, *Canadian journal of phys.* **48** , 1288 (1970).

21. R. P. McEachran, A. D. Stauffer, and L. E. M. Campbell, *J. Phys.* B **13**, 1281 (1980).

22. T. J. Murphy and C. M. Surko, *J. Phys.* B **23**, L727 (1990).

23. P. G. Coleman, T. C. Griffith, G. R. Heyland and T. L. Killeen, *J. Phys.* B **8**,1734 (1975).

24. G. L. Wright, M. Charlton, T. C. Griffith and G. R. Heyland, *J. Phys.* B **18**, 4327 (1985).

Positron Annihilation In Xe Atoms

M. M. Abdel-Mageed, H. S. Zaghloul, M. Abdel-Aziz

Physics Department, Faculty of Science, Ain Shams University, Cairo, Egypt

Abstract. The annihilation parameters Z_{eff} of positrons in xenon atoms are calculated. The elastic cross sections for positron scattering from Xe atoms reported at low energy, below positronium (Ps) formation threshold. The bound wavefunctions of electrons are generated from Cowan program. The least-squares variational method (LSVM) is used for determining the scattering wavefunction. The present results of annihilation parameters and cross sections are consisted with simiemprical and other theoretical results.

INTRODUCTION

An important and interesting feature of low-energy positron collisions with atoms and molecules is the possibility of annihilation of the positron with one of the electrons in the target. This has been the subject of extensive experimental and theoretical work [1–11]. Only elastic scattering or direct annihilation are possible when the positron incident with energy less than the positronium formation threshold.

Positron annihilation is a good tool to give useful information about matter. It is important in the study of metals [12], alloys [13], polymers [14] and super-conductors [15]. Particularly, it has been confirmed that positron annihilation life time technique is one of the most effective nondestructive tools for providing valuable information about the electronic structure and defects of materials and determination of the Fermi surface in the single crystal which is useful in band structure calculation [16, 17].

A unique aspect of these interactions is the emission of γ-rays when a positron annihilates with an electron. This signal provides information about the interaction, and it is the basis of many types of measurements. For example, this signal has been used to characterize defects and interfaces in solids [18, 19].

When positrons annihilate with atoms they annihilate predominantly with valence electrons because of the repulsive potential exerted on the positron by the nuclei. However, a small fraction of the positrons can tunnel through this repulsive potential and annihilate with inner-shell electrons. An important implication of the inner-shell annihilation is the emission of Auger electrons and consequent formation of doubly ionized atoms. The positron-induced Auger electron emission has been observed in condensed matter [20].

We have applied least squares variational method to positron-xenon interaction at low energy below positronium formation threshold to determine s-wave scattering wavefunction. This scattering wavefunction will be used to calculate annihilation parameter Z_{eff} at each orbit of the target as well as the elastic cross section.

DETAILS OF CALCULATION

Annihilation into two γ-rays is far more probable than into three γ-rays. Assuming that the positrons are unpolarized, the annihilation rate in a gas is given by

$$\lambda = \pi \, r_0^2 \, c \, \rho \, Z_{eff}(k) \tag{1}$$

where $\mathbf{r_0} = \mathbf{e}^2 / (\mathbf{mc}^2)$ is the classical radius of the electron, c the speed of light, ρ is the density of electrons per atom available to the positron for annihilation and k is the positron momentum. Z_{eff} is the effective number of electrons in the target system. The value of Z_{eff}, which varies with the momentum of the positron, is a measure of the probability of the positron being at the same position as one of the target electrons . It is calculated from the elastic scattering wavefunction for the positron - target system as follows:

CP888, *Modern Trends in Physics Research,*
Second International Conference on Modern Trends in Physics Research—MTPR-06,
edited by L. El Nadi
© 2007 American Institute of Physics 978-0-7354-0354-9/07/$23.00

$$Z_{eff}(k) = \left\langle \Psi(x, r_N; k) \left| \sum_{i=1}^{N} \delta(r_i - x) \right| \Psi(x, r_N; k) \right\rangle \tag{2}$$

where $\Psi(x, r_N; k)$ is the scattering wavefunction, including all partial waves, for the system made up of the incident positron with wave vector k and the target atom. x and r_N stand for the position vectors of the positron and the target (composed of N electrons), respectively. The error in Z_{eff} is only of first order in the error in Ψ, whereas the error in the elastic scattering phase shifts is usually of second order in the error [11]. A good agreement between the calculated value of Z_{eff} and an accurate experimental value, derived from measurements of the annihilation rate, is therefore an important test of the quality of the scattering wavefunction..

The least-Squares Variational Method and Trial Wavefunction

For s-wave scattering process, the variational treatment starts by defining a trial wavefunction $\left| \Psi_t^n(x, r_N; k) \right\rangle$. It consists of two multiplicative wavefunction

$$\left| \Psi_t^n(x, r_N; k) \right\rangle = \left| \Phi_T(r_N) \right\rangle \left| \Psi_{Sc}^n(x; k) \right\rangle, \tag{3}$$

where $\left| \Phi_T(r_N) \right\rangle$ represents the target in its ground-state and $\left| \Psi_{Sc}^n(x; k) \right\rangle$ is the positron scattering wavefunction which is composed of the angular part ($Y_{0,0} = \sqrt{1/4\pi}$) multiplied by the radial part ($\Psi_P^n(x; k)$). We have

$$\left| \Psi_P^n(x; k) \right\rangle = a^n \left| \hat{S}(x; k) \right\rangle + b^n \left| \hat{C}(x; k) \right\rangle + \sum_{i=1}^{n} d_i \chi_i(x), \tag{4}$$

n refers to the dimension of the Hilbert-space part of the trial wavefunction representing all possible virtual states of quantum mechanical system composed of the positron and the target. $\left| \hat{S}(x; k) \right\rangle$ and $\left| \hat{C}(x; k) \right\rangle$ are the regular and irregular parts of the scattering wavefunction, respectively. $\left| \hat{C}(x; k) \right\rangle$ should contain a cut-off function to avoid the singularity at the origin. This cut-off function will tend to zero at the origin and to unity at infinity. $\left| \Psi_P^n(x; k) \right\rangle$ has to satisfy the boundary conditions:

$$\left| \Psi_P^n(0) \right\rangle = \left| 0 \right\rangle$$

$$\left| \Psi_P^n(x; k) \right\rangle \xrightarrow{x \to \infty} a^n \left| \hat{S}(x; k) \right\rangle + b^n \left| \hat{C}(x; k) \right\rangle \tag{5}$$

where $\left| \chi_i(x) \right\rangle$ is a quadratic integrable wavefunction. a^n, b^n and d_i are variational parameters. In this case the reactance matrix R_{11} contains a single element which is identical with the tangent of the s-wave scattering phase shift η_0 and is calculated by

$$R_{11} = \tan \eta_0 = b^n / a^n \tag{6}$$

The s-wave elastic scattering trial wavefunction for the system may be written in abbreviated form as:

$$\left| \Psi_t^n \right\rangle = \left| S \right\rangle + R_{11} \left| C \right\rangle + \left| \phi_n \right\rangle \tag{7}$$

where S is the regular part ;

$$S = \hat{S} . \Phi_T (r_N) = \frac{1}{\sqrt{4\pi}} \; Sinc\, \beta . \Phi_T (r_N) \; , \tag{8}$$

Where, $Sinc\, \beta = \dfrac{Sin\, \beta}{\beta}$, $\beta = (k\,x)$. The function C, consists of a cut-off function and the irregular part of the asymptotic solution. It has the form

$$C = \hat{C} . \Phi_T (r_N) = \frac{1}{\sqrt{4\pi}} \; (1-e^{-\alpha x})(\, Cosc\, \beta \,) \, \Phi_T (r_N) \; , \tag{9}$$

where $Cosc\, \beta = \dfrac{Cos\, \beta}{\beta}$ and α is an adjustable (free) parameter which is selected from the values that give

a plateau of R_{11} (see ref. [24] P.73). $\Phi_T (r_N)$ is the target ground state wavefunction which can be expressed as a Slater determinant of mutually orthonormal one-electron wavefunction u_i in the form:

$$\Phi_T (r_N) = \frac{1}{\sqrt{N}} \; det \left[u_1 (r_1) \, u_2 (r_2) \, u_3 (r_3) \, ... u_z (r_z) \right] \tag{10}$$

where N is the total number of electrons. According to the central field model, $u_i (r_i)$ can be expressed as

$$u_i (r_i) = \frac{1}{r_i} R_{n_i \ell_i} Y_{\ell_i m_i} (\hat{r}_i) \zeta (\sigma) \, , \tag{11}$$

where $R_{n_i \ell_i}$ is the radial wavefunction, which is the solution of the equation:

$$\left[-\frac{d^2}{dr_i^2} + \frac{\ell_i (\ell_i +1)}{r_i^2} + V_i (r_i) \right] R_{n_i \ell_i} = \varepsilon_i \, R_{n_i \ell_i} \tag{12}$$

where $V_i (r_i)$ is the assumed potential energy function for the field in which the atomic electron i moves. These functions are generated from Cowan program using Hartree-Fock-Slater method with Hartree plus statistical exchange approximated potential.

$Y_{\ell_i m_i} (\hat{r}_i)$ is the usual spherical harmonics and $\zeta (\sigma)$ stands for the spin wavefunction of the orbit i such that n_i , ℓ_i and m_i are the corresponding principal, orbital and magnetic quantum The Hilbert-space part $\left| \phi_n (x, r_N) \right\rangle$ possesses the form

$$\left| \phi_n (x, r_N) \right\rangle = \Phi_T (r_N) . \sum_{i=1}^{n} d_i \left| \chi_i (x) \right\rangle = \sum_{i=1}^{n} d_i \left| \phi_i \right\rangle \tag{13}$$

where $\chi_i = x^i e^{-\alpha x}$ and $\phi_i = \chi_i \Phi_T$ $\tag{14}$

The next step in the variational treatment is to select a proper test-wave function $\left| \phi_s \right\rangle$ and define the functional

$$\left\langle \phi_s \left| H - E \right| \Psi_t^n \right\rangle = V \tag{15}$$

The linear variational parameters R_{11} and d_i are chosen according to the following variational principle:

$$\partial V^2 = 0 \tag{16}$$

Thus, they are chosen following to a least-squares variational principle in which all projections of the vector $(H-E)\left|\Psi_t^n\right\rangle$ on $\left|\phi_s\right\rangle$ are minimum. The test wavefunction $\left|\phi_s\right\rangle$ is constructed by:

$$\left|\phi_s\right\rangle = \left\{\ \left|S\right\rangle, \left|C\right\rangle, \left|\phi_j\right\rangle;\ j=1,2,.....n\right\}. \tag{17}$$

In this case we have the system of projections

$$\left(S\,|\,S\right) + R_{11}\left(S\,|\,C\right) + \sum_{i=1}^{n} d_i\ \left(S\,|\,\phi_i\right) = V_1$$

$$\left(C\,|\,S\right) + R_{11}\left(C\,|\,C\right) + \sum_{i=1}^{n} d_i\ \left(C\,|\,\phi_i\right) = V_2 \tag{18}$$

$$\left(\phi_j\,|\,S\right) + R_{11}\left(\phi_j\ |\,C\right) + \sum_{i=1}^{n} d_i\ \left(\phi_j\ |\,\phi_i\right) = V_{j+2}\ ;\ j=1,2,...n$$

and the LSVM implies :

$$\partial\sum_{j=1}^{n+2} V_j^2\ =\ 0 \tag{19}$$

which means that the sum of the projections of $\left(H-E\right)\left|\Psi_t^n\right\rangle$ on the test function space ϕ_s are minimum. The minimization of $\sum_{j=1}^{n+2} V_j^2\ =\ 0$ guarantees that the vector $\left(H-E\right)\left|\Psi_t^n\right\rangle$ has a minimum length. The variational parameters are obtained by applying this variational principle. The matrix elements required for the employment of the LSVM, namely $\left(S\,|\,S\right)$, $\left(S\,|\,C\right)$, $\left(S\,|\,\phi_i\right)$, $\left(C\,|\,S\right)$, $\left(C\,|\,C\right)$, $\left(C\,|\,\phi_i\right)$, $\left(\phi_j\,|\,S\right)$, $\left(\phi_j\,|\,C\right)$, and $\left(\phi_j\,|\,\phi_i\right)$, have the general form [24]:

$$\left(g\,|\,f\right) = \left\langle g\,|\,E-H\,|\,f\right\rangle = \left\langle g\,|\,\hat{H}\,|\,f\right\rangle = \left\langle g\,|\,\hat{f}\right\rangle$$

$$= \int_0^\pi Sin\theta\,d\theta \int_0^{2\pi} d\varphi \int_0^\infty x^2 g\,\hat{f}\,dx \tag{20}$$

Where θ is the angle between x and the Z-axis and φ is the azimuthal angle. The operator $\hat{H}$ possesses the form

$$\hat{H} = \left(E-H\right), \tag{21}$$

The total Hamiltonian (in Rydberg units) of positron-target atom system has the form

$$H = H_T - \nabla_x^2 + V_{int}\left(r,x\right) + V_{pl}(r) \tag{22}$$

where H_T being the Hamiltonian for the target atom, ∇_x^2 is the kinetic energy operator for the incident positron, and $V_{int}\left(r,x\right)$ stands for the static interaction potential between the positron and the target. The polarization potential $V_{pl}(r)$ is given the form [21]

$$V_{pl}(r) = -\alpha_d(1-\exp(-r^6/r_c^{\ 6})\,/\,2r^4 \tag{23}$$

The experimental values for the dipole polarizability $\alpha_d = 27.61$, and the cut off parameter $r_c = 1.63$ [21].

The total energy E of the system may be written, in Rydberg, as

$$E = E_T + k^2\,, \tag{24}$$

where $E_T\ and\ k^2$ are the energy of the target and the kinetic energy of the incident positrons, respectively.

$V_{int}(r,x)$ is the interaction potential between the incident positron and the target and is given by $V_{int}(r,x) = \dfrac{2Z}{x} - \displaystyle\sum_{i}^{N} \dfrac{2}{|x-r_i|}$ \hfill (25)

RESULTS AND DISCUSSION

The computation of the annihilation parameters was started by the calculation of the orbital wavefunctions and energies of the target atoms using Cowan computer code (program RCN32). These wavefunctions are used for the calculation of the positron-Xe atom interacting potential

The construction of the matrix elements $(S|S)$, $(S|C)$, $(S|\phi_i)$, $(C|S)$, $(C|C)$, $(C|\phi_i)$, $(\phi_j|S)$, $(\phi_j|C)$, and $(\phi_j|\phi_i)$ are used in the LSVM program (LLSQ) with starting value of the free parameter (non linear parameter) α and certain value of n (the dimension of the Hilbert-space part of the trial wavefunction). The optimum value of α is obtained from the plateau curve (the stationary behavior for R11 with α). We have found that this corresponds to $\alpha = 1.45$. The electron and positron wavefunctions are used to compute the annihilation Z_{eff}. for valance and inner subshells. The total Z_{eff} are found by adding the contribution of different subshells.

At low energies the dominant contribution to annihilation parameter Z_{eff} is from s-wave scattering. The dependence of the total Z_{eff} on the positron momentum, k, over the range $0 \leq k \leq 1.5$ which are plotted, together with the total Z_{eff}, in figure1. This figure shows that our theoretical calculations have the same behavior as that calculated by semi-empirical model [7] and polarized orbital method [30]. The annihilation parameter $Z_e Z_{eff ff}$ reveal an initial decrease as positron energy is increased from zero, instead of Z_{eff} continuing to decrease, it starts to increase at positron energies just below the positronium formation threshold.

We have calculated the annihilation parameter Z_{eff} at each inner-shell of the Xe-atom . The contribution of them to the annihilation parameter Z_{eff} are shown in figure (2).

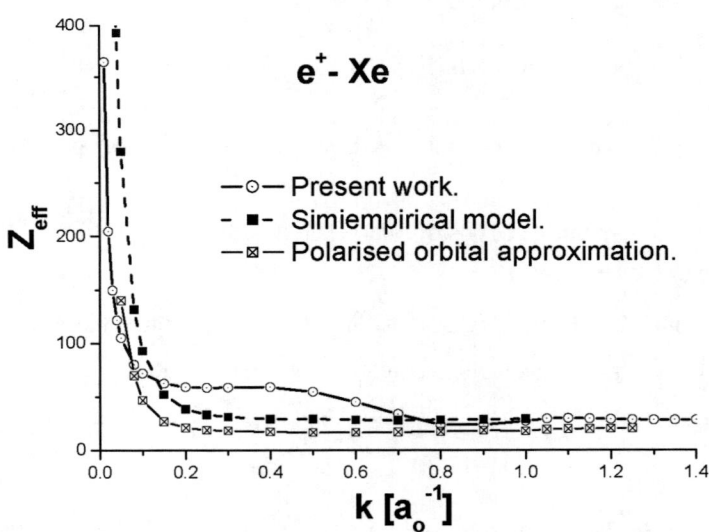

FIGURE 1. The theoretical dependence of Z_{eff} on the positron momentum.

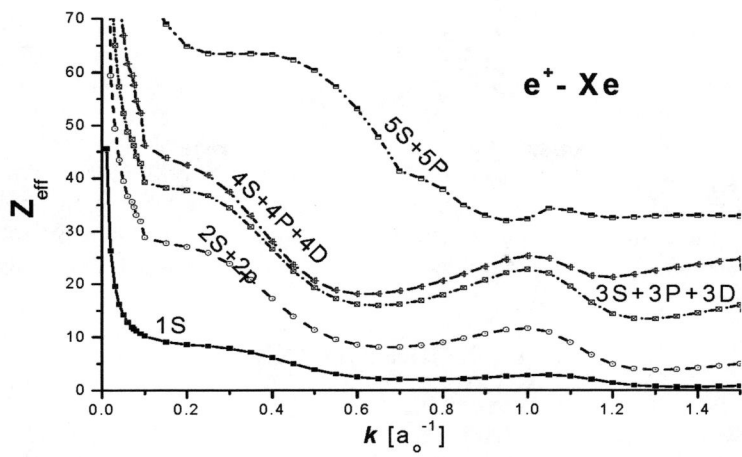

FIGURE 2. The momentum dependence of the annihilation parameter Z_{eff} of positron with different shells of positron – xenon atoms.

Figure (2) shows that positrons annihilate predominantly with electrons of upper shells. For the principal quantum numbers $n=3$ and $n=4$ the annihilation parameter Z_{eff} is roughly proportional to the number of electrons in the s, p, and d subshells, namely, 2, 6, and 10, respectively.

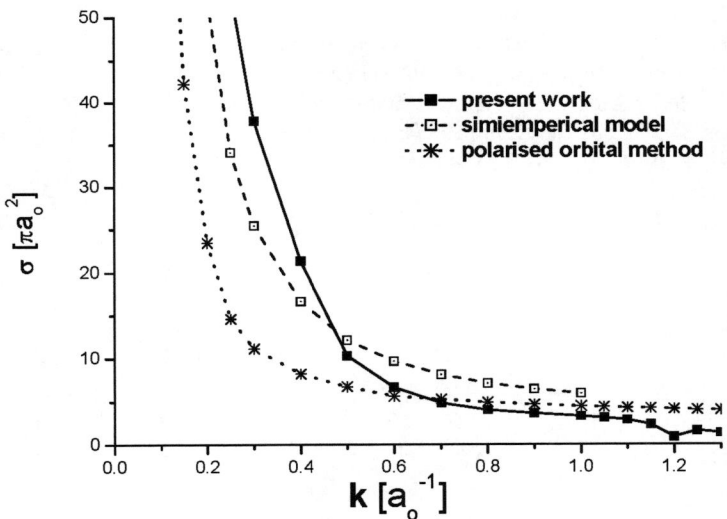

FIGURE 3. The s-wave cross section for positron-xenonn scattering

The s-wave elastic scattering cross-section σ_{el} (πa_0^2) of positron by xenon atom is drawn in figure (3). The figure demonstrates the monotonic decrease of σ_{el} as the energy of the incident positron increases. The s-wave elastic scattering cross section (in πa_0^2 units) is related to the phase shift by the following relation

$$\sigma_{el} = \frac{4}{k^2} Sin^2(\eta_0) \tag{26}$$

CONCLUSION

In this work we have show how the least-squares variational method (LSVM) can be developed for scattering wave function which used to calculate the annihilation rate of positrons in xenon atom as well as the scattering cross section. Hartree-Fock-Slater method is used to generate the orbital wavefunctions for xenon atom The contributions of Z_{eff} has calculated numerically for valance and inner subshells of Xe-atom. It is found an agreement with a previously calculated values.

REFERENCES

1. S. A. Novikov, M. W. J. Bromley and J. Mitroy, Phys.Rev.**A69**, 052702 (2004).
2. E. A. G. Armour, D. J. Baker and M. Plummer, J. Phys. B: At. Mol. Opt. Phys. **23**,3057 (1990).
3. R. J. Drachman Phys. Rev. **179**,237 (1969).
4. Iwata K, Greaves R G, Murphy T J, Tinkle M D and Surko C M 1995 Phys. Rev. A **51** 473.
5. McEachran R P, Morgan D L, Ryman A G and Stauffer A D 1977 J. Phys. B: At. Mol. Phys. **10** L663.
6. G. F. Gribakin, Phys.Rev.A **61** 022720. (2000).
7. J..Mitroy and I. A. Ivanov, Phys.Rev.A **65**,042705 (2002).
8. Koji Iwata, G. F.Gribakin, R. J. Greaves and C. M. Surko, Phys. Rev. Lett. 79,1 (1997).
9. D. H. Modison and W. N. Shelton, Phys.Rev.A7,499 (1973).
10. R. Ferrell, Review of Modern Physics **28** 308(1956) ; P.A.Fraser, Adv.At.Mol..Phys.**4**,63-107 (1968).
11. P. Van Reeth, J. W. Humberston, Koji Iwata, R.G.Greaves and C.M.Surko, J. Phys. B: at. Mol. Opt. Phys. **29** No **12**, L465-L471. (28 June 1996).
12. D. M. Schrader and Y. C. Jean, editors, Positron and Positronium Chemistry, Amsterdam, Elsevier(1988).
13. M. Dimanchev, A. Shofan, T. Troev and J. Serna, Material Science Forrum **105** (1992).
14. Q. Deng and Y. C. Jean, Journal of Polymer Science B **30**,1359(1992).
15. C. S. Sunder, A. Bharathi, W. Y. Ching and Y. C. Jean, Phys.Rev.B **42**, 2193 (1990).
16. K. O. Jensen and A. Weiss, Phys.Rev. **B41**, 3928 (1990).
17. J. Liuand and Y. C. Jean, American Chemical Society **28**, 5774 (1995).
18. Z. Tang *et al.*, Phys. Rev. Lett. **78**, 2236 (1997).
19. J. P. Peng *et al.*, Phys. Rev. Lett. **76**, 2157 (1996).
20. Koji Iwata, G.F.Gribakin,R.G. Greaves,and C. M. Surko,Pys. Rev. Lett .79,1(1997).
21. J. Mitroy and M. W. J. Bromley Pys. Rev. A **67**,034502 (2003).

Quantitative Elemental Analysis of Biological Samples by Energy Dispersive X-ray Fluorescence Spectrometry

H. T. Mohsen, N. F. Zahran and A. I. Helal

Nuclear Research Center, Atomic Energy Authority, P. 0. Box 13759, Cairo, Egypt

Abstract. Energy dispersive XRF (EDXRF) measures the energy of the emitted x-ray by collecting the ionization products induced in a solid state semiconductor detector (SiLi). On their pathway through the sample the exciting radiation and emitted radiation are attenuated. Attenuation of emitted radiation depends on many parameters, like sample thickness, homogeneity, humidity and density. In the present study, EDXRF method for the quantitative analysis of some biological samples using standard reference materials (SRM) has been developed.

INTRODUCTION

The use of EDXRF as a direct nondestructive method for multi-elemental analysis of biological and foodstuff samples has been increased over the last few years [1,2]. Atomic spectroscopy, including FAAS (Flame Atomic Absorption Spectrometry), ETAAS (Electrothermal Atomic Absorption Spectrometry), ICP-OES (Inductively Coupled Plasma Optical Emission Spectrometry) and ICP-MS (Inductively Coupled Plasma Mass Spectrometry) are the usual used technique for elemental determination of those kind of samples [3]. However, these techniques imply a prior total destruction of the matrix by mineral acids, which may lead to problems of contaminants by reactants employed or disturbances of the measured concentration by element losses due to incomplete solubilization and/or evaporation [4]. Moreover, the method of matrix destruction used strongly depends on the chemical composition of the sample and on the element to be determined [5]. Simplicity of sample preparation, minimum manipulation, speed and the chance of analyzing some elements such as sulfur that can hardly be determined by other techniques have promoted EDXRF as a useful alternative to conventional spectroscopic techniques. In general, EDXRF quantitative analysis is carried out by the calibration curve method, obtained with many calibrators.

However, for some applications (such as plant specimens) it is difficult to get sufficient certified standards, with similar matrices to the samples, to achieve a good spread of data points over the range of each element to be determined [6]. The aim of this work was to study the applicability of a quantitative EDXRF method based on a several standard reference materials (SRM) and synthetic calibrators from similar matrices.

In EDXRF quantitative analysis, sample preparation is very important to reduce and eliminate many sources of uncertainty. Good grinding, drying and pressing all samples in the same weight and geometry introduce good quality quantitative results. Also working in a vacuum atmosphere eliminates the effect of x-ray attenuation in intensity during its path to the detector.

EXPERIMENTAL

Measurements

The EDXRF quantitative elemental analyses were done using (JEOL JSX-3222) Elemental Analyzer installed in Central Laboratory for Elemental and Isotopic Analysis – Atomic Energy Authority. It is equipped with x-ray tube of Rh anode and 127 $\square$m Be window. The characteristic x-ray radiation was measured using a Si(Li) detector with energy resolution of 149 eV at 5.9 keV and 1000 cps. Detectable elements are from Na to U. Primary x-rays are applied to the spacemen from its underside through a primary collimator for limiting the irradiated region of the specimen as showing in figure (1). Characteristic x-rays are also collimated by a secondary collimator for eliminating x-rays scattering from objects other than the specimen and adjusting x-ray intensity. The specimen chamber is under vacuum to prevent decreasing in x-ray intensity. Table (1) illustrates the experimental conditions for EDXRF applied for all samples under investigation

CP888, *Modern Trends in Physics Research,*
Second International Conference on Modern Trends in Physics Research—MTPR-06,
edited by L. El Nadi
© 2007 American Institute of Physics 978-0-7354-0354-9/07/$23.00

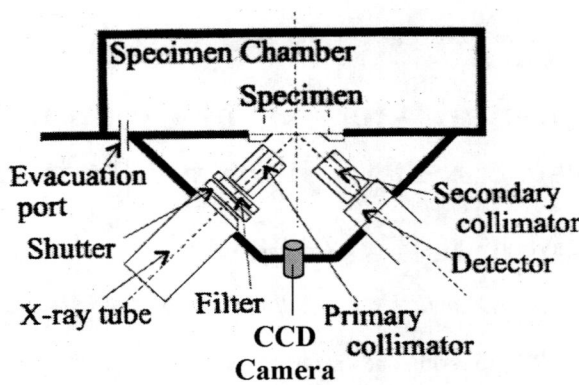

FIGURE 1. EDXRF optical system including x-ray tube and Si(Li) detector in the specimen chamber.

TABLE 1. Experimental conditions for EDXR.

Parameter	Value
Energy	30 keV
X-ray	2500 CPS
Acquisition time	600 sec life time
Primary collimator	4 mm
Secondary collimator	7 mm
Atmosphere	Vacuum
Energy calibration	Al-Mg alloy

Samples

The samples analyzed with the EDXRF included SRM's used for calibration are powder samples of 0.25g weight. All samples are pressed hydraulically up to 5 Ton (with area of 10 mm^2) after drying. The samples are grain samples (maize 1 & maize 2) and milk samples (milk 1 & milk 2). SRM used are illustrated in the following table.

TABLE 2. SRM's used for EDXRF quantitative analysis

Sample	Name
Whey powder	IAEA-155
Milk powder	IAEA-153
Hay powder	IAEA-V 10
Rice flour	NIST 1568a

To increase number of points on the calibration curve, pure graphite was added to Whey (IAEA-155) SRM with two ratios 1:1 and 2:1.

RESULTS AND DISCUSSION

X-ray fluorescence technique is a comparative method, this means that in order to analyze a material under investigation one has first to calibrate the measurement arrangement with the use of standard samples with known concentrations of the wanted elements. Also the standard samples and the samples under investigation must be from the same matrix. Quantitative results are introduced for maize and milk samples by calibration curves using the above SRM's (Table 2) and two pressed pellets of whey powder with additive pure graphite 1:1 and 2:1. The choice of graphite as additive because it is not detectable in EDXRF (JEOL JSX-3222). Addition of pure graphite increases number of calibration samples from only one SRM. This gives EDXRF quantitative analysis method for solid samples the ability to use only one SRM to make a calibration curve. Milk powder SRM (IAEA-153) is also measured as unknown samples to check the quality of the results. Table (3) illustrates elemental concentrations in ppm for the unknown maize samples. Table (4) illustrates elemental concentrations in ppm for the unknown milk samples. Table (5) illustrates elemental concentrations in ppm for milk SRM with good agreement between measured and certified concentration.

TABLE 3. EDXRF elemental concentrations in ppm for maize samples.

Element	Maize 1	Maize 2
Br	21.63	22.28
Ca	1719	1551.8
Fe	94.07	99.9
Mg	1505	1080
Rb	19.22	19.19
Zn	96.63	95.84

TABLE 4. EDXRF elemental concentrations in ppm for milk samples.

Elements	Milk 1	Milk 2
Br	12.29	17.75
Ca	12105	6390
Fe	11.42	10.89
K	20046	21670
Mg	978	1835
Rb	13.37	14.1
Zn	40.71	40.39

TABLE 5. Elemental concentrations in ppm for Milk SRM (IAEA-153).

Elements	Certified Concentration in ppm	95 % Confidence Interval in ppm	EDXRF results in ppm
Br	12.32	11.01 - 13.64	12.49
Ca	12870	12540 -13170	13855.00
Fe	2.53	166 - 3.47	7.9072
K	17620	16480 - 18760	16240.00
Mg	1060	1000 - 1150	2484.33
Rb	14.03	12.27 - 16.10	13.21
Zn	39.56	37.66 - 41.23	39.78

CONCLUSION

Sample preparation is very important to get accurate EDXRF quantitative results. The addition of pure graphite to solid SRM's can be used as synthetic calibrators in EDXRF quantitative elemental analysis method.

REFERENCES

1. C. Vazquez, N. Barbaro, S. Lopez, X-ray Spectrum. **32**, 57 (2003).
2. P.J. Potts, A.T. Ellis, P. Kregsamer, J. Marshall, C. Streli, M. West, P. Wobrauscjek, J. Anal. At. Spectrom. 19, 1397 (2004).
3. C. Baffi, M. Bettinelli, G.M. Beone, S. Spezia, Chemosphere **48**, 299 (2002).
4. M. Hoeing, H. Baeten, S. Vanhentenrijk, E. Vassileva, Ph. Quevauviller, Anal. Chim. Acta **358** 85 (1998).
5. H. Polkowska-Mortrenko, B. Danko, R. Dybczynski. A. Koster-Ammerlaan, P. Bode, Anal. Chim. Acta **408**, 89 (2000).
6. J. Omote, H. Kohno, K. Toda, Anal. Chim. Acta **307**, 117 (1995).

The Periodicity of the Alfven Mode Waves in the Earth's Magnetosheath

A. Tawfik[1], M. A. Amer[1], O. M. Shalabiea[1,2], and M. S. El-Nawawy[1,3]

1-Astronomy Dept., Faculty of Science, Cairo University, Cairo, Egypt
2- Physics Department, University of Sebha, Sebha, P.O. Box 625, Libya
3- Space program of Egpyt, NARSS.

ABSTRACT. The magnetosheath is a region located between the bow shock and the Earth's magnetopause. The importance of the Earth's magnetosheath in space physics is related to its role as a mediator between the solar wind and terrestrial magnetosphere. The magnetosheath should play an important role in the transmission of matter and energy observed in the solar wind through to the magnetopause. Several theoretical models have been developed to understand the plasma and magnetic field properties in the magnetosheath. These models differ in the role that the magnetic forces play in them. The interaction of the magnetosheath with the magnetopause can be described in terms of waves. These waves in general can take various forms, such as oscillation, discontinuities or standing waves and infinitesimal increments. The function of these waves is to reconfigure the solar wind flow and its frozen-in magnetic field from the solar wind state to the state specified by the magnetopause boundary condition. The main characteristics and positions of the wave modes in the Earth's magnetosheath are briefly summarized. In this paper we estimated the periodicity of the wave modes across the Earth's magnetosheath using Eigen modes of the global magnetosphere and ideal gas model. In the ideal gas model, we found two values of the periodicity of the Alfven waves corresponding to two different positions in the magnetosheath. The first is, (~20.4 min) at the magnetopause, while the second is (~ 42.36 min) at the bow shock. For the Eigen modes of the global magnetosphere, the periodicity is about (~16.59 min). From our calculations, we concluded that the Alfven waves could propagate near the inner magnetosheath (at the magnetopause) with shorter periodicity than at the bow shock. These waves should play the required role to explain the observed transfer of matter and energy from the solar wind into the magnetopause.

INTRODUCTION

The Earth is a magnetized object. The terrestrial magnetic field is strong forming a huge magnetosphere weakly loaded with plasma. The solar wind is weakly magnetized flow of matter with a significant kinetic pressure interacting with the terrestrial magnetosphere. The magnetosheath is a region located between the bow shock and the Earth's magnetopause. The importance of the Earth's magnetosheath in space physics is related to its role as a mediator between the solar wind and the terrestrial magnetosphere (Song et. al, 1999). According to the ideal MHD which considers the plasma as infinitely conducting, the solar wind should never penetrate the magnetosphere and no matter transfer should be possible through the frontier separating the two objects. This is not the observed case. Therefore, detailed studies of the physical processes occurring in the magnetosheath are required to explore the conditions for such transfer to occur.

There are many disturbances in the magnetosheath, which can be described in terms of waves. In the (MHD) theory three different wave modes exist, called the fast, intermediate and slow modes (e. g., Kantrowitz and Petschek, 1966). The interaction of the magnetosheath with the magnetopause can be described in terms of these waves. The important role of these waves is to reconfigure the solar wind flow and its frozen-in magnetic field from the solar wind state to the state specified by the magnetopause boundary condition.

Waves can be generated by the processes of the solar wind-bow shock interactions. The waves in the magnetosheath can modify the magnetopause structure and also affect the magnetosphere as a whole. In addition to the shock related waves, the interaction between the magnetosheath and the magnetopause can also lead to the excitation of waves, some of which end up in the magnetosheath. These waves could play a role in flow diversion or the formation of the boundary layer.

CP888, *Modern Trends in Physics Research,*
Second International Conference on Modern Trends in Physics Research—MTPR-06,
edited by L. El Nadi
© 2007 American Institute of Physics 978-0-7354-0354-9/07/$23.00

The main aim of this work is to calculate the periodicity of the Alfven mode waves in the Earth's magnetosheath in order to explore its in the transfer of matter through the magnetopause. The paper is organized as follow: In Section (2) we summarized the main characteristics and position of the most dominant wave modes that can propagate in the Earth's magnetosheath. While in Section (3) we calculate the periodicity of Alfven mode waves. The results and discussion are given in Section (4). Conclusions are given in Section (5).

MAIN CHARATERSITIC WAVE MODES

1. The main characteristics

The wave modes in the Earth's magnetosheath can be classified into two major types: First, is the compressional wave mode. Second, is the transverse or Alfven Ion Cyclotron (AIC) wave modes.

(i) The compressional wave modes:

Regular compressional waves with periods about 20 sec are perhaps the most recognizable feature of the terrestrial magnetosheath, observed downstream of the quasi-perpendicular shocks (Tsurutani et al., 1982; Song et al., 1992b; Anderson et al., 1994).

Belmont and Rezeau (2003) calculated the general dispersion relation for the compressional modes (in ideal MHD) as follows:

$$(\frac{\omega}{k})^4 - (V_A^2 + c_s^2)(\frac{\omega}{k})^2 + V_A^2 c_s^2 \cos^2 \theta = 0 \qquad \text{(with } \cos\theta = \frac{k_{\text{II}}}{k} \text{)} \qquad (1)$$

where, ω, k, c_s, θ are the angular frequency, the wave number, the sound speed, and the angle between the wave number and the magnetic field, respectively.

TABLE 1. The main characteristics of the fast mode waves

Definition	The fast mode is a compressional mode that propagates at frequencies higher than the proton gyrofrequency as well as lower (Rezeau et al., 1999).
The main characteristics	It propagates very fast in any direction in the magnetosheath. Its phase speed is found to be grater or equal to the Alfven velocity, of about 3.5 V_{A0}. It is observed, with a right-hand (RH) polarization with respect to the magnetic field (Rezeau et al., 1999). There is a positive correlation between the plasma density ρ and magnetic field B upstream of the bow shock (Yan and Lee, 1994). Fast waves of zero frequency ($\alpha=0$) can transfer into Alfven or slow modes by "mode conversion". For ($\alpha\neq0$) "mode coupling" is dominant (fast and Alfven mode are existed); where a part of the energy incident fast mode can be transferred into the shear Alfven form (Rezeau and Belmont, 2001). The dispersion relation (Eq.1) showed that the magnetic field and the density are in-phase for the fast mode waves.
Sources	The fast-mode wave can travel from the bow shock to the magnetopause in 1.5 t_A (Alfven time), and travel from the magnetopause back to the bow shock in about 5 t_A (Yan and Lee, 1994).

$$V_A = \frac{B_0}{\sqrt{\mu_0 \rho}} \qquad \text{, is the Alfven velocity.} \qquad (2)$$

Where, B_0, ρ, μ_0 are the magnetic field strength, the mass density, and the permeability of free space, respectively.

As mentioned before, the compressional modes can be classified as: Fast, Slow, and Mirror mode waves. The main characteristic of the dominant wave modes are summarized in Tables (1, 2, and 3), respectively.

TABLE 2. The main characteristics of the slow mode waves.

Definition	The slow mode wave is a wave with a finite propagation speed and the same characteristic perturbation relations as the MHD slow mode (Song and Russell, 1997).
The main characteristics	It has velocity $\leq$ the Alfven velocity. Its total pressure ($P_{particle} + P_{magnetic}$) ~ constant, across the background field. It carries energy predominately along the background field. It propagates along B at the sound speed and reduces plasma-pressure gradients (Kivelson and Russell, 1995). The observed waves do not convict with the magnetosheath plasma flow; they seem to stand against the magnetosheath flow. The wavelength (λ_n) in the direction normal to the magnetopause is typically 200 to 5000 km, while the wavelength (λ_t) in the tangential direction is much larger than (λ_n). A steepened plasma density profile with a shock-like structure also appears in the enhanced density region (Yan and Lee, 1994). The dispersion relation in Equation (1) indicated that the magnetic filed and the density, are in opposite direction for the slow mode waves.
Sources	1- Disturbances in the solar wind may generate the waves. 2- The movement or oscillations of the magnetopause may create MHD waves in the magnetosheath (Yan and Lee, 1994).

TABLE 3. The main characteristics of the mirror mode waves.

Definition	The mirror mode waves are 3D structures with their major axis nearly perpendicular to the magnetopause surface (Hubert et al., 1998).
The main characteristics	They are linearly polarized and compressive, with fluctuations anti-correlated with density fluctuations (Tsurutani et al., 1982). Their total pressure is constant across the event (Tsurutani et al., 1982). For small temperature anisotropy, the mirror instability has a largest growth rate in the direction nearly perpendicular to the magnetic field (Gary, 1992). It is a non-oscillating ($\omega_r = 0$) instability with the wave vector in a direction highly oblique to the magnetic field (Leckband et al., 1995). The maximum growth rate of the mirror mode occurs at oblique angles.
Sources	The mirror instability develops from the slow mode branch of the MHD dispersion relation when account is taken of finite temperature anisotropy (Song et al., 1992a). This instability can be generated by an increase in anisotropy either through anisotropic heating at the bow shock (Lee et al., 1988) or by adiabatic field compression in the magnetosheath (Crooker and Siscoe, 1977, Hubert, 1994).

(ii)   The transverse or Alfven (AIC) wave modes:

The identifications of linear AIC waves rely on a comparison of (B ${}_{\Pi}$ and B_T). The plasma density and pressure are found to be in phase. The main characteristics of the Alfven mode waves are summarized in Table (4).

TABLE 4. The main characteristics of the Alfven mode waves

Definition	The Alfven waves are traversal ion waves. They are non-dispersive waves with Alfven velocity (Kivelson and Russell, 1995).
The main Characteristics	The plasma pressure and magnetic field magnitude are constant, across the incident Alfven wave (at time t=0), while By and Bz change sinusoidally (Yan et al., 1994).
	No density perturbation is associated with the magnetic field perturbation, and no perturbation of the magnetic field magnitude.
	The Alfven mode has a maximum growth rate along the magnetic field which is larger than the maximum growth rate of the mirror mode (Balikhin et al., 2001).
	It is a right-hand polarized wave. Gleaves and Southwood (1990) observed a strong linearly polarized Alfven wave, below fcp /10, propagating along the inward shock normal and across the magnetic field.
	The plasma density and pressure are found to be in-phase.
	Lacombe et al. (1995) showed that some wave properties depend on the shape of the distribution functions of electrons and protons.
	The magnetic components are mainly polarized perpendicular to the static magnetic field, which is a characteristic of the Alfven waves (Anderson et al., 1991, Song et al., 1993).
	Belmont and Rezeau (2003) showed that the dispersion relation for the Alfven waves in the general case is given by: $\omega^2 - k_{11}^2 V_A^2 = 0$, where the direction of perturbation of the magnetic field direction is perpendicular to B and k.
Sources	1- Omidi et al. (1994) referred to magnetosheath frontier regions as the source of the propagating AIC modes.
	2- The observed AIC waves propagate along the ambient field in both the upstream and downstream directions. This implies that the bow shock is not the only source of AIC waves. AIC waves are locally growing in the magnetosheath (Lacombe et al., 1995).

2. The position

Denton et al. (1998) summarized the best fitting of different modes as shown in Table (5).

TABLE 5. The best fitting modes of the three regions of the magnetosheath.

Event	Position	Best Fitting Mode	Second Best Mode
1	Middle magnetosheath	Q-⊥ Mir	Q-⊥ Alf
2		Q-⊥ Mir	Q-⊥ Alf
3		Q-⊥ Mir	Alf/Q-∥ Mg S
4	Outer magnetosheath	Alf/Q-∥ Mg S	Q-⊥ Mir
5		Alf/Q-∥ Mg S	Mir
6		Alf/Q-∥ Mg S	Q-⊥ Mir
7	Inner Magnetosheath	Q-∥ Mir	Q-⊥ Mir
8		Q-⊥ Mir	Alf/Q-∥ Mg S

Q-∥ (quasi-parallel) corresponds to $5° < \theta_{kB} \le 20°$; Q-⊥ (quasi-perpendicular) corresponding to $60° < \theta_{kB} \le 85°$; MgS, magnetosonic, Alf, Alfven; AIc, ion acoustic; and Mir, mirror. Where, θ_{kB} is the angle between k and Bo.

CONDITIONS FOR MATTER AND ENERGY TRANSFER ROM THE SOLAR WIND TO THE MAGNETOSPHERE

The magnetosphere is a big (magnetic) spring. The outermost part of this spring is the magnetosheath. The magnetosheath is a bound region, consists of several mode waves. The approximate mass of this region can be calculated (just sonic part). If we take this mass and push it inside the equilibrium point, the magnetosphere pushes back. Then its mass will move out past the equilibrium point and the solar wind slows this down and pushes it against the magnetosphere again. In this case, the solar wind is considered as the restoring force of this spring.

For simplicity, we calculate the periodicity of an Eigen mode wave (e.g. the Alfven waves). We study two cases taking into account the shape of the magnetosheath; first is the spherical case, and the second is the non-spherical case. Then we compare the results.

1. Eigen modes of the global magnetosphere (The spherical case):

The fundamental frequency can be estimated from energy considerations (see Alfven and Falthammar, 1963). Consider a sphere of infinitely conductive fluid in a homogenous magnetic field B_o. Following Lundquist (1952), we suppose that the sphere is slightly deformed (see Fig. 1) so that it becomes a prolate spheroid with the half-axes R ($1+\alpha$) and R ($1- 0.5\ \alpha$), and then left free to oscillate.

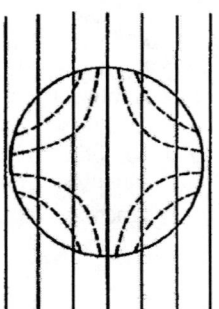 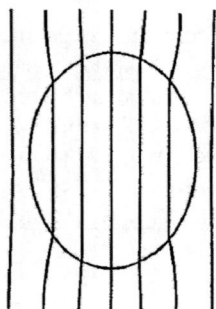

FIGURE 1. Hydromagnetic oscillations of a sphere. Original field lines and stream lines (dashed curves) are shown in the left Fig., and the deformed state to the right. (Alfven-Falthammar, 1963).

The excess magnetic energy at maximum deformation is:

$$W_m = \frac{\alpha^2 B_0^2 R^3}{4\mu}$$

(3)

Where, μ, α, Bo, and R are the permeability, the deformation factor, the magnetic field strength, and the radius, respectively.

If it is assumed that the oscillation is harmonic and characterized by an angular frequency ω, the kinetic energy is found to be:

$$W_{Kin} = \frac{1}{5}\rho\pi\alpha^2 R^5\omega^2$$

(4)

At the instant when the body is spherical and the excess magnetic energy vanishes. Equating W_m and W_{Kin} we

obtain:

$$\omega = \frac{\sqrt{5}}{R}\frac{B_0}{(\mu_0\rho)^{1/2}}$$

(5)

The Alfven velocity V_A which propagates along a magnetic field line is "elastic wave" arising from the Maxwell stress of the field line. It can be expressed as:

$$V_A = \frac{B_0}{\sqrt{\mu_0 \rho}}$$

(6)

The terrestrial magnetic field is well approximated by a dipole field. Hence, the field strength at a distance R is approximately given by:

$$B(R) = B_e \left(\frac{R_e}{R}\right)^3$$

(7)

We will discuss two different positions of the waves. The first is at the magnetopause (R~10 Re), where the magnetic field is given by:

$$B(10\ R_e) \approx 3 \times 10^{-8}\ T$$

Therefore, the Alfven velocity at the magnetopause is equal to:

$$V_A = 1.46 \times 10^5\ \text{m.s}^{-1}.$$

(8)

Substitute this value of V_A, in equation (5) we get:

$$\omega = \frac{\sqrt{5}}{R} \frac{B_0}{(\mu_0 \rho)^{1/2}} \approx 5.129 \times 10^{-3} s^{-1}$$

(9)

Then, we could obtain the periodicity at the magnetopause as given by:

$$P_{magnetopause} = \frac{2\pi}{\omega} \approx 1224.825\, s \approx 20.4\ \text{min}$$

(10)

The second position is at the bow shock (12 R_e). The magnetic field is calculated using equation (8) as:

$$B(12\ R_e) \approx 1.74 \times 10^{-8}\ T$$

(11)

The Alfven velocity is given by:

$$V_A = 8.46 \times 10^4\ \text{m.s}^{-1}.$$

(12)

Using this value of V_A, and substitute in Equation (5) to get:

$$\omega = \frac{\sqrt{5}}{R} \frac{B_0}{(\mu_0 \rho)^{1/2}} \approx 2.47 \times 10^{-3} s^{-1}$$

(13)

Then, the periodicity at the bow shock is given by:

$$P_{Bowshock} = \frac{2\pi}{\omega} \approx 2542\, s = 42.36\ \text{min}$$

(14)

2. Wave propagation in the solar wind (non-spherical case) :

The origin of the restoring force is essential to get the spring constant. Let us first consider compression and expansion of an ideal gas as an analogy.

Imagine an ideal gas in a box which has a piston of mass on one side. For simplicity, we assume that the box is a perfect thermal insulator: so there is no energy exchange between the gas and the external environment in which the box is placed. The piston is at rest at a position x =0, because the pressure of the gas in the box balances with an external pressure (atmospheric pressure). Thus,

$$P(0) = P_0$$

(15)

Where P (0) is the gas pressure when the piston is at x =0 and P_0 is the external pressure.

If we push the piston, the gas is compressed and the piston moves to the position x. The volume of the gas is given by:

$$V(x) = V_0 \frac{L}{L-x}$$

(16)

Where, V(0) is the volume when the piston is at x =0, and L is the length of the box along the x-axis. Since we assume that there is no energy exchange between the gas and the external environment, the state of the gas changes adiabatically.

The pressure P(x) of the gas when the piston is at a position x is calculated from:

$$P_o V_o = P(x) V^{\gamma}(x) \tag{17}$$

Where, γ is an adiabatic constant (which is the ratio of the specific heats Cp/Cv, for constant pressure and volume). Namely, the gas pressure P(x) is given by:

$$P(x) = P_0 (\frac{L-x}{L})^{\gamma} \tag{18}$$

The force $F(x)$ needed to put the piston at x is calculated from the balance of the forces exerted on the piston, i.e.

$$SP_0 + F(x) = SP_0 (\frac{L-x}{L})^{\gamma} \tag{19}$$

Where, S is the surface area of the piston. Using the approximation x $\ll$ L, we get:

$$(\frac{L-x}{L})^{\gamma} \cong 1 - \frac{x}{\gamma L} \tag{20}$$

Thus, we get

$$F(x) \cong -\frac{\gamma P_0 S}{L} x \tag{21}$$

Using equation (16), one can notice that $F(x)$ is proportional to the displacement x from the equilibrium position x = 0. This indicates that the piston makes harmonic oscillation around the equilibrium position x = 0, if we remove the force $F(x)$ when the piston is at the position x.

The equation of motion of the piston whose mass M is expressed by

$$M \ddot{x} = -\frac{\gamma P_0 S}{L} x \tag{22}$$

Thus, the angular frequency ω for the oscillation of the piston is given by:

$$\omega = \sqrt{\frac{\gamma P_0 S}{M L}} \tag{23}$$

Now, let us go into the magnetosheath. The external pressure is now the solar wind pressure. The restoring force is the magnetic pressure (Maxwell stress tensor, in general). Magnetic field lines are like rubber string, but these strings repel each other. Because of the repulsion the magnetic field lines have a "pressure" which resists against external force, the solar wind pressure in our case. Namely, we can regard the magnetosheath as a "magnetic field gas", whose pressure (magnetic pressure) is given by:

$$P_{mo} = \frac{B_0^2}{8\pi} \tag{24}$$

Thus $P(0)$ in Equation (21) may be replaced by P_{m0}, where B_0 is the field strength in the "ordinary" solar wind intensity. In our problem S in Equation (22) is the area and L is the thickness of the magnetosheath. Thus we obtain

$$\omega \cong \sqrt{\frac{B_0^2 S}{8\pi \gamma M L}} \tag{25}$$

Ignoring γ, which is of order 1, M is the mass of the magnetosheath and is given by

$$M = \rho S L \tag{26}$$

Substituting this M into Equation (25) we get

$$\omega = \frac{B_0}{L} \sqrt{\frac{1}{8\pi\rho}} = \frac{B_0}{L} \sqrt{\frac{1}{2\mu_0 \rho}} \tag{27}$$

Where, L, is the length of the magnetosheath = 2- 3 Re (see, Farris et al., (1991) $\approx 1.9 \times 10^7$ m. B_0, is the magnetosheath magnetic field strength = 30 – 40 nT $\approx 35 \times 10^{-9}$ T. and ρ, is the density $\approx 3.35 \times 10^{-20}$ Kg.m^{-3}. Then,

$$\omega = 0.0063098 \ s^{-1}$$

(28)

Hence,

$$P_{non-spherical} = \frac{2\pi}{\omega} \approx 995.76 \, s \approx 1 \times 10^3 \, s = 16.59 \ min$$

(29)

This result is very similar to the result of the first case 1.5×10^3 s, given by Equation (10).

RESULTS AND DISCUSSION

Our results are summarized in Table (6)

TABLE 6. The periodicity of Alfven waves for two different cases.

The case	Spherical		Non-spherical
The position (x)	At magnetopause (~10 Re)	At bow-shock (~12 Re)	
The Alfven velocity (V_A)	$V_A = 1.46 \times 10^5$ m. s^{-1}	$V_A = 8.46 \times 10^4$ m. s^{-1}	$V_A = 1.7 \times 10^5$ m.s^{-1}
The periodicity (P) in minutes	~ 20.4 min	~ 42.36 min	~ 16.59 min

As we have seen from Table (6), there are two values of the periodicity of the Alfven waves in the spherical case corresponding to two different positions of the waves in the magnetosheath. The first position is at the magnetopause (at R $\approx$ 10 R$_e$, V$_A \approx 1.46 \times 10^5$ m. s^{-1}) with periodicity (~20.4 min). The second position is at the bow shock (At R $\approx$ 12 R$_e$, V$_A \approx 8.46 \times 10^4$ m. s^{-1}) with periodicity (~ 42.36 min). On the other hand, in the non-spherical case the periodicity is even shorter than the spherical two cases. It is nearly about (~16.59 min). The plasma frequency by the two cases is of the same order of magnitude. Therefore one can conclude that conditions are met for a matter transfer provided that a mode conversion from pressure to Alfven is possible. It has to be noticed that, only the frequency matching has been considered and that a full treatment of the problem involves polarization and wave number analysis, coupling problem and a quantitative assessment of the transfer allowed by these processes is still to be done.

Based on the results, we concluded that the Alfven waves could exist near the magnetopause with shorter periodicity than at the bow shock. This means that as we go far from the center of the Earth, the periodicity becomes long while the oscillation is faster near the Earth's surface. This is indicated that the resulting waves are generated as a result of the magnetosphere oscillation.

We have also seen that the periodicity at the magnetopause is close to the non-spherical case, which confirming that the Alfven waves could exist in the inner magnetosheath too. This result is consistent with the observational results of Rezeau and Belmont (2001) that in the presence of a sufficient magnetic field rotation, the magnetosheath waves are converted into a trapped Alfven waves in the magnetopause boundary. Our results in the spherical and non-spherical (at the bow-shock and magnetopause) case also meet the suggestion of Denton et al. (1998), that the Alfven waves could exist in the inner and outer parts of the magnetosheath.

It is important to note that the periodicity in the non-spherical is (~16.59 min), which has nearly the same duration of the slow mode waves (~15 min) observed by Yan and Lee (1994). They found that the slow-mode waves are generated through the interaction between the bow shock and various MHD waves (fast-mode, Alfven-mode, or mirror-mode waves). From the above results we concluded that the Alfven wave's periodicity is equal to or greater

than the periodicity of the slow mode waves. These Alfven waves could be produced as a result of the oscillations of the magnetosphere as a spring. The frequencies are not completely stable; it is also not surprising to see the other parameters vary a bit from one interval to another. If the frequency was stable we could expect the polarization properties to be stable.

ACKNOWLEDGMENT

We thank C. T. Russell for suggesting the problem, L. Rezeau and T. Yamamoto for fruitful discussion.

REFERENCES

1. Alfven, H. and Falthammar, C.-G, Cosmical electro-dynamics, Fundamental principles, p.90, Oxford University press, 1963.
2. Anderson, B. J., and S. A. Fuselier, Magnetic pulsations from 0.1 to 0.4 H_z and associated plasma properties in the Earth's subsolar magnetosheath and plasma depletion layer, J. Geophys. Res., **98**, 1461 (1991).
3. Anderson, B. J., S. A. Fuselier, S. P. Gary, R. E. Denton, Magnetic spectral signatures in the Earth's magnetosheath and plasma depletion layer, J. Geophys. Res., **99**, 5877 (1994).
4. Balikhin, M. A., S. Schwartz, S. N. Walker, H. St. C. K. Alleyne, M. Dunlop, and H. Luhr, Dual-spacecraft observations of standing waves in the magnetosheath, J. Geophys. Res., **106**, 25395 (2001).
5. Belmont, G. and Rezeau, L., Introduction a la physique des plasmas, RI- CETP/01/2003.
6. Crooker, N. U., and G. L. Siscoe, A mechanism for pressure anisotropy and mirror instability in the dayside magnetosheath, J. Geophys. Res., **82**, 185 (1977).
7. Denton, R. E., M. R. Lessard, J. W. LaBelle, and S. P. Gary, Identification of low-frequency magnetosheath waves, J. Geophys. Res., **103**, 23661 (1998).
8. Farris, M. H., S. M. Petrinec and C. T. Russell, The thickness of the magnetosheath: Conservation the polytrophic index, Geophys. Res. Lett., **18**, 1821 (1991).
9. Gary, S. P., The mirror and ion cyclotron anisotropy instabilities, J. Geophys. Res., **97**, 8519 (1992).
10. Gleaves, D. G. and D. J. Southwood, Phase delays in transverse disturbances in the Earth's magnetosheath, Geophys. Res. Letts., **17**, 2249 (1990).
11. Hubert, D., C. Lacombe, C. C. Harvey, M. Moncuquet, C. T. Russell, and M. F. Thomsen, Nature, properties, and origin of low frequency waves from an oblique shock to the inner magnetosheath, J. Geophys. Res., **103**, 26783 (1998).
12. Kantrowitz, A., and H. E., Petschek, MHD characteristics and shock waves, in plasma physics in theory and application, edited by W. B. Kunkel, p.148, McGraw-Hill, New York, 1966.
13. Kaufmann, R. L., J. T. Horng and A. Wolfe, Large amplitude hydromagnetic waves in the inner magnetosheath, J. Geophys. Res., **75**, 4666 (1970).
14. Kivelson, M. G., and C. T. Russell, Introduction to space physics, Cambridge University press, 1995.
15. Lacombe, C., G. Belmont, D. Hubert, C. C. Harvey, A. Mangeney, C. T. Russell, J. T. Gosling and S. A. Fuselier, Density and magnetic field fluctuations observed by ISEE 1-2 in the quiet magnetosheath, Ann. Geophysicae, **13**, 343 (1995).
16. Lee, L. C., C. P. Price and C. S. Wu, A study of mirror waves generated downstream of a quasi-perpendicular shock, J. Geophys. Res., **93**, 247 (1988).
17. Lee, L. C., Yan, M. and Hawkins, J. G., A study of slow mode structure in front of the dayside magnetopause, Geophys. Res. Lett., **19**, 381 (1991).
18. Leckband, J. A., D. Burgess, F. G. E. Pantellini, and S. J. Schwartz, Ion distributions associated with mirror waves in the Earth's magnetosheath, Adv. Space Res., **15**, 345 (1995).
19. Lundquist, S., Studies in Magneto-hydrodynamics, ibid. 5, 297, 1952.
20. Omidi, N., A. O'Farrell and Krauss-Varban, Sources of magnetosheath waves and turbulence, Adv. Space Res., **14**, 45 (1994).
21. Rezeau, L., G. Belmont, N. Cornilleau-Wehrlin, F. Reberac, and C. Briand, Spectral law and polarization properties of the low-frequency waves at the magnetopause, Geophys. Res. Lett., **26**, 651 (1999).
22. Rezeau, L., and G. Belmont, Magnetic turbulence at the magnetopause, a key problem for understanding the solar wind / magnetosphere exchanges, Space Science Reviews, **95**, 427 (2001).
23. Song, P., C. T. Russell and M. F. Thomsen, Waves in the inner magnetosheath: A case study, Geophys. Res. Lett. **19**, 2191 (1992).
24. Song, P., C. T. Russell, and C. Y. Huang, Wave properties near the subsolar magnetopause: Pc 1 waves in the sheath transition layer, J. Geophys. Res., **98**, 5907 (1993).
25. Song, P., C. T. Russell, What do we really know about the magnetosheath?, Adv. Space Res., **20**, 747 (1997).
26. Song, P., C. T. Russell, T. I. Gombosi, J. R. Spreiter, S. S. Stahara and X. X. Zhang, On the processes in the terrestrial magnetosheath, J. Geophys. Res., **104**, 345 (1999).
27. Tsurutani, B. T., E. J. Smith, R. R. Anderson, K. W. Ogilvie, J. D. Scudder, D. N. Baker and S. J. Barne, lion roars and non-oscillatory drift mirror waves in the magnetosheath, J. Geophys. Res., **87**, 6060 (1982).
28. Yan, M., and L. Lee, Generation of slow-mode waves in front of the dayside magnetopause, Geophys. Res. Lett, **21**, 629 (1994).

Lorentz Force Effects on the Orbit of a Charged Artificial Satellite: A New Approach

Yehia A. Abdel-Aziz

National Research Institute of Astronomy and Geophysics, Helwan, Cairo, Egypt
e-mail: yehia@nriag.sci.eg

Abstract: A charged artificial satellite moving relative to a magnetic field accelerates in a direction perpendicular to its velocity and the magnetic field due to the Lorentz force. The geomagnetic field is considered as a multipole potential field and the satellite electrical charged is supposed to be constant. The study is provided to compute Lorentz force acceleration of a charged satellite in Earth's magnetic field as a function of orbital elements of the satellite. Periodic perturbations in the orbital elements of the satellite are derived using Lagrange planetary equations. Numerical results for a chosen satellites orbit shows the most effects of Lorentz force are in semi major axis, eccentricity, and the longitude of the satellite, but there aren't any effects of the force on the inclination and the argument of the perigee of the satellite elements.

INTRODUCTION

The important quantity which determines the magnitude of the effect is the satellite's electrical charge. The surface of a satellite is charged to a negative potential (see Al'pert et al., 1964) and in the first approximation behaves like a spherical condenser with respect to the ionosphere vicinity.

The relationship between the plasma environment and spacecraft potential must be taken into account for the importance of mission role and spacecraft configuration in evaluating absolute and differential charging effects. The build-up of large potentials on spacecraft relative to ambient plasma is not, of itself, a serious electrostatic discharged design concern. However, such charging enhances surface contamination, which degrades thermal properties. These and other charging effects can produce potential differences between spacecraft surface or between spacecraft surface and spacecraft ground. When a breakdown threshold is exceeded, an electrostatic discharge can occur. Vehicle torquing or wobble can also be produced when multiple discharging occur (Carolyn K. Purvis et al. 1984).

Anderson (et al. 1994) examined the relationship between the plasma environment and spacecraft potential for the Dynamics Explorer 2 (DE 2) spacecraft in an attempt to improve the accuracy of ion drift measurements by the retarding potential analyzer (RPA). The geomagnetic field had considerable effect on the spacecraft potential due to magnetic field confinement of the electrons as well as to the (Lorentz force) V×B electric field resulting from the movement of the spacecraft across magnetic field lines. They derived an algorithm for determining the spacecraft potential (at the location of the RPA on the spacecraft) for any point of the DE 2 orbit.

Juhasz Antal and Mihaly Horanyi (1997) studied the Degrading objects in orbit around the Earth as well as solid rocket motors generate micron- and submicron-sized space debris. They shown, the motion of these particles is dictated by gravity, solar radiation pressure, and electromagnetic forces, since these grains collect electrostatic charges and become vulnerable to the electric and magnetic fields in the Earth's magnetosphere. They shown that magnetosphere effects tend to reduce the lifetime of these grains either by forcing them onto elliptical orbits to collide with the Earth or by swiftly ejecting them into the interplanetary space.

Several attempts were already made to assess the effects of Lorentz force to show in principle its negligible value with respect to some other, pre-important effects (e.g. Sehnal, 1969). However, the necessary high precession of orbital determination of some proposed space experiments (Ciufolini, 1987), requires a full knowledge of the electrodynamical effects connected with the Lorentz force, which we shall try to study in detail.

VokRouhlicky (1989) determined the orbital effects of the Lorentz force on the motion of an electrically charged artificial satellite moving in the Earth's magnetic field. In his case the influence of the geomagnetic field manifests itself predominantly by Lorentz force.

Peek, (2005) found out the components of the Lorentz force in the spherical coordinate system.

CP888, *Modern Trends in Physics Research,*
Second International Conference on Modern Trends in Physics Research—MTPR-06,
edited by L. El Nadi
© 2007 American Institute of Physics 978-0-7354-0354-9/07/$23.00

He evaluated the use of Lorentz force as a means of orbit control for finite bodies, including small spacecraft.

In the present work, the components of the Lorentz force acceleration of a charged satellite in the Earth's magnetic field are computed as a function of orbital elements of the satellite in two directions (radial and tangential). Periodic perturbations in orbital elements of the satellite due to the Lorentz force are derived using Lagrange planetary equations. Numerical applications for the LAGOS Satellite are introduced.

EQUATIONS of MOTION

Let us consider the magnetic field of the earth to be given as a multipole potential field and the satellite electrical charged is supposed to be constant.

The components of the disturbing force (Lorentz force) we are looking for will be the components of a vector

$$F_L = QV \times B \tag{1}$$

where V is the velocity of the satellite in the orbit, B is the vector of the magnetic field intensity of the Earth, Q is the satellite's electrical charge. The components of the vector F_L, from equation (1) in the direction $\hat{r}$ (the direction of the position vector) and $\hat{t}$ (the tangential direction, which normal to r and the orbital angular momentum)

As in Peck (2005) Augmented with the Lorentz Force, Newton's law of gravitation for a satellite of mass m moving in r^{-2} gravitational field of the Earth with point mass M becomes

$$m \frac{^N d^2}{dt^2} r = -m \frac{\mu}{r^2} \hat{r} + Q \left(\frac{^N d}{dt} r - \boldsymbol{\omega}_e \times r \right) \times B , \tag{2}$$

where, the superscript N indicates a derivative taken with respect to a Newtonian, or inertial frame, r is the vector position (magnitude r and direction $\hat{r}$) of the satellite relative to the system barycentre, $\mu = MG$ where G is the universal gravitational constant, $\boldsymbol{\omega}_e$ is the Earth's angular velocity vector fixed frame with the respect to an inertial frame. This expression acknowledges that it is the satellite's velocity relative to the magnetic field $V = \frac{^N d}{dt} r - \boldsymbol{\omega}_e \times r$ that determines the Lorentz force. In the simplest model, the earth's magnetic field rotates with Earth. Br relativity, this time-varying magnetic field represents an electric field, which is the means by which work can be done on the Lorentz force.

In a frame E the rotates with the Earth, the equation of motion in terms of the relative velocity $V = \frac{^E d}{dt} r$ and a gravitational potential Φ_{gr} is

$$\frac{^E d}{dt} V = -\nabla \Phi_{gr} + \frac{Q}{m} V \times B - 2 \boldsymbol{\omega}_e \times V + 2 \boldsymbol{\omega}_e \times (\boldsymbol{\omega}_e \times r) \tag{3}$$

where, dividing through by m introduces the commonly used charge per mass $\frac{Q}{m}$ as a parameter that determines the scale of the Lorentz force. Following Schaffer and Burns (1994), we project this equation onto V

$$\frac{^E d}{dt} V \cdot V = \frac{1}{2} \frac{d}{dt} V^2 = -\nabla \Phi_{gr} \cdot \frac{^E d}{dt} r + \frac{d}{dt} \left(\frac{1}{2} r^2 \omega_e^2 \sin^2 \theta \right) \tag{4}$$

where θ is a coordinate of a spherical coordinate system (r, θ, ϕ) with origin at the Earth's centre and associated with an E-fixed basis. Integrating between t_1 and t_2 shows that the total mechanical energy in the rotating fram H is constant:

$$H(t_2) - H(t_1) = 0, \tag{5}$$

Schaffer and Burns (1994) point out that this function is the appropriate Hamiltonian in the non-canonical variables $(r, p = m\dot{r})$ as demonstrated by Littlejohn (1982, 1979), who used these coordinates in a perturbation theory for highly charged particles in slowly varying electromagnetic fields.

The existence of this constant Hamiltonian in a rotating frame suggests that only the rotation of the magnetic field, which causes the co-rotational electrical field, can do work in the inertial frame. By way of illustration, we consider the osculating elements of a restricted two- body orbit whose angular momentum is aligned with the Earth's magnetic moment (i.e., a magnetic equatorial orbit). The energy E per unit satellite mass in inertial frame is given by

$$E = -\frac{\mu}{2a} \,, \tag{6}$$

where a is the semimajor axis of the sateillite's orbit. Its time rate of change is

$$\dot{E} = \dot{a}\frac{\mu}{2a^2} = \frac{^{N}\mathrm{d}}{\mathrm{d}t}r \cdot dF_L \,, \tag{7}$$

a familiar result in which the energy depends entirely on the semimajor axis. Work done if and only if the semimajor changed. The perturbing force dF per unit mass is

$$dF_L = F_r\hat{r} + F_t\hat{t} \tag{8}$$

In the basis of proposed directions of the Lorenzt force, then

$$\frac{^{N}\mathrm{d}}{\mathrm{d}t}r = \dot{r}\hat{r} + r\dot{\phi}\hat{t} \tag{9}$$

where ϕ is the true anomaly of the orbit. Therefore,

$$\dot{a}\frac{\mu}{2a^2} = \left(\dot{r}\hat{r} + r\dot{\phi}\hat{t}\right)\left(F_r\hat{r} + F_t\hat{t}\right) \tag{10}$$

or

$$\dot{a}\frac{\mu}{2a^2} = \dot{r}F_r + r\dot{\phi}F_t \,. \tag{11}$$

At the magnetic equator $B = B\hat{r} \times \hat{t}$. In this case the radial component of the Lorentz force is

$$F_r = \hat{r} \cdot \frac{Q}{m}\left(\frac{^{N}\mathrm{d}}{\mathrm{d}t}r - \omega_e \times r\right) \times B = \frac{Q}{m}Br\left(\dot{\phi} - \omega_e\right), \tag{12}$$

And the tangential component of the Lorentz force is

$$F_t = \hat{t} \cdot \left[\frac{Q}{m}\left(\frac{^{N}\mathrm{d}}{\mathrm{d}t}r - \omega_e \times r\right) \times B\right] = \frac{QB}{m}\dot{r} \,. \tag{13}$$

The signs of Q and B are important her: positive charge (due to, say electron emission) and negative (southward) magnetic fields cause a loss in orbital energy, as does negative charge in a positive (northward) magnetic field. Only when Q and B are of opposite sign is the energy change positive. Although this result would indicate that greater Lorentz force is available at higher altitude, in fact B drops off approximately with r^{-3}; so that net effect of increased altitude is deleterious.

We shall make use of the formulas

$$r\dot{\phi} = a\sqrt{1-e^2}\,\dot{E}, \qquad\qquad \dot{r} = ae\sin E\,\dot{E},$$

$$\sin\phi = \sqrt{1-e^2}\,\frac{a}{r}\sin E, \qquad \cos\phi = \frac{a}{r}\left(\cos E - e\right) \tag{14}$$

where E is the eccentric anomaly of the satellite's orbit, e is the eccentricity of the orbit.
Using the relations in (14), we can rewrite the components of the Lorentz force $\boldsymbol{F}_L$ in radial and tangential directions as the following:

$$F_r == \frac{Q}{m} B\left[a\sqrt{1-e^2}\,\dot{E} - r\omega_e \right], \tag{15}$$

$$F_t == \frac{Q}{m} Bae\sin E\,\dot{E}. \tag{16}$$

The Lagrange Planetary equations:

$$\frac{da}{dt} = \frac{2}{na}\left(\frac{\partial F}{\partial M} \right),$$

$$\frac{de}{dt} = \frac{\sqrt{1-e^2}}{na^2 e}\left(\sqrt{1-e^2}\,\frac{\partial F}{\partial M} - \frac{\partial F}{\partial \omega} \right),$$

$$\frac{dI}{dt} = \frac{1}{na^2\sqrt{1-e^2}\,\sin I}\left(\cos I\,\frac{\partial F}{\partial \omega} - \frac{\partial F}{\partial \Omega} \right),$$

$$\frac{d\omega}{dt} = \frac{1}{na^2}\left(-\frac{\cos I}{\sqrt{1-e^2}\,\sin I}\,\frac{\partial F}{\partial I} - \frac{\sqrt{1-e^2}}{e}\,\frac{\partial F}{\partial e} \right), \tag{17}$$

$$\frac{d\Omega}{dt} = \frac{1}{na^2\sqrt{1-e^2}\,\sin I}\,\frac{\partial F}{\partial I},$$

$$\frac{dM}{dt} = n - \frac{1-e^2}{na^2 e}\,\frac{\partial F}{\partial e} - \frac{2}{na}\,\frac{\partial F}{\partial a},$$

are used to derive the perturbations in the orbital elements, where $n,\, I,\, M,\, \omega,\, \Omega$ are the mean motion, inclination, mean anomaly, argument of perigee, ascending node of the satellite's orbit respectively.
Therefore, the periodic perturbations in the Keplerian elements due to the Lorentz force are in the following shape

$$\Delta a = \frac{da}{dt} = \left(\frac{Q}{m}\right)B\left(\frac{2ae}{nr}\right)\sin E\left[\sqrt{1-e^2} - \omega_e + \cot E\,\dot{E} + e\sin E \right],$$

$$\Delta e = \frac{de}{dt} = \left(\frac{Q}{m}\right)B\left(\frac{1-e^2}{nr}\right)\sin E\left[\sqrt{1-e^2} - \omega_e + \cot E\,\dot{E} + e\sin E \right],$$

$$\Delta I = \frac{dI}{dt} = 0,$$

$$\Delta\omega = \frac{d\omega}{dt} = -\left(\frac{Q}{m}\right)B\left(\frac{\sqrt{1-e^2}}{na^2e}\right)\left[-\frac{2ae}{\sqrt{1-e^2}}\dot{E} + \frac{a\sqrt{1-e^2}}{n}\left(\frac{ae}{r}\sin^2 E - \cos E\right)+\right.$$

$$\left. + \omega_e a\cos\phi + a\sin E\,\dot{E} + \frac{a^2e}{r}\sin E\cos E\,\dot{E} + \frac{a^2e^2}{nr}\sin^3 E - \frac{ae}{n}\sin E\cos E\right],$$ (18)

$$\Delta\Omega = \frac{d\Omega}{dt} = 0,$$

$$\Delta M = \frac{dM}{dt} = n - \left(\frac{Q}{m}\right)B\left\{\left(\frac{1-e^2}{na^2e}\right)\left[-\frac{2ae}{\sqrt{1-e^2}}\dot{E} + \frac{a\sqrt{1-e^2}}{n}\left(\frac{ae}{r}\sin^2 E - \cos E\right)+\right.\right.$$

$$\left. + \omega_e a\cos\phi + a\sin E\,\dot{E} + \frac{a^2e}{r}\sin E\cos E\,\dot{E} + \frac{a^2e^2}{nr}\sin^3 E - \frac{ae}{n}\sin E\cos E\right]$$

$$\left. -\frac{2}{na}\left[\sqrt{1-e^2}\dot{E} + \frac{3}{2}n^2a^3\sqrt{1-e^2}\dot{E} - \omega_e\frac{r}{a} + e\sin E\,\dot{E} + \frac{3}{2}n^2a^3e\sin E\,\dot{E}\right]\right\}.$$

RESULTS and DISCUSSION

As we already stated, The LAGEOS orbital elements were taken as a basis for numerical examples. The electrical charge of the satellite was taken as $Q = -3\times10^{-11}C$, (see VokRouhlicky, 1989). The elements of the LAGEOS satellite are summarized in the table 1.

TABLE 1. Parameter and elements of LAGEOS satellite under the computation.

Semi-major axis	12270 km
Eccentricity	0.004
Inclination	109.9 deg
Mass of satellite	407 kg
Charge	3X10^{-11} C

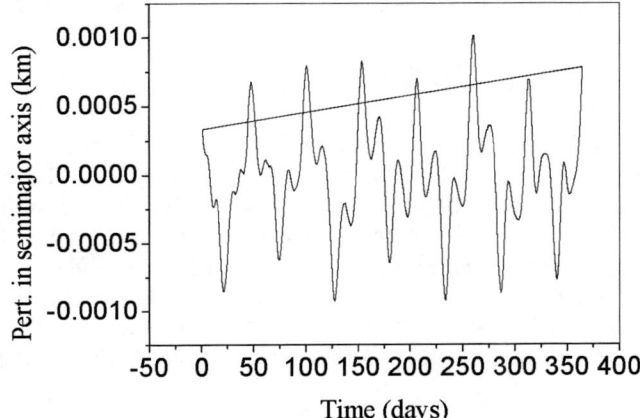

FIGURE 1. Perturbations in the semimajor axis due to Lorentz force.

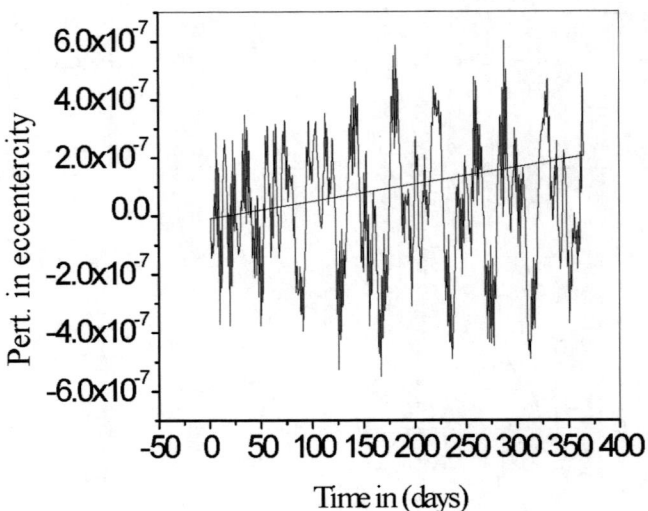

FIGURE 2. Perturbations in the eccentricity due to Lorentz force.

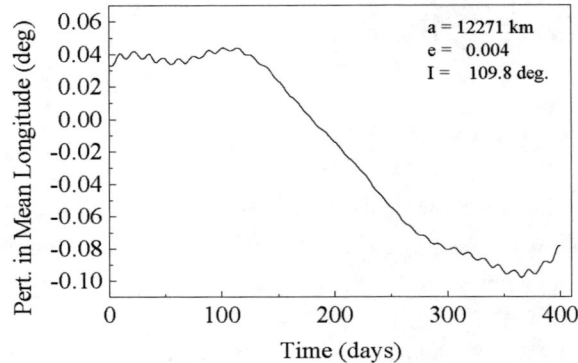

FIGURE 3. Perturbations in the longitude of the satellite orbit due to Lorentz force.

Figure 1, presents the perodic perturbations in the semi-major axis due to the Lorentz force, notice that Lorentz force change the semi-major axis between – 0.001and 0.001 km in a period of 400 days, it may help us to explain the reason of the shift in LAGEOS position about 1 millimetre per day. Figure 2, presents the effects of the Lorentz force on the eccentriicity, which descirbe these efeects as a periodic perturbations in the eccentricity in a level between - -6×10^{-7} and 6×10^{-7} after 400 days. Figure 3, presents the influence of the Lorenzt force on the mean longitude of the satellite $\lambda = \omega + \Omega + M$, which has secular perturbations from 0.04 to -0.1 deg after 400 days.

CONCLUSION

The components of the Lorentz force acceleration of a charged satellite in the Earth's magnetic field are computed as a function of the orbital elements of the satellite in two directions (radial and tangential). Periodic perturbations in the orbital elements of the satellite due to the Lorentz force are derived using Lagrange planetary equations. Numerical applications for the LAGOS Satellite are introduced. These Numerical results shown that the most effects of Lorentz force are in the semimajor axis, eccentricity, and the longitude of the satellite, but there aren't any effects of the Lorentz force on the inclination and the argument of perigee of the orbit. The Lorentz force that derived in this work may help us to explain the reason of the shift in LAGEOS position about 1 millimetre per day.

REFERENCES

1. Anderson, P. C.; Hanson, W. B.; Coley, W. R.; J. Geophy Res, **99**, 3985 (1994)
2. Al'pert Ja. et al., Artificial satellite in rare plasma, Moscow , (1964).
3. Ciufolini, I., Celest. Mech. **40**, 19 (1987).

4. Carolyn K. Purvis, Henry B., Whittlesey, A.C. and Stevens, N. John, NASA Technical paper 2361, (1984).

5. Juhász, Antal and Horányi, Mihály, J Geophy Res, **102**, 7246 (1997).

6. Littlejohn, R. G. , J Math Phy, **23** , 747, (1982).

7. Littlejohn, R. G, A J of Math Phy, **20**, 2458 (1979).

8. Peck Mason A., AIAA 2005-5995 Guidance, Navigation, and control conference and Exhibit 15-18 August 2005, San Francisco, California (2005).

9. Schaffer L. and Burns J., J Geophy Res, **99**, 211 (1994).

10. Sehnal L., SAO reporet No. 271 (1969).

11. VokRouhlicky, D, Celest. Mech. **46**, 85 (1989).

Radionuclides And Heavy Metals Pollution Of The Bottom Sediments Of The Reservoir At Dzierzno Duze Near The Upper Silesia Region, Poland

Kelany A.[1], Tuszynski M.[2], Kostecki M.[3]

[1]*Physics Department, Faculty of Science, Helwan University, Cairo, Egypt.*
Email: kelany12@yahoo.com
[2]*Medicine Physics Department, Institute of Physics, University of Silesian, Katowice,*
[3]*Polish Academy of Science, Zabrze, Poland.*

ABSTRACT. The Dzierzno Duze lake is big artificial reservoir on Klodnica river which goes through the Silesia region, the most industrial area in Poland. The mine waters and other industrial pollution were collected in the sediments in this lake for years. There are plan to give people of the region clean and beautiful lake for leisure activities, so, the global chemical, biological and radiological measurements were done for the reservoir. These samples have been taken from water, bottom sediments (7 points). The concentration of radionuclides was measured by high sensitive gamma spectrometry technique with semiconductor detector HPGe type. The heavy metals were evaluated by ASA method [1].

INTRODUCTION

Mankind and all forms of life on earth have been unavoidable exposed to radiation from natural sources. Exposure to natural sources varies little from year to year, and involves the whole world population to the same extent. The exposure dose from natural sources depends mainly on place of residence, and altitude. For most world population, the range of individual effective dose from natural sources is between one half and two times the average global value at sea level, which is 2.4 mSv per year [2]. Study of natural radiation background and exposure of human beings are of great importance, not only for practical reasons but also for radiological impact of nuclear activities. Firstly the background may change with the development of the nuclear industry (including nuclear power), the wide use of radiation and radioactive isotopes and the development of other technologies. It is necessary to determine the baseline of natural radiation and radioactivity so as to distinguish man-made contamination in time and take appropriate measures to protect the environment [3]. Secondly, the accumulation of information on natural radiation is of great value for drawing up rules and regulations on radiation protection; some countries have taken the dose from natural as reference for radiation protection standards [3]. Lastly, some natural radionuclides are trace elements, which play important roles in the fields of meteorology, hydrology, geology and astronomy [3]. In addition, there have been many attempts to relate health effects to both this basic exposure and to the exposure added from medical uses, weapons tests fallout and nuclear power generation. These uses of the data make it desirable to define the exposure in more details than in the past and to indicate the range of natural exposure expected under various common conditions.

The of sediment samples showed that the mean values of ^{226}Ra and ^{228}Ac were 49.3 and 68 Bq/kg respectively with values ranges between 42 to 57 Bq/kg for ^{226}Ra and 45 to 57 Bq/kg for ^{228}Ac [4]. In the study of low-level radioactivity in the marine environment of the South Spain they found that the average activity of ^{238}U radionuclide in sediment was 282.9 Bq/kg with minimum value of 42 Bq/kg and maximum value of 1100 Bq/kg [5].

The aim of the work was the determination and evaluation of the radioactive isotopes in the sediments and living plants of the Dzierżno Duże water reservoir at upper Silesia in Poland. Application of the work is give people living at this region for time spending (recreation, fishing).

CP888, *Modern Trends in Physics Research,*
Second International Conference on Modern Trends in Physics Research—MTPR-06,
edited by L. El Nadi

EXPERIMENTAL

Sampling and sample preparation

Our work was to determine that kind of radioactive isotopes present in sediments, what is their concentration, the pollution living plants with radioactive elements, besides we want to determine from where the radioactive elements are in the investigated lakes (natural, artificial, technical development and so on) samples from bottom sediment in spite of sort and amount of radionuclides and other chemical elements and biological substances in the lake of Dzierżno Duże.

The reservoir is 6 km wide and the average depth is 5m with maximum depth of 25m, the area of the lake is 615 ha and the volume of water is 93.5 mln m^3. The level of water is 203 m above sea level [6], the location of the reservoir in the Upper Silesia district see Figure 1. The lake is divided onto two parts by underwater dam. its artificial water reservoir (the Odra reservoir) in the west part of Upper Silesia industrial region, near Gliwice. It was created in old sand mining in 1963. The lake is a reservoir on Kłodnica river, which goes through the Silesia region, which is the most industrial area in Poland. The mine waters and other industrial pollution were collected in the sediments in this lake for years.

The depth distribution of radioactive substances has been studied in cores collected in seven points of the lake Dzierżno Duże. These samples have been taken from seven points along the whole reservoir and were divided by 5 cm layers up to 25 cm. The collection of 19 samples was done, the mass of the samples ranged from 12 g to 47 g with and average of 30 g, for that we was used special boxes with small volumes, and the energy and efficiency calibration procedures was done for each volume. Some points was possible to take only one sample of sediments (layer 0-5cm), because of missing wider bottom sediment, there is a flowing of water at the inlet and the outlet of the lake. At the middle of the reservoir it was possible to take five layers of 5cm wide sediments, from 0-25cm depths.

The position of the sampling are present in the map of Figure 2, the sampling points No. 1, 2 and 3 are in first part and No. 4, 5, 6 and 7 in the second part. These samples have been taken from seven points along the whole reservoir and were divided by 5 cm layers up to 25 cm. The collection of 19 samples was done and the characteristics of this was shown in Table 1, the mass of the samples ranged from 12 g to 47 g with and average of 30 g, for that we was used special boxes with small volumes, and the energy and efficiency calibration procedures was done for each volume. In point 1 and 2 (the near Kłodnicy reservoir) and point 7 it was possible to take only one sample of sediments (layer 0-5cm), because of missing wider bottom sediment, there is a flowing of water at the inlet and the outlet of the lake. At the middle of the reservoir it was possible to take five layers of 5cm wide sediments, from 0-25cm depths (points 4 and 5).

Gamma Spectrometry

HPGe semiconductor detector, type Canberra, coaxial with FWHM of 2KeV and efficiency of 30%, at site of Silesian Technical University, Department of Physics, Gliwice, Poland. The gamma identification was carried between 100 keV to 3 MeV, the analysis was done by using comparison method. A 8192 Channels must be displayed on the MCA screen before a successful data transmission to the computer can be made. Over the energy range of interest, (usually 60 keV to 2000 keV). The energy range must be adequately covered by calibration points so that interpolation between the points is accurate. It is recommended to take calibration points approximately every 50 keV from 60 keV to 300 keV and approximately 200 keV from 300 keV to 1400 keV and at least one calibration point between 1400 keV and 2000 keV. This spacing of calibration points is adequate to define the selected energy range, particularly the region from 100 keV to about 2000 keV. it may be necessary to calibrate with a specific radionuclide when the purpose of the measurement is to determine that specific radionuclide more precisely.

Through the detail analysis of the spectra, the observed isotopes from ^{238}U-series, (^{214}Bi, ^{214}Pb, ^{234}Th, ^{206}At and ^{226}Ra) and the ^{232}Th-series (^{228}Ac, ^{212}Bi, ^{212}Pb and ^{208}Tl), the natural radioactive isotope ^{40}K and artificial isotopes of ^{137}Cs. The calculated of specific activity of samples were present in Table 1, and in histograms shown in Figure 3. (A, B, C and D).

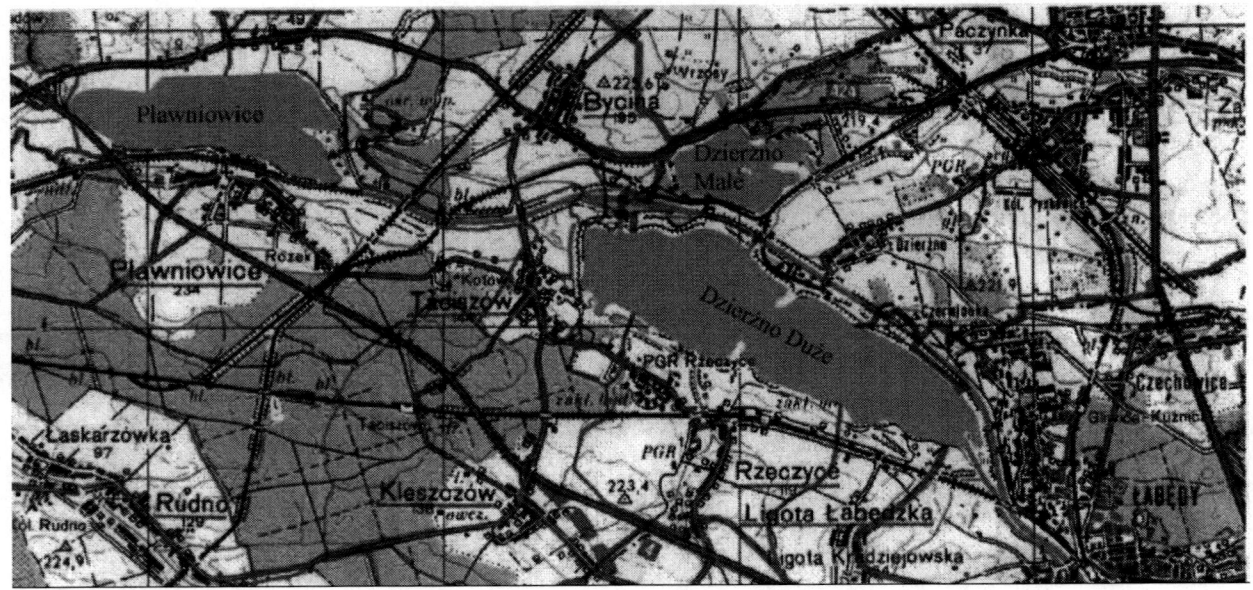

FIGURE 1. The western water reservoir collection in the Upper Silesia industrial district

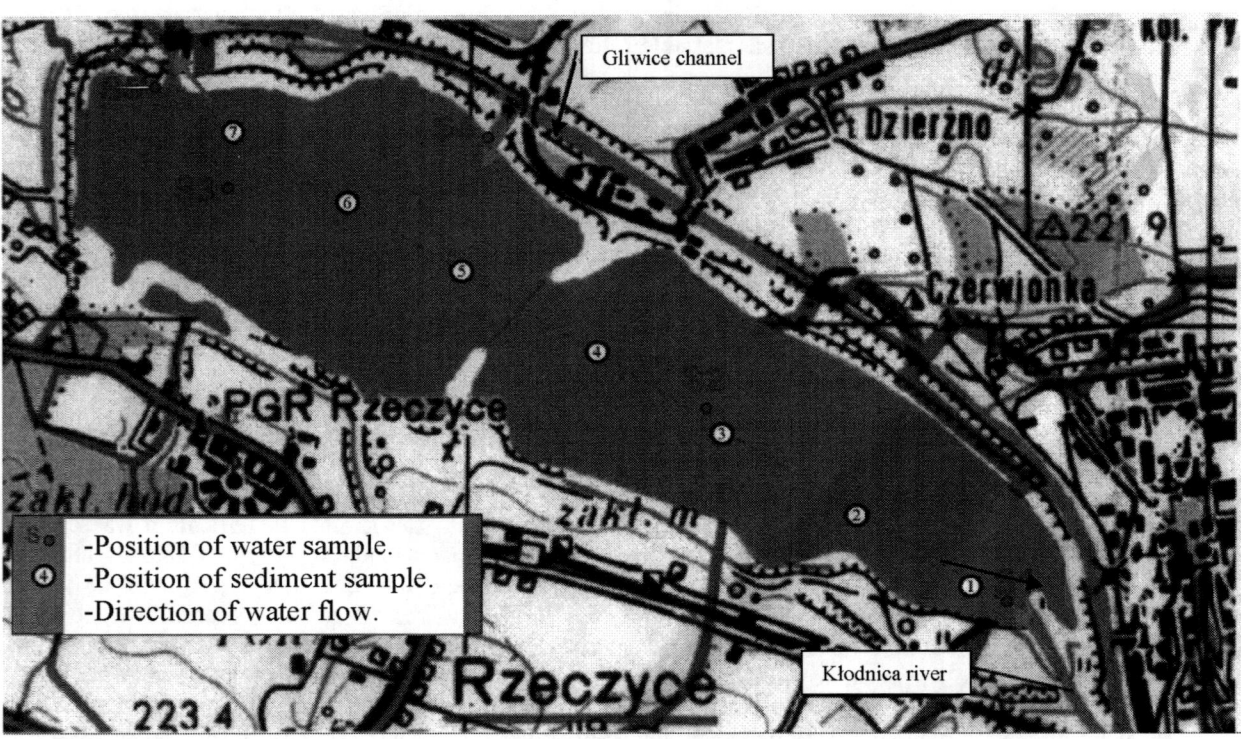

FIGURE 2. Reservoir of Dzierżno Duże (near Gliwice) – Positions of the samples.

TABLE 1. The characteristic and specific activity of sediment samples from the reservoir of Dzierżno Duże.

SAMPLE No.	Points No.	Layer No.	Depth [cm]	Mass [g]	Density [g/cm³]	^{40}K [Bq/kg]	^{137}Cs [Bq/kg]	^{238}U [Bq/kg]	^{232}Th [Bq/kg]
1	1	1	0-5	29.20	**0.449**	380.40±19.5	13.69±3.7	74.84±8.65	30.16±5.49
2	2	1	0-5	42.30	0.651	355.30±18.85	17.23±4.15	**67.48±8.21**	31.65±5.63
3	3	1	0-5	39.40	0.606	**336.20±18.34**	16.39±4.05	77.73±8.82	**29.06±5.39**
4	3	2	5-10	35.80	0.551	375.60±19.38	12.72±3.57	75.71±8.70	35.55±5.96
5	3	3	10-15	36.10	0.555	379.20±19.47	13.86±3.72	87.72±9.37	40.38±6.35
6	3	4	15-20	31.80	0.489	362.70±19.04	13.61±3.69	75.85±8.71	61.10±7.82
7	4	1	0-5	32.80	0.505	484.70±22.02	13.12±3.62	118.29±10.88	74.72±8.64
8	4	2	5-10	34.90	0.537	483.40±21.99	32.99±5.74	147.59±12.15	59.09±7.69
9	4	3	10-15	37.00	0.569	450.26±21.22	53.42±7.31	153.37±12.38	43.66±6.61
10	4	4	15-20	**13.80**	0.578	474.00±21.77	11.73±3.42	134.97±11.62	36.53±6.04
11	4	5	20-25	34.60	0.532	494.60±22.24	10.50±3.24	137.32±11.72	34.06±5.84
12	5	1	0-5	12.60	0.548	**540.30±23.24**	27.21±5.22	181.19±13.46	**107.53±10.37**
13	5	2	5-10	42.20	0.649	504.80±22.47	65.59±8.10	164.61±12.83	75.46±8.69
14	5	3	10-15	18.00	0.783	440.10±20.98	**10.11±3.18**	**184.06±13.57**	54.10±7.36
15	5	4	15-20	15.40	0.670	493.00±22.20	12.16±3.49	144.17±12.01	31.52±5.61
16	5	5	20-25	**47.10**	0.725	527.70±22.97	47.66±6.90	101.00±10.05	46.20±6.80
17	6	1	0-5	16.40	0.713	534.40±23.12	54.17±7.36	153.80±12.40	100.06±10.00
18	6	2	5-10	20.70	**0.900**	442.70±21.04	22.10±4.70	152.13±12.33	56.81±7.54
19	7	1	0-5	19.40	0.843	470.50±21.69	**63.86±7.99**	172.76±13.14	74.97±8.66
Average				29.45	0.624	448.94±21.19	26.95±5.19	126.56±11.25	53.82±7.34

- Points No. are shown on the map of Figure 2.
- The minimum and maximum values are given in bold

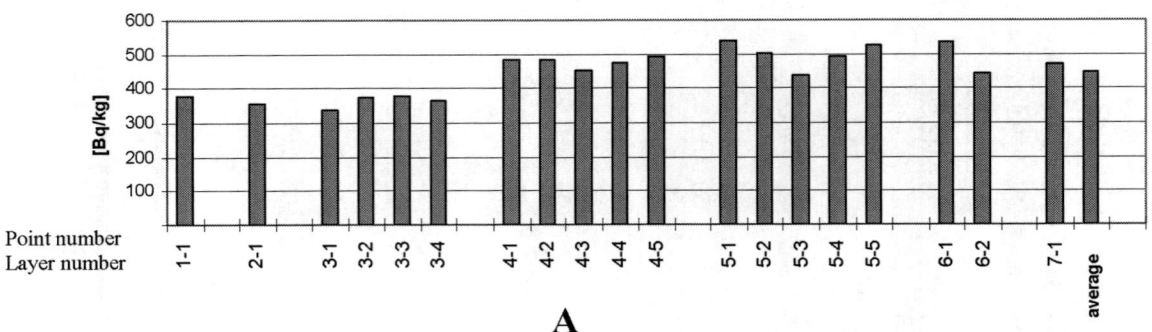

A

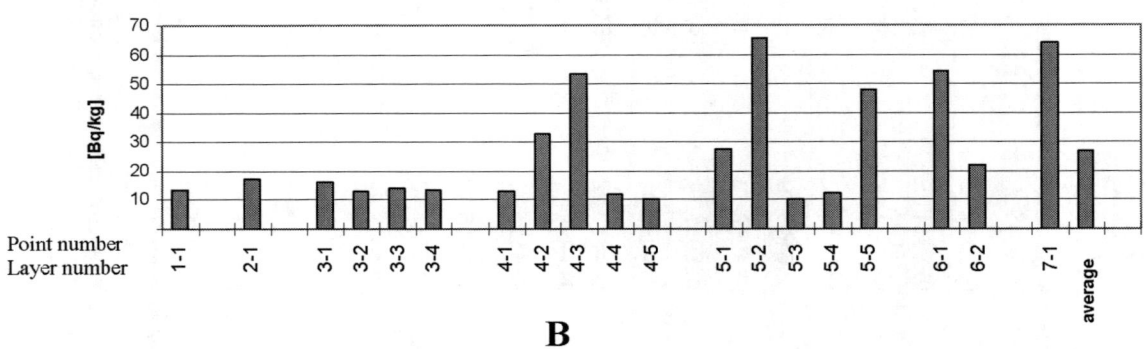

B

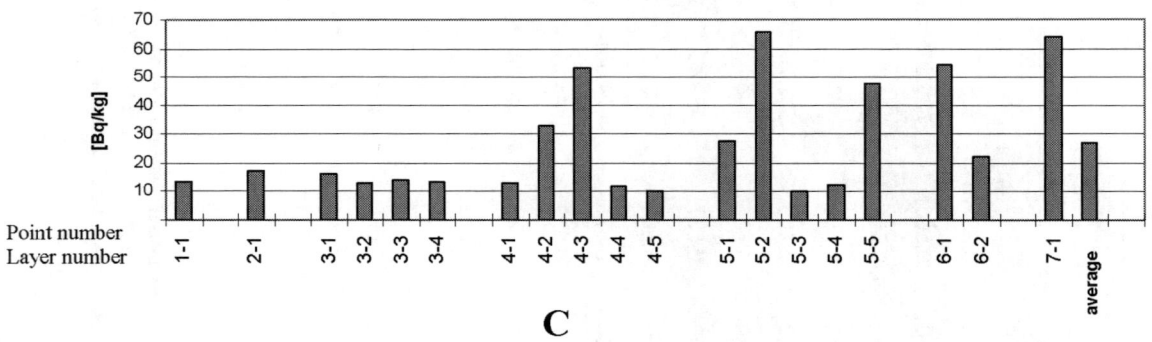

C

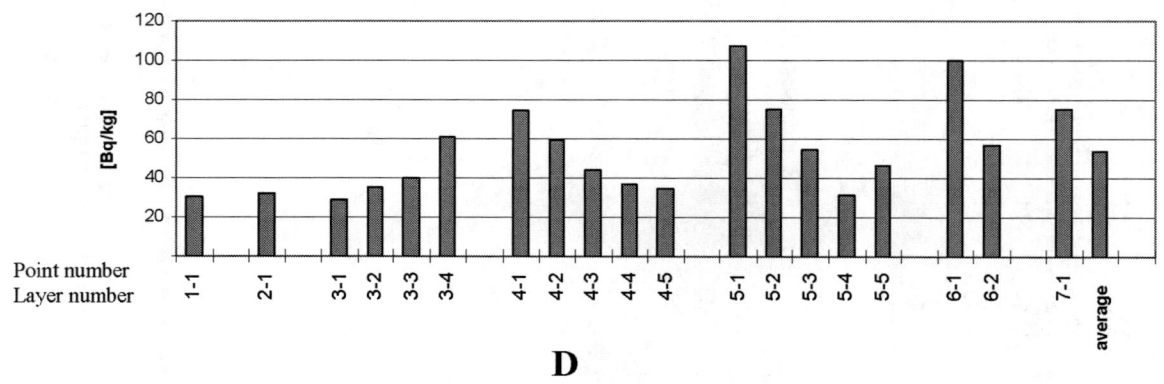

D

FIGURE 3. The concentration of radioisotopes in sediment samples from the reservoir of Dzierzno Duze.
A= Specific activity of ^{40}K [Bq/kg] B= Specific activity of ^{137}Cs [Bq/kg]
C= Specific activity of ^{238}U [Bq/kg] D= Specific activity of ^{232}Th [Bq/kg]

Calculation Of The Activity

The activity of radionuclides in measurement sample was evaluated by the comparing method, to obtain the activity of γ-ray emitting from the samples we are use this formula;

$$A_x = \left(\frac{J_x - J_{bx}}{J_s - J_{bs}} \right) A_s \frac{M_s}{M_x} \tag{1}$$

Where A_s -activity of the standard sample in Bq/kg,
A_x -activity of the measured sample in Bq/kg,
M_s -mass of standard sample in kg,
M_x -mass of measured sample in kg,
J_s -count rate per 1000 minutes for standard sample,
J_x -count rate per 1000 minutes for measured sample,
J_{bs} -count rate per 1000 minutes for background standard sample,
J_{bx} -count rate per 1000 minutes for background measured sample.

Calculation Of The Uncertainty For The Activity

The calculation of activity depends on many different parameters, the uncertainty for this activity is calculated by using the "error propagation law "of Gauss,

$$\Delta A_x = A_x \left[\frac{\Delta J_x^2 + \Delta J_{bx}^2}{(J_x - J_{bx})^2} + \frac{\Delta J_s^2 + \Delta J_{bs}^2}{(J_s - J_{bs})^2} + \frac{\Delta A_s^2}{A_s^2} \right]^{1/2} \tag{2}$$

Where ΔA_s -uncertainty of activity of the standard sample,
ΔA_x -uncertainty of activity of the measured sample,
ΔJ_s -uncertainty of count rate per 1000 minutes for standard sample,
ΔJ_x -uncertainty of count rate per 1000 minutes for measured sample,
ΔJ_{bs} -uncertainty of count rate per 1000 minutes for background for standard sample,
ΔJ_{bx} -uncertainty of count rate per 1000 minutes for background measured sample.

RESULTS AND DISCUSSION

The depth distribution of radioactive substances has been studied in cores collected in seven points of the lake Dzierżno Duże. The gamma spectroscopy was involves into radionuclide measurement. The heavy metals (Pb, Cu, Mn, Cr, Cd) concentration was obtained by the ASA method[1]. There was shown appearance of radionuclides in bottom sediments from uranium progeny with mean value Aav=127.5 Bq/kg, thorium progeny (Aav=53.3 Bq/kg), the natural radionuclide of potassium (Aav=433.8 Bq/kg) and the artificial radioisotope of caesium (Aav=25.4 Bq/kg). Besides the change of concentration of radioisotopes along the reservoir, and with the deep of sediments were analysed. Mean concentration of heavy metals in sediments in mg/kg were as follows - Pb: 128.8, Cd: 14.1, Cr: 48.4, Cu: 92.9, Hg: 1.8. The strong correlation between concentration of heavy metals (Pb, Cu, Mn, Cr, Cd) and uranium and thorium progeny was observed, Figure 4., shows the average concentration of the radionuclides and heavy metals in the sediments of Dzirżnu Duże lake and Figure 5., shows the coexistence of the heavy metals concentration (A) with the ^{238}U-series and (B) ^{232}Th-series concentrations. Table 2., shows the amount of basic pollution of water in the reservoir Dzierżno Duże from inlet and outlet.

The highest concentration of the caesium is in the layer 5-10 cm what may be connected with the Chernobyl accident. The second higher concentration is observed in core 20-25 cm which may be associated with bomb-derived fallout in the 1960s. There was shown appearance of radionuclides from uranium-radium progeny, thorium progeny, the natural radionuclide of potassium and the artificial radioisotope of caesium. The mean activity concentrations of radioelements is: ^{40}K = 449 Bq/kg, ^{137}Cs = 26 Bq/kg, ^{238}U = 133 Bq/kg, ^{232}Th = 55 Bq/kg. Besides the change of concentration of radioisotopes along the reservoir and with the deep of sediment was analysed. The higher concentration of radioisotopes at the end of the lake is observed generally.

Activity of ^{40}K

The specific activity of the isotope of ^{40}K with natural concentration ca 0.0117% in soils, with very long $T_{1/2}$ =1.277*10^9 year, was calculated on the presence of energy 1.461 MeV γ-line. The results are shown in Table 1. and Figure 3.(A). This concentration is in the range of 336.2-540.3 Bq/kg, with an average 448.94 Bq/kg. Concentration of ^{40}K is roughly stable for each measurement points and doesn't change with depth. The slightly increase of concentration of ^{40}K was observed in the second part of reservoir. Such results may be connected with transportation of ^{40}K along the reservoir. The observed concentration values are typical for soils and sands [7], and are not radiological hazards.

Activity of ^{238}U-series

The concentration of ^{238}U and progeny was investigated, after the equilibrium state was achieved. The results are shown in Table 1. and Figure 3.(C). It is observed an increase in the concentration of ^{238}U along the reservoir. In the second part of the reservoir, the concentration of ^{238}U is twice higher than at the entrance of the reservoir (increase from ca 75 Bq/kg to ca 170 Bq/kg). It may be connected with slow stream of water and gradual deposition of sediments.

The changes of ^{238}U with depth are not clearly shown. The average concentration of ^{238}U is equal to 126.56 Bq/kg, this value is almost five times higher than in soils and it is similar as granite rocks [7].

The conclusion is that, the water in the reservoir come from the nearest mines. The water from mines always has isotopes from U and Th-progeny. The values of ^{238}U concentration are not radiological hazards, according to norms [8]. The measured values are less than the limit of concentration for ^{238}U (1000 Bq/kg).

Activity of ^{232}Th-series

The concentration of ^{232}Th and its progeny was measured after the achieving of the equilibrium state, the results are shown in Table 1 and Figure 3(D). The changes of ^{232}Th concentration are similar to ^{238}U and in the second part of the reservoir, the concentration is three times higher than in the first part (ranged from ca 30 Bq/kg to ca 100 Bq/kg), in this case was shown changes of the concentration with depth, the changes of the concentration indicate on the strong flow of water and deposition of ^{232}Th isotopes in sediments (see results for point 3). The different results are observed in points No. 4 and 5, around the under water dam and in point No. 6 the concentration of ^{232}Th decrease with depth. In these points of sampling the lowest value was observed on the level 25 cm and it is half of that for the upper level, average value received for all results is 53.82 Bq/kg and this is twice higher than in typical soils, and lower than in typical granite rocks and in buildings blocks [7]. Similar as for ^{238}U, we conclude that the reservoir was supported by water from the nearest mines, which are rich in U and Th isotopes, for many years. As in ^{238}U, the concentration of ^{232}Th is not radiological hazards.

Activity of ^{137}Cs

Naturally not occur in the environment, it is artificial and create during destruction of uranium in atomic bomb and nuclear power station, $T_{1/2}$ for ^{137}Cs is 30 years. A large amount of this isotope was escaped to atmosphere during nuclear tests especially in 1961 and 1962 and after the Chernobyl accident in year 1986 [9]. The radioactive of cesium is easy to detect through the nuclear methods of detection, even a very small concentration of it. It makes that the cesium was observed in mostly of all surface samples of soils and sediments.

The cesium, ^{137}Cs is detected through gamma line, 661.62 keV. The result of the cesium investigation was shown in Table 1. and Figure 3.(B). For this artificial isotope, the big dynamic of concentration is observed. In the first part of the reservoir, the changes of the concentration with depth are observed in point No. 4.

The observed changes of ^{137}Cs concentration in the first and second part of the lake are determined probably by links with Chernobyl accident and nuclear tests in the upper atmosphere during 1960's. Many investigators observed it before [9,10].

From the radiological point of view detected concentrations of ^{137}Cs is not dangerous of living people and not increase the natural gamma radiation.

TABLE 2. The Amount of Basic Pollution of Water in The Reservoir Of Dzierżno Duże from Inlet and Outlet.

Units /Definition	Inlet Amount	Class	Outlet Amount	Class	I	II	II
mg N-NH4/l	7.21	Upper of limit	7.47	Upper of limit	1	3	6
mg N-NO2/l	0.34	Upper of limit	0.31	Upper of limit	0.02	0.03	0.06
mg N-NO3/l	4.19	I	3.68	I	5	7	15
mg N/l	18.08	Upper of limit	17.05	Upper of limit	5	10	15
mg O2/l	350.62	Upper of limit	75.03	I	25	70	100
mg Cl/l	1043.6	Upper of limit	892.03	III	250	300	400
mg SO4/l	430.27	Upper of limit	383.10	Upper of limit	150	200	250
mg P-PO4/l	0.318	II	0.412	Upper of limit	0.2	0.6	1
mg P/l	0.676	Upper of limit	0.765	II	0.1	0.25	0.4
mg Na/l	665.83	Upper of limit	569.3	Upper of limit	100	120	150
mg K/l	33.83	Upper of limit	31.35	Upper of limit	10	12	15
mg CaCO3/l	700.80	Upper of limit	619.2	Upper of limit	350	550	700
mg Cr/l	0.01	I	0.01	III	0.05	0.1	0.1
mg Zn /l	0.05	I	0.02	I	0.2	0.2	0.2
mg Cd/l	0.00	I	0.00	I	0.005	0.03	0.1
mg Cu/l	0.04	I	0.03	I	0.05	0.05	0.05
mg Pb/l	0.05	I	0.03	I	0.05	0.05	0.05
mg Mn/l	0.608	III	0.448	III	0.1	0.3	0.8
mg Fe/l	1.33	II	0.12	I	1	1.5	2
Alkaline PH	7.63	I	7.74	I	6.5 to 8.5	6.5 to 9	6 to 9
Conductivity μs	4246	Upper of limit	3726	Upper of limit	800	900	1200
Suspension mg/l	413.6	Upper of limit	57.8	Upper of limit	20	30	50

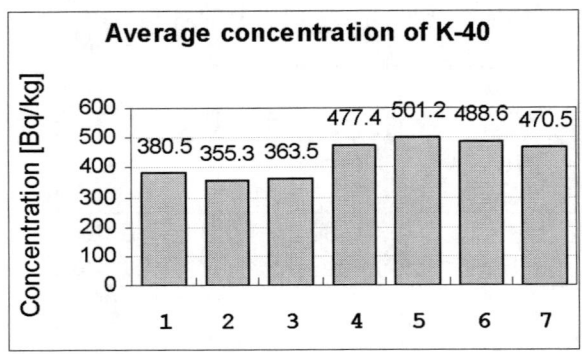

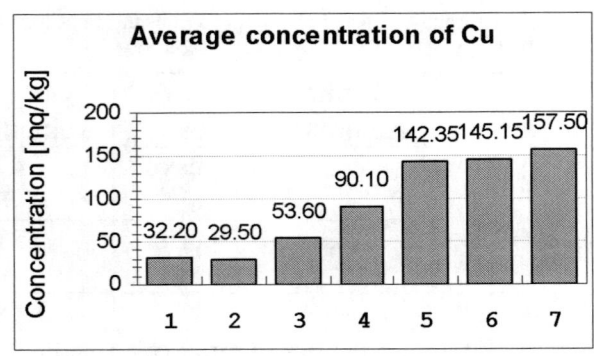

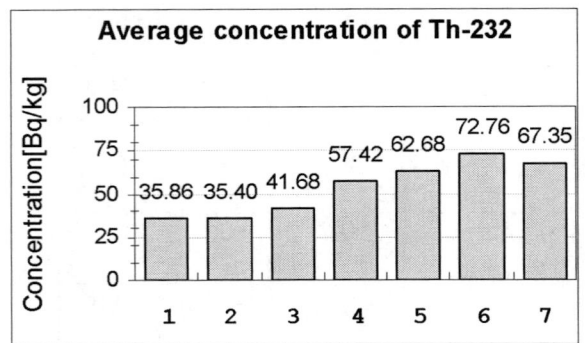

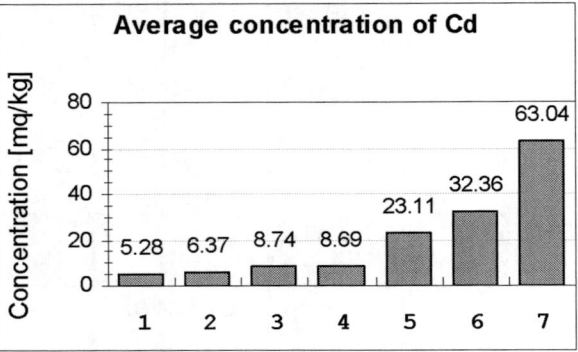

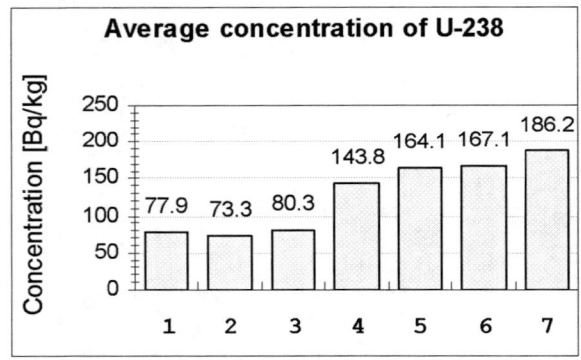

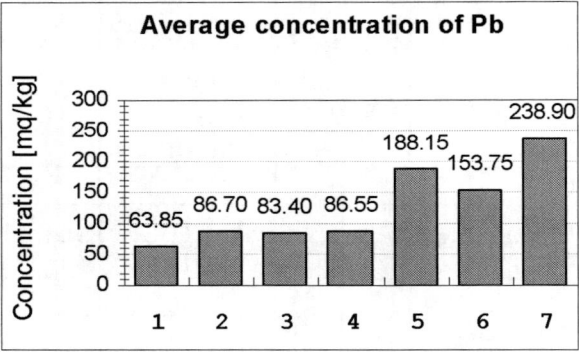

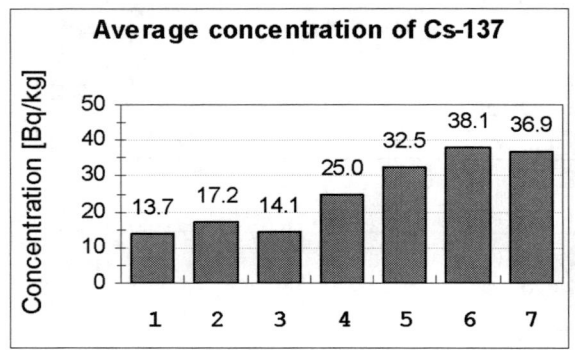

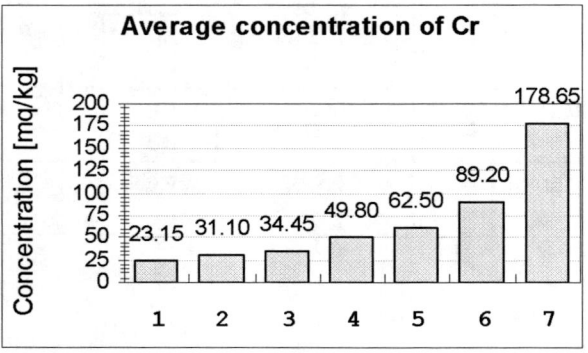

FIGURE 4. The average concentration of the radionuclides and heavy metals in the sediments of Dzirżnu Duże lake.

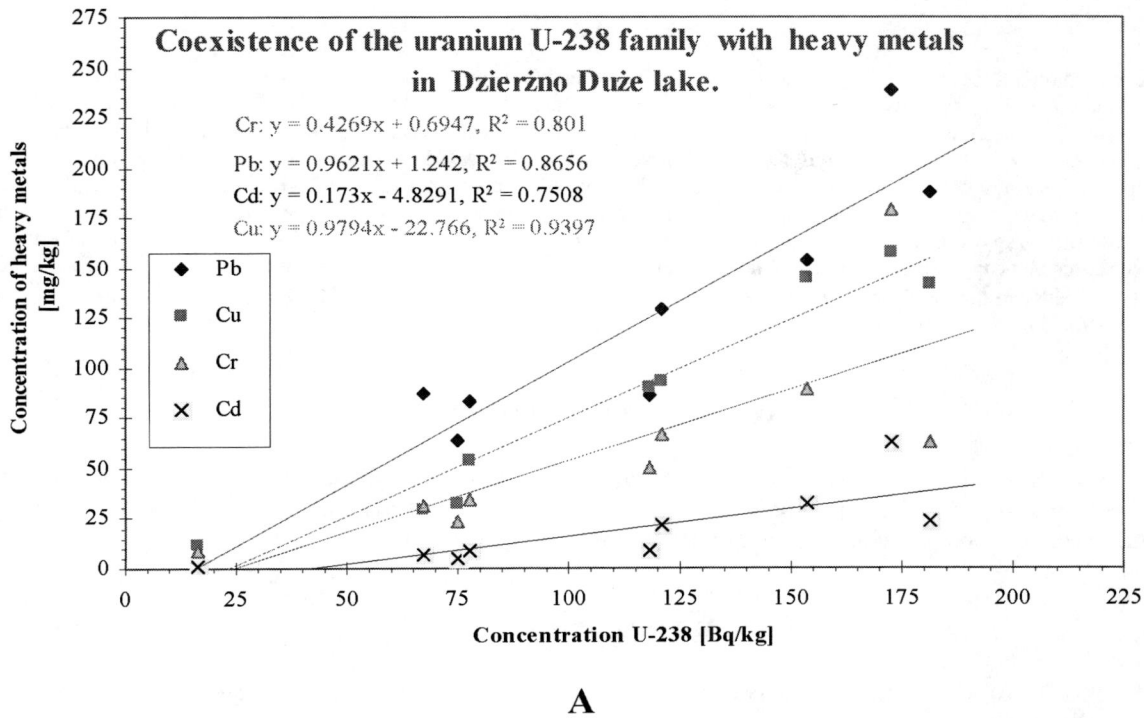

A

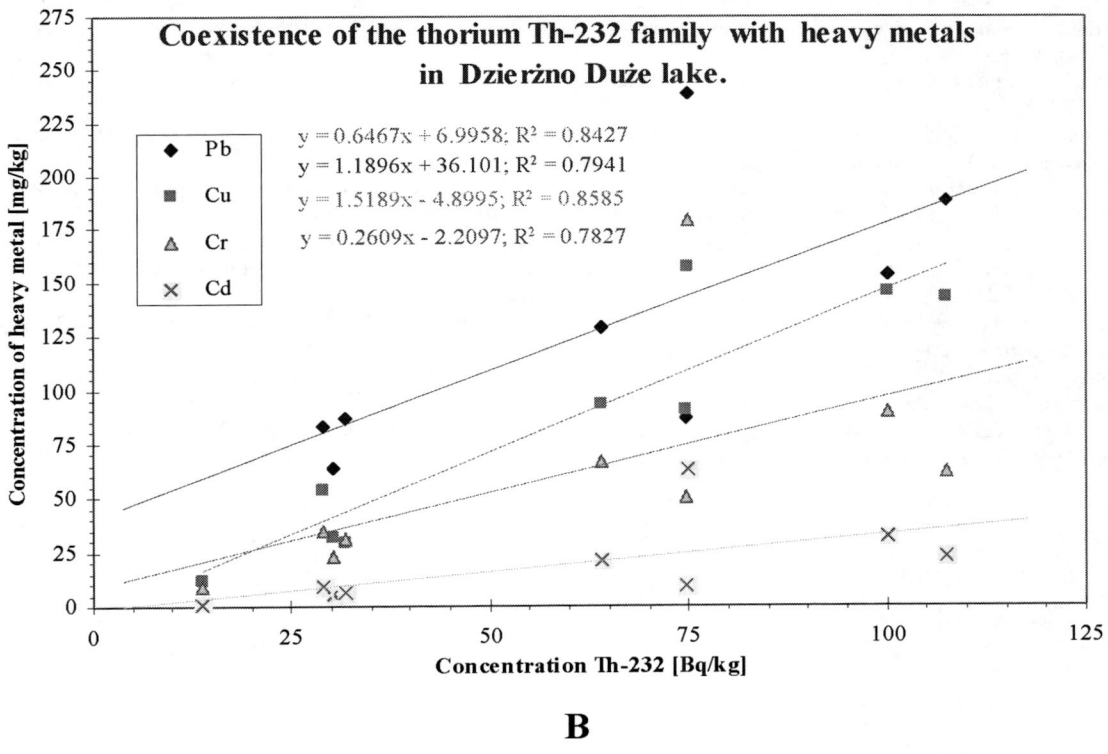

B

FIGURE 5. The coexistence of heavy metals concentration with (A) the ^{238}U-series and (B) ^{232}Th-series concentrations.

CONCLUSION

There was shown appearance of radionuclides in bottom sediments from uranium progeny with mean value activity of 127.5 Bq/kg, thorium progeny of 53.3 Bq/kg, the natural radionuclide of potassium of 433.8 Bq/kg and the artificial radioisotopes of caesium of 25.4 Bq/kg. Besides the change of the concentration of radioisotopes along the reservoir and with the deep of sediment was analyzed. High concentration of Zn, Pb, Cu, Cd and other heavy metals in the atmosphere of the Upper Silesia industrial region have led to the considerable and, in the course of time, increasing heavy metals pollution of the bottom sediments of the Dzierzno Duze lake situated in the vicinity. Mean concentration of heavy metals in sediments in mg/kg were as follows Pb: 128.8, Cd: 14.1, Cr: 48.4, Cu: 92.9, Hg: 1.8. The strong correlation between concentration of heavy metals (Mn, Pb, Cu, Cd, Cr) and uranium and thorium progeny was observed. There are good correlation between the heavy metals concentration with the ^{238}U-series and ^{232}Th-series concentrations , are due to the chemical reactions and physical adsorptions between them.

ACKNOWLEDGMENTS

I have taken collaboration in this field with Dr Marek Tuszyński from Medicine Physics Department, Institute of Physics, University of Silesia in Katowice,Poland. I would like to thank him very much for sampling of interesting sites, discussions and help in interpretation of measurements results.

REFERENCES

1. T. Kotani and H. Akai, "KKR-ASA method in exact exchange-potential band-structure calculations", Physical Review B (Condensed Matter), **54**, 16502 (1996).
2. UNSCEAR (1988): "Sources, Effects and Risks of Ionizing Radiation-United Nation Scientific Committee on the Effects of Atomic Radiation". *Report to the General Assembly,* with annexes. United Nation Sales Publication, United Nations, eds, New York.
3. Ziqiang P., Yin Y. and Migqiang G. : "Natural radiation and radioactivity in China". *Radiation Protection Dosimetery* **24** , 29 (1988).
4. Carreiro M. C. V. and Segueira M. M. A. (1988): *"^{226}Ra and ^{228}Ac in freshwater ecosystem"*. LNETI/Departamento de proteccacaoe Seguranca Radiologica Estrada Nacional 10, ed., 2685 Sacavem, Portugal.
5. Mangon G., Martinez-Aguire A. and Garsia-Leon M.: *"Low-level radioactivity studies in the marine environment of the south of Spain"*. Facullad de Fisica, Universidad de Sevilla, ed., 41080 Sevilla, Spain.
6. Mileska M. I. (1983): *"Geographic and Topographic Dictionary of Poland"* PWN, ed., Warsaw.
7. Hughes J.S., Shaw K. B. and O'Riodan M. C. (1998): "Radiation Exposure of the UK Population - 1998 Review", NRPB 100, {NRPB-R227} National Radiological Protection Board, Chilton.
8. Disposition of the Head of National Atomic Agency, from 28 Aug. 1997 about "Definite accidents in which activities with sources of ionization rays not permission", *Monitor Polski* No. 59, pos. 569.
9. He Qingping, Walling D. E. and Owens P.N. : Interpreting the ^{137}Cs profiles observed in several small lakes and reservoirs in southern England. *Chemical Geology* **129**: 115 (1996).
10. Owens P.N., Walling D. E. and He Qingping : The Behavior of Bomb-Derived Caesium-137 Fallout in Catchment Soils. *Journal of Environmental Radioactivity* **323** 169 (1996).

Author Index

A

Abbas, W. A., 186
Abd Al-Azeem, M. F., 36
Abd El-Aziz, M., 177
Abdel-Aziz, M., 358, 366
Abdel-Aziz, Y. A., 385
Abdel-Baset, T. A., 93
Abdel-Kader, M. H., 220, 225
Abdellatif, G., 197
Abdel-Mageed, M. M., 358, 366
Abdelsalam, A., 249, 305, 314, 323
Abou El-Ela, F. M., 86, 93, 101
Abu-Elmagd, M. S. M., 294
Afaneh, F., 14
Ahmed, M. A., 122, 182, 288
Ahmed, M. M., 294, 323, 345
Ahmed, R., 42, 182
Ahmed, S. M. M. S., 220
Ahmed, S. M. S., 225
Akbarzadeh, H., 42
Al-Aloosy, A. S., 128, 138
Aleem, Fazal-e-, 42, 182, 241
Aleem, H., 274
Ali-Mossa, N., 323
Al-Sherbini M., S., 231
Aly, S. H., 46, 78
Amer, M. A., 376
Ammar, Kh., 59
Ashosh, M., 138
Ashoush, M., 128
Azim M., O. A., 213
Azzawe, A. J. M. Al, 278

B

Badawy, B. M., 249, 305, 314
Badawy, W. A., 29, 110
Badr, Y. A., 160, 177, 186

D

de Feraudy, H., 258
Doaa, M. H., 345

E

E-El Gamal, Y., 197
El-Ahdal, M. A., 288
El-Alfi, G., 78
El-Dek, S. I., 122
El-Din, M. Z., 231
EL-Falaky, E., 337
Elkashef, N., 36
Elkholy, R., 323
El Nadi, L. M., 167, 173
El–Nagdy, M. S., 249, 305, 314, 323
El-Nawawy, M. S., 376
El-Nozahy, A. M., 220, 225
Eloker, M. M., 36
El-Okr, M. M., 128
Elokr, M. M., 132, 138
Elokr, R., 132
El-Okr, R. M., 128
El Sayed, K., 207, 231
El Sherbini, Th. M., 152
El Sherief, R. M., 110
El-Wazzan, N., 46
El Zenki, G., 138

F

Fadl-Allah, S. A., 110
Faridi, A., 274
Farrag, A., 53
Fazal-e-, H., 274

H

Hamid, A. S., 294
Hammam, M., 78
Harith, M. A., 197, 231
Hassab Elnaby, S. I., 160
Hassan, M. A., 132
Hassebo, Y., 207
Hegazy, H., 152
Helal, A. I., 373
Helmerson, K., 0
Hussain, T., 301

I

Imam, H., 197

K

Kamal, M. M., 288
Kamel, M. A., 345
Kelany, A., 392
Khalifa, I. A., 220, 225
Khashaba, E. M., 314

ERRATA

Errata: Modern Trends in Physics Research

[Aip Conference Proceeding – March 17, 2005—Volume 748, Issue 1]

I. ATOMIC, MOLECULAR AND CONDENSED MATTER PHYSICS

An Application of error Reduction and Harmonic Inversion Schemes to the Semi
Classical Calculation of Molecular Vibrational Energy Levels (Pages 3-14)
by H.S Taylor

Table1 was published within the reference section by error

Page	paragraph	Line	should
14	TABLE1	1-18	be transferred to page 11 on top of figure 3.

Errata: Modern Trends in Physics Research

[AIP Conference Proceeding – March 17, 2005—Volume 748, Issue 1]

I. ATOMIC, MOLECULAR AND CONDENSED MATTER PHYSICS

Real Time Monitoring of Growing Nanoparticles by Insito Small Angle Grazing Incidence X-Ray Scattering (Pages 63-71)
by Gilles Renaud

In advertently some corrections could not be included in the final release due to author problems on his end. Since both the print and online versions contain these errors, we herewith provide corrections to this article.

Page	paragraph	Line	Should
67	FIGURE 4	Upper left	Include the notation B
68	FIGURE 5	CAPTION 9	(C) Particle shapes fit truncate Octahedron yielding d=12.6± 4 nm, log-normal Direction of 6nm FWHM, D=21±0.2 nm, $h_{(001)}$=5.8±0.1 nm and total height h = 7.9±0.1 nm (D) 2D image simulated with the above parameters Compares very well with the experimental image

CP888, *Modern Trends in Physics Research,*
Second International Conference on Modern Trends in Physics Research—MTPR-06,
edited by L. El Nadi
© 2007 American Institute of Physics 978-0-7354-0354-9/07/$23.00

Errata: Modern Trends in Physics Research

[AIP Conference Proceeding – March 17, 2005—Volume 748, Issue 1]

II. CHEMICAL PHYSICS, LASERS AND ELECTRONICS

Laser Spectroscopy and Cavity QED With Nanometric Gas Cells (Page 215-221)
By Martial Ducloy

The abstract of the article is missing it should read:

Abstract. Recent advances in fabrication of extremely thin cells (ETC) of dilute vapor with thickness L typically spanning from L = 50 nm to 1 μm, open new prospects for sup-Doppler spectroscopy, and permits the detection of atom surface and cavity QED effects in an unexplored range of distance. When vapor cell is short enough to make the mean free path anisotropic (for atoms flying from wall to wall) the transient build up of the resonant interaction with light is responsible for a specific enhancement of the response of the slowest atoms (i.e.atoms moving quasi parallel to the to the wall) this provides the principle of a novel method for Doppler free spectroscopy, applicable to a variety of situations (Velocity dependent optical pumping, Linear absorption, two photon transition, etc…

CP888, *Modern Trends in Physics Research,*
Second International Conference on Modern Trends in Physics Research—MTPR-06,
edited by L. El Nadi
© 2007 American Institute of Physics 978-0-7354-0354-9/07/$23.00

Errata: Modern Trends in Physics Research

[AIP Conference Proceeding – March 17, 2005—Volume 748, Issue 1]

III. NUCLEAR, PARTICLE AND ASTROPHYSICS.

Neutrino Recent Developments (Pages 331- 349)
by Riazuddin

In advertently some corrections could not be included in the final release due to publishing problems. Since both the print and online versions contain these errors, we herewith provide corrections to this article.

Figure 5. was published after the acknowledgement by error
Figure 5. on page 348, it should be transferred to page 347 before the CONCLUSION.

CP888, *Modern Trends in Physics Research,*
Second International Conference on Modern Trends in Physics Research—MTPR-06,
edited by L. El Nadi
© 2007 American Institute of Physics 978-0-7354-0354-9/07/$23.00

Errata: Modern Trends in Physics Research

[AIP Conference Proceedings -- March 17, 2005 -- Volume 748, Issue 1]

III NUCLEAR, PARTICLE AND ASTROPHYSICS

Theory and Design of Thermionic Electron Beam Guns (pages 376-386)

Inadvertently some corrections could not be included in the final release due to authors problems on their end. Since both the print and online versions contain these errors, we herewith provide corrections to this article.

Page	Paragraph	Line	Should read/ refer
377	4	1	Electron beams are ..... **may be read as** .... Electron beams are generated by thermionic, field and photo emission effects depending on the application devices [2].
377	5	1	Thermionic emission ..... **may be read as** .... Thermionic emission is the escape of electrons from a heated surface [2]
378	1	1	The application .... **may be read as** .... The application [2] .....
378	5	last	Electron beam pulse .... **may be read as** .... Electron beam pulse trains intended for the generation of microwave power in future linear colliders [2]
379	2	4	(field emission source) .... **may be read as** .... (field emission source) [2].
383	1	2	[11] .... **may be read as** [2, 11].
384	1	1	*Electron Beam Welding* ..... **may be read as** .... *Electron Beam Welding* [2] .... .
386	-	-	Reference 2 may be read as...... S.W. Shultz, Proceedings of ebeam 2002. Int. Conf. on High Power Electron Beam Technology, USA, 1, (2002) 1; C. E. Hill, Ion & Electron Sources, CERN, PS-94-36, (1994); A.V.Crewe et al., Rev.Sci. Instr. 39(4) (1968) 576-583; http://www4.nau.edu/microanalysis/Microprobe/Column-ElectronGun.html; M. Iqbal, "Review of Electron Beam Sources "CHEP-Report, March 2004 and references given therein.

CP888, *Modern Trends in Physics Research,*
Second International Conference on Modern Trends in Physics Research—MTPR-06,
edited by L. El Nadi
© 2007 American Institute of Physics 978-0-7354-0354-9/07/$23.00

411

Errata: Modern Trends in Physics Research

[AIP Conference Proceedings -- March 17, 2005 -- Volume 748, Issue 1]

III NUCLEAR, PARTICLE AND ASTROPHYSICS

Beowulf Supercomputers: Scope and Trends (pages 401-405)

Inadvertently some corrections could not be included in the final release due to authors problems on their end. Since both the print and online versions contain these errors, we herewith provide corrections to this article.

Page	Paragraph	Line	Should read/ refer
401	3	last	The first such machine .... *may be read as* the first such machines to win a Gorden Bell price/performance prize [4].
402	1	2	perform billions of calculations in a second. *may be read as* perform billions of calculations in a second [4].
402	2	3	[7,8] *may be read as* [4, 7, 8] &
402	4	1	A typical setup will .... *may be read as* A typical setup will be to configure the nodes for a private Beowulf network [4]
403	1	last	netconfig command for network setup on the node1. *may be read* as netconfig command for network setup on the node1.[4].
403	2	2	Next we will set up the NFS....... *may be read as* Next we will set up the NSF [4]
403	5	1	For parallel programming...... *may be read as* For parallel programming [4] ......
403	last	3	The purpose of the batch system...... *may be read as* The purpose of the batch system [4]........
404	3	6	[17] *may be read as* [4, 17].
404	4	4	[20] *may be read as* [4,20].
405	-	-	Reference 4 may be read as Becker D.J .... Conference on parallel processing, 1995, pp 11-14; http://climate.ornl.gov/~forrest/osdj-2000-11]; http://loki-www.lanl.gov/papers/sc98/, www.cacr.caltech.edu/beowulf/tutorial/building.html, http://www.aspsys.com/clusters/disciplines/default.aspx/chemistry.aspx, www.aamu.edu/MechanicalEngineering/hpcl_cluster.htm, Scientific American 285 (2), pp. 72-79, M. Ahmed, "Burraq Cluster" CHEP Publication, January 2004 and references given therein.

CP888, *Modern Trends in Physics Research,*
Second International Conference on Modern Trends in Physics Research—MTPR-06,
edited by L. El Nadi
© 2007 American Institute of Physics 978-0-7354-0354-9/07/$23.00